BIOACTIVES AND PHARMACOLOGY OF LEGUMES

Cover photo credits:

Left: Phanera purpurea tree (alibangbang) with fruits (Philippines). Photo by Obsidian Soul. https://commons.wikimedia.org/w/index.php?curid=84386087

Middle: Camel Thorn flower - Alhagi maurorum. Photo by Eitan Ferman. 3.0, https://commons.wikimedia.org/w/index.php?curid=11228928. https://creativecommons.org/licenses/by-sa/3.0/

Right: Pterocarpus marsupium. Photo by Silviculture South. https://commons.wikimedia.org/w/index.php?curid=66618249. https://creativecommons.org/licenses/by-sa/4.0/deed.en

AAP Focus on Medicinal Plants

BIOACTIVES AND PHARMACOLOGY OF LEGUMES

Edited by
T. Pullaiah, PhD

AAP APPLE ACADEMIC PRESS

First edition published 2024

Apple Academic Press Inc.
1265 Goldenrod Circle, NE,
Palm Bay, FL 32905 USA

760 Laurentian Drive, Unit 19,
Burlington, ON L7N 0A4, CANADA

CRC Press
6000 Broken Sound Parkway NW,
Suite 300, Boca Raton, FL 33487-2742 USA

4 Park Square, Milton Park,
Abingdon, Oxon, OX14 4RN UK

Library and Archives Canada Cataloguing in Publication

Title: Bioactives and pharmacology of legumes / edited by T. Pullaiah, PhD.
Names: Pullaiah, T., editor.
Series: AAP focus on medicinal plants.
Description: First edition. | Series statement: AAP focus on medicinal plants | Includes bibliographical references and index.
Identifiers: Canadiana (print) 20230142265 | Canadiana (ebook) 20230142281 | ISBN 9781774911266 (hardcover) | ISBN 9781774911273 (softcover) | ISBN 9781003304555 (ebook)
Subjects: LCSH: Materia medica, Vegetable. | LCSH: Legumes. | LCSH: Bioactive compounds.
Classification: LCC RS431.M37 B56 2023 | DDC 615.3/21—dc23

Library of Congress Cataloging-in-Publication Data

CIP data on file with US Library of Congress

ISBN: 978-1-77491-126-6 (hbk)
ISBN: 978-1-77491-127-3 (pbk)
ISBN: 978-1-00330-455-5 (ebk)

AAP Focus on Medicinal Plants

ABOUT THE SERIES

This new book series, edited by T. Pullaiah, focuses on bioactives and pharmacology of medicinal plants.

For millennia, medicinal plants have been a valuable source of therapeutic agents, and still many of today's drugs are based on plant-derived natural products or their derivatives. Bioactive compounds typically occur in small amounts, and they have more subtle effects than nutrients. Bioactive compounds influence cellular activities that modify the risk of disease and help to alleviate disease symptoms. The bioactive compounds have potentially important health benefits, and these compounds can act as antioxidants, enzyme inhibitors and inducers, inhibitors of receptor activities, and inducers and inhibitors of gene expression among other actions. A wide array of biological activities and potential health benefits of medicinal plants have been reported, which include antiviral, antibacterial, antifungal, antioxidant, anticancer, anti-inflammatory, antidiabetic, hepatoprotective, cardioprotective, nephroprotective properties as well as other protective effects on the liver, kidney, heart, and nervous system.

The volumes aim to be comprehensive desk references on bioactives and pharmacology of all the medicinal plants. They will also be important sourcebooks for the development of new drugs.

Book Series Editor

Prof. T. Pullaiah
Department of Botany
Sri Krishnadevaraya University,
Anantapur 515003, A.P., India
Email: pullaiah.thammineni@gmail.com

Book in the Series

- Bioactives and Pharmacology of Lamiaceae
- Bioactives and Pharmacology of Legumes
- Bioactives and Pharmacology of Medicinal Plants (set of 2 volumes)
- Biomolecules and Pharmacology of Medicinal Plants, 2-volume set

- Frankincense–Gum Olibanum: Botany, Oleoresin, Chemistry, Extraction, Utilization, Propagation, Biotechnology, and Conservation
- Phytochemical Composition and Pharmacy of Medicinal Plants, 2-volume set
- Phytochemistry and Pharmacology of Medicinal Plants, 2-volume set

Other Books from AAP by Dr. T. Pullaiah

Ethnobotany of India, 5-volume set:
Editors: T. Pullaiah, PhD, K. V. Krishnamurthy, PhD, and Bir Bahadur, PhD

- Volume 1: Eastern Ghats and Deccan
- Volume 2: Western Ghats and West Coast of Peninsular India
- Volume 3: North-East India and the Andaman and Nicobar Islands
- Volume 4: Western and Central Himalaya
- Volume 5: Indo-Gangetic Region and Central India

Global Biodiversity, 4-volume set:
Editor: T. Pullaiah, PhD

- Volume 1: Selected Countries in Asia
- Volume 2: Selected Countries in Europe
- Volume 3: Selected Countries in Africa
- Volume 4: Selected Countries in the Americas and Australia

Handbook of Research on Herbal Liver Protection: Hepatoprotective Plants
T. Pullaiah, PhD, and Maddi Ramaiah, PhD

Bio-Inspired Technologies for the Modern World: Emerging Trends and Applications
R. Ramakrishna Reddy, PhD, and T. Pullaiah, PhD

Biodiversity of Hotspots–36 volumes
Editor: T. Pullaiah, PhD

About the Editor

T. Pullaiah, PhD
Former Professor, Department of Botany, Sri Krishnadevaraya University, Andhra Pradesh, India

T. Pullaiah, PhD, is a former Professor at the Department of Botany at Sri Krishnadevaraya University in Andhra Pradesh, India, where he has taught for more than 35 years. He has held several positions at the university, including Dean, Faculty of Biosciences, Head of the Department of Botany, Head of the Department of Biotechnology, and Member of Academic Senate. He was the President of the Indian Botanical Society (2014), President of the Indian Association for Angiosperm Taxonomy (2013), and Fellow of the Andhra Pradesh Academy of Sciences. He was awarded the Panchanan Maheshwari Gold Medal, the Prof. P. C. Trivedi Medal, the Dr. G. Panigrahi Memorial Lecture Award of the Indian Botanical Society, and Prof. Y. D. Tyagi Gold Medal of the Indian Association for Angiosperm Taxonomy, and the Best Teacher Award from Government of Andhra Pradesh. Under his guidance, 54 students obtained their doctoral degrees. He has authored 52 books, edited 23 books, and published over 330 research papers, including reviews and book chapters. His books include *Advances in Cell and Molecular Diagnostics* (published by Elsevier), *Camptothecin*, and *Camptothecin producing Plants* (Elsevier) *Ethnobotany of India* (5 volumes published by Apple Academic Press), *Global Biodiversity* (4 volumes, Apple Academic Press), *Red Sanders: Silviculture and Conservation* (Springer), *Genetically Modified Crops* (2 volumes, Springer), *Monograph on Brachystelma and Ceropegia in India* (CRC Press), *Flora of Andhra Pradesh* (5 volumes), *Flora of Eastern Ghats* (4 volumes), *Flora of Telangana* (3 volumes), *Encyclopedia of World Medicinal Plants* (7 volumes, 2nd edition), and *Encyclopedia of Herbal Antioxidants* (3 volumes).

He was also a member of the Species Survival Commission of the International Union for Conservation of Nature (IUCN). Professor Pullaiah received his PhD from Andhra University, India, attended Moscow State University, Russia, and worked as Post-Doctoral Fellow during 1976–1978.

Contents

Contributors .... *xv*

Abbreviations .... *xxi*

Preface .... *xxv*

1. ***Saraca asoca*** **(Roxb.) W.J.de Wilde: Phytochemical and Pharmacological Perspectives 1**
   Thadiyan Parambil Ijinu, Neenthamadathil Mohandas Krishnakumar, Parameswaran Sasikumar, Pathissery John Sarlin, and Palpu Pushpangadan
2. **Bioactive Compounds and Pharmacological Activities of *Mucuna pruriens* L. 21**
   M. Indira, D. Lavanya, G. Pallavi, Liya Siby and S. Krupanidhi
3. **Bioactives and Pharmacology of *Acacia ataxacantha* DC: A Review 37**
   N. K. Sethiya, V. Walia, S. K. Chaudhary, and Yogesh Chand Yadav
4. **A Review on Bioactives and Pharmacology of *Acacia auriculiformis* A. Cunn. ex Benth. 47**
   N. K. Sethiya, V. Walia, S. K. Chaudhary, Yogesh Chand Yadav, and S. Bhargava
5. **Phytochemistry and Pharmacology of Red Bead Seed Tree *Adenanthera pavonina* L.. .... 65**
   Digambar N. Mokat and Tai D. Kharat
6. **Phytochemistry and Pharmacology of *Abrus precatorius* L. 79**
   Avinash Patil, Sonali Patil, and Darshana Patil
7. **Biomolecules and Pharmacology of *Alhagi maurorum* Medik (Family: Fabaceae) 101**
   N. V. L. Sirisha Mulukuri and Pasala Praveen Kumar
8. **A Systematic Review of *Astragalus membranaceus* (Fisch.) Bge. (Fabaceae) 113**
   Pasala Praveen Kumar and Silvia Netala
9. **Chemical Composition and Biological Properties of *Bauhinia purpurea* L. Benth. (Family: Fabaceae) 129**
   Meghna Adhvaryu and Bhasker Vakharia

10. Medicinal Properties of the *Bowdichia virgilioides* Kunth: An Evidence-Based Review..141

Maria Danielma dos Santos Reis, Jamylle Nunes de Souza Ferro, Felipe Lima Porto, Rafael Vrijdags Calado, Tayhana Priscila Medeiros Souza, and Emiliano Barreto

11. Phytoconstituents and Biological Activities of *Caesalpinia mimosoides* Lam...151

Pradeep Bhat, Harsha V. Hegde, Savaliram G. Ghane, and Santoshkumar Jayagoudar

12. A Review on Phytochemistry and Pharmacology of *Cassia fistula* L.....161

Lepakshi M. D. Bhakshu, K.Venkata Ratnam, and R.R. Venkata Raju

13. Bioactives and Pharmacology of *Derris scandens* (Roxb.) Benth.. 191

Jitendra R. Patil, Savaliram G. Ghane, and Ganesh C. Nikalje

14. Remedial Potential of Bioactive Compounds from *Macrotyloma uniflorum* (Lam.) Verdc. 201

Chetna Faujdar and Priyadarshini

15. Chemical Composition and Biological Properties of *Pterocarpus marsupium* Roxb. (Family: Fabaceae)..................................213

Santhivardhan Chinni and Ravilla Jyothsna

16. Biomolecules and Therapeutics of *Senna siamea* (Lam.) H. S. Irwin & Barneby (Syn.: *Cassia siamea* Lam.).................................225

Victorien Dougnon, Brice Boris Legba, Esther Deguenon, Jerrold Agbankpe, Hornel Koudokpon, Jean Robert Klotoe, Honoré Bankole and Jacques Dougnon

17. Bioactives and Pharmacology of *Senna sophera* (L.) Roxb.....................235

Ambika Viswanathan Pillai and V. Suresh

18. *Stryphnodendron adstringens* (Mart.) Coville: Bioactive Compounds and Pharmacological Actions 247

Maria Danielma dos Santos Reis, Jamylle Nunes de Souza Ferro, Felipe Lima Porto, Rafael Vrijdags Calado, Tayhana Priscila Medeiros Souza, and Emiliano Barreto

19. *Tamarindus indica* and Its Bio-activities: An Important Fruit Tree......261

Poornananda Madhava Naik and Vinayak Upadhya

20. Pharmacological Significance of *Uraria picta* (Jacq.) DC. in the Prevention and Treatment of Diseases....................................271

Chetna Faujdar, Abhishek Negi, and Priyadarshini

21. *Acacia ferruginea* DC. (Safed Khair): Bioactive Compounds and Pharmacological Activity...279

Neetesh K. Jain, Yogesh Chand Yadav, Sumeet Dwivedi, and Pankaj Yadav

22. *Acacia modesta* Wall. (Phulai): Bioactive Compounds and Pharmacological Activity.....287

Yogesh Chand Yadav, Neetesh K. Jain, Sumeet Dwivedi, and Pankaj Yadav

23. Bioactive Compounds and Pharmacological Activity of *Acacia macrostachya* Rchb. ex DC. and *Acacia leucophloea* (Roxb.)Willd.....295

Neetesh K. Jain, Sumeet Dwivedi, and Yogesh Chand Yadav

24. Bioactives and Pharmacology of *Flemingia strobilifera* (L.) W.T. Aiton.....303

Sandip Kisan Gavade and Manoj Madhwanand Lekhak

25. Biomolecules and Pharmacology of *Flemingia vestita* Benth. ex Baker.....317

Sandip Kisan Gavade and Manoj Madhwanand Lekhak

26. *Psoralea corylifolia*: A Herbal Drug with Promising Metabolites and Pharmacology Activity.....327

Supriya Thale and Nutan Malpathak

27. *Albizia lebbeck* (L.) Benth.: *Shirish* Tree with Its Phytochemistry, Ethnobotany and Pharmacology.....339

Digambar N. Mokat and Tai D. Kharat

28. Phytochemical and Pharmacological Activities of *Senna hirsuta* (L.) H. S. Irwin & Barneby.....353

K. S. Shanthi Sree, B. Kavitha, A. Suvarna Latha, and P. Lakshmi Padmavathi

29. Bioactive Metabolites and Pharmacological Activities of *Vigna radiata* (L.) Wilczek (Mung Bean).....359

Desai Krishna, Nainesh R. Modi, and Anjali Shukla

30. Bioactive Compounds and Pharmacological Activities of *Acacia dealbata* Link and *Acacia mearnsii* De Wild. (Wattle).....373

Yogesh Chand Yadav, Pankaj Yadav, Pradeep Kumar, Neetesh K. Jain, and Sumeet Dwivedi

31. An Insight of Phytochemical and Pharmacological Prospective of *Senna auriculata* (L.) Roxb.....381

Lepakshi M. D. Bhakshu, K. Venkata Ratnam and R. R. Venkata Raju

32. Phytochemical and Pharmacological Profile of *Clitoria ternatea* L.....417

Ajay Neeraj, R. Y. Hiranmai, Supriya Vaish, and Sunil Soni

33. Secondary Compounds Profile and Bioactive Properties of *Sesbania sesban* (L.) Merr.....433

J. M. Sasikumar, Sibbala Subramanyam, K. N. Jayaveera, Nafyad I. Batu, and Meseret C. Egigu

34. A Review on Bioactives and Pharmacological Activities of *Senna occidentalis* L. 447

K. S. Shanthi Sree, A. Suvarna Latha, P. Lakshmi Padmavathi, D. Bharathi, and B. Kavitha

35. *Indigofera tinctoria* L., Phytochemical and Pharmacological Aspects 461

Thadiyan Parambil Ijinu, Ragesh Raveendran Nair, Vasantha Kavunkal Hridya, Valpasseri Purakkat Akhilesh, and Palpu Pushpangadan

36. Bioactives and Pharmacology of *Pterolobium hexapetalum* (Roth) Santapau & Wagh 473

B. Kavitha and N. Yasodamma

37. Bioactives and Pharmacology of *Aeschynomene aspera* L. and *Aeschynomene indica* L. 485

B. Kavitha and N. Yasodamma

Index 497

Contributors

Meghna Adhvaryu
Department of Botany, Government Science College, Vankal, Mangrol, Surat, Gujarat, India;
E-mail: meghna.adhvaryu@gmail.com

Jerrold Agbankpe
Research Unit in Applied Microbiology and Pharmacology of Natural Substances,
Research Laboratory in Applied Biology, Polytechnic School of Abomey-Calavi,
University of Abomey-Calavi, Godomey, Benin

Valpasseri Purakkat Akhilesh
Amplicon BioLabs, Kinfra Techno-Industrial Park, Malappuram 673635, Kerala, India;
E-mail: vpakhilesh@gmail.com

Honoré Bankole
Research Unit in Applied Microbiology and Pharmacology of Natural Substances,
Research Laboratory in Applied Biology, Polytechnic School of Abomey-Calavi,
University of Abomey-Calavi, Godomey, Benin

Emiliano Barreto
Laboratory of Cell Biology, Federal University of Alagoas, Brazil

Nafyad I. Batu
Department of Biology, Bonga University, Bonga, P.O. Box 334, Ethiopia

Lepakshi M. D. Bhakshu
Department of Botany, PVKN Government College (A), Chittoor, Andhra Pradesh 517002, India;
E-mail: bakshu@pvkngcchittoor.info; lmbakshu@gmail.com

S. Bhargava
Faculty of Pharmacy, DIT University, Dehradun, Uttarakhand, India

D. Bharathi
Department of Biosciences and Sericulture, Sri Padmavati Mahila Visvavidyalayam, Tirupati,
Andhra Pradesh 517502, India

Pradeep Bhat
ICMR, National Institute of Traditional Medicine, Nehru Nagar, Belagavi, Karnataka 590010, India

Rafael Vrijdags Calado
Laboratory of Cell Biology, Federal University of Alagoas, Brazil

S. K. Chaudhary
Institute of Bioresources and Sustainable Development, Imphal, Manipur, India

Santhivardhan Chinni
Department of Pharmacology, Raghavendra Institute of Pharmaceutical Education and Research,
Anantapuramu, Andhra Pradesh, India; E-mail: chinni0143@gmail.com

Esther Deguenon
Research Unit in Applied Microbiology and Pharmacology of Natural Substances, Research Laboratory in Applied Biology, Polytechnic School of Abomey-Calavi, University of Abomey-Calavi, Godomey, Benin

Jacques Dougnon
Research Unit in Applied Microbiology and Pharmacology of Natural Substances, Research Laboratory in Applied Biology, Polytechnic School of Abomey-Calavi, University of Abomey-Calavi, Godomey, Benin

Victorien Dougnon
Research Unit in Applied Microbiology and Pharmacology of Natural Substances, Research Laboratory in Applied Biology, Polytechnic School of Abomey-Calavi, University of Abomey-Calavi, Godomey, Benin; E-mail: victorien.dougnon@gmail.com

Sumeet Dwivedi
UIP, Faculty of Pharmacy, Oriental University, Indore, Madhya Pradesh, India

Meseret C. Egigu
School of Biological Sciences and Biotechnology, College of Natural and Computational Sciences, Haramaya University, Haramaya, P.O. Box 138, Ethiopia

Chetna Faujdar
Department of Biotechnology, Jaypee Institute of Information Technology, Noida, Uttar Pradesh 201309, India

Jamylle Nunes de Souza Ferro
Laboratory of Cell Biology, Federal University of Alagoas, Brazil

Sandip Kisan Gavade
Department of Botany, Dattajirao Kadam Arts, Science and Commerce College, Ichalkaranji, Maharashtra, India; E-mail: skgavadenogra@gmail.com

Savaliram G. Ghane
Plant Physiology Laboratory, Department of Botany, Shivaji University, Kolhapur, Maharashtra 416004, India

Harsha V. Hegde
ICMR, National Institute of Traditional Medicine, Nehru Nagar, Belagavi, Karnataka 590010, India

R. Y. Hiranmai
School of Environment and Sustainable Development, Central University of Gujarat, Gandhinagar 382030 Gujarat, India

Vasantha Kavunkal Hridya
Chemical Sciences and Technology Division, CSIR-National Institute for Interdisciplinary Science and Technology, Thiruvananthapuram 695019, Kerala, India; E-mail: vkhridya@gmail.com

Thadiyan Parambil Ijinu
Amity Institute for Herbal and Biotech Products Development, Thiruvananthapuram 695005, Kerala, India; E-mail: ijinutp@gmail.com

M. Indira
Department of Biotechnology, Vignan's Foundation for Science, Technology and Research, Vadlamudi 522213, Andhra Pradesh, India; E-mail: indu221007@gmail.com

Neetesh K. Jain
OCPR, Faculty of Pharmacy, Oriental University, Indore, Madhya Pradesh, India

Santoshkumar Jayagoudar
Department of Botany, G. S. S. College & Rani Channamma University P. G. Centre, Belagavi, Karnataka 590006, India; E-mail: santoshjayagoudar@gmail.com

K. N. Jayaveera
Department of Chemistry, Jawaharlal Nehru Technological University, Anantapur, Andhra Pradesh 515002, India

Ravilla Jyothsna
Department of Pharmacology, Raghavendra Institute of Pharmaceutical Education and Research, Anantapuramu, Andhra Pradesh, India

B. Kavitha
Department of Botany, Rayalaseema University, Kurnool, Andhra Pradesh 518007, India; E-mail: kavithab15@gmail.com

Tai D. Kharat
Department of botany, Savitribai Phule Pune University, Pune 411007, Maharashtra, India

Jean Robert Klotoe
Research Unit in Applied Microbiology and Pharmacology of Natural Substances, Research Laboratory in Applied Biology, Polytechnic School of Abomey-Calavi, University of Abomey-Calavi, Godomey, Benin

Hornel Koudokpon
Research Unit in Applied Microbiology and Pharmacology of Natural Substances, Research Laboratory in Applied Biology, Polytechnic School of Abomey-Calavi, University of Abomey-Calavi, Godomey, Benin

Desai Krishna
Department of Botany, Bioinformatics and Climate Change Impacts Management, Gujarat University, Ahmedabad, Gujarat 38009, India

Neenthamadathil Mohandas Krishnakumar
Department of Biosciences, Rajagiri College of Social Sciences, Ernakulam 683104, Kerala, India; E-mail: krishnakumarnmohandas@gmail.com

S. Krupanidhi
Department of Biotechnology, Vignan's Foundation for Science, Technology and Research, Vadlamudi 522213, Andhra Pradesh, India

Pradeep Kumar
Chaudhary Sughar Singh College of Pharmacy, Jaswant Nagar Etawah 206130 Uttar Pradesh, India

Pasala Praveen Kumar
Department of Pharmacology, Santhiram College of Pharmacy, Kurnool, Andhra Pradesh, India; E-mail: praveenpharmaco@gmail.com

A. Suvarna Latha
Department of Biosciences and Sericulture, Sri Padmavathi Mahila Visvavidyalayam, Tirupati, Andhra Pradesh, India

D. Lavanya
Department of Biotechnology, Vignan's Foundation for Science, Technology and Research, Vadlamudi 522213, Andhra Pradesh, India

Brice Boris Legba

Research Unit in Applied Microbiology and Pharmacology of Natural Substances, Research Laboratory in Applied Biology, Polytechnic School of Abomey-Calavi, University of Abomey-Calavi, Godomey, Benin

Manoj Madhwanand Lekhak
Angiosperm Taxonomy Laboratory, Department of Botany, Shivaji University, Kolhapur, Maharashtra, India

Nutan Malpathak
Department of Botany, Savitribai Phule Pune University, Maharashtra, Pune 411007, India; E-mail: mpathakster@gmail.com

Nainesh R. Modi
Department of Botany, Bioinformatics and Climate Change Impacts Management, Gujarat University, Ahmedabad, Gujarat 38009, India; E-mail: nrmodi@gujaratuniversity.ac.in

Digambar N. Mokat
Department of Botany, Savitribai Phule Pune University, Pune 411007, Maharashtra, India; E-mail: parimalsai@gmail.com

N. V. L. Sirisha Mulukuri
Department of Natural Products, Nitte College of Pharmaceutical Sciences, Bangalore, Karnataka, India; E-mail: sirishamulukuri056@gmail.com

Poornananda Madhava Naik
Department of Botany, Karnatak University, Dharwad, Karnataka 580003, India

Ragesh Raveendran Nair
Department of Botany, NSS College Nilamel, Kollam 691535, Kerala, India; E-mail: ragshrnair87@gmail.com

Ajay Neeraj
School of Environment and Sustainable Development, Central University of Gujarat, Gandhinagar 382030 Gujarat, India; E-mail: ajayneeraj.hansraj92@gmail.com

Abhishek Negi
Department of Biotechnology, Jaypee Institute of Information Technology, Noida, Uttar Pradesh 201309, India

Silvia Netala
Department of Pharmacognosy, Shri Vishnu College of Pharmacy, Bhimavaram, West Godavari, Andhra Pradesh, India; E-mail: silvia@svcp.edu.in

Ganesh C. Nikalje
Department of Botany, SevaSadan's R. K. Talreja College of Arts, Science and Commerce, Affiliated to University of Mumbai, Ulhasnagar, Maharashtra421003, India; E-mail: ganeshnikalje7@gmail.com/g.nikalje@ssrkt.edu.in

P. Lakshmi Padmavathi
Department of Biosciences and Sericulture, Sri Padmavathi Mahila Visvavidyalayam, Tirupati, Andhra Pradesh, India

G. Pallavi
Department of Biotechnology, Vignan's Foundation for Science, Technology and Research, Vadlamudi 522213, Andhra Pradesh, India

Avinash Patil
Department of Botany-Biotechnology, B. K. Birla College of Arts, Science & Commerce (Autonomous), Kalyan (West), Maharashtra, India; E-mail: dravinashpatil@rediffmail.com

Darshana Patil
Department of Botany, Smt. C.H.M. College, Ulhasnagar-03, Maharashtra, India

Jitendra R. Patil
Department of Botany, SevaSadan's R. K. Talreja College of Arts, Science and Commerce, Affiliated to University of Mumbai, Ulhasnagar, Maharashtra 421003, India

Sonali Patil
Department of Bioanalytical Sciences, B. K. Birla College of Arts, Science & Commerce (Autonomous), Kalyan (West), Maharashtra, India

Ambika Viswanathan Pillai
Post Graduate and Research Department of Botany, Govt. Victoria College, Palakkad, Kerala, India

Felipe Lima Porto
Laboratory of Cell Biology, Federal University of Alagoas, Alagoas, Brazil

Priyadarshini
Department of Biotechnology, Jaypee Institute of Information Technology, Noida, Uttar Pradesh 201309, India; E-mail: priyadarshini@jiit.ac.in/priya.juit@gmail.com

Palpu Pushpangadan
Amity Institute for Herbal and Biotech Products Development, Thiruvananthapuram 695005, Kerala, India; E-mail: palpuprakulam@yahoo.co.in

R. R. Venkata Raju
Department of Botany, Sri Krishnadeveraya University, Ananthapuramu, Kerala 515003, India

K. Venkata Ratnam
Department of Botany, Rayalaseema University, Kurnool, Kerala 518007, India

Maria Danielma dos Santos Reis
Laboratory of Cell Biology, Federal University of Alagoas, Brazil; E-mail: danielma.reis@icbs.ufal.br

Pathissery John Sarlin
Department of Zoology, Fatima Mata National College, Kollam 691001, Kerala, India; E-mail: sarlinpoly@yahoo.com

J. M. Sasikumar
School of Biological Sciences and Biotechnology, College of Natural and Computational Sciences, Haramaya University, Haramaya, P.O. Box 138, Ethiopia; E-mail: jmsashikumar@gmail.com

Parameswaran Sasikumar
Drug Testing Laboratory, Department of Rasasastra and Bhaishajyakalpana, Government Ayurveda College, Thiruvananthapuram 695001, Kerala, India

K. S. Shanthi Sree
Department of Biosciences and Sericulture, Sri Padmavathi Mahila Visvavidyalayam, Tirupati, Andhra Pradesh, India; E-mail: shanthi4botany@gmail.com

N. K. Sethiya
Faculty of Pharmacy, DIT University, Dehradun, Uttarakhand, India; E-mail: neeraj.sethiya@dituniversity.edu.in

Anjali Shukla
Department of Botany, Bioinformatics and Climate Change Impacts Management, Gujarat University, Ahmedabad, Gujarat 38009, India

Liya Siby
Department of Biotechnology, Vignan's Foundation for Science, Technology and Research, Vadlamudi 522213, Andhra Pradesh, India

Sunil Soni
School of Environment and Sustainable Development, Central University of Gujarat, Gandhinagar 382030 Gujarat, India

Tayhana Priscila Medeiros Souza
Laboratory of Cell Biology, Federal University of Alagoas, Alagoas, Brazil

Sibbala Subramanyam
Department of Pharmaceutical Chemistry, College of Health and Medical Sciences, Haramaya University, Haramaya, Ethiopia

V. Suresh
Post Graduate and Research Department of Botany, Govt. Victoria College, Palakkad, Kerala, India; E-mail: sureshmagnolia@gmail.com

Supriya Thale
Department of Botany, V. G. Vaze College of Arts, Science and Commerce (Autonomous), Maharashtra, Mumbai 400081, India; E-mail: supriyathale@vazecollege.net; s.sups1983@gmail.com

Vinayak Upadhya
Department of Forest Products and Utilization, College of Forestry (University of Agricultural Sciences, Dharwad), Banavasi Road, Sirsi, Karnataka 581401, India; E-mail: sirsivinayak@gmail.com

Supriya Vaish
School of Environment and Sustainable Development, Central University of Gujarat, Gandhinagar 382030 Gujarat, India

Bhasker Vakharia
Bhuma Research in Ayurvedic & Herbal Medicine (BRAHM); E-mail: bhasker.vakharia@gmail.com

V. Walia
Department of Pharmacology, SGT College of Pharmacy, SGT University, Gurugram, Haryana, India; E-mail. vaibhav.walia00@gmail.com

Pankaj Yadav
UIP, Faculty of Pharmacy, Oriental University, Indore, Madhya Pradesh, India

Yogesh Chand Yadav
Faculty of Pharmacy, UP University of Medical Science, Saifai, Etawah, Uttar Pradesh, India; E-mail: drycy31@gmail.com

N. Yasodamma
Department of Botany, Sri Venkateswara University, Tirupati, Andhra Pradesh, India; E-mail: yasodanpalli@gmail.com

Abbreviations

AAE	ascorbic acid equivalent
ABTS	2,2′-azino-bis(3-ethylbenzothiazoline-6-sulfonic acid
AChE	anticholinesterase
AIA	adjuvant-induced arthritis
ALP	alkaline phosphatase
ALAT	alanine aminotransaminase
ALT	alanine aminotransferase
ALT	acetylcholinesterase
AST	aspertate aminotransferase
AMACR	α-Methyl Acyl CoA racemose
AMAM	aqueous methanolic extract
Anti-TMV	against tobacco mosaic virus
APS	Astragalus polysaccharide
ASA	acetylsalicylic acid
ASAT	aspartate aminotransaminase
AUUP	Amity University Uttar Pradesh
BHA	butylated hydroxyanisole
BHT	butylated hydroxyl toluene
BP	benzo (a) pyrene
BW	body weight
CCl_4	carbon tetrachloride
CP	cyclophosphamide
CS	cultivated species
CT	*Clitoria ternatea*
CUMS	chronic unpredictable mild stress
CuO NPs	copper oxide nanoparticles
DE	diethyl ether
DN	diabetic nephropathy
DPPH	2-diphenyl-1-picrylhydrazyl
DZSS	*D. scandens, Zingibercassumunar, Suregadamultiflora,* and *Siphonodoncelastrineus*
EA	ethyl acetate
EAC	Ehrlich Ascitis Carcinoma

EPG	eggs per gram
ER	estrogen receptor
FBG	fasting blood glucose
FRAP	ferric reducing antioxidant power
FST	forced swimming test
GEMO	genoprotective effects in non-cellular genomodifier capacity
GI	glycemic index
HbA1c	glycated hemoglobin
HDA	heat denatured
HFD	high fat diet
HK	hexokinase
HRBC	human red blood cell
IBR	infectious bovine rhinotracheitis
ICMR	Indian Council of Medical Research
ILF	intestinal intraluminal fluid
LDH	lactate dehydrogenase
LPS	lipopolysaccharide
MBC	minimum bactericidal concentration
MBC	mung bean coat
MBS	mung bean sprout
MBSC	mung bean seed coat
MDA	malonyl dialdehyde
MDR	multidrug resistant
MEC	minimum effective concentration
MF	fraction of methanol
MGC	microglial cell
MFC	minimum fungal concentration
MIC	minimum inhibitory concentrations
MMP	matrix metalloproteinase
MRSA	methicillin-resistant *S. aureus*
MS	market species
NFκB	nuclear factor kappa B
NO	nitric oxide
NOS	nitric oxide synthase
NSD	nonsun dried
OA	osteoarthritis
PBMC	peripheral blood mononuclear cells
PCM	paracetamol
PE	petroleum ether

PEPCK	phosphoenolpyruvate carboxykinase
PTZ	pentylenetetrazole
RBEF	Ritnand Balved Education Foundation
ROS	reactive oxygen species
SABE	bark extracts of *S. auriculata*
SALE	leaf extract of *S. auriculata*
SAME	*S. auriculata* leaf methanol extract
SD	sun dried
SI	selective index
SMART	somatic mutation and recombination test
SPF	sun protection factor
STZ	streptozotocin
TFC	total flavonoid content
TNF-α	tumor necrosis factor-α
TPC	total phenolic content
TRAP	telomerase repeated amplification protocol
UC	ulcerative colitis
VF	ventricular fibrillation
VT	ventricular tachycardia

Preface

Legumes belong to the third largest flowering plant family Fabaceae. The family is very important to humans and animals as it is the source of vegetable protein pulses. It also contains vegetables and cooking oils. In addition, it is also a source of many medicines. Since it is one of the largest family, we thought of bringing a separate volume on bioactives and pharmacology of selected plants of the legume family.

A variety of bioactives and therapeutics from legumes have been reported, which include antiviral, antimicrobial, antioxidant, anticancer, anti-inflammatory, and antidiabetic, hepatoprotective, nephroprotective, and cardioprotective activities. In this book, bioactives and pharmacology of some selected medicinal plants of Fabaceae are given. A brief introduction is given for each species. Under each species, bioactive compounds are listed, and their chemical structures are given. This is followed by pharmacological activities. All the published literature on pharmacological activities on that species is reviewed.

I hope that this will be a sourcebook for the development of new drugs.

I am thankful to all the authors who contributed to the review chapters. I thank them for their cooperation and erudition.

I request that readers give their suggestions for improvement of the coming volumes and a future edition.

CHAPTER 1

Saraca asoca (Roxb.) W.J.de Wilde: Phytochemical and Pharmacological Perspectives

THADIYAN PARAMBIL IJINU[1,2*],
NEENTHAMADATHIL MOHANDAS KRISHNAKUMAR[3],
PARAMESWARAN SASIKUMAR[4], PATHISSERY JOHN SARLIN[5], and
PALPU PUSHPANGADAN[1]

[1]*Amity Institute for Herbal and Biotech Products Development, Thiruvananthapuram 695005, Kerala, India*

[2]*Naturæ Scientific, Kerala University Business Innovation and Incubation Centre, Karyavattom Campus, Thiruvananthapuram 695581, Kerala, India*

[3]*Department of Biosciences, Rajagiri College of Social Sciences, Ernakulam 683104, Kerala, India*

[4]*Drug Testing Laboratory, Department of Rasasastra and Bhaishajyakalpana, Government Ayurveda College, Thiruvananthapuram 695001, Kerala, India*

[5]*Department of Zoology, Fatima Mata National College, Kollam 691001, Kerala, India*

[*]*Corresponding author E-mail: ijinutp@gmail.com*

ABSTRACT

Saraca asoca is an important tropical medicinal plant mainly found in the Western Ghats of India. It is a small evergreen tree, belongs to the family Fabaceae. It is commonly known as the "Ashoka". Traditionally,

Bioactives and Pharmacology of Legumes. T. Pullaiah, PhD (Ed.)

it is used in the treatment of various gynaecological disorders, dyspepsia, diseases of the blood, biliousness, tumors, abdominal enlargement, colic, piles, ulcers, and bone fractures. More than 30 compounds have been reported from the different parts of *S. asoca*, which mainly consist of flavonoids, tannins, steroids, fatty acids, and glycosides. The flowers and bark of *S. asoca* can be considered a good source of gallic acid and ellagic acid. The extracts, fractions, and isolated compounds from different parts of the *S. asoca* showed antioxidant, antiinflammatory, analgesic, antipyretic, anticancer, antiproliferative, anti-mutagenic, genoprotective, antiulcer, radioprotective, cardioprotective, antidiabetic, antidepressant, antiosteoporosis, anthelmintic, antimicrobial, uterotonic, immunomodulatory, wound-healing, hepatoprotective, neuroprotective, aphrodisiac, and larvicidal activities.

1.1 INTRODUCTION

Saraca asoca (Roxb.) W.J.de Wilde (Syn.: *Jonesia asoca* Roxb., *Jonesia confusa* Hassk., *Saraca confusa* (Hassk.) Backer, *Saraca indica* sensu Bedd. and *Saraca asoca* (Roxb.) Willd.) belongs to the family Fabaceae. Vernacular names of this plant include ashok, ashoka, anganapriya (Hindi), ashoka, asokah, gatasokah, hemapuspa, kankeli, tamrapallava, anganapriya, apashoka, chakraguchha, chitra, dohali, doshahari, gandapushpa, gatasoka, kantacharandohada, karnapura, kelika, krimikaraka, madhupuspah, nata, palladru, pindapushpa, prapallava, raktapallava, rama, rogitaru, shhaya, shokaharta, shokanasha, smaradhivasah, sokanasanah, strinirikshanadohada, subhaga, vajula, vamanghrighataka, vamankayatana, vanjuladrumah, vichitrah, vishoka, vitashoka (Sanskrit), asokam, asoka, hemapushpam, vanculam, vanjulam (Malayalam), asavu, ashoka pattai, asogam, asoka pattai, acoka maram, acokaturu, acoku, acunam, anagam, ankanappiriyam, ayil, ayil pattai, cacupam, cakam, cakkarakuccam, caricam, caripam, centu, ceyalai, cincupam, cittirakaranam, cittiram, cokam, emaputbam, iravattam, kakoli, kakorinimaram, kamukam, kankeli, kannapuram, karcokam, kattutonimaram, kelikomaram, kopikamaram, malaikarunai, malaikkoran, mati, matu, nettilinkam, nettulinkam, pallavattirumaram, pintaputpam, pintitam, pirakipattikam, pitam, ravatam, sasubam, tamirapallavam, tamirapalliyamaram, tokali, vacciram, vanculam, vantiputpam, vicokam, yukapatrikai (Tamil), abhanga, achanga, akshatha, aksunkara, asage, ashoka, ashoka mara, ashokada mara, ashuge, ashunkar, asokada, gandha pushpi, jaasundi, kankelli, kempuchinnadeke gida, kengalimara, kusgemara, seethe mara, achenge, akshath, asaje, asanke, hashige, jassoondie, kenkalimara (Kannada), ashokamu, seethamma asokamu, vanjalamu, asogam, asokamanu (Telugu), ashoka, tamada ashok, asok, jasundi (Marathi), a so ka, my nan med, mya-nan-med (Tibetan), ashok chal (Arabic), and asoka (English).

It is a middle-sized evergreen tree. Its bark is dark brown to gray or almost black. Leaves are paripinnate, glabrous, leaflets 4–5 pairs, 10–20 cm long, margin slightly wavy. Flowers are bisexual, fragrant, orange scarlet in terminal or axillary racemes. Pods are flat, leathery, black, compressed, tapering at both ends. Seeds 4–8, slightly compressed, shiny and glabrous. It is a very beautiful tree when having flowers. The purplish pink new leaves appear at the intervals (Sasidharan, 2004, 2011). It is native to Indo-Malaysian region and Sri Lanka. In India, it is confined to the moist zones of the Western and Eastern Ghats (Peninsular India), Central and Eastern Himalayan foot hills to an altitude of 750 m (Pushpangadan et al., 2016). In Ayurveda, it is used for treating dyspepsia, diseases of the blood, biliousness, tumors, abdominal enlargement, colic, piles, ulcers, and bone fractures. The flowers pounded in water are used to treat hemorrhagic dysentery and the dried flowers for diabetes. It is considered to be an excellent uterine tonic, and useful for the treatment of syphilis and biliousness (Pushpangadan et al., 2016). It is traditionally used for various gynecological disorders (Pradhan et al., 2009; Varaprasad et al., 2011), including menorrhagia (tender leaves), oligomenorrhea (bark), and dysmenorrheal (bark) (Rajith et al., 2012). Flowers have been traditionally used against syphilis, hyperdipsia, burning sensation, hemorrhoids, dysentery, scabies in children, and inflammation (Prathapan et al., 2012). The Kani tribes of the Western Ghats used dried flowers of *S. asoca* along with leaves of *Lawsonia inermis* boiled in coconut oil as a topical application for treating eczema and scabies (Anitha et al., 2008).

1.2 PHYTOCHEMICAL CONSTITUENTS

More than 30 compounds have been reported from the different parts of *S. asoca* which mainly consist of flavonoids, tannins, steroids, fatty acids, and glycosides (Singh et al., 2015). Flavonoids such as (-)-epicatechin, epiafzelechin-(4β-8)-epicatechin, and procyanidin B2, and lignan glycosides such as lyoniside, nudiposide, 5-methoxy-9-β-xylopyranosyl-(-)-isolariciresinol, icariside E3, and schizandriside, together with β-sitosterol glucoside, were isolated from the *S. asoca* bark methanolic extract (Sadhu et al., 2007; Pandey et al., 2011; Shirolkar et al., 2013). Yadav et al. (2015, 2019) identified the presence of gallic acid, β-guanine, indolyl methyl glucosinolate, trimethyl apigenin, tyramine, β-xanthine, gallic acid hexoside, hypophyllanthin, phloridzin, lignin, galloyl-isorhamnetin, myoinositol, cellotriose, 17-decarboxy betanin, lyoniside, procyanidin gallate from the bark. The wax obtained from bark includes n-alkanes (C20–C35), esters (C34–C60), primary alcohols (C20–C30), and n-octacasanol (Joshi, 2004; Rangari, 2012; Yadav et al., 2013). Gahlaut et al. (2013) reported the quantitative estimation of β-sitosterol present in bark, leaves, and flowers of the *S. asoca* using high-performance liquid chromatography coupled with mass spectrometry. The results depicted that the bark was found to have higher quantity of β-sitosterol compared with other parts. The flower essential oil is rich in E,E-α-farnesene (41.2%), hexadecanoic acid (15.3%), methyl salicylate (9.5%), and Z-lanceol (6.6%) (Joshi, 2016).

Ahmad et al. (2015) first reported the isolation of flavanol glycosides such as 3',5-dimethoxy epicatechin, 3'-deoxyepicatechin-3-O-β-D-glucopyranoside, 3'-deoxycatechin-3-O-α-L-rhamnopyranoside, and epigallocatechin along with known flavanols such as leucopelargonidin, leucocyanidin, epicatechin, catechin, lyoniside, and gallocatechin. Saha et al. (2012) first reported the occurrence of gallic acid in the methanolic extract of *S. asoca* using HPTLC analysis. The flower contains fatty acids such as oleic, linoleic, palmitic, and stearic acids along with flavonoids and its glucosides quercetin, kaempferol-3-0-β-D-glucoside, quercetin-3-0-β-D-glucoside, apigenin-7-0-β-D-glucoside, pelargonidin 3,5-diglucoside, cyanidin 3,5-diglucoside. Seed and pod contains oleic, linoleic, palmitic, and stearic acids; catechol, (-) epicatechol, and leucocyanidin (Jain, 1968; Rastogi, 2003; Sadhu et al., 2007). Mittal et al. (2013) carried out high-performance liquid chromatography coupled with quadrupole time-of-flight mass spectrometer for the identification of phenolic and other compounds in the water extracts of different parts of *S. asoca*. It is found that there are 34 catechin derivatives, 34 flavonoids, and 17 other compounds identified through the comparison with standards and characteristic base peaks as well as other daughter ions. The identified known compounds include (-)-epicatechin, leucopelargonidin 3-O-glucoside, epiafzelechin-(4β->8)-epicatechin, procyanidin B2, epicatechin-(4β->8)-epiafzelechin, icariside E3, 3-O-β-D-glucopyranosyl sitosterol, lyoniside, nudiposide, and schizandriside.

1.3 PHARMACOLOGICAL STUDIES

The biological properties of extracts, fractions, and isolated compounds from different parts of the *S. asoca* are discussed below.

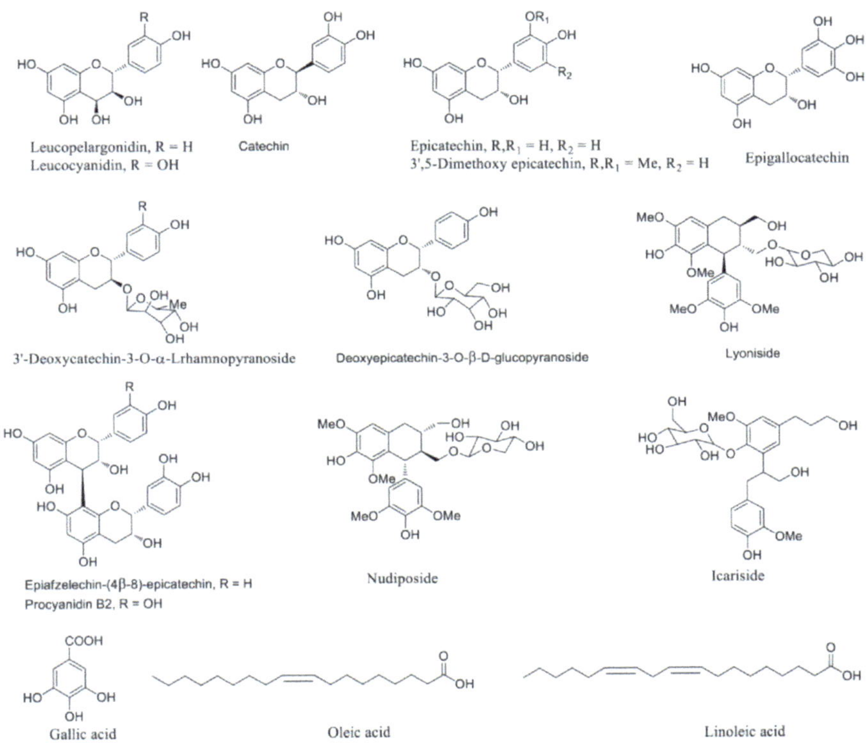

1.3.1 Antioxidant Activity

Ashour et al. (2020) carried out the antioxidant screening of total ethanol, petroleum ether, chloroform, ethyl acetate, and methanol extracts of *S. asoca* leaves. 2,2-Diphenyl-1-picrylhydrazyl assay showed a dose-dependent increase in the free radical scavenging effect. Total ethanol and ethyl acetate extracts exhibited maximum percentage inhibition in 2,2-diphenyl-1-picrylhydrazyl assay as 69.69% and 68.18%, respectively. The IC_{50} value of the alcoholic extract of the bark is found to be 38.5 μg/mL. The nitric oxide radical scavenging assay revealed the dose-dependent radical scavenging activity of the extract (IC_{50} 29.1 μg/mL). The extract also showed the lipid peroxidation inhibition effect with IC_{50} 66 μg/mL (Yadav et al., 2015). In another study, Panchawat and Sisodia (2010) showed the *S. asoca* stem bark hydro-alcoholic extract exhibited a significant antioxidant effect in scavenging 2,2-diphenyl-1-picrylhydrazyl radicals with an IC_{50} of 193.88 μg/mL. Mir et al. (2012) found the 2,2-diphenyl-1-picrylhydrazyl radical scavenging potential of ethanol extract of *S. asoca* bark with an IC_{50} value 45 μg/mL. Gallic acid and quercetin containing flower extract and ellagic acid containing bark extract showed 2,2-diphenyl-1-picrylhydrazyl radical scavenging activity with IC_{50} values of 6.83 μg/mL and 6.6 μg/mL, respectively (Saha et al., 2013). Pal et al. (2014) found that the fresh and dried flower ethanol extracts of *S. asoca* exhibited 81.34% and 58.20% of 2,2-diphenyl-1-picrylhydrazyl radical scavenging activity and water extracts of fresh and dried flowers showed 79.88% and 23.82% of radical scavenging activity.

1.3.2 Anti-Inflammatory Activity

The leaf ethanol extract of *S. asoca* exhibited a potent anti-inflammatory effect in Wistar rats at 200 mg/kg dose. The extract reduced the paw edema significantly compared with the standard diclofenac (10 mg/kg) (Shelar et al., 2010). The ethanolic extract of *S. asoca* bark was evaluated for anti-inflammatory activity in carrageenan-induced hind paw edema in Long Evans rat. The ethanol extract showed a statistically significant, dose-dependent reduction in paw edema. The maximum inhibition of edema (64.14%) was observed at the fifth hour of the study at 300 mg/kg dose (Sharif et al., 2011). Ahmad et al. (2016) reported two anti-inflammatory glycosides from the methanol extract of *S. asoca* bark. The compounds 3'-deoxyepicatechin-3-*O*-β-D-glucopyranoside and 3'deoxycatechin-3-*O*-α-L-rhamnopyranoside exhibited a significant dose-dependent anti-inflammatory property. These two compounds exhibited a maximum anti-inflammatory effect at 200 μg/mL dose in the whole blood assay and showed 52% and 33% of inhibition respectively in carrageenan-induced paw edema in rats.

Evaluation of the anti-arthritic effect of the methanol extract of *S. asoca* bark was carried out by adjuvant-induced arthritis in rats. The methanol extract at 50, 100, and 200 mg/kg doses significantly reduced the paw thickness and levels of both plasma and liver lysosomal enzymes. The extract treatment significantly reduced the protein-bound carbohydrates, urinary collagen content, and pro-inflammatory cytokine levels. The near normalized histopathological architecture of the joints further supported the above results (Saravanan et al., 2011). The anti-arthritic effect of the acetone extract of *S. asoca* seeds was studied in Freund's adjuvant-induced arthritic rats. The acetone extract (300 and 500 mg/kg) decreased prostaglandin levels in blood. The extract significantly decreased paw edema, ankle joint inflammation, and hydrxyproline and glucosamine concentrations in urine. The histopathological studies of liver, stomach, kidney, and joints supported the results. The normal radiological images of the joint further confirmed its anti-arthritic effect (Gupta et al., 2014).

1.3.3 Analgesic Activity

Verma et al. (2010) evaluated the analgesic activity of petroleum ether, chloroform, methanol, and water extracts of the whole parts of *Saraca asoca* at doses 200 and 400 mg/kg, by using the tail immersion method and the formalin-induced pain method in albino mice. The extracts exhibited dose-dependent analgesic activity. Methanol extract (400 mg/kg) showed highest activity and produced 52.64% and 43.30% inhibition of formalin-induced pain response in the first and second phases, respectively, compared with pentazocine (10 mg/kg), which produced 61.30% and 52.38% of inhibition of the pain response. *S. asoca* bark aqueous and alcoholic extracts showed a significant analgesic effect at dose 300 mg/kg (Mohod et al., 2014).

1.3.4 Antipyretic Activity

The seed acetone extract of *S. asoca* was assessed for the antipyretic effect using Brewer's yeast-induced pyrexia in Wistar rats. The extract at 500 mg/kg dose showed a significant antipyretic effect compared with the control group (Sasmal et al., 2012).

1.3.5 Anticancer Activity

The flowers of *S. asoca* showed 50% in vitro cytotoxicity effect on Dalton's lymphoma ascites and Sarcoma-180 tumor cells at 38 μg and 54 μg, respectively. It did not affect the normal lymphocytes but showed percentile activity for lymphocytes derived from leukemia patients (Varghese et al., 1992). The flavonoid fraction from *S. asoca* flowers reported to have a chemopreventive effect on 7, 12-dimethyl benzanthracene-induced second-stage skin carcinogenesis in mice models. The pretreatment of the fraction (10 mg/kg) resulted in a significant reduction in the number of tumors per mouse and the percentage of tumor-bearing mice. The latency period for the appearance of the first tumor was delayed by the pretreatment. There was a significant reduction in the expression of ornithine decarboxylase, the key enzyme that promotes stage of 2-stage skin cancer (Cibin et al., 2012). *S. asoca* alcoholic extract inhibited the proliferation of breast cancer cells MCF-7 (estrogen receptor (ER) positive) and MDA-MB-231 (ER negative) revealing its anti-proliferative effect. The effect of the extract was more prominent in MCF-7 cells with IC_{50} 73.6 μg/mL compared to MDA-MB-231 cells with IC_{50} 128 μg/mL (Yadav et al., 2015). A major flavonoid compound, catechin, was isolated from the methanol extract of *S. asoca* bark evaluated for its chemo-preventive potential in benzene-induced toxicity of acute myeloid leukemia mice. After the treatment, the catechin-treated group exhibited a significant increase in p53 and p21, caspase 11, myeloperoxidase, bcl2, and CYP2E1 (cytochrome P_{450} family 2 subfamily E member 1). The study also revealed the reduced fragmentation of DNA in the catechin-administered group compared with the control indicating the modulatory effect of the compound on the cell cycle. The results suggested that catechin significantly attenuated benzene-induced secondary acute myeloid leukemia in the bone marrow of experimental animals (Mukhopadhyay et al., 2017). The flavonoid fraction of *S. asoca* showed significant cytotoxicity against Dalton's lymphoma ascites and Sarcoma-180 tumor cells (Kaur and Misra, 1980). Saracocide isolated from *S. asoca* exhibited a significant DNA topoisomerase IB inhibition effect (Mukherjee et al., 2012).

1.3.6 Antiproliferative, Anti-Mutagenic, and Geno-Protective Activities

The effect of *S. asoca* bark on estradiol-induced keratinized metaplasia in rat uterus endometrium was studied by Shahid et al. (2015). The stem bark methanolic extract significantly reduced estradiol-induced endometrial thickening and the serum levels of estrogen (82.9 pg/mL) compared with estrogen control (111.2 pg/mL) revealing the antiproliferative and anti-keratinizing effects of the extract. The *S. asoca* bark ethanolic extract was evaluated for the antimutagenic effect in *Salmonella* strains (TA97A, TA98, TA100, and TA102) in the presence or absence of metabolic activation.

The antigenotoxic activity was studied against cyclophosphamide in male Swiss albino mice. The results showed that the extract possessed significant antimutagenic and antigenotoxic properties (Nag et al., 2015).

1.3.7 Antiulcer Activity

S. asoca flower aqueous suspension treatment against pylorus ligation-induced and aspirin-induced gastric ulcers significantly reduced basal gastric secretion and causes stimulation of mucous secretion and endogenous gastric mucosal prostaglandin synthesis in rats. The aqueous suspension (10–30 mg/kg) significantly inhibited the occurrence of acute gastric mucosal lesions (Maruthappan and Shree, 2010). Madhuri et al. (2013) reported the anti-ulcerogenic property of the aqueous extract of *S. asoca* bark at 250 mg/kg and 500 mg/kg doses in aspirin-induced gastric ulcers in albino rats. The results indicated that the extract significantly decreased the ulcers in a dose-dependent manner. The ethanolic leaf extract of *S. asoca* was investigated for its antiulcer activity against ethanol, pylorus ligature, and indomethacin-induced ulcer models in albino rats. It is evident from the results that the extract (200 and 400 mg/kg) significantly reduced ulcer index. There was a significant reduction in total acidity gastric volume and increase in pH compared with the standard drug ranitidine (80 mg/kg) (Verma et al., 2020). The aqueous extract (500 mg/kg) of *S. asoca* bark exhibited a significant protective effect on aspirin-induced gastric ulcer in rats (Lakshmi et al., 2013).

1.3.8 Radio-Protective Activity

The radiation mitigating property of the hydro-alcoholic extract of *S. asoca* bark against gamma radiation in mice was reported (Rao et al., 2010). There was a significant increase in the mean survival time in mice and the extract (400 mg/kg) increased antioxidant enzyme levels and lowered lipid peroxidation. The hydro-alcoholic extract of *S. asoca* bark exhibited radio-protective activity against X-ray-induced cellular damage in Chinese hamster lung fibroblast (V79) cells. The treatment resulted in the increase in cell viability. The in vitro treatment of the extract (50 μg/mL) also reversed radiation-induced increase in intracellular reactive oxygen species and mitochondrial depolarization and decreased antioxidant enzyme levels in V79 cells (Das et al., 2015). The *S. asoca* leaf ethanolic extract exhibited a protective effect against gamma-irradiation-induced renal damage in female albino rats.

The extract (200 mg/kg) significantly ameliorated the renal function test, high sensitivity C-reactive protein, interleukin-1β, angiotensin-converting enzyme, tumor necrosis factor-α, glutathione, and malondialdehyde levels (Mohamed et al., 2021).

1.3.9 Cardio-Protective Activity

Swamy et al. (2013) studied the cardio-protective effect of the alcoholic extract of *S. asoca* bark against cyclophosphamide-induced cardio-toxicity. The extract at 400 mg/kg dose produced significant dose-dependent protection against cyclophosphamide-induced changes by increasing the body weight and decreasing the heart weight and relative heart weight. The extract (200 and 400 mg/kg) significantly reversed the status of cardiac biomarkers such as creatine kinase, creatine kinase isoenzyme MB, lactate dehydrogenase, oxidative enzymes, and lipid profile. The findings were supported by the histopathological studies that the extract treatment reduced the severity of cellular damage of the myocardium. The biochemical parameters and electrocardiogram pattern also supported the results.

1.3.10 Antidiabetic Activity

The hypoglycemic effect of the *S. asoca* bark methanolic extract was investigated in normal and streptozotocin-induced diabetic rats. The extract at 400 mg/kg exhibited a significant hypoglycemic effect in rats (Preethi et al., 2010). The administration of petroleum ether, chloroform, and methanol extracts of *S. asoca* leaves for 21 days caused a significant decrease in blood glucose levels in diabetic mice. The treatment of the extracts (250 and 500 mg/kg) improved the body weight and altered biochemical parameters of streptozotocin-induced diabetic mice. The histopathological studies of liver, pancreas, and kidney supported the antihyperglycemic effect of the extracts (Kumar et al., 2012). Prathapan et al. (2012) reported the antiglycation and inhibitory potential of flavonoid fraction of *S. asoca* flowers against α-glucosidase, α-amylase, and low-density lipoprotein oxidation. The fraction significantly inhibited α-glucosidase (EC_{50} 63.62 μg/mL) and α-amylase (EC_{50} 360.35 μg/mL) in a dose-dependent manner. The flavonoid fraction (50 μg/mL) inhibited the production of advanced glycation end products (72.89%). The low-density lipoprotein oxidation was also inhibited significantly with an EC_{50} value of 40.11 μg/mL. The hypoglycemic, hypolipidemic, and antioxidant potential of *S. asoca* leaves ethanol extract was evaluated in streptozotocin-induced diabetic rats. The extract treatment (200 and 400 mg/kg) normalizes the elevated lipid profile and it reduced glucose levels in a dose-dependent manner. The extract also reduced the diabetes-induced renal oxidative stress (Jain et al., 2013). Mishra and Vijayakumar (2014) carried out the evaluation of ethanolic extract of *S. asoca* flowers for its antihyperglycemic activity in streptozotocin-nicotinamide-induced diabetic rats. The extract at 200 mg/kg dose improved the elevated blood glucose levels in diabetic rats. The decreased levels of antioxidant enzymes such as glutathione, glutathione peroxidase, glutathione reductase, superoxide dismutase, and catalase were also reversed by the extract treatment. The extract and different bioactive fractions of *S. asoca* flowers were evaluated for the in vitro aldose reductase inhibitory effect in high glucose-induced cataract in goat lens. The ethyl acetate fraction lowered the lens opacity. The treatment with the fraction delayed cataract progression in streptozotocin-induced diabetic rats. The fraction also inhibited rat lens aldose reductase enzyme and prevented cataractogenesis. The ethyl acetate fraction exhibited higher percentage of inhibition of aldose reductase with an IC_{50} value of 4.69 μg/mL (Somani and Sathaye, 2015). The methanol extract of *S. asoca* leaf showed significant inhibition of protein tyrosine phosphatase enzyme, a drug target for the treatment of type 2 diabetes (Kalakotla et al., 2014).

1.3.11 Antidepressant Activity

The antidepressant effects of ethanolic extracts of *S. asoca* bark were assessed by reserpine-induced hypothermia and forced swim tests in rats. It was observed that the extract at doses 100, 200, and 400 mg/kg exhibited a significant effect in a dose-dependent manner (Shetty et al., 2008). The effect of the ethanolic bark extract of *S. asoca* on chronic unpredictable mild stress (CUMS) was evaluated in male Sprague-Dawley rats by the forced swim test, the open field test, and the sucrose preference test. The extract (100 mg/kg) exhibited significant reduction in the immobility time in the forced swim test and increased sucrose consumption. The extract also increased the levels of endogenous antioxidant levels compared with the control. The possible mechanism behind the alleviation of the symptoms associated with depression might be achieved by reducing the CUMS-induced oxidative stress and reactive oxygen species in the brain (Gill et al., 2018a).

1.3.12 Anti-Osteoporosis Activity

Anti-osteoporotic activity of *S. asoca* bark was evaluated in dexamethasone-induced osteoporotic female Wistar rats. The extract (200 mg/kg) treatment resulted in an increase in bone density, bone biomechanical strength of the femur and increased the mean ash value of the femur compared with the osteoporotic control group (Thakur et al., 2016).

1.3.13 Anthelmintic Activity

The leaf ethanolic and methanolic extracts of *S. asoca* at 1, 2.5, and 5% concentrations were used to evaluate the anthelmintic activity against piperazine citrate, standard drug. The study showed that the ethanolic extracts were relatively more potent as an anthelmintic agent (Sarojini et al., 2011). The anti-helmintic effect of the methanol extract of *S. asoca* bark was studied at 50 and 100 mg/mL doses. The dose 100 mg/mL exhibited significant anti-helmintic activity compared with the control drug albendazole (10 mg/mL) (Deepti et al., 2012).

1.3.14 Antimicrobial Activity

Antibacterial effect of the alcoholic and aqueous extracts from the aerial parts and the in vitro raised calli of *S. asoca* were analyzed by the agar well diffusion method and the micro dilution method. The alcoholic extract of all the explants and calli exhibited antibacterial property and the calli-derived extracts showed comparable results to the extracts from the explants. The minimum inhibitory concentration of the extracts ranged from 0.039 to 1.25 mg/mL (Shahid et al., 2007). The methanolic and aqueous extracts of the *S. asoca* stem bark were evaluated for the antimicrobial effects against the bacterial strains of *Escherichia coli*, *Pseudomonas aeruginosa*, *Staphylococcus aureus*, *Bacillus cereus*, *Proteus mirabilis*, *Salmonella typhimurium*, *Klebsiella pneumoniae*, and *Streptococcus pneumoniae* and the fungal strains of *Cryptococcus albidus* and *Candida albicans*. The extracts showed antimicrobial effects with minimum inhibitory concentration ranging from 0.5 to 2% for bacterial strains and from 1 to 3% for fungal strains (Sainath et al., 2009). Hot water extracts of *S. asoca* leaves and regenerated bark were found to possess an antimicrobial effect and the minimum inhibitory concentration was recorded to be 0.266–0.533 mg/

mL against bacteria. The hot water leaf extract showed a better antifungal effect with a minimum inhibitory concentration of 1.065 mg/mL. The compounds (+)-catechin and (+)-epicatechin and their biosynthesis-related compound were upregulated in the extracts, which might be the reason for its antimicrobial effect (Shirolkar et al., 2013). There are many literature reports establishing the antibacterial properties of methanol, ethanol, acetone, and aqueous extracts of *S. asoca* dried bark, flower buds, and leaves (Singh et al., 2015). Jain et al. (2018) studied the antimicrobial mechanism of *S. asoca* using the differential proteomics and metabolic profile of *P. aeruginosa*. The *S. asoca* fraction treatment resulted in the upregulated DNA topological and metabolic processes. It further inhibited cellular component biogenesis and response to chemical stimuli impairing the cell membrane function and quorum-sensing system in *P. aeruginosa*. Pal et al. (1985) reported the antibacterial activity of the extracts prepared from flower and flower buds of *S. asoca*.

1.3.15 Uterotonic Activity

The ethanolic extract of *S. asoca* was assessed for estrogenic potential in ovariectomized female Wistar albino rats. The extract at 400 mg/kg was found to be effective compared with ovariectomized control (Swar et al., 2017). Satyavati et al. (1970) reported the oxytocic activity of phenolic glycoside (P2) isolated from *S. asoca* plant. Mitra et al. (1999) evaluated the uterine tonic effect of *S. asoca* containing herbal preparation U-3107 in Wistar rats. The administration of U-3107 significantly elevated both wet and dry uterine weights in rats compared with the control group. The estrogen levels were also increased significantly and the treatment did not affect the regular estrus cycle.

1.3.16 Immunomodulatory Activity

Saracin isolated from *S. asoca* was found to be mitogenic for human lymphocytes. The treatment with saracin induced interleukin 2 secretion from human peripheral blood mononuclear cells. The compound has high affinity for CD8+ cells and it induced apoptosis in activated T-lymphocytes (Ghosh et al., 1999).

1.3.17 Wound-Healing Activity

Deepti et al. (2012) evaluated the wound-healing effect of *S. asoca* bark in Wistar rats. The methanol extract as 1% (w/w), 4% (w/w) ointment, and nitrofurazone 0.2% (w/w) ointment was taken as the standard drug. The highest dose of the extract, 4% (w/w), showed potent wound-healing activity by significantly increasing the tensile strength and wound closure, and decreasing the period of epithelialization.

1.3.18 Hepatoprotective Activity

The hepatoprotective property of the methanolic and hydroalcoholic extracts of the bark of *S. asoca* was evaluated using the carbon tetrachloride-induced hepatotoxicity model in rats. Pretreatment with the methanolic extract (200 and 400 mg/kg) significantly reduced the elevated levels of liver marker enzymes such as serum glutamic oxloacetic transaminase, serum glutamic pyruvic transaminase, and alkaline phosphatase (Arora et al., 2015).

1.3.19 Neuroprotective Activity

The methanolic extract of *S. asoca* bark exhibited potent in vitro neuroprotection toward the differentiated cells and in cell viability assessment. The methanolic extract (100 mg/kg) reduced doxorubicin-induced elevated acetylcholine esterase activity and prevented the depletion in episodic memory (Cheruku et al., 2019).

1.3.20 Aphrodisiac Activity

The methanolic extract of *S. asoca* bark at a dose of 100 mg/kg showed improvement in sexual behavior characterized by an increase in both mount frequency and intromission frequency in male Wistar rats. The extract did not affect sperm count, motility, and sperm morphology (Gill et al., 2018b).

1.3.21 Larvicidal Activity

The leaf petroleum ether extract and bark chloroform extract of *S. asoca* were found effective against the larvae of *Culex quinquefasciatus* with LC_{50} 228.9 ppm and 291.5 ppm, respectively (Mathew et al., 2009).

1.3.22 Toxicity Study

Petroleum ether, chloroform, methanol, and water extracts of *S. asoca* bark did not produce any toxic symptoms of mortality up to 2000 mg/kg. The biochemical parameters, antioxidant enzyme status, and histopathological studies did not exhibit any significant alterations and toxic effects of the extract (Yadav et al., 2015). Different solvent extracts of *S. asoca* leaves were screened for in vitro cytotoxic activity (Ashour et al., 2020). The total ethanol extract showed good activity against MCF-7, HepG-2, and HCT-116 cells ($IC_{50} \leq 30$ μg/mL). The ethyl acetate extract was cytotoxic to HCT-116, MCF-7, HeLa, and HepG-2 with IC_{50} values 0.038, 3.29, 10.14, and 19.21 μg/mL, respectively. The cytotoxic activity of the petroleum ether extract was prominent against HeLa and MCF-7 with IC_{50} values 7.6 and 23.15 μg/mL, respectively. The chloroform extract showed the effect against MCF-7, HeLa, and HepG-2 with the IC_{50} values 2.29, 10.37, and 28.7 μg/mL, respectively. The methanol extract exhibited a cytotoxic effect against MCF-7 with IC_{50} 1.33 μg/mL.

ACKNOWLEDGMENTS

The authors express their sincere thanks to Dr. Ashok K. Chauhan, Founder President, Ritnand Balved Education Foundation (RBEF) and Amity Group of Institutions, and Dr. Atul Chauhan, Chancellor, Amity University Uttar Pradesh (AUUP) for facilitating this work. Thadiyan Parambil Ijinu has received Young Scientist Fellowship from the Department of Science and Technology, Government of India (SP/YO/413/2018).

KEYWORDS

- **traditional medicinal use**
- **gallic acid**
- **ellagic acid**
- **antioxidant activity**
- **anti-inflammatory activity**
- **anticancer activity**

REFERENCES

Ahmad, F.; Misra, L.; Tewari, R.; Gupta, P.; Mishra, P.; Shukla, R. Anti-Inflammatory Flavonol Glycosides from *Saraca asoca* Bark. *Nat. Prod. Res.* **2015,** *23*, 1–4.

Ahmad, F.; Misra, L.; Tewari, R.; Gupta, P.; Mishra, P.; Shukla, R. Anti-Inflammatory Flavanol Glycosides from *Saraca asoca* Bark. *Nat. Prod. Res.* **2016,** *30*, 489–492.

Anitha, B.; Mohan, V. R.; Athiperumalsami, T.; Sutha, S. Ethnomedicinal Plants Used by the Kanikkars of Tirunelveli District, Tamil Nadu, India to Treat Skin Diseases. *Ethnobot. Leafl.* **2008,** *12*, 171–180.

Arora, B.; Choudhary, M.; Arya, P.; Kumar, S.; Choudhary, N.; Singh, S. Hepatoprotective Potential of *Saraca ashoka* (Roxb.) De Wilde Bark by Carbon Tetrachloride Induced Liver Damage in Rats. *Bull. Faculty Pharm. Cairo Uty.* **2015,** *53*, 23–28.

Ashour, W. S.; Mohammed, R. S.; Zeid, A. H. A.; Sabry, O. M.; Hawary, S. S. E. *Saraca indica* L. Leaves, Phytochemical, In Vitro Antioxidant and Cytotoxic Activities. *Egypt. J. Chem.* **2020,** *63*, 3779–3790.

Cheruku, S. P.; Chamallamudi, M. R.; Ramalingayya, G. V.; Biswas, S.; Gourishetti, K.; Nandakumar, K.; Devkar, R.; Mallik, S. B.; Nampoothiri, M.; Kumar, N. Neuroprotective Potential of Methanolic Extract of *Saraca asoca* Bark Against Doxorubicin-Induced Neurotoxicity. *Pharmacogn. Mag.* **2019,** *15*, 309–316.

Cibin, T. R.; Devi, D. G.; Abraham, A. Chemoprevention of Two-Stage Skin Cancer In Vivo by *Saraca asoca. Integr. Cancer Ther.* **2012,** *11*, 279–286.

Das, S.; Kumar, R.; Rao, B. S. S. Radio-Modifying Potential of *Saraca indica* Against Ionizing Radiation: An In Vitro Study Using Chinese Hamster Lung Fibroblast (V79) Cells. *Cell Biol. Int.* **2015,** *39*, 1061–1072.

Deepti, B.; Rani, T. S.; Srinivasa, B. P. Evaluation of Anti-Helmintic and Wound Healing Potential of *Saraca asoca* (Roxb.) Bark. *Pharmacogn. J.* **2012,** *4*, 40–45.

Gahlaut, A.; Shirolkar, A.; Hooda, V.; Dabur, R. β-Sitosterol in Different Parts of *Saraca asoca* and Herbal Drug Ashokarista: Quali-Quantitative Analysis by Liquid Chromatography-Mass Spectrometry. *J. Adv. Pharm. Technol. Res.* **2013,** *4*, 146–150.

Ghosh, S.; Majumder, M.; Majumder, S.; Ganguly, N. K.; Chatterjee, B.P. Saracin: A Lectin from *Saraca indica* Seed Integument Induces Apoptosis in Human T-Lymphocytes. Arch. *Biochem. Biophys.***1999,** *371*, 163–168.

Gill, M.; Kinra, M.; Rai, A.; Chamallamudi, M. R.; Kumar, N. Evaluation of Antidepressant Activity of Methanolic Extract of *Saraca asoca* Bark in a Chronic Unpredictable Mild Stress Model. *Neuroreport* **2018a,** *29*, 134–140.

Gill, M.; Rai, A.; Kinra, M.; Sumalatha, S.; Rao, C. M.; Cheruku, S. P.; Devkar, R.; Kumar, N. Chemically Characterized Extract of *Saraca asoca* Improves the Sexual Function in male Wistar Rats. *Andrologia* **2018b,** *50*, e13037.

Gupta, M.; Sasmal, S.; Mukherjee, A. Therapeutic Effects of Acetone Extract of *Saraca asoca* Seeds on Rats with Adjuvant-Induced Arthritis via Attenuating Inflammatory Responses. *ISRN Rheumatol.* **2014,** 959687.

Jain, A.; Chaudhary, J.; Sharma, S.; Saini, V. Hypolipidemic, Hypoglycemic and Antioxidant Potential of *Saraca asoca* Ethanolic Leaves Extract in Streptozotocin Induced- Experimental Diabetes. *Int. J. Pharm. Pharmaceut. Sci.* **2013,** *5*, 302–305.

Jain, P.; Nale, A.; Dabur, R. Antimicrobial Metabolites from *Saraca asoca* Impair the Membrane Transport System and Quorum-Sensing System in *Pseudomonas aeruginosa. Arch. Microbiol.* **2018,** *200*, 237–253.

Jain, S. K. *Medicinal Plants*; National Book Trust: New Delhi, 1968; p 124.

Joshi, R. K. E,E-α-Farnesene Rich Essential Oil of *Saraca asoca* (Roxb.) Wilde Flower. *Nat. Prod. Res.* **2016,** *30*, 979–981.

Joshi, S. G. *Medicinal Plants*; Oxford and IBH Publishing Co Pvt Ltd., 2004; pp 123–124.

Kalakotla, S.; Mohan, G. K.; Rani, M. S.; Divya, L.; Pravallika, P. L. Screening of *Saraca indica* (Linn.) Medicinal Plant for Antidiabetic and Antioxidant Activity. *Der. Pharm. Lett.* **2014,** *6*, 227–233.

Kaur, J. D.; Misra, K. Biological and Pharmacological Activity of *Saraca asoca*: Anticancer Activity. *J. Indian Chem. Soc.* **1980,** *57*, 1243.

Kumar, S.; Narwal, S.; Kumar, D.; Singh, G.; Narwal, S.; Arya, R. Evaluation of Antihyperglycemic and Antioxidant Activities of *Saraca asoca* (Roxb.) De Wild Leaves in Streptozotocin Induced Diabetic Mice. *Asian Pac. J. Trop. Med.* **2012,** *1*, 170–176.

Lakshmi, K. N. V.C.; Madhuri, M.; Anwar, S.; Ali, L.; Bhimavarapu, P.; Shaik, A. Evaluation of Antiulcerogenic Activity of Various Extracts of *Saraca indica* Bark on Aspirin Induced Gastric Ulcers in Albino Rats. *Int. J. Res. Pharm. Chem.* **2013,** *3*, 753–758.

Madhuri, M.; Anwar, S.; Ali, L.; Bhimavarapu, P.; Shaik, A. Evaluation of Antiulcerogenic Activity of Various Extracts of *Saraca indica* Bark on Aspirin Induced Gastric Ulcers in Albino Rats. *Int. J. Res. Pharm. Chem.* **2013,** *3*, 743–758.

Maruthappan, V.; Shree, K. S. Antiulcer Activity of Aqueous Suspension of *Saraca indica* Flower Against Gastric Ulcers in Albino Rats. *J. Pharm. Res.* **2010,** *3*, 17–20.

Mathew, N.; Anitha, M. G.; Bala, T. S. L.; Sivakumar, S. M.; Narmadha, R.; Kalyanasundaram, M. Larvicidal Activity of *Saraca indica, Nyctanthes arbor-tristis*, and *Clitoria ternatea* Extracts Against Three Mosquito Vector Species. Parasitol. Res. **2009,** *104*, 1017–1025.

Mir, M. A.; Sawhney, S. S.; Kumar, S. Comparative Study Upon the Antioxidant Potential of *Saraca indica* and *Pterospermum acerifolium*. *J. Chem. Pharmaceut. Res.* **2012,** *4*, 4716–4720.

Mishra, S. B.; Vijayakumar, M. Anti-Hyperglycemic and Antioxidant Effect of *Saraca asoca* (Roxb.) De Wilde Flowers in Streptozotocin Nicotinamide Induced Diabetic Rats: A Therapeutic Study. J. Bioanal. Biomed. **2014,** *12*, 1–5. 2013

Mitra, S. K.; Gopumadhavan, S.; Venkataranganna, M. V.; Sarma, D. N. K.; Anturlikar, S. D. Uterine Tonic Activity of U-3107, a Herbal Preparation in Rats. *Indian J. Pharmacol.* **1999,** *31*, 200–203.

Mittal, A.; Kadyan, P.; Gahlaut, A.; Dabur, R. Nontargeted Identification of the Phenolic and Other Compounds of *Saraca asoca* by High Performance Liquid Chromatography-Positive Electrospray Ionization and Quadrupole Time-of-Flight Mass Spectrometry. *ISRN Pharm.* **2013,** 1–10.

Mohamed, M.A.H.; Mohammed, H.S.; Mostafa, S.A.; Ibrahim, M.T. Protective effects of *Saraca indica* L. Leaves Extract (Family Fabaceae) Against Gamma Irradiation Induced Injury in the Kidney of Female Albino Rats. *Environ. Toxicol.* **2021,** *36*, 506–519.

Mohod, S. P.; Jangde, C. R.; Narnawore, S. D.; Rant, S. Experimental Evaluation of Analgesic Property of Bark of Skin of *Saraca indica* (Ashoka) and *Shorea robusta* (Shal). *J. Appl. Pharmaceut. Sci.* **2014,** *4*, 62–65.

Mukherjee, T.; Chowdhury, S.; Kumar, A.; Majumder, H. K.; Jaisankar, P.; Mukhopadhyay, S. Saracoside: A New Lignan Glycoside from *Saraca indica*, a Potential Inhibitor of DNA Topoisomerase IB. *Nat. Prod. Commun.* **2012,** *7*, 767–769.

Mukhopadhyay, M.K.; Shaw, M.; Nath, D. Chemopreventive Potential of Major Flavonoid Compound of Methanolic Bark Extract of *Saraca asoca* (Roxb.) in Benzene -Induced Toxicity of Acute Myeloid Leukemia Mice. *Pharmacogn. Mag.* **2017,** *50*, 216–223.

Nag, D.; Ghosh, M.; Mukherjee, A. Antimutagenic and Genoprotective Effects of *Saraca asoca* Bark Extract. *Toxicol. Ind. Health.* **2015,** *31*, 696–703.

Pal, S. C.; Maiti, A. P.; Chatterjee, B. P.; Nandy, A. Antibacterial Activity of Flowers and Flower Buds of *Saraca indica* Linn. *Indian J. Med. Res.* **1985,** *82*, 188–189.

Pal, T. K.; Bhattacharyya, S.; Dey, A. Evaluation of Antioxidant Activities of Flower Extract (Fresh and Dried) of *Saraca indica* Grown in West Bengal. *Int. J. Curr. Microbiol. App. Sci.* **2014,** *3*, 251–259.

Panchawat, S.; Sisodia, S. S. In Vitro Antioxidant Activity of *Saraca asoca* Roxb. De Wilde Stem Bark Extracts from Various Extraction Processes. *Asian J. Pharm. Clin. Res.* **2010,** *3*, 231–233.

Pandey, A. K.; Ojha, V.; Yadav, S.; Sahu, S. K. Phytochemical Evaluation and Radical Scavenging Activity of *Bauhinia variegata*, *Saraca asoca* and *Terminalia arjuna* Barks. *Res. J. Phytochem.* **2011,** *5*, 89–97.

Pradhan, P.; Joseph, L.; Gupta, V.; Chulet, R.; Arya, H.; Verma, R.; Bajpai, A. *Saraca asoca* (Ashoka): A Review. *J. Chem. Pharm. Res.* **2009,** *1*, 62–71.

Prathapan, A.; Nampoothiri, S. V.; Mini, S.; Raghu, K. G. Antioxidant, Antiglycation and Inhibitory Potential of *Saraca asoca* Flowers Against the Enzymes Linked to Type 2 Diabetes and LDL Oxidation. *Eur. Rev. Med. Pharmacol. Sci.* **2012,** *16*, 57–65.

Preethi, F.; Fernandes, J.; Pricilla, K. Hypoglycemic Activity of *Saraca indica* Linn. Barks. *J. Pharm Res.* **2010,** *3*, 491–493.

Pushpangadan, P.; George, V., Sreedevi, P.; Ijinu, T. P.; Anzar, S.; Bincy, A. J. *Saraca asoca* (Roxb.) De Wilde. In *Plants for Health and Nutritional Security*; Amity Institute for Herbal and Biotech Products Development: Thiruvananthapuram, India, 2016; pp 377–380,

Rajith, N. P.; Ambily, D. V.; Dan, V. M.; Sreedevi, P.; George, V.; Pushpangadan, P. A Survey on Ethnomedicinal Plants Used for Menstrual Disorders in Kerala. *Indian J. Tradit. Knowl.* **2012,** *11*, 453–460.

Rangari, V. D. *Pharmacognosy and Phytochemistry*, Vol. II, 2nd ed.; Career Publication: Nashik, 2012; pp 269–271.

Rao, B. S. S.; Rao, N.; Archana, P. R.; Gayathri, P.; Shetty, P. Antioxidant and Radiation Antagonistic Effect of *Saraca indica*. *J. Environ. Pathol. Toxicol. Oncol.* **2010,** *29*, 69–79.

Rastogi, V. D. *Pharmacognosy and Phytochemistry*; Career Publication: Nashik, 2003; pp 269–270.

Sadhu, S. K.; Khatun, A.; Phattanawasin, P.; Ohtsuki, T.; Ishibashi, M. Lignan Glycosides and Flavonoids from *Saraca asoca* with Antioxidant Activity. *J. Nat. Med.* **2007,** *61*, 480–482.

Saha, J.; Mitra, T.; Gupta, K.; Mukherjee, S. Phytoconstituents and HPTLC Analysis in *Saraca asoca* (Roxb.) Wilde. *Int. J. Pharm. Pharm. Sci.* **2012,** *4*, 96–99.

Saha, J.; Mukherjee, S.; Gupta, K.; Gupta, B. High-Performance Thin-Layer Chromatographic Analysis of Antioxidants Present in Different Parts of *Saraca asoca* (Roxb.) De Wilde. *J. Pharm. Res.* **2013,** *7*, 798–803.

Sainath, R.S.; Prathiba, J.; Malathi, R. Antimicrobial Properties of the Stem Bark of *Saraca indica* (Caesalpiniaceae). *Eur. Rev. Med. Pharmacol. Sci.* **2009,** *13*, 371–374.

Saravanan, S.; Babu, N.P.; Pandikumar, P.; Ignacimuthu, S. Therapeutic Effect of *Saraca asoca* (Roxb.) Wilde on Lysosomal Enzymes and Collagen Metabolism in Adjuvant Induced Arthritis. *Inflammopharmacol.* **2011,** *19*, 317–325.

Sarojini, N.; Manjari, S. A.; Kanti, C. C. Phytochemical Screening and Anthelmintic Activity Study of *Saraca indica* Leaves Extracts. *Int. Res. J. Pharm.* **2011,** *2*, 194–197.

Sasidharan, N. *Biodiversity Documentation for Kerala, Part 6: Flowering Plants*; Kerala Forest Research Institute: Kerala, India, 2004.

Sasidharan, N. *Flowering Plants of Kerala*, Ver. 2.0 (DVD), Serial Number 698329520, Kerala Forest Research Institute, Thrissur, Kerala, India, 2011.

Sasmal, S.; Majumdar, S.; Gupta, M.; Mukherjee, A.; Mukherjee, P.K. Pharmacognostical, Phytochemical and Pharmacological Evaluation for the Antipyretic Effect of the Seeds of *Saraca asoca* Roxb. *Asian Pac. J. Trop. Biomed.* **2012,** *2*, 782–786.

Satyavati, G. V.; Prasad, D. N.; Sen, S. P.; Das, P. K. Oxytocic activity of a pure phenolic glycocide (P2) from, *Saraca indica* Linn. (Ashoka): A Short Communication. *Indian J. Med. Res.* **1970,** *58*, 660–663.

Shahid, A. P.; Salini, S.; Sasidharan, N.; Padikkala, J.; Raghavamenon, A. C.; Babu, T. D. Effect of *Saraca asoca* (Asoka) on Estradiol Induced Keratinizing Metaplasia in Rat Uterus. *J. Basic Clin. Physiol. Pharmacol.* **2015,** *26*, 509–515.

Shahid, M.; Shahzad, A.; Malik, A.; Anis, M. Antibacterial Activity of Aerial Parts as Well as In Vitro Raised Calli of the Medicinal Plant *Saraca asoca* (Roxb.) De Wilde. *Can. J. Microbiol.* **2007,** *53*, 75–81.

Sharif, M. K.; Hossain, M.; Uddin, M. E.; Farooq, A. O.; Islam, M. A.; Sharif, M. M. Studies on the Anti-Inflammatory and Analgesic Efficacy of *Saraca asoca* in Laboratory Animals. *Arch. Pharm. Pract.* **2011,** *2*, 47–52.

Shelar, D. B.; Shirote, P. J.; Naikwade, N. S. Anti-Inflammatory Activity and Brine Shrimps Lethality Test of *Saraca indica* Linn Leaves Extract. *J. Pharma. Res.* **2010,** *3*, 2004–2008.

Shetty, P.; Krishnamoorthy, M.; Vijayanarayana, K.; Subramanyam, E.; Satyanarayana, D. Antidepressant Activity of Bark of *Saraca indica* Linn. *Asian J. Chem.* **2008,** *20*, 1075–1080.

Shirolkar, A., Gahlaut, A., Chhillar, A. K.; Dabur, R. Quantitative Analysis of Catechins in *Saraca asoca* and Correlation with Antimicrobial Activity. *J. Pharm. Anal.* **2013,** *3*, 421–428.

Singh, S.; Krishna, A. T. H.; Kamalraj, S.; Kuriakose, G. C.; Valayil, G. M.; Jayabaskaran, C., Phytomedicinal Importance of *Saraca asoca* (Ashoka): An Exciting Past, an Emerging Present and a Promising Future. *Curr. Sci.* **2015,** *109*, 1790–1801.

Somani, G.; Sathaye, S. Bioactive Fraction of *Saraca indica* Prevents Diabetes Induced Cataractogenesis: An Aldose Reductase Inhibitory Activity. *Pharmacogn. Mag.* **2015,** *11*, 102–110.

Swamy, H. M. V.; Patel, U. M.; Koti, B. C.; Gadad, P. C.; Patel, N. L.; Thippeswamy, A. H. M. Cardioprotective Effect of *Saraca indica* Against Cyclophosphamide Induced Cardiotoxicity in Rats: A Biochemical, Electrocardiographic and Histopathological Study. *Indian J. Pharmacol.* **2013,** *45*, 44–48.

Swar, G.; Shailajan, S.; Menon, S. Activity Based Evaluation of a Traditional Ayurvedic Medicinal Plant: *Saraca asoca* (Roxb.) De Wilde Flowers as Estrogenic Agents Using Ovariectomized Rat Model. *J. Ethnopharmacol.* **2017,** *195*, 324–333.

Thakur, R.; Pawar, R.; Ahirwar, B. Evaluation of *Saraca indica* for the Management of Dexamethasone-Induced Osteoporosis. *J. Acute Med.* **2016,** *6*, 7–10.

Varaprasad, N.; Suresh, A.; Suresh, V.; Kumar, N. S.; Rajendar, A.; Madeshwaran, M. Anti-Pyretic Activity of Methanolic Extract of *Saraca asoca* (Roxb.) De Wild leaves. *Intern. J. Pharma Res. Dev.* **2011,** *3*, 202–207.

Varghese, C. D.; Nair, S. C.; Panikkar, K. R. Potential Anticancer Activity of *Saraca asoca* Extracts Towards Transplantable Tumors in Mice. *Indian J. Pharmaceut. Sci.* **1992,** *54*, 37–40.

Verma, A.; Jana, G. K.; Chakraborty, R.; Sen, S.; Sachan, S.; Mishra, A. Analgesic Activity of Various Leaf Extracts of *Saraca indica* Linn. Der. *Pharmacia. Lettre*. **2010,** *2*, 352–357.

Verma, P.; Paswan, S. K.; Vishwakarma, V. K.; Saxena, P.; Rao, C. V.; Shrivastva, S. Evaluation of the Antiulcer Activity of Ethanolic Leaves Extracts of *Saraca indica* Against Ethanol, Pylorus Ligature and Indomethacin Induced Ulcer in Albino Rats. *Curr. Bioact. Compd.* **2020,** *16*, 1191–1196.

Yadav, G., Garg, V. K., Thakur, N.; Khare, P. Locomotor Activity of Methanolic Extract of *Saraca indica* Bark. *Adv. Biol. Res.* **2013,** *7*, 1–3.

Yadav, N. K.; Saini, K. S.; Hossain, Z.; Omer, A.; Sharma, C.; Gayen, J. R.; Singh, P.; Arya, K. R.; Singh, R. K. *Saraca indica* Bark Extract Shows In Vitro Antioxidant, Anti-Breast Cancer Activity and Does Not Exhibit Toxicological Effects. *Oxidative Med. Cell. Longev.* **2015,** Article ID 205360.

Yadav, S. S.; Woo, A. A.; Choure, K. Review of Medicinally Important Plant Species *Saraca asoca* (Roxb.). *Int. J. Adv. Res.* **2019,** *7*, 154–166.

CHAPTER 2

Bioactive Compounds and Pharmacological Activities of *Mucuna pruriens* L.

M. INDIRA*, D. LAVANYA, G. PALLAVI, LIYA SIBY and S. KRUPANIDHI

Department of Biotechnology, Vignan's Foundation for Science, Technology and Research, Vadlamudi-522213, Andhra Pradesh, India

Corresponding author. E-mail: indu221007@gmail.com

ABSTRACT

Mucuna pruriens is a tropical legume commonly known as velvet bean or magic bean. It is an annual climbing shrub reaches up to 15 m length. *M. pruriens* plant belongs to the family Fabaceae, which is widely distributed in tropical and subtropical countries. Every part of the plant has medicinal value used as antiparkinsonian, antioxidant, anti-inflammatory, antidiabetic, sexual enhancing, antimicrobial, antiviral, anticholesterolemic and also acts against nerve disorders. The seed is the excellent source of dopamine precursor L-Dopa which is L-3,4-dihydroxy phenyl alanine. The legume consists of high amount of lipids, carbohydrates, minerals, amino acids and fiber. The legume is used as multipurpose for many applications. The mucuna plant is also used as green manure, fixes atmospheric nitrogen and improves soil fertility. Due to the presence of micronutrients, it is used as a vegetable for nourishment. The chapter discusses the geographical distribution of *M. pruriens* plant, bioactive compounds and its chemical structures. Further it emphasized on various pharmacological activities, nutritional value and medicinal value of the plant. The bioactive compounds present in the plant can be used as a source for the development of potential lead molecules to treat the different ailments.

Bioactives and Pharmacology of Legumes. T. Pullaiah, PhD (Ed.)

2.1 INTRODUCTION

Mucuna pruriens L., usually known as velvet bean or dopa bean, belongs to the family Fabaceae. It is an annual climbing legume native to Asia, South America, and Africa (Kumar and Saha, 2013). The common English names are velvet bean, cowage, cowitch, and Lyon bean (Lampariello et al., 2012). It is widely used as an ayurvedic medicine for treating various ailments (Verma et al., 2014). Every part of the plant contains phytochemical compounds having medicinal properties (Shanmugavel and Krishnamoorthy, 2018). There is a huge demand in the Indian market and it is the authentic drug used for Parkinson's disease (Kavitha and Thangamani, 2014). All parts of the plant have rich nutrients and medicinally important biochemical compounds (Divya et al., 2017). The leaves are trifoliate, flowers are white, lavender, or purple, 10 cm long seed pods with hairs that cause itching when contacted with the skin. The itching is due to the presence of a protein mucuna present in hairs (Kavitha and Thangamani, 2014). The seeds contain high starch content, protein, and fiber. The plant is widely used as an anti-Parkinson's, antihypertensive, antiviral, antimicrobial, anticancer, antihypertensive, antioxidant, and antidiabetic (DeFilipps and Krupnick, 2018). It also plays an important role against nervous disorders, rheumatoid arthritis, aging, hypoglycemia, and male infertility (Pathania et al., 2020). The plant is commonly distributed in Asia, Africa, Tropical America, West Indies, USA, and Pacific islands (DeFilipps and Krupnick, 2018).

2.2 BIOACTIVE COMPOUNDS

The phytochemicals present in *M. pruriens* are phytosterols, carbohydrates, alkaloids, glycosides, tannins, flavonoids, saponins, reducing sugars, and steroids (Murthy et al., 2016). The phytochemical constituents are: L-dopa, glutathione, lecithin, β-sitosterol, 6-methoxyharman, gallic acid, indole-3-alkylamines-N, N-dimethyl tryptamine, oleic acids, stearic acid, palmitic acids, and linoleic acids (Verma et al., 2014). The bioactive compounds and their phytochemical class and pharmacological activities are listed in Table 2.1 and the chemical structures are listed in Figure 2.1.

Mucuna seeds contain L-dopa (3-(3,4 dihydroxyl phenyl)-L-alanine), lecithin, glutathione, β-sitosterol, and gallic acid (Pathania et al., 2020). The leaves, roots, seeds, and pods contain indole-3-alkylamine-N, N-dimethyl tryptamine (Kumar and Saha, 2013). The compound 6-methoxyharman is an indole alkaloid present only in leaves and serotonin in pods (Divya et

TABLE 2.1 Bioactive Compounds and Their Pharmacological Activities.

Name of the bioactive compound	Molecular formula	Phytochemical class	Pharmacological activity	References
5-Hydroxy Tryptamine	$C_{10}H_{12}N_2O$	Tryptamines	Antidepressant activity	Rana and Galani (2014)
Phytosterols	$C_{29}H_{50}O$	Stigmastanes	Anti-cholesterol activity	Ratnawati and Widowati (2011)
2'-Hydroxy genistein	$C_{15}H_{10}O_6$	Isoflavo-2-enes	Lowers blood pressure and kidney toxicity	Palanisamy and Venkataraman (2013)
Ambroxol Hydrochloride	$C_{13}H_{19}Br_2ClN_2O$	Phenyl methylamines	Mucolytic agent	Jahan et al. (2010)
Arachidic acid	$C_{20}H_{40}O_2$	Fatty acids and conjugates	Modulate cyclooxygenase-2, that plays a role in inflammation	Akgun et al. (2017)
Cardenolide	$C_{23}H_{34}O_2$	Steroid lactones	Used for heart failure	Fung et al. (2010)
HydroxythioAcetildenafil	$C_{25}H_{34}N_6O_3S$	Indolecarboxylic acids	Male infertility	Mireille et al. (2017)
Coumarin	$C_9H_6O_2$	Coumarins	Anticoagulant agents	Verma et al. (2014)
Linoleic acid	$C_{18}H_{34}O_3$	linoleic acid	Treating diseases of the heart and blood vessels	Tavares et al. (2015)
Eupatin	$C_{18}H_{16}O_8$	Flavones	Anti-inflammatory	Motta et al. (2013)
Gallic acid	$C_7H_6O_5$	Benzene	Anti-Parkinson's disease & anti-amyloid fibril formation	Rai et al. (2017)
Genistein	$C_{15}H_{10}O_5$	Isoflavo-2-enes	Lowers blood pressure & kidney toxicity	Palanisamy and Venkataraman (2013)
Glutathione	$C_{10}H_{17}N_3O_6S$	Amino acids, Peptides and Analogues	Antioxidant activity	Obogwu et al. (2014)
Levodopa	$C_9H_{11}NO_4$	Amino acids, Peptides and Analogues	Anti-Parkinson's disease	Lieu et al. (2012)
Nicotine	$C_{10}H_{14}N_2$	Alkaloids	Neuroprotective activity	Kasture et al. (2013)

TABLE 2.1 *(Continued)*

Name of the bioactive compound	Molecular formula	Phytochemical class	Pharmacological activity	References
Octadeca-9,12-dienoic acid	$C_{18}H_{32}O_2$	Linoleic acids	Treating diseases of the heart and blood vessels	Tavares et al. (2015)
Oleic acid	$C_{18}H_{34}O_2$	Fatty acids and conjugates	Treating diseases of the heart and blood vessels	Tavares et al. (2015)
Palmitic acid	$C_{16}H_{32}O_2$	Fatty acids and Conjugates	Used as precursor for the synthesis of antipsychotic drug paliperidone palmitate	Tavares et al. (2015)
Tetradecanoic acid	$C_{14}H_{28}O_2$	Fatty acids and conjugates	Reduces blood sugar levels	Bhaskar et al. (2011)
Tryptamine	$C_{10}H_{12}N_2$	Tryptamines	Serotonergic hallucinogens	Lampariello et al. (2012)
Linoleic acid	$C_{18}H_{32}O_2$	Linoleic acids	Octadecanoic acid, Biosynthesis of Prostaglandins	Lampariello et al. (2012)
Bufotenine	$C_{12}H_{16}N_2O$	Tryptamine alkaloid and tertiary amine	Affects cardiovascular function	Szabo (2003)
6-methoxy harman	$C_{13}H_{12}N_2O$	B-carboline alkaloid	antidepressant	Ferraz et al. (2019)
-3-carboxy-1,1-dimethyl-6,7-dihydroxy-1,2,3,4-tetrahydroisoquinoline	$C_{12}H_{15}NO_4$	Iso quinoline alkaloid	Antioxidant activity	Verma et al. (2014)
-3-carboxy-1,1-dimethyl-7,8-dihydroxy-1,2,3,4-tetrahydroisoquinoline	$C_{11}H_{13}NO_4$	Iso quinoline alkaloid	Antioxidant activity	Verma et al. (2014)

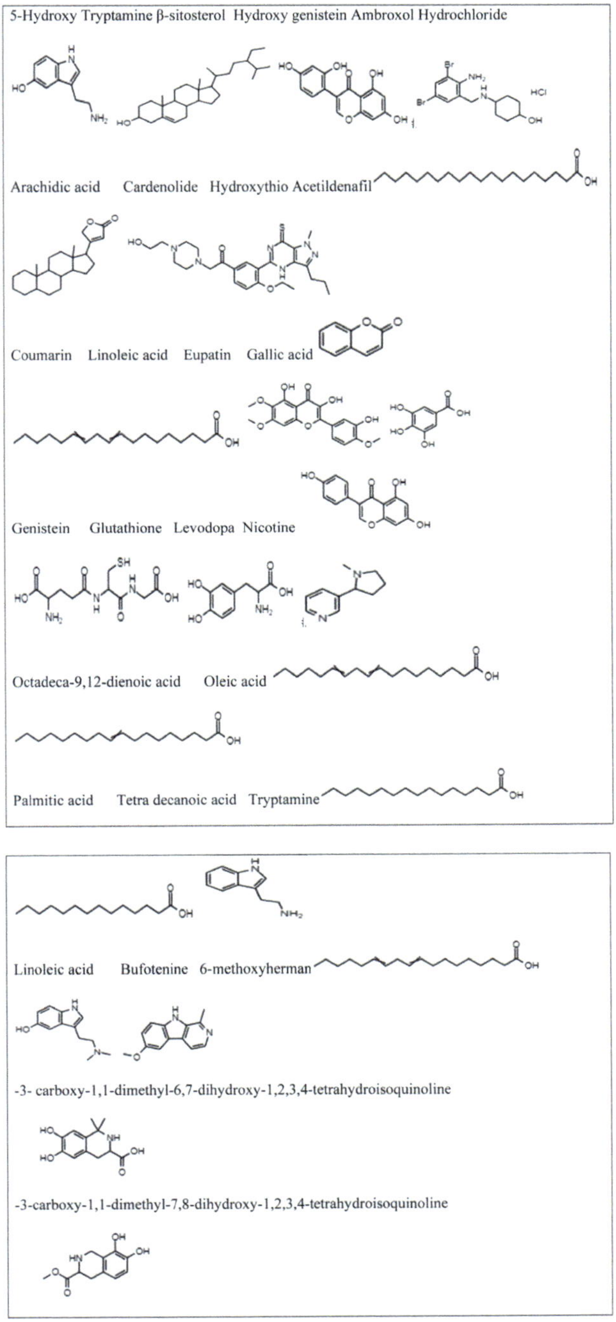

FIGURE 2.1 Chemical structures of bioactive compounds of *M. pruriens*.

al., 2017). The seeds consist of fatty acids that include stearic acid, linoleic acid, oleic acid, palmitic acid, and myristic acid (Kumar and Saha, 2013). The GC-MS analysis data revealed the occurrence of oleic acid, hexadecenoic acid, octadecanoic acid, ascorbic acid, and squalene (Bhaskar et al., 2011). The seeds also consist of iso quinoline alkaloids including -3-carboxy-1,1-dimethyl-7,8-dihydroxy-1,2,3,4-tetrahydroisoquinoline, -3-carboxy-1,1-dimethyl-6,7-di hydroxy-1,2,3,4-tetrahydroisoquinoline, and -3-carboxy-1,2,3,4-tetrahydroisoquinoline (Misra and Wagner, 2004; Natarajan et al., 2012; Yadav et al., 2017).

Levodopa is a precursor compound of the brain neurotransmitter dopamine which is an amino acid precursor of catecholamines (Ovallath and Sulthana, 2017). Dopamine cannot enter the brain without crossing a barrier called blood–brain barrier (Haddad et al., 2018). Levodopa additionally can cross the blood–brain barrier and act locally by using the remaining healthy dopamine-producing neurons to synthesize dopamine. Hence, it restores dopamine and reduces symptoms of Parkinson's disease (Ovallath and Sulthana, 2017). The other alkaloids present in this plant are prurienine, prurienimine, pruriemidine, 1,2,3,4-tetra hydroxyquinoline and they are selective to opioid receptors (Lampariello et al., 2012). In phytosterols, the β-sitosterol plays a role in the ordering of fatty acid chains in cell membrane. The phytosterols reduce cholesterol levels in the human body (Pathania et al., 2020).

The *M. pruriens* flour and seeds consist of 43.12% and 43.4% total proteins, starch 37.19% and 33.33%, lipids 7% and 7.6%, and fiber 5.64% and 2.36%, respectively. The minerals present in flour and seeds are iron, phosphorous, and potassium (Tavares et al., 2015). The amino acids present in *Mucuna pruriens* are glycine, leucine, alanine, methionine, glutamic acid, arginine, lysine, serine, histidine, cytosine, proline, tyrosine, and threonine. It also contains other amino acids (Deokar et al., 2016).

2.3 PHARMACOLOGICAL ACTIVITIES

2.3.1 Neuroprotective Role in Parkinson's Disease

The neuroprotective role of seed powder was evaluated in Parkinson's disease patients clinically. In this study different doses (15 g and 30 g) of seed powder were given to Parkinson's disease patients in a randomized manner at weekly intervals and related to the standard drug L-dopa/carbidopa. Following 4 h of drug ingestion in patients, the dose 30 g seed powder showed rapid onset of action and decrease in dyskinesias compared with the standard drug (Katzenschlager et al., 2004). *M. pruriens* seed extract was evaluated for its therapeutic effect in Parkinson's disease. The Swiss albino mice were taken for evaluating the neuroprotective role and found a significant decrease in tyrosine hydroxylase enzyme activity in the brain (Singh et al., 2016).

2.3.2 Antidiabetic Activity

The antidiabetic activity of *M. pruriens* seed extract was investigated in diabetes wistar rats induced by streptozocin. The rats fed with seed extract had improved cholesterol and insulin levels in serum compared with the control. The beta cell mass in islets of pancreas was increased and a decrease in necrotic changes was observed. The liver hepatocytic cells were restored and reduction in necrotic changes was observed (Rajesh et al., 2016).

The hypoglycemic activity was evaluated for the aqueous extract of *M. pruriens* in normal, glucose-loaded, and streptozocin-induced diabetic rat models. The rats fed with seed extract at a dose of 100 mg/kg and 200 mg/kg of body weight reduced the blood glucose levels in normal rats after oral administration of glucose for 2 h. In diabetic rats, oral administration of seed extract reduced the blood glucose levels after 21 days (Bhaskar et al., 2008).

2.3.3 Fertility Enhancement

Mucuna pruriens seed powder was evaluated for stress and fertility enhancement in infertile men. 5 g/day seed powder was given orally to the 60 infertile subjects and the sperm samples were collected before and after treatment. In case of the subjects with psychological stress, the sperm count decreased. The serum seminal plasma lipid profiles and cortisol levels were elevated. The ascorbic acid and seminal plasma glutathione levels decreased, whereas the catalase and superoxide dismutase activities were reduced. The *M. pruriens* treated test group improved the motility and count of sperm cells. The seminal plasma lipid peroxide levels and psychological stress were reduced. The ascorbic acid, SOD, catalase, and seminal plasma glutathione levels were restored (Shukla et al., 2010). Abraham (2011) studied the role of fertility in male albino rats using mucuna seed powder. The results revealed the increase in sperm cells and decrease in the abnormal sperm count. The histological studies also revealed the improvement in spermicides that were densely packed in seminiferous tubules in treated mice. Noah et al. (2014) investigated the role of fertility in female Sprague Dawley rats. After oral administration of seed extract ovulation, reproductive hormones levels, estrous cycle, and oxidation stress in ovaries of rats were studied. This study reported increased levels of FSH and LH in serum which in turn enhanced the fertility and number of oocytes in the ovulation process.

2.3.4 Hepatoprotective Activity

The hepatoprotective activity was evaluated for hydroethanolic extract of *M. pruriens* leaves and also in-vivo antioxidant activity was observed in rat models (Obogwu et al., 2014). The ethanolic seed extract was evaluated for hepatoprotective activity in albino wistar rats against paracetamol-induced hepatotoxicity. The biochemical parameters of hepatic damage like SGPT, SGOT, bilirubin, antioxidant, and alkaline phosphatase levels were evaluated in both test and control groups. The seed extract treated group was restored their biochemical marker levels to normal levels (Kumar et al., 2014).

2.3.5 Antiviral Activity

Phytosterols are naturally occurring sterol molecules having similar structure of cholesterol. According to many studies phytosterols have antiviral activity by the interaction with the lipid raft components of the cell membrane. The lipid raft components are rich in cholesterol, glycosyl phosphatidyl inositol-anchored proteins, glycosphingolipids, and signaling proteins and contribute to cell-to-cell communication. When phytosterols bind to the lipid rafts of the cell membrane, it results in interaction between them that leads to destabilization of lipid raft shape. Hence, phytosterol could decrease the adhesion of virus to the host cells (Kiani et al., 2020).

2.3.6 Antimicrobial Activity

Ogundare and Olorunfemi (2007) evaluated the antimicrobial activity against indicator organisms. The solvents ethanol, methanol, and acetone extracts of *M. pruriens* showed low antimicrobial activity against selected indicator organisms namely *Staphylococcus aureus*, *Salmonella typhi, S. pneumoniae,*

K. pneumoniae, P. mirabilis, E. coli, Aspergillus niger, Candida albicans, and *Fusarium solani*. The MIC value for the extracts ranged between 24 mg/mL and 25 mg/mL. Pandey and Pandey (2016) evaluated the alcoholic extract of the *M. pruriens* leaves against *Pseudomonas aeruginosa, Staphylococcus* sp., *Bacillus subtilis*, and *Escherichia coli*. The extract at a concentration of 200 mg/kg exhibited antibacterial activity against all strains and streptomycin used as a control. The concentration 800 mg/mL displayed good scavenging activity.

2.3.7 Antivenom Activity

The antivenom activity was evaluated for *M. pruriens* extract against snake venoms in *in-vitro* observation. The *M. pruriens* extract was used to produce the antibodies and the anti-*M. pruriens* extract was used to neutralize snake venom in *in-vitro* against *Calloselasma rhodostoma* venom and *Naja sputatrix* venom. The indirect ELISA assay cross-reactions were reported between anti-*Mucuna pruriens* extract IgG and poisonous snake venoms (Tan et al., 2009). Fung et al. (2011) evaluated the pharmacological activity of seed extract against snake venom in male Sprague Dawley rat models. The seed extract was protective against *Naja sputatrix* venom given at a dose of 0.45 μg/g.

2.3.8 Antitumor Activity

M. pruriens methanolic extract was investigated for antitumor activity in Swiss albino mice. Mice were fed with methanolic extract at a dose of 125 and 250 mg/kg body weight once in a day for 14 days. Tumor was inoculated in rats and after 24 h of inoculation the parameters such as viable cell count, tumor volume, and packed cell volume were observed in treated and control group. The decrease in the weight of the tumor and restored hemoglobin levels to normal in the test group of animals compared with the control group were observed (Rajeshwar et al., 2005a). The peptide fractions of *Mucuna pruriens* showed activity against liver cancer, and hepatitis C virus. The peptide fractions (5-10 KDa) collected by ultrafiltration showed anti-hepatoma activity against the liver cancer cell (HepG2 and QGy7703) by the MTT assay. The functions 5-10 KDa and >10 KDa showed activity against hepatitis C virus (Taghizadeh et al., 2021).

2.3.9 Anti-Inflammatory Activity

The cotton pellet implantation model and the carrageenan-induced paw edema model were used for the anti-inflammatory activity of the aerial parts of the *M. pruriens* in Wistar rats. The ethanolic extract reduced the size of the paw edema and weight of the cotton pellet and the results were compared with control group (Bala et al., 2011). The *Mucuna pruriens* (50 mg/kg) showed decreased inflammation at a rate of 45% and 38.8% for acute inflammation and chronic inflammation, respectively, in albino rats compared with standard drug 10 mg/kg diclofenac at a rate of 9.8% and 62% respectively (Uchegbu et al., 2016).

2.3.10 Aphrodisiac Effect

The aphrodisiac activity was investigated for seeds of *M. pruriens* at a dose of 75 mg/kg body weight given as an aqueous suspension in male albino rats. The parameters such as mount latency and mount frequency were evaluated. The seed powder increased the sexual activity in rats (Kumar et al., 1994).

2.3.11 Antioxidant Activity

Tripathi and Upadhyay (2002) evaluated the antioxidant activity of alcoholic extract using rat liver homogenate. The antioxidant effect was checked on oxidation of glutathione content, iron-induced lipid peroxidation, and its interaction with superoxide and hydroxyl radicals. The $FeSO_4$-induced lipid peroxidation was inhibited and the rate of oxidation of glutathione content was not changed. This indicates that the extract has anti-lipid peroxidation activity. Rajeshwar et al. (2005b) evaluated in-vitro lipid peroxidation activity for *M. pruriens* methanolic extract. The IC_{50} value for plant extract is 217.25 μg/mL compared with controls ascorbate 41 μg/mL and quercetin 19.75 μg/mL. Kumar et al. (2010) evaluated antioxidant activity of various extracts of whole plant using DPPH assay, super oxide anion scavenging activity, and iron chelating activity. In case of DPPH assay, the ethyl acetate extract showed an IC_{50} value of 420 μg/mL and for control rutin 480 μg/mL. The ethyl acetate extract was effective in superoxide radical scavenging activity than petroleum ether and methanol extract. In iron chelating activity, the methanol extract was more effective compared with ethyl acetate and petroleum ether extract.

2.3.12 Anti-Helminthic Activity

Okoli et al. (2015) evaluated anti-helminthic activity of crude and solvent extracts of *M. pruriens* on *Pheretima posthuma* and *Ascaris suum* eggs and larval stages by *in-vitro*. The crude and Soxhlet extract showed difference in the time for paralysis and death of the worms that result in both ovicidal and larvicidal activity.

2.3.13 Mucuna pruriens as Food and Feed Supplement

The seeds are used as a source of food by tribal of Asia and Africa continents. The ethnic group of Nigerian people use leaves and pods as vegetables and seeds as condiments. Mature seeds are toasted and used to prepare soup and sauce. The seeds are used as thickeners for soup, several food items, and in beverages by Nigerian people. *M. pruriens* seeds are used to prepare fermented products namely Mucuna fortified weaning food, Mucuna tempe, and Mucuna condiment through bacterial and fungal fermentation using *Bacillus* sp. and *Rhizopus oligosporus*, respectively. Mucuna plant is used as a diary livestock feed supplement to increase the milk yield, weight, and growth of the animals, laying hens, and broilers (Sridhar and Bhat, 2007). The ensiling of Mucuna seed in varying proportions of maize grains is used as a feed supplement in ruminant animals. The lactating goats are used in the feeding trail and the milk yields, doe weight, feed intake, and kid weights are monitored. The ensiling decreased the crude protein content but increased the palatability. The feed intake increased and the doe weight decreased. The silage fermentation content and toxicity were decreased by 10-47 % (Matenga et al., 2003).

2.3.14 Nutritional Properties

Mucuna pruriens consists of rich source of carbohydrates, crude proteins, crude lipids, and dietary fiber. The fatty acids present in high levels are linolenic acid, stearic acid, oleic acid, and linoleic acid. The behenic acid is the antinutritional fatty acid present in *M. pruriens*. The minerals are potassium, sodium, magnesium, calcium, phosphorous, zinc, manganese, copper, and iron. The amino acids are proline, arginine, threonine, aspartic acid, serine, alanine, glycine, glutamic acid, lysine, cysteine, methionine, valine, isoleucine, tyrosine, leucine, phenyl alanine, histidine, and tryptophan (Janardhanan, 2000; Kala and Mohan, 2010).

2.3.15 Antinutritional Properties

The antinutritional factors present in *Mucuna pruriens* are phytic acid, trypsin inhibitors, hydrogen cyanide, tannins, oligosaccharides, and phytohemagglutinins. Phytic acid present in the seeds lowers the bioavailability of essential minerals, forms complexes with proteins, and inhibits the digestion of proteins. The hydrogen cyanide and trypsin inhibitor levels are low in *M. pruriens*. The phytohemagglutinin activity was higher with respect to "A" blood group of human erythrocytes and low levels of activity with respect to `"O" blood group of human erythrocytes (Janardhanan, 2000; Kala and Mohan, 2010).

2.3.16 Mitochondrial Dysfunction and DNA Damage

The *M. pruriens* seed extract was evaluated for DNA damage and mitochondrial dysfunction in epididymal spermatozoa of hyperglycemic rats. The male wistar albino rats are divided into diabetes-induced streptozocin rats, diabetic rats administered with seed extract (200 mg/kg), normal rats with seed extract (200 mg/kg) and control group. For 60 days the seed extract was administered daily once and after the 60th day, the animals were sacrificed and the sperms were collected from epididymis. The parameters antioxidant activity, lipid peroxidation, ROS, DNA damage, mitochondrial membrane potential, and chromosomal integrity were evaluated. In case of streptozocin-induced diabetes rats the sperm count, viability, and motility were increased compared with the normal rats. The lipid peroxidation and DNA damage were increased in sperm cells. Mitochondrial membrane potential, enzymatic and nonenzymatic functions were affected. In case of diabetic rats administered with seed extract, reduced LPO levels, well-preserved DNA, recovery in antioxidant levels, MMP, and mitochondria functions were shown. The seed extract reduced the sperm damage induced by oxidative stress (Suresh et al., 2013).

KEYWORDS

- **alkaloids**
- **antimicrobial**
- **antioxidant**
- **antiviral**
- **dopamine**
- **L-Dopa**
- ***Mucuna pruriens***

REFERENCES

Abraham, J. Effect of *Mucuna pruriens* Seeds on Fertility of Male Albino Rats *Rattus norvegicus*. *J. Pharm. Res.* **2011,** *4* (1), 233–236.

Akgun, B.; Sari, A.; Ozturk, S.; Erol, F. S.; Ozercan, I. H.; Ulu, R. Effects of *Mucuna pruriens* on Free Fatty Acid Levels and Histopathological Changes in the Brains of Rats Fed a High Fructose Diet. *Med. Principles Pract.* **2017,** *26* (6), 561–566.

Bala, V.; Debnath, A.; Shill, A. K.; Bose, U. Anti-Inflammatory, Diuretic and Antibacterial Activities of Aerial Parts of *Mucuna pruriens* Linn. *Intern. J. Pharmacol.* **2011,** *7* (4), 498–503.

Bhaskar, A.; Nithya, V.; Vidhya, V. G. Phytochemical Evaluation by GC-MS and Antihyperglycemic Activity of *Mucuna pruriens* on Streptozotocin Induced Diabetes in Rats. *J. Chem. Pharm. Res.* **2011,** *3* (5), 689–696.

Bhaskar, A.; Vidhya, V. G.; Ramya, M. Hypoglycemic Effect of *Mucuna pruriens* Seed Extract on Normal and Streptozotocin-Diabetic Rats. *Fitoterapia* **2008,** *79* (7–8), 539–543.

DeFilipps, R. A.; Krupnick, G. A. The medicinal Plants of Myanmar. *PhytoKeys* **2018,** 102, 1–341.

Deokar, G.; Deore, P.; Kshirsagar, S. Phytochemistry and Pharmacological Activity of *Mucuna pruriens*: A Review. *Pharm. Biol. Evaluation.* **2016,** *3*, 50–59.

Divya, B. J.; Suman, B.; Venkataswamy, M.; Thyaga Raju, K. The Traditional Uses and Pharmacological Activities of *Mucuna pruriens* (L) DC: A Comprehensive Review. *Indo Am. J. Pharm. Res.* **2017,** *7* (01), 7516–7525.

Ferraz, C. A. A.; de Oliveira Júnior, R. G.; Picot, L.; da Silva Almeida, J. R. G.; Nunes, X. P. Pre-Clinical Investigations of β-carboline Alkaloids as Antidepressant Agents: A Systematic Review. *Fitoterapia* **2019,** *137*, 104196. DOI: 10.1016/j.fitote.2019.104196.

Fung, S. Y.; Tan, N. H.; Sim, S. M.; Marinello, E.; Guerranti, R.; Aguiyi, J. C. *Mucuna pruriens* Linn. Seed Extract Pretreatment Protects Against Cardiorespiratory and Neuromuscular Depressant Effects of *Naja sputatrix* (Javan Spitting Cobra) Venom in Rats. *Indian J. Exp. Biol.* **2011,** *49*, 254–259.

Fung, S. Y.; Tan, N. H.; Sim, S. M. Protective effects of *Mucuna pruriens* Seed Extract Pretreatment Against Cardiovascular and Respiratory Depressant Effects of *Calloselasma rhodostoma* (Malayan Pit Viper) Venom in Rats. *Trop. Biomed.* **2010,** *27* (3), 366–372.

Haddad, F.; Sawalha, M.; Khawaja, Y.; Najjar, A.; Karaman, R. Dopamine and Levodopa Prodrugs for the Treatment of Parkinson's Disease. *Molecules* **2018,** *23* (1), 40. DOI: 10.3390 /molecules23010040

Jahan, Y.; Mahmood, T.; Bagga, P.; Kumar, A.; Singh, K.; Mujahid, M. Future Prospects of Cough Treatment; Herbal Medicines V/S Modern Drugs. *Intern. J. Pharma. Res. Sci.* **2010,** *6* (9), 3689–3697.

Janardhanan, V. V. K. Nutritional and Anti-Nutritional Composition of Velvet Bean: An Under-Utilized Food Legume in South India. *Int. J. Food Sci. Nutr.* **2000,** *51* (4), 279–287.

Kala, B. K.; Mohan, V. R. Nutritional and Anti-Nutritional Potential of Three Accessions of Itching Bean (*Mucuna pruriens* (L.) DC var. *pruriens*): An Under-Utilized Tribal Pulse. *Int. J. Food Sci. Nutr.* **2010,** *61* (5), 497–511.

Kasture, S.; Mohan, M.; Kasture, V. *Mucuna pruriens* Seeds in Treatment of Parkinson's Disease: Pharmacological Review. *Orient. Pharm. Exp. Med.* **2013,** *13* (3), 165–174.

Katzenschlager, R.; Evans, A.; Manson, A.; Patsalos, P. N.; Ratnaraj, N.; Watt, H.; Lees, A. J. *Mucuna pruriens* in Parkinson's Disease: A Double Blind Clinical and Pharmacological Study. *J. Neurol. Neurosurg. Psych* **2004,** *75* (12), 1672–1677.

Kavitha, C.; Thangamani, C. Amazing Bean *Mucuna pruriens*: A Comprehensive Review. *J. Med. Plants Res.* **2014,** *8* (2), 138–143.

Kiani, A. K.; Dhuli, K.; Anpilogov, K.; Bressan, S.; Dautaj, A.; Dundar, M.; Bertelli, M. Natural Compounds as Inhibitors of SARS-CoV-2 Endocytosis: A Promising Approach Against COVID-19. *Acta bio-medica: Atenei Parmensis* **2020,** *91* (13-S), e2020008. DOI: 10.23750/abm.v91i13-S.10520.

Kumar, C. H.; Ramesh, A.; Mohan, G. K. Hepato Protective and Antioxidant Effects of *Mucuna pruriens* Against Acetaminophen-Induced Hepatotoxicity in Albino Wister Rats. *Res. J. Pharm. Technol.* **2014,** *7* (1), 70–73.

Kumar, D. S.; Muthu, A. K.; Smith, A. A.; Manavalan, R. In Vitro Antioxidant Activity of Various Extracts of Whole Plant of *Mucuna pruriens* (Linn). *Intern. J. Pharm Tech Res.* **2010,** *2* (3), 2063–2070.

Kumar, K. A.; Srinivasan, K. K.; Shanbhag, T.; Rao, S. G. Aphrodisiac Activity of the Seeds of *Mucuna pruriens*. *Indian Drugs* **1994,** *31* (7), 321–327.

Kumar, P.; Saha, S. An Updated Review on Taxonomy, Phytochemistry, Pharmacology and Toxicology of *Mucuna pruriens*. *J. Pharmacogn. Phytochem.* **2013,** *2* (1), 306–314.

Lampariello, L. R.; Cortelazzo, A.; Guerranti, R., Sticozzi, C.; Valacchi, G. The Magic Velvet Bean of *Mucuna pruriens*. *J. Trad. Complem. Med.* **2012,** *2* (4), 331–339.

Lieu, C. A.; Venkiteswaran, K.; Gilmour, T. P.; Rao, A. N.; Petticoffer, A. C.; Gilbert, E. V.; Subramanian, T. The Anti Parkinsonian and Anti Dyskinetic Mechanisms of *Mucuna pruriens* in the MPTP-Treated Nonhuman Primate. *Evidence-Based Complem. Altern. Med.* **2012,** Article ID 840247. DOI: 10.1155/2012/84024.

Matenga, V. R.; Ngongoni, N. T.; Titterton, M.; Maasdorp, B. V. Mucuna Seed as a Feed Ingredient for Small Ruminants and Effect of Ensiling on Its Nutritive Value. *Trop. Subtrop. Agroecosyst.* **2003,** *1* (2–3), 97–105.

Mireille, K. P.; Desire, D. D. P.; Pierre, K. Male Sexual Disorders in Patients with Parkinson Disease: Treatment with Natural Remedies. *Adv. Tissue Eng. Regen. Med.* **2017,** *3* (2), 355–360.

Misra, L.; Wagner, H. Alkaloidal Constituents of *Mucuna pruriens* Seeds. *Phytochemistry* **2004,** *65* (18), 2565–2567.

Motta, E. V. S.; Pinto, N. C. C.; Duque, A. P. N.; Mendes, R. F.; Bellozi, P. M. Q.; Scio, E. Antioxidant, Antinociceptive and Anti-Inflammatory Activity of *Mucuna pruriens* (L.) DC. Leaves *Revista Brasileira de Plantas Medicinais* **2013,** *15* (2), 264–272.

Murthy, S. N.; Sangvikar, S.; Malgaonkar, M. M.; Sharma, C.; Kulkarni, Y. R. In Vitro Physico-Chemical, Phytochemical and Fluroscence Assessment of *Mucuna* sps. *IOSR J. Biotech. Biochem.* **2016,** *2* (2), 1–10.

Natarajan, K.; Narayanan, N.; Ravichandran, N. Review on "Mucuna"—The Wonder Plant. *Int. J. Pharm. Sci. Rev. Res.* **2012,** *17* (1), 86–93.

Noah, O. T.; Chinwe, G. S.; Ayodele, O. A. Fertility Enhancing Potential of *Mucuna pruriens* Seeds in Female Sprague-Dawley Rats. *J. Adv. Med. Med. Res.* **2014,** *4* (16), 3148–3157.

Obogwu, M. B.; Akindele, A. J.; Adeyemi, O. O. Hepatoprotective and In Vivo Antioxidant Activities of the Hydroethanolic Leaf Extract of *Mucuna pruriens* (Fabaceae) in Antitubercular Drugs and Alcohol Models. *Chinese J. Nat. Med.* **2014,** *12* (4), 273–283.

Ogundare, A. O.; Olorunfemi, O. B. Antimicrobial Efficacy of the Leaves of *Dioclea reflexa, Mucuna pruriens, Ficus asperifolia* and *Tragia spathulata*. *Res. J. Microbiol.* **2007,** *2* (4), 392–396.

Okoli, B. J.; Ayo, R. G.; Habila, J. D.; Ndukwe, G. I. In-Vitro Anthelmintic Activity of *Mucuna pruriens* (DC) and *Canarium schweinfurthii* (Engl) on Acarissuum. *J. Emerg. Trends Eng. Appl. Sci.* **2015,** *6* (7), 236–241.

Ovallath, S.; Sulthana, B. Levodopa: History and Therapeutic Applications. *Ann. Indian Acad. Neurol.* **2017,** *20* (3), 185–189.

Palanisamy, N.; Venkataraman, A. C. Beneficial Effect of Genistein on Lowering Blood Pressure and Kidney Toxicity in Fructose-Fed Hypertensive Rats. *Br. J. Nutr.* **2013,** *109* (10), 1806–1812.

Pandey, J.; Pandey, R. Study of Phytochemical and Antimicrobial Activity of Alcoholic Extract of *Mucuna pruriens* (L.) Leaves. *Intern. J. Appl. Res.* **2016,** *2* (2), 219–222.

Pathania, R.; Chawla, P.; Khan, H.; Kaushik, R.; Khan, M. A. An Assessment of Potential Nutritive and Medicinal Properties of *Mucuna pruriens*: A Natural Food Legume. *3 Biotech.* **2020,** *10*, 1–15.

Rai, S. N.; Birla, H.; Singh, S. S.; Zahra, W.; Patil, R. R.; Jadhav, J. P.; Singh, S. P. *Mucuna pruriens* Protects Against MPTP Intoxicated Neuroinflammation in Parkinson's Disease Through NF-κB/pAKT Signaling Pathways. *Front. Aging Neurosci.* **2017,** *9*, 421. doi.org/10.3389/fnagi.2017.00421.

Rajesh, R.; Singh, S. A.; Vaithy, K. A.; Manimekalai, K.; Kotasthane, D.; Rajasekar, S. S. The Effect of *Mucuna pruriens* Seed Extract on Pancreas and Liver of Diabetic Wistar Rats. *Intern. J. Current Res. Rev.* **2016,** *8* (4), 61–67.

Rajeshwar, Y.; Gupta, M.; Mazumder, U. K. Antitumor Activity and In Vivo Antioxidant Status of *Mucuna pruriens* (Fabaceae) Seeds Against Ehrlich Ascites Carcinoma in Swiss Albino Mice. *Iran. J. Pharmacol. Therap.* **2005a,** *4* (1), 46–53.

Rajeshwar, Y.; Gupta, M.; Mazumder, U. K. In-Vitro Lipid Peroxidation and Antimicrobial Activity of *Mucuna pruriens* Seeds. *Iran. J. Pharma. Therap.* **2005b,** *4*, 32–35.

Rana, D. G.; Galani, V. J. Dopamine Mediated Antidepressant Effect of *Mucuna pruriens* Seeds in Various Experimental Models of Depression. *Ayu.* **2014,** *35* (1), 90–97.

Ratnawati, H.; Widowati, W. Anti-Cholesterol Activity of Velvet Bean (*Mucuna pruriens* L.) Towards Hypercholesterolemic Rats. *Sains Malaysiana* **2011,** *40* (4), 317–321.

Shanmugavel, G.; Krishnamoorthy, G. Nutraceutical and Phytochemical Investigation of *Mucuna pruriens* Seed. *Pharma Innov. J.* **2018,** *7* (11), 273–278.

Shukla, K. K.; Mahdi, A. A.; Ahmad, M. K.; Jaiswar, S. P.; Shankwar, S. N.; Tiwari, S. C. *Mucuna pruriens* Reduces Stress and Improves the Quality of Semen in Infertile Men. *Evidence-Based Complem. Altern. Med.* **2010,** *7* (1), 137–144.

Singh, B.; Pandey, S.; Verma, R.; Singh, S.;Mahdi, A. A. Neuroprotective Role of *Mucuna pruriens* in Parkinson's Disease Model System. *Intern. J. Res. Dev. Pharm. Life Sci.* **2016,** *5* (6), 2397–2404.

Sridhar, K. R.; Bhat, R. Agro Botanical, Nutritional and Bioactive Potential of Unconventional Legume-Mucuna. *Livestock Res. Rural Dev.* **2007,** *19* (9), 126–130.

Suresh, S.; Prithiviraj, E.; Lakshmi, N. V.; Ganesh, M. K.; Ganesh, L.; Prakash, S. Effect of *Mucuna pruriens* (Linn.) on Mitochondrial Dysfunction and DNA Damage in Epididymal Sperm of Streptozotocin Induced Diabetic Rat. *J. Ethnopharmacol.* **2013,** *145* (1), 32–41.

Szabo, N. J. Indole Alkylamines in Mucuna Species. *Trop. Subtrop. Agroecosyst.* **2003,** *1* (2–3), 295–307.

Taghizadeh, S. F.; Azizi, M.; Asili, J.; Madarshahi, F. S.; Rakhshandeh, H.; Fujii, Y. Therapeutic Peptides of *Mucuna pruriens L.*—Anti-Genotoxic Molecules Against Human Hepatocellular Carcinoma and Hepatitis C Virus. *Food Sci. Nutr.* **2021,** *00*, 1–7.

Tan, N. H.; Fung, S. Y.; Sim, S. M.; Marinello, E.; Guerranti, R.; Aguiyi, J. C. The Protective Effect of *Mucuna pruriens* Seeds Against Snake Venom Poisoning. *J. Ethnopharmacol.* **2009,** *123* (2), 356–358.

Tavares, R. L.; Silva, A. S.; Campos, A. R. N.; Schuler, A. R. P.; de Souza Aquino, J. Nutritional Composition, Phytochemicals and Microbiological Quality of the Legume, *Mucuna pruriens. Afr. J. Biotech.* **2015,** *14* (8), 676–682.

Tripathi, Y. B.; Upadhyay, A. K. Effect of the Alcohol Extract of the Seeds of *Mucuna pruriens* on Free Radicals and Oxidative Stress in Albino Rats. *Phytother. Res.* **2002,** *16* (6), 534–538.

Verma, S. C.; Vashishth, E.; Singh, R.; Pant, P.; Padhi, M. M. A Review on Phytochemistry and Pharmacological Activity of Parts of *Mucuna pruriens* Used as an Ayurvedic Medicine. *World J. Pharma. Res.* **2014,** *3* (5), 138–158.

Yadav, M. K.; Upadhyay, P.; Purohit, S.; Pandey, B. L.; Shah, H. Phytochemistry and Pharmacological Activity of *Mucuna pruriens*: A Review. *Intern. J. Green Pharm.* **2017,** *11* (2), 69–73.

CHAPTER 3

Bioactives and Pharmacology of *Acacia ataxacantha* DC: A Review

N. K. SETHIYA[1*], V. WALIA[2], S. K. CHAUDHARY[3], and YOGESH CHAND YADAV[4]

[1]*Faculty of Pharmacy, DIT University, Dehradun, Uttarakhand, India*

[2]*Department of Pharmacology, SGT College of Pharmacy, SGT University, Gurugram, Haryana, India*

[3]*Institute of Bioresources and Sustainable Development, Imphal, Manipur, India*

[4]*Faculty of Pharmacy, UP University of Medical Science, Saifai, Etawah, Uttar Pradesh, India*

**Corresponding author. E-mail: neeraj.sethiya@dituniversity.edu.in*

ABSTRACT

Acacia ataxacantha (*A. ataxacantha*) DC is an African traditional medicinal species of *Acacia*. The current book chapter provide a details overview on traditional uses, phytoconstituents investigated and pharmacology of *A. ataxacantha* after through scrutiny of available literature. The plant was documented for prevention and treatment of tooth decay, dysentery, bronchitis, excessive cough, joint pain, headaches, pneumonia, bleeding, abscesses, hyperthermic convulsions, chickenpox, yellow fever, pain reliever, toothache and skin sores in traditional system. Phytoconstituents such as α-amyrenol, lupeol, betulinic acid, betulinic acid-3-trans-caffeate, acthaside, lupenol, α-spinasterol and n-tetradecanyl-3-methoxy-4-hydroxy-trans-cinnamate was identified and separated. Further, plant was found to be safe and effective

Bioactives and Pharmacology of Legumes. T. Pullaiah, PhD (Ed.)

as antibacterial, antifungal, anti-diabetic, anti-inflammation, antipyretic, antioxidant and gastro-intestinal protection.

3.1 INTRODUCTION

Genus *Acacia* consisting about 1400 species belongs to family Fabaceae. *Acacia ataxacantha* DC. is an African species of *Acacia*, commonly known as Flame thorn and characterized by numerous hooked prickles and conspicuous red pods. It is widespread species of Africa. *A. ataxacantha* is a multi-stemmed, large, thorny, and scrambling woody shrub with a tendency for the scrambling shoots due to recurved prickles. It attains height up to 5–10 m with trunk diameter of ~300 mm. Morphologically, leaves are alternate, with spine and carries pinnae (5–12 pairs). Further, twigs contain short spines (pointing toward down), fruit pods are flattened (brownish red in color) and flowers (clustered with terminal spikes) are off-white in color (Lynette and Barbara, 1981; Maroyi, 2018). In traditional medicine plant is documented for tooth decay, dysentery, bronchitis, excessive cough, joint pain, headaches, pneumonia, bleeding, abscesses, hyperthermic convulsions, chickenpox, yellow fever, pain reliever, toothache, skin sores, and human immunodeficiency virus, and as a major ingredient of many traditional herbal mixture including "dzovheyo" (Burkill, 1997; Kereru et al., 2007; MacDonald et al., 2010; Cheikhyoussef et al., 2011; Oladunmoye and Kehinde, 2011; Sani and Aliyu, 2011; Hedimbi and Chinsembu, 2012; Kareji, 2013; Akapa et al., 2014a, b, Maroyi, 2018).

3.2 BIOACTIVES

A. ataxacantha has been reported for various secondary metabolites mainly tannins, flavonoids, and gums (Amoussa et al., 2014). In this context, α-Amyrenol [(3β)-Urs-12-en-3-ol) ($C_{30}H_{50}O$; m/z = 426.7) **[1]** was isolated from root bark extracts of *A. ataxacantha* (Aba et al., 2015). Further, lupeol ($C_{30}H_{50}O$; m/z=426.31) **[2]**, betulinic acid ($C_{30}H_{48}O_3$; m/z =456.7) **[3]**, and betulinic acid-3-trans-caffeate ($C_{39}H_{54}O_6$; m/z=618) **[4]** were reported from the ethyl acetate extract of *A. ataxacantha* stem bark (Amoussa et al., 2016a). In addition, acthaside (7-hydroxy-2-methyl-6-[β-galactopyranosyl-propyl]-4H-chromen-4- one) ($C_{19}H_{24}O_9$; m/z=396.14243) **[5]** was isolated from ethyl acetate extract of *A. ataxacantha* (Amoussa et al., 2016b). Ahmadu et al., (2018) isolated two triterpenoid lupenol ($C_{30}H_{50}O$; m/

z=426.31) **[2]** and α-spinasterol ($C_{29}H_{48}O$; m/z=412.28) **[6]** from stem bark of the *A. ataxacantha* extract in chloroform. Recently, n-Tetradecanyl-3-methoxy-4-hydroxy-trans-cinnamate (tetradecanyl ferulate) ($C_{24}H_{38}O_4$; m/z= 391.2298); a ferulic acid derivative **[7]** and lupenol **[2]** was further isolated from stem bark chloroform extract of *A. ataxacantha* stem bark (Ahmadu et al., 2019).

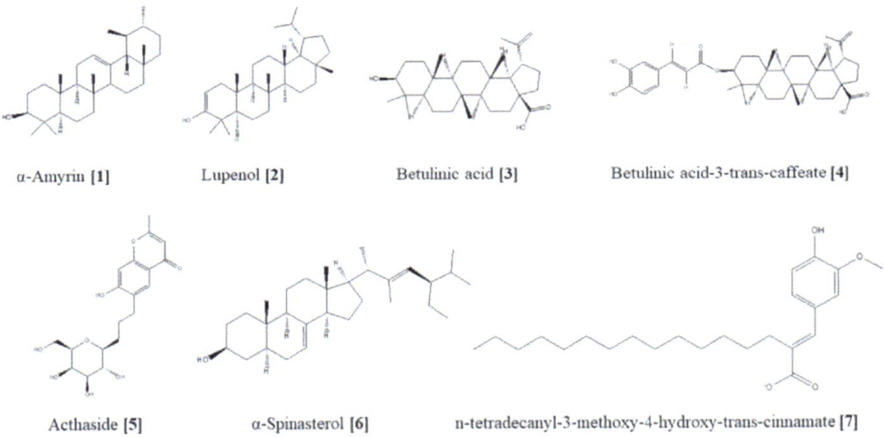

FIGURE 3.1 Chemical structure of different chemical compound isolated from *Acacia ataxacantha*

3.3 PHARMACOLOGY

A wide range of pharmacological activities of *A. ataxacantha* leaves, barks, and roots extracts have been reported.

3.3.1 Toxicity Studies

Oral acute toxicity study conducted on the crude hydroethanol extract of *A. ataxacantha* barks in rats suggested that it is safe and nontoxic at 2000 mg/kg body weight (bw) and all the animals were found to be physically active without any deaths. Further, *A. ataxacantha* (2000 mg/kg/bw) did not produce any significant changes in haematological parameters, biochemical parameters, and bw of animals (Amoussa et al., 2015a). Another study involving the administration of *A. ataxacantha* leaves extract (10, 100, 1000, 1600, 2900, and 5000 mg/kg/bw) in rats suggested that the oral median lethal dose (LD_{50}) of the extract is greater than 5000 mg/kg/bw (Abbas et al., 2017). However, in a recent study involving acute and chronic toxicity on the extract resulted to increases both in the liver (alanine transaminase, aspartate transaminase, and alkaline phosphatase) and kidney (creatinine, urea, and sodium ion) parameters at the dose of 400 mg/kg/bw. It was further suggested that acute administration of the extract is safe, but use for prolong period use may produce harmful effect on the kidney, liver, and stomach (Abbas et al., 2018).

3.3.2 Antibacterial Activities

It has been reported that the dichloromethane, ethyl acetate, hexane, hydroalcohol, and methanol extract of *A. ataxacantha* barks exhibited antibacterial activity against *Enterococcus faecalis*, *Escherichia coli*, *Pseudomonas aeruginosa*, *Staphylococcus epidermidis*, and *Staphylococcus aureus* in broth microdilution method. *A. ataxacantha* extracts exhibited minimum inhibitory concentrations (MIC) of about 0.3 to 5 mg/mL (Amoussa et al. 2014). In another study, the chloroform, ethyl acetate, methanol, and petroleum ether root extracts and one of the isolated compound, that is, α-amyrenol was evaluated against *Bacillus subtilis*, *Corynebacterium ulcerans*, *E. coli*, *Proteus mirabilis*, *Klebsiella pneumoniae*, *Salmonella enteritidis*, *Salmonella typhi*, *S. aureus*, *Streptococcus faecalis*, *Streptococcus pneumoniae*, *P. aeruginosa*, and *Streptococcus pyogenes* using well diffusion method. The result of study illustrated the MIC ranged from 2.5 to 10 mg/mL and minimum bactericidal concentration (MBC) ranged from 2 to 20 mg/mL for extracts, while MIC and MBC values of α-amyrenol were found to be 12.5–25 mg/mL and 25–50 mg/mL, respectively. However, none of the extract showed any activity against *C. ulcerans*, *P. mirabilis*, and *S. faecalis* (Aba et al., 2015). Further, the antibacterial activity has been reported by the betulinic acid-3-trans-caffeate isolated from the barks extract of *A. ataxacantha* against *P. aeruginosa*, *E. faecalis*, *S. aureus*, and *S. epidermidis* in microdilution and disc diffusion assay method exhibited MIC values 12.5–50 μg/mL and MBC values 25–50 μg/mL (Amoussa et al., 2016a). Further, 7-hydroxy-2-methyl-6-[β-galactopyranosyl-propyl]-4Hchromen-4-one 4 isolated from ethyl acetate extract of *A. ataxacantha* barks showed antibacterial activity and shown MIC value of 25–50 μg/mL and MBC values 25–50 μg/mL, respectively (Amoussa et al., 2016b).

3.3.3 Antifungal Activities

Antifungal activities of chloroform, methanol, ethyl acetate, and petroleum ether extract of *A. ataxacantha* roots and its active constituent α-amyrenol was evaluated against *Candida albicans*, *Candida tropicalis,* and *Candida krusei* in well diffusion method. It was observed that none of the extract was active against *C. tropicalis.* Further, MIC values of 5–10 mg/mL and minimum fungal concentration (MFC) of 10–20 mg/mL obtained for extract. However, MIC and MFC values for α-amyrenol was found to be 25 and 50 mg/mL, respectively (Aba et al., 2015). In another study, the betulinic acid-3- trans-caffeate isolated from *A. ataxacantha* barks extract showed antifungal activity against *C. albicans* in microdilution and disc diffusion assay method and exhibited MIC value of 12.5 μg/mL and MFC value of 25 μg/mL (Amoussa et al., 2016a). In addition, *A. ataxacantha* barks extract showed antifungal activities against *Aspergillus clavatus*, *Aspergillus fumigatus*, *Aspergillus flavus*, *Aspergillus nidulans*, *Aspergillus parasiticus,* and *Aspergillus ochraceus* in agar diffusion method. It was found that *A. ataxacantha* bark extracts inhibited sporulation and mycelial growth and showed the percentages inhibition ranging from 33.8 to 99.7% and 5.4 to 53.0%, respectively. Further acthaside, that is, 7-hydroxy-2-methyl-6-[β-galactopyranosylpropyl]-4H-chromen-4-one isolated from ethyl acetate extract of *A. ataxacantha* showed activity against *C. albicans* and exhibited MIC value of 25 μg/mL and MFC value of 25 μg/mL (Amoussa et al., 2016b).

3.3.4 Antidiabetic Activities

It has been reported that the oral administration of *A. ataxacantha* barks extract (125 mg/kg/bw) for 7 days showed antidiabetic activity in streptozotocin-induced diabetic rats (Arise et al., 2014). Further, *A. ataxacantha* root extract reduced the blood glucose in streptozotocin-induced hyperglycaemic rats to levels comparable to reference drug metformin after 7 days of treatment (Arise et al., 2016).

3.3.5 Antiinflammatory Activities

A. ataxacantha leaves extract (200 and 400 mg/kg/bw) showed antiinflammatory activities in rats in carrageenan and albumin-induced paw edema model of inflammation. It was observed that the treatment with the extracts reduced the edema after 3rd hour in a dose-dependent manner (Abbas et al., 2017).

3.3.6 ANTIPYRETIC ACTIVITIES

Hydroalcohol extract of *A. ataxacantha* leaves was evaluated for antipyretic activity against the yeast-induced pyrexia and the rat body temperature was recorded after every 20 minutes of interval time for 60 minutes. It was found that there was not any significant difference in the mean rectal temperature obtained after treatment with various doses of extract when compared with the negative control group. It was further suggested that the extract does not possess any significant antipyretic activity (Abbas et al., 2017).

3.3.7 Antioxidant Activities

A. ataxacantha barks extract exhibited antioxidant activity in 2, 2-diphenyl-1-picrylhydrazyl (DPPH) radical scavenging and ferric reducing antioxidant power assays. The extracts exhibited antioxidant activities (0.7–92.6%) in DPPH assay, while result obtained from the ferric reducing antioxidant power assay of the extracts varied from 120.3 to 1273.6 μmol of ascorbic acid equivalent (AAE) per gram. The strongest ferric reducing ability was found in ethyl acetate extract with value of 1273.6 μmol AAE/g followed by methanol (849.1 μmol AAE/g), hydroalcohol (816.7 μmol AAE/g), dichloromethane (489.4 μmol AAE/g), and n-hexane (120.3 μmol AAE/g) (Amoussa et al., 2015b). It has been further reported that the betulinic acid-3-trans-caffeate, lupeol, betulinic acid, and 7-hydroxy-2- methyl-6-[β-galactopyranosyl-propyl]-4Hchromen-4-one isolated from the barks extract of *A. ataxacantha* also showed antioxidant activity in DPPH assay. The IC_{50} value for betulinic acid-3-transcaffeate is 3.6 μg/mL; for lupeol is 16.8 μg/mL; for betulinic acid is 25.2 μg/mL and for 7-hydroxy-2-methyl-6-[β-galactopyranosyl-propyl]-4Hchromen-4-one is 3.6 μg/mL (Amoussa et al., 2016a, b).

3.3.8 Gastro-Intestinal Role

It has been reported that the ethanol extract of *A. ataxacantha* administered (500 and 1000 mg/kg) intraperitoneally (i.p.) reduces the ulcer in dose-dependent manner. Further, *A. ataxacantha* extract treatment produced no pharmacologic effect at lower concentration (10 and 100 mg/mL/i.p.), while relaxes the ileum at the concentration of 200 mg/mL/i.p. suggesting the ulcer protective and antimotility effect (Abbas et al., 2019). In another study, on the peptic ulcers in the indomethacin and stress, induced ulcer model methanol leaves extract of *A. ataxacantha* (100 and 200 mg/kg/bw) results in the reduction of the ulcer index in a dose-dependent manner and normalized the gastric pH (Akapa et al., 2014a). In addition, *A. ataxacantha* has been reported to show the laxative properties in the loperamide-induced constipation in rats. During the study, constipated control rats received normal saline, while the treated constipated rats group were given *A. ataxacantha* (100, 200, and 400 mg/kg/bw) extract for 10 days. It has been observed that the administration of *A. ataxacantha* (100, 200, and 400 mg/kg) extract for 10 days produced significant laxative activities and reduced loperamide-induced constipation in a dose-dependent manner (Akapa et al., 2014b). This line of evidences shows the potential role of the *A. ataxacantha* in the treatment of GIT disorders.

KEYWORDS

- **acacia**
- **bioactives**
- **phytoconstituents**
- **antibacterial**
- **antidiabetic**

REFERENCES

Aba, O. Y.; Ezuruike, I. T.; Ayo, R. G.; Habila, J. D.; Ndukwe, G. I. Isolation, Antibacterial and Antifungal Evaluation of α-Amyrenol from the Root Extract of *Acacia ataxacantha* DC. *Sch. Acad. J. Pharm.* **2015,** *4* (2), 124–131.

Abbas, M. Y.; Ejiofor, J. I.; Yakubu, M. I. Acute and Chronic Toxicity Profiles of the Methanol Leaf Extract of *Acacia ataxacantha* D. C. (Leguminosae) in Wistar Rats. *Bull. Fac. Pharm. Cairo Univ.* **2018,** *56*, 185–189.

Abbas, M. Y.; Ejiofor, J. I.; Yaro, A. H.; Yakubu, M. I.; Anuka, J. A. Anti-Inflammatory and Antipyretic Activities of the Methanol Leaf Extract of *Acacia ataxacantha* (Leguminosae) in Mice and Rats. *Bayero J. Pure Appl. Sci.* **2017,** *10* (1), 1–5.

Abbas, M. Y.; Yakubu, M. I.; Aliyu, I. M.; Yakubu, R. A. Gastrointestinal Tract Profile of Laboratory Animals Treated with Ethanol Extract of *Acacia ataxacantha* (Leguminosae) D. C. Stem Bark. *J. Pharm. Bioresour.* **2019,** *16* (2), 152–157.

Ahmadu, A. A.; Agunu, A.; Lawal, B. A. Ferulic Acid Ester from the Stem Bark of *Acacia ataxacantha. Afr. J. Biomed. Res.* **2019,** *22*, 214–217.

Ahmadu, A. A.; Agunu, A.; Myrianthopoulos, V.; Fokialakis, N. Chemical Constituents of the Stem Bark of *Acacia ataxacantha* (Fabaceae). *Trop. J. Nat. Prod. Res.* **2018,** 2(8), 380–382.

Akapa, T. C.; Arise, R. O.; Olajide, O. J.; Ikusemoro, I. T. Ulceroprotective Potentials of Methanolic Extract of *Acacia ataxacantha* Leaves in Indomethacin and Stress Induced Gastric Ulcer Models. *Int. J. Biochem. Res. Rev.* **2014a,** *4* (4), 312–321.

Akapa, T. C.; Obidola, S. M.; Philip, F. O. Loperamide Induced Constipated Wistar Rats: Laxative Role of Aqueous Extract of *Acacia ataxacantha* leaves. *World J. Pharm. Pharma. Sci.* **2014b,** *3* (12), 189–199.

Amoussa, A. M. O.; Bourjot, M.; Lagnika, L.; Vonthron Sénécheau, C.; Sanni, A. Acthaside: A New Chromone Derivative from *Acacia ataxacantha* and Its Biological Activities. *BMC Complem. Altern. Med.* **2016b,** *16*, 506.

Amoussa, A. M. O.; Lagnika, L.; Bourjot, M.; Vonthron Sénécheau, C.; Sanni, A. Triterpenoids from *Acacia ataxacantha* DC: Antimicrobial and Antioxidant Activities. *BMC Complem Altern Med* **2016a,** *16*, 284.

Amoussa, A. M. O.; Lagnika, L.; Sanni, A. *Acacia ataxacantha* (Bark): Chemical Composition and Antibacterial Activity of the Extracts. *Int. J. Pharm. Pharm. Sci.* **2014,** *6* (11), 138–141.

Amoussa, A. M. O.; Lagnika, L.; Tchatchedre, M.; Laleye, A.; Sanni, A. Acute Toxicity and Antifungal Effects of *Acacia ataxacantha* (Bark). *Int. J. Pharmacog. Phytochem. Res.* **2015a,** *7* (4), 661–668.

Amoussa, A. M. O.; Sanni, A.; Lagnika, L. Antioxidant Activity and Total Phenolic, Flavonoid and Flavonol Contents of the Bark Extracts of *Acacia ataxacantha. J. Pharmacog. Phytochem.* **2015b,** 4(2), 172–178.

Arise, R. O.; Akapa, T.; Adigun, M. A.; Yekeen, A. A.; Oguntibeju, O. O. Normoglycaemic, Normolipidaemic and Antioxidant Effects of Ethanolic Extract of *Acacia ataxacantha* Root in Streptozotocin—Induced Diabetic Rats. *Nat. Sci. Biol.* **2016,** *8* (2), 144–150.

Arise, R. O.; Ganiyu, A. I.; Oguntibeju, O. O. Lipid Profile, Antidiabetic and Antioxidant Activity of *Acacia ataxacantha* Bark Extract in Streptozotocin-Induced Diabetic Rats. In *Antioxidant Antidiabetic Agents and Human Health*; Oguntibeju, O. O., Ed.; In Tech, 2014. http://dx.doi.org/10.5772/57151

Burkill, H. M. The Useful Plant of West Tropical Africa. *R Bot. Garden Kew, UK* **1997,** *3*, 178–179.

Cheikhyoussef, A.; Shapi, M.; Matengu, K.; Ashekele, H. M. Ethnobotanical Study of Indigenous Knowledge on Medicinal Plant Use by Traditional Healers in Oshikoto Region, Namibia. *J. Ethnobiol. Ethnomed.* **2011,** *7*, 1–11.

Hedimbi, M.; Chinsembu, K. C. Ethnomedicinal Study of Plants Used to Manage HIV/AIDS-Related Disease Conditions in the Ohangwena Region, Namibia. *Int. J. Med. Pl. Res.* **2012,** *1* (1), 4–11.

Kareji, A. E. Evaluation of Herbal Drugs Used to Treat Fungal and Bacterial Diseases in Kenya. *Int. J. Herb. Med. Plants* **2013,** *1* (4), 85–87.

Kereru, P. G.; Kenji, G. M.; Gachanga, A. N.; Keriko, J. M.; Mungai, G. Traditional Medicines Among EMBU and Mbeere People of Kenya. *Afr. J. Complement. Alt. Med.* **2007,** *4*(1), 75–86.

Lynette, D.; Barbara, J. *Acacia, a Field Guide to the Identification of the Species of Southern Africa*; 1st ed.; Centaur, 1981; p 121.

MacDonald, I; Joseph, O. E.; Harriet, M. E. Documentation of Medicinal Plants Sold in Markets in Abeokuta, Nigeria. *Trop. J. Pharm. Res.* **2010,** *9* (2), 110–118.

Maroyi, A. Review of Ethnopharmacology and Phytochemistry of *Acacia ataxacantha. Trop. J. Pharm. Res.* **2018,** *17* (11), 2301–2308.

Oladunmoye, M. K.; Kehinde, F. Y. Ethnobotanical Survey of Medicinal Plants Used in Treating Viral Infections Among Yoruba Tribe of South Western Nigeria. *Afr. J. Microbiol. Res.* **2011,** *5* (19), 2991–3004.

Sani, H. D.; Aliyu, B. S. A Survey of Major Ethno Medicinal Plants of Kano North, Nigeria, Their Knowledge and Uses by Traditional Healers. *Bayero J. Pure Appl. Sci.* **2011,** 4(2), 28–34.

CHAPTER 4

A Review on Bioactives and Pharmacology of *Acacia auriculiformis* A. Cunn. ex Benth.

N. K. SETHIYA[1*], V. WALIA[2], S. K. CHAUDHARY[3], YOGESH CHAND YADAV[4], and S. BHARGAVA[1]

[1]*Faculty of Pharmacy, DIT University, Dehradun, Uttarakhand, India*

[2]*Department of Pharmacology, SGT College of Pharmacy, SGT University, Gurugram, Haryana, India*

[3]*Institute of Bioresources and Sustainable Development, Imphal, Manipur, India*

[4]*Faculty of Pharmacy, UP University of Medical Science, Saifai Etawah, Uttar Pradesh, India*

Corresponding author. E-mail: neeraj.sethiya@dituniversity.edu.in

ABSTRACT

Acacia auriculiformis A. Cunn. ex Benth. is a medicinally important tree, with a marked geographical distribution. The current chapter focuses on the several significant medicinal components, contributing to its remarkable usage. Hence, in brief the identified metabolically significant bioactive components are categorized in carbohydrates, fatty acids, flavonoids, glycosides and tannins. The chapter explored a wide range of traditional and pharmacological usages of *A. auriculiformis* from available literature. The detailed description of evidences, proves the scientific importance of *A. auriculiformis* as future remedies to develop several value added products.

Bioactives and Pharmacology of Legumes. T. Pullaiah, PhD (Ed.)

4.1 INTRODUCTION

Acacia auriculiformis A. Cunn. ex Benth. is a resilient, deciduous, vigorously growing evergreen tree belongs to family Fabaceae (Leguminosae). It attains height up to 50 ft, leaves are 4–10 cm long and 2–3 cm wide (alternate, simple, falcate, blade-like, and sickle shaped), flowers are greenish-white colored (clustered) and pods are 1 cm wide thick twisted (coiled). The distribution pattern is wider in various parts of world including Australia, Indonesia, New Guinea, Queensland, and India. Interestingly, popular by so many common names such as Darwin black wattle, Black wattle, Tan wattle, Earpod wattle, Australian wattle, Australian babul, Australian acacia, Acacia, Akashmani, and Kasia. In India, it was first recorded in 1946 and currently widely distributed on roadsides, railway lines sides and in parks. Traditionally, it has been documented in the treatment of rheumatism, pains, inflammation, sore eyes, aches, itching, allergy, rashes, and malaria. In addition, it serves as source of fuel wood, paper production industries, furniture, toys, and tools industries (Kirtikar and Basu, 1984; Kushalapa, 1991; Ghosh et al., 1993; Langeland and Burks, 1998; Orwa et al., 2009; Sathya and Siddhuraju, 2012; Pullaiah, 2006; Sharma et al., 2016, 2017).

4.2 BIOACTIVES

A. auriculiformis was found to contain several metabolites including natural coloring pigments (Chakraborty et al., 2020), carbohydrates, fatty acids, flavonoids, glycosides, and tannins (Sharma et al., 2015, 2017; Rangra et al., 2019a).

4.2.1 Carbohydrates

A. auriculiformis is identified as rich source of carbohydrates such as D-glucuronic acid (1), methylglucuronic acid (2), D-galactose (3), L-rhamnose (4), L-arabinose (5), D-xylose (6), and D-glucose (7) (Anderson, 1978; Ray et al., 1989). The chemical structure of various carbohydrates is depicted in Figure 4.1.

4.2.2 Fatty Acids

The seeds of *A. auriculiformis* were studied by gas-liquid chromatography for composition consisting fatty acid with reference to epoxy acid ~ Epoxy

18:1 (8). The fatty acid contains was found to be 4.9% and Figure 4.1 depicted the chemical structure of epoxy acid (Chowdhury et al., 1983).

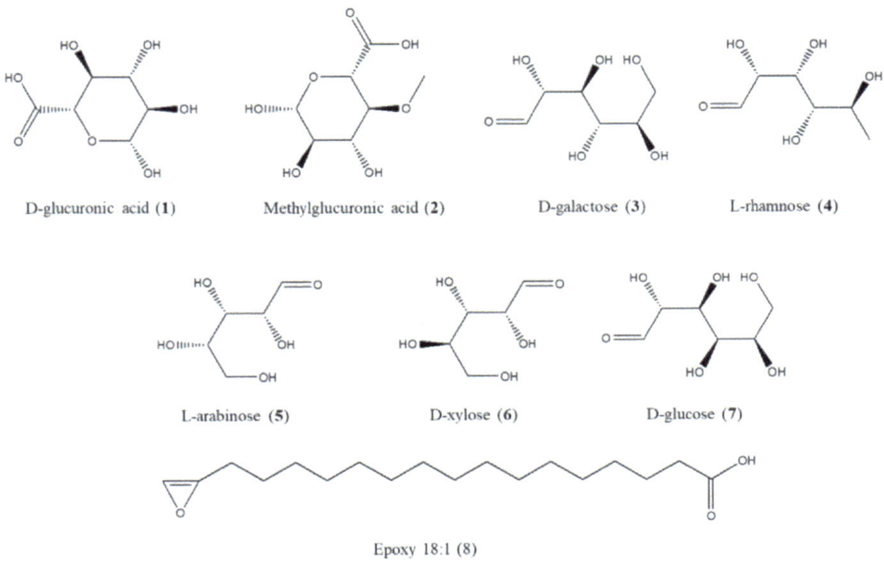

FIGURE 4.1 Chemical structure of carbohydrates and fatty acid.

4.2.3 *Flavonoids and Tannins*

An earlier study depicted isolation of flavan-3,4-diol from the heartwood by adopting the paper ionophoresis technique (Drewes and Roux, 1966). Auriculoside (I, Glc= β-D-glucopyranosyl) or 7,3′,5′-trihydroxy-4′-methoxyflavan 3′- glucoside (9) was isolated from aerial part of *A. auriculiformis* (Dhawan et al., 1980; Sahai et al., 1980). Barry et al. (2005) identified and purified three flavonoids 2,3-trans-3,4′,7,8-tetrahydroxyflavanone (10), teracacidin (11), and 4′,7,8,-trihyroxyflavanone (12) from heartwood of *A. auriculiformis*. Quercetin (13) and epicatechin (14) were reported from the barks of *A. auriculiformis* (Kaur et al., 2014). Anderson (1978) reported the presence of red color Anthocyanidins from barks such as leucodelphinidins (15) and leucocyanidins (16). Tannins ranges from 12 to 16% were reported in the barks of *A. auriculiformis* and content of tannins was found greater when compared with younger trees (Sastri, 1950). The chemical structure of various flavonoids is depicted in Figure 4.2.

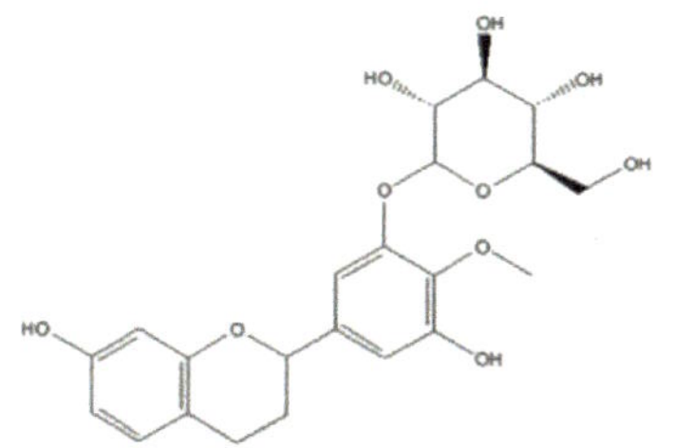

7,3′,5′-trihydroxy-4′-methoxyflavan 3′- glucoside (9)

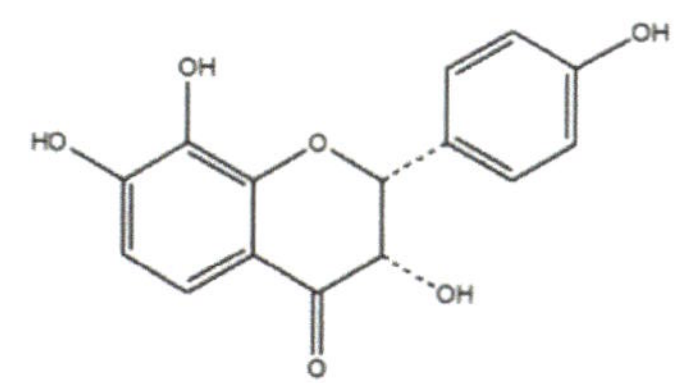

2,3-trans-3,4',7,8-tetrahydroxyflavanone (10)

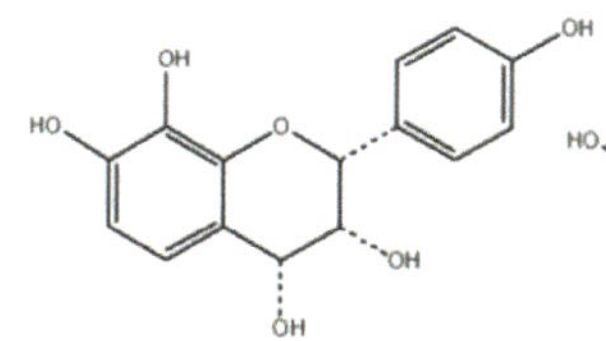

Teracacidin (11)

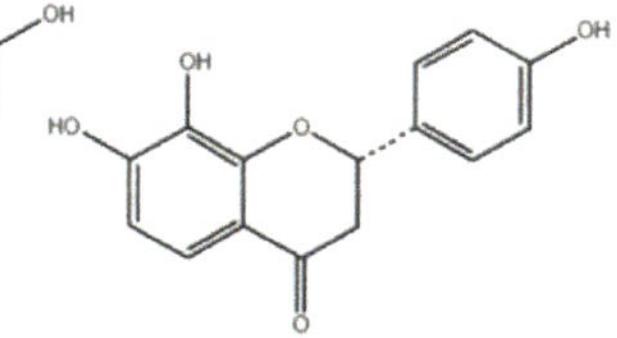

4',7,8, -trihyroxyflavanone (12)

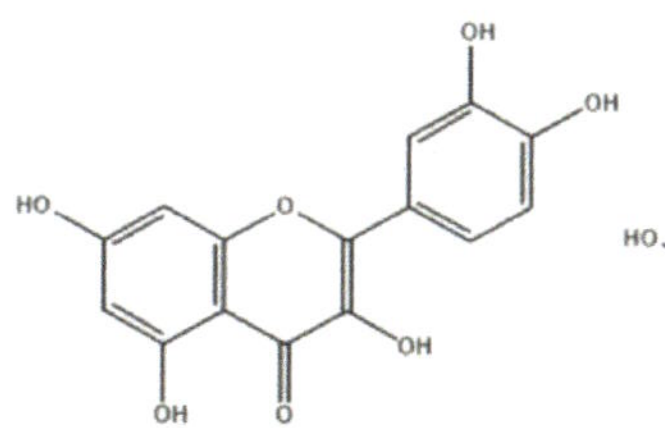

Quercetin (13)

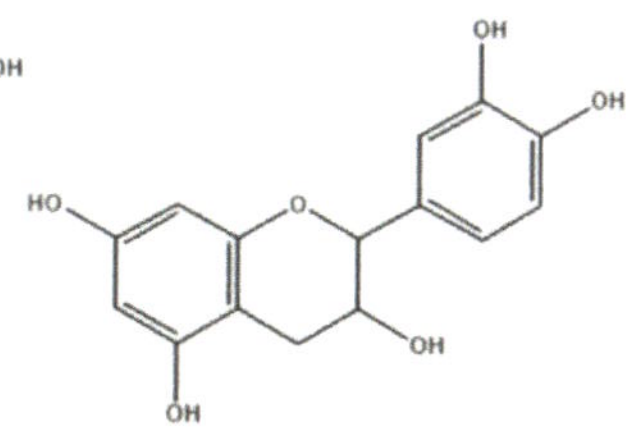

Epicatechin (14)

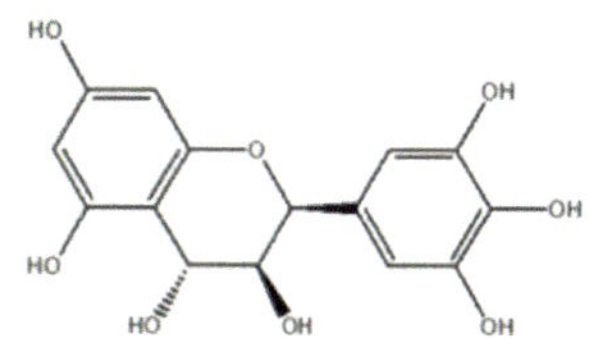

Leucodelphinidins (15)

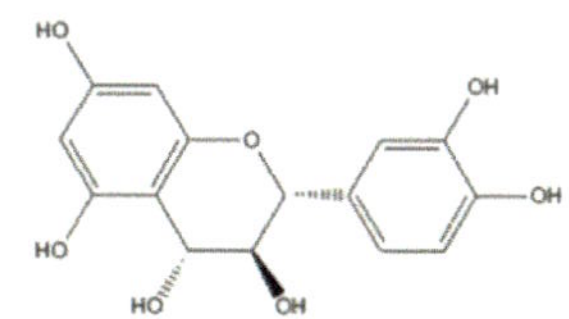

Leucocyanidins (16)

FIGURE 4.2 Chemical structure of flavonoids.

4.2.4 Glycosides

A triterpenoid trisaccharide [acacic acid lactone-3-O-b-D-glucopyranosyl (1→6) - [α- L-arabinopyranosyl (1→2)] – β - D-glucopyranoside (17)] was isolated from the seeds (Mahato et al., 1989). Subsequently, isolation of partial characterized mixture of triterpenoid saponins having aglycon acacic acid lactone (19) structure namely Acaciasides A and B (18) was reported from *A. auriculiformis* seeds (Pakrashi et al., 1991). Further, A triterpenoid saponin was isolated and characterized as 3-O-{[b-D-xylopyranosyl (1→3)-b-D-xylopyranosyl (1→4)-a-L-rhamnopyranosyl (1→2)]- [a-L-rhmanopyranosyl (1→4)]-b-D-glucopyranosyl}-3,16,21- trihydroxyolean-12-en-28-oic acid (20) from the hydroalcohol extract (Uniyal et al., 1992). Apart from these, two novel compounds namely Acaciasides A and B (acylated triterpenoid bisglycosides saponins) have also been separated and isolated from fruits and funicles (Mahato et al., 1992; Ghosh et al., 1993). Subsequently, Proacaciasides I and II (21) glycosides along with Acaciasides A and B triterpenoid saponins, were identified and isolated from the fruits and funicles of *A. auriculiformis* (Mahato, 1996a, 1996b). Isolation of triterpenoid saponins proacaciaside-I, proacaciaside-II, acaciaside, and acaciamine (22) was also reported from the fruits of *A. auriculiformis* (Garai and Mahato, 1997). A triterpenoid saponin 3-O-[α-L-rhamnopyranosyl-(1→4)]-α-L-arabinopyranosyl-(1→6)-β-D-galactopyranosyl-3β, 16α, 21β, 22α, 28-pentahydroxy-olean-12-ene (23) and one known compound corosolic acid (24) has been isolated and identified from methanol extract of plant stems by adoption many color reactions, chemical degradation method, and spectral analysis (Asati and Yadava, 2014). The chemical structure of various flavonoids is shown in Figure 4.3.

4.3 PHARMACOLOGY

4.3.1 Toxicity Studies

The *A. auriculiformis* roots and barks hydroalcohol extracts has the LD_{50} value lies in the range of 500–1000 mg/kg (Nakanishi et al., 1965). The toxicity signs include writhing, palpitation, gasping, body limb bone, decreased respiratory rate, and death. Following intraperitoneal administration, the LD_{50} value was found to be 3741.7 mg/kg in another study. However, the complete mortality was observed at 4000 mg/kg extract (Ghosh et al., 1993).

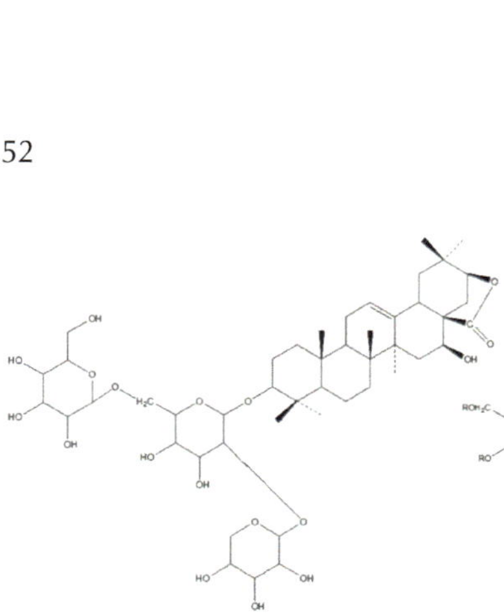

Acacic acid lactone-3-O-b-D-glucopyranosyl (1→6) - [a- L-arabinopyranosyl (1→2)] – b - D-glucopyranoside (**17**)

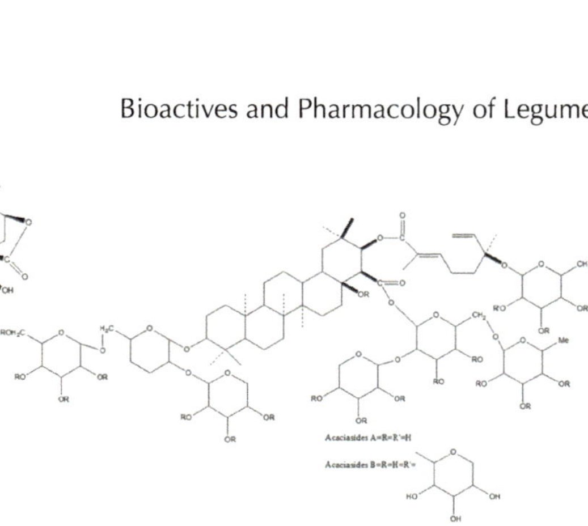

Acaciasides A and B (**18**)

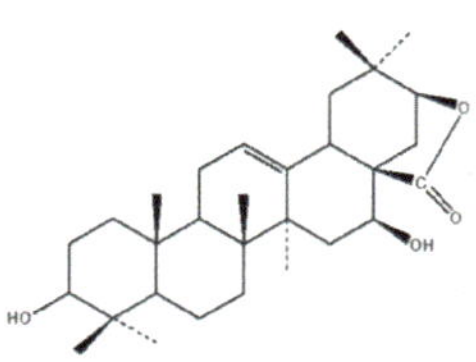

Acacic acid lactone (**19**)

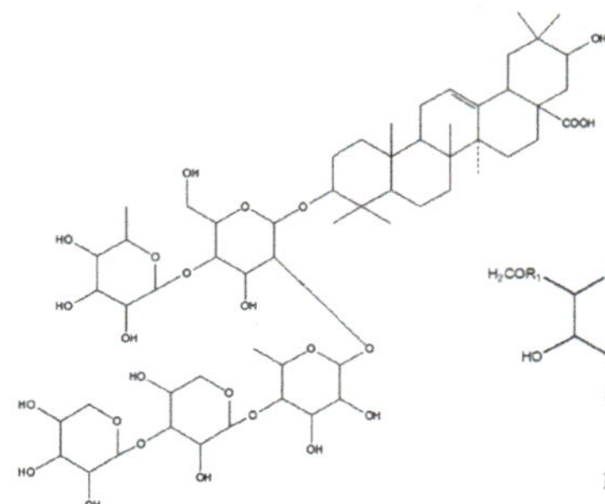

3-O-{[b-D-xylopyranosyl(1→3)-b-D-xylopyranosyl(1→4)-a-L-rhamnopyranosyl(1→2)]- [a-L-rhmanopyranosyl (1→4)]-b-D-glucopyranosyl}-3,16,21- trihydroxyolean-12-en-28-oic acid (**20**)

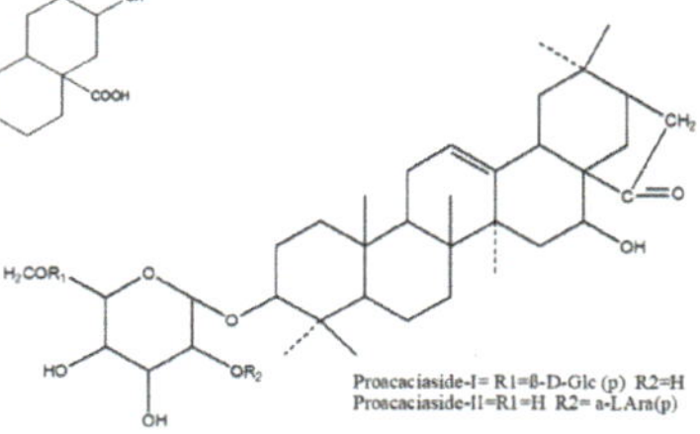

Proacaciaside I and II (**21**)

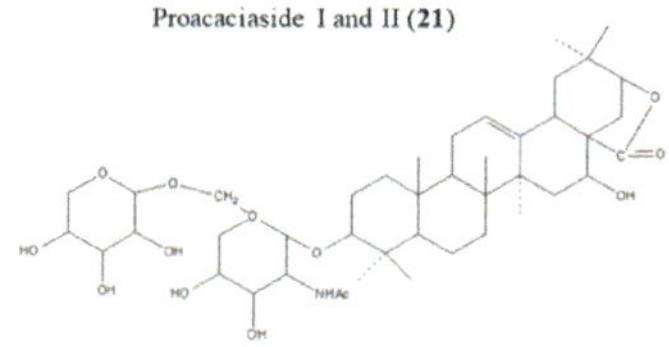

Acaciamine (**22**)

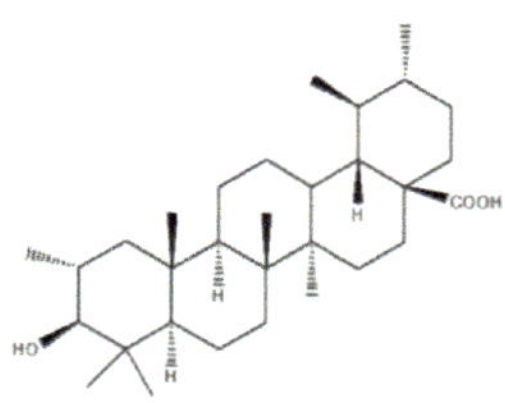

Corosolic acid (**23**)

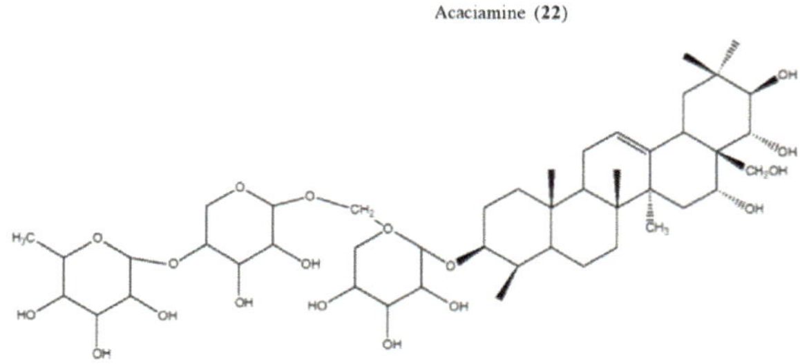

3-O-[α-L-rhamnopyranosyl-(1 → 4)]-α-L-arabinopyranosyl-(1 → 6)-β-D-galactopyranosyl-3β, 16α, 21β, 22α, 28-pentahydroxy-olean-12-ene (**24**)

FIGURE 4.3 Chemical structure of glycosides.

In another study, it was suggested that the ethanol extract of leaves is safe up to 2000 mg/kg (Sharma et al., 2014). *A. auriculiformis* roots and barks extract 12% (w/w) applied on the back portion with hair cleared skin of mice suggested to have no change in appearance or general behavior and body weight loss etc. The ointment containing extract was found to be safer for maximum selected dose of 12% (w/w) (Singh and Sharma, 2014). Further, in another study highest cytotoxic activity of methanol fractions of leaves and barks of *A. auriculiformis* was observed through lethality bioassay performed on brine shrimp with LC_{50} value of 0.55 and 0.79 μg/mL, respectively. In addition, the LC_{50} value of ethyl acetate fraction of leaves was 0.95 μg/mL, when compared with standard vincristine sulfate, which was 0.52 μg/mL (Sravanthi et al., 2014).

4.3.2 Antioxidant Activity

The antioxidant activity was performed using DPPH scavenging assay on petroleum ether, ethyl acetate, chloroform, ethanol, and water extracts of *A. auriculiformis* leaves (25–150 μg/mL) was reported that the ethanol extract is most active in DPPH radical scavenging activity. The order of activity was found to be highest in ethanol, then chloroform, then ethyl acetate, then petroleum ether, and at last in water (Kumar et al., 2017). It was further suggested that *A. auriculiformis* extracts possess dose-dependent activity in reducing power, DPPH, and hydroxyl radical assays. Finally, 1 mg/mL final reaction mixture was observed to possess 91.3–94.1% peroxidation inhibiting activity level (Loganayaki et al., 2011). The mild antioxidant activity was reported in the ethanol extract of *A. auriculiformis* flowers and leaves using DPPH radical scavenging assay [IC_{50} of flowers (152 ± 13) μg/mL and leaves (161 ± 30) μg/mL] (Chew et al., 2011). It has been also reported that barks and pods extracts showed higher quenching capacity on DPPH and ABTS, lower quenching capacity on OH^- and equivalent quenching on O^{2-} and NO (Sathya and Siddhuraju, 2012). The methanol fruit extract of *A. auriculiformis* showed the antioxidant potential in DPPH assay and the observed IC_{50} is 0.031 mg/mL, EC_{50} 1.28 mg/mg, ARP 79.5 and ASE/mL is 1.58, respectively (Prakash et al., 2011a). Similar result was again reproduced and reported in another work by same group (Prakash et al., 2011b). The ethanol barks extract of *A. auriculiformis* possess antioxidant activity at 900 μg/mL (Sravanthi et al., 2014). However, the comparative study performed on leaves and barks extract of *A. auriculiformis* suggested that the ethyl acetate barks extract showed highest DPPH scavenging assay activity when compared with methanol leaves extract (Urmi et al., 2013). Further, the antioxidant activity of *A. auriculiformis* was found to be equivalent to the *Acacia mangium* in DPPH radical scavenging activity method (Mihara et al., 2005). In addition, antioxidant activity of methanol extract/fractions of barks powder of *A. auriculiformis* was tested by DPPH, relative reducing power, and hydroxyl radical (site-specific and nonsite specific) assays. The methanol extract/fractions were further partitioned with ethyl acetate and water. The result of scavenging activity suggested that the both ethyl acetate and water fraction exhibit good scavenging activity in DPPH (72.0%; 57.2%), reducing power (1.76%; 1.52%), site-specific (88.0%; 82.6%), and nonsite specific (93.6%; 83.47%) hydroxyl radical scavenging assay. An increasing (and decreasing) order of polarity of solvents at the maximum concentration (1–100 μg/mL) was observed when, compared with crude extract (Singh et al. 2007a). Similarly, percentage of inhibitions observed were 71.2%, 73.66%, 83.37%, 75.63%, and 72.92% at maximum concentration (10–150 μg/mL) in DPPH, chelating power, lipid peroxidation, site-specific, and nonsite specific deoxyribose scavenging assays, respectively, on water fraction of ethyl acetate barks extract (Singh et al. 2007b). However, acetone extract/fractions of barks at concentration range of 10–700 μg/mL, exhibit maximum inhibitory percentages, that is, 72.3% (DPPH assay), 91.7% (deoxyribose assay), 1.63% (reducing power assay), 83.3% (chelating power assay), and 70.9% (lipid peroxidation assay) (Singh et al. 2007c). It was further reported that soaking of plant material followed by autoclaving with various solutions including sodium bicarbonate, ash, palm sugar and water, fermentation and dry heating processing improve reducing power and free radical-scavenging activity, when compared with unprocessed method (Sathya and Siddhuraju, 2013a). In another study, Nandi et al. (2004) reported that both Acaciasides A and B can generate superoxide anions and initiate lipid peroxidation.

4.3.3 Antidiabetic Activity

A. auriculiformis might be the potential treatment for the diabetes mellitus. When acetone extracts of *A. auriculiformis* barks and pods were evaluated for their inhibitory potential against α-amylase and α-glucosidase enzymes suggest that both barks and pods extracts possess dual inhibitory potential against the enzymes. The phenolic compounds obtained from the barks and pods of *A. auriculiformis* are responsible for antidiabetic activity was depicted (Sathya and Siddhuraju, 2012). Further, *A. auriculiformis* barks and empty pods extracts were found to improve alloxan-induced type II diabetes in another study (Sathya and Siddhuraju, 2013b).

4.3.4 Memory Enhancing Activity

A. auriculiformis leaves extract has shown to improve the learning and memory in rats in both rewarded alteration and passive avoidance tests. In addition, *A. auriculiformis* inhibit acetylcholinesterase (AChE) enzyme in dose dependent and produce improvement in dementia (Sharma et al., 2014).

4.3.5 CNS Suppression Activity

A. auricufiformis extract contains auriculoside responsible for the CNS depressant activity (Sahai et al., 1980). Further, in mice during barbiturate potentiation test, CNS depressant activity of *A. auriculiformis* extract was observed (Dhawan et al., 1980; Sahai et al., 1980).

4.3.6 Wound Healing Activity

It has been reported that the ointments compounded with both ethanol and aqueous extracts of *A. auriculiformis* stem barks showed the wound healing activity due to the presence of various phytoconstituents like phenolic constituents, flavonoids, and tannins, respectively. The wound healing activity of *A. auriculiformis* is evident in both incision and excision wound models. Which were further characterized by increased wound contraction rate, reduction of epithelialization period, tensile strength, hydroxyproline content, granulation tissue, and formation of collagen fiber (Singh and Sharma, 2014).

4.3.7 Antimutagenic and Anticarcinogenic Activity

A. auriculiformis barks extracts possess the antimutagenic property. It has been reported that both chloroform and acetone extracts of *A. auriculiformis* possess antimutagenic activity against both indirect-acting [2-amino-fluorene (2AF)] and direct-acting [4-nitro-o-phenylenediamine (NPD) or sodium azide] mutagens and anticarcinogenic activity against the chemical carcinogen 7, 12-dimethyl-benz [a]anthracene (DMBA) (Kaur et al., 2002). Further, in another study, the acetone extract of barks and fruits of *A. auriculiformis* was found to exhibit significant antimutagenic activities (Arora et al., 2003).

4.3.8 Hepatoprotective Activity

A. auriculiformis barks and pods extracts were found to exert protective effect on liver cells against paracetamol intoxicated liver injury. *A. auriculiformis* barks and pods extracts treatment improved the liver function as an evidence from biochemical markers such as aspartate aminotransferase, alanine transaminase, alkaline phosphatise, total bilirubin, and total protein. This study further suggested that *A. auriculiformis* barks and pods extracts are good therapeutic candidate in the treatment of the liver injury (Sathya and Siddhuraju 2013b).

4.3.9 Antiinflammatory Activity

A. auriculiformis exhibited the anti-inflammatory actions due to the presence of high phenolic and sterol content. *A. auriculiformis* exerted anti-inflammatory effects in both carrageenan and formalin-induced inflammatory assays. *A. auriculiformis* methanol and petroleum ether leaves extracts (400 mg/kg) showed the highest percent inhibition of 84.88 and 82.12, respectively, in the carrageenan-induced rat paw edema model and highest percent inhibition of 65.68 and 63.34 in formalin-induced rat paw edema model of inflammation (Rangra et al., 2019b).

4.3.10 Antibacterial Activity

The hydroalcohol extract of *A. auriculiformis* roots and barks (10 mg/mL) was tested for antibacterial activity against *Staphylococcus aureus*, *Bacillus subtilis*, *Escherichia coli,* and *Proteus vulgaris*. The hydroalcohol extract of *A. auriculiformis* roots and barks (1–4.9 mg/mL) was found to inhibit *B. subtilis* significantly than other strain (Nakanishi et al., 1965). *A. auriculiformis* leaves extracts (water, chloroform, petroleum ether, ethyl acetate, and ethanol) showed antibacterial activities against *B. subtilis*, *E. coli*, *Proteus mirabilis*, *Pseudomonas aeruginosa*, *Salmonella enterica*, *Shigella flexneri*, *Vibrio cholerae,* and *S. aureus* in agar disc diffusion assay method. In this context the mean of minimum inhibitory concentration was found to be 2, 3, 6, 5, and 1 mg/mL, from ethanol, ethyl acetate, chloroform, petroleum ether, and water extract, respectively, against tested bacterial stains (Kumar et al., 2017). The antimicrobial activity of *A. auriculiformis* was due to mainly the presence of two acylated biglycoside saponins, Acaciasides A and B.

Acaciasides A and B have been further shown to possess the antibacterial activity against the *Bacillus megaterium*, *Salmonella typhimurium* and *P. aeruginosa* at concentration of 700 μg/mL in another study (Mandal et al., 2005). The ethanol extracts of *A. auriculiformis* flowers and the leaves also showed the antibacterial activity against the Gram-positive bacteria, that is, *Micrococcus luteus*, *Bacillus cereus*, methicillin-sensitive *S. aureus*, two strains of methicillin-resistant *S. aureus* but no activity against the Gram-negative bacteria, that is, *E. coli*, *P. aeruginosa*, *Klebsiella pneumoniae*, and *Enterobacter aerogenes* using disc diffusion (Kirby-Bauer) method. However, no activity was observed against Gram-negative strain due to the presence of permeability outer membrane barrier (Chew et al., 2011). Similarly, 2 and 6 mg/mL, methanol extract of *A. auriculiformis* phyllodes concentration inhibited actively both Gram-positive (*S. pyogenes* and *S. aureus*) and Gram-negative (*E. coli*) bacteria (Pennacchio et al. 2005). Further, the antimicrobial activities of methanol and ethanol extract of *A. auriculiformis* seeds pods (5, 10, 15, and 20 μg/mL) against the several Gram positive and Gram-negative bacteria including *S. aureus*, *Rhodo coccus*, *Listeria monocytogenes*, *B. subtilis*, *E. coli*, *Shigella dynsenteriae*, *Salmonella typhi*, *K. pneumoniae*, *P. aeruginosa*, S. *entrica* (serovar *typhimurium* and *arizona*), *Vibrio cholerae*, and *Acinetobacter boumanii* using the standard nutrient agar media was investigated. A study on silver nanoparticles synthesized from *A. auriculiformis* aqueous extract of pods was tested against Gram-positive (*Staphylococcus* spp. and *Bacillus cereus*) and Gram-negative organisms (*Klebsiella* spp. and Wild-type *E. coli* BW 25113) exhibit significant result through minimum inhibitory concentration method results (Nalawade et al., 2014).

The ethanol extract was observed to be more potent and bioactive than methanol extracts against both Gram-negative and Gram-positive bacteria except *L. monocytogenes*, *P. aeruginosa*, *K. pneumoniae*, and *A. boumanii* (Chaki et al., 2015). However, ethanol extract of *A. auriculiformis* exhibited antibacterial activity against MDR strains of *K. pneumoniae* and *E. coli* (Rao et al., 2018).

4.3.11 Antifungal Activity

The hydroalcohol extract of *A. auriculiformis* roots and barks (10 mg/mL) showed no growth inhibition against *Aspergillus niger*, *Candida albicans*, *Penicillium luteum*, and *Mucor spinescens* in agar streak-dilution method in an earlier study (Nakanishi et al., 1965). Further, the antifungal activity has been shown by both aqueous and ethanol extract of *A. auriculiformis* leaves against *C. albicans* (Rao et al., 2018). The ethanol extract of *A. auriculiformis* barks showed antifungal activity against the two fungal stains, that is, *A. niger* and *C. albicans*. Further at a concentration of 100 mg/mL the highest antifungal activity was exhibited against *A. niger* (20.62 ± 0.17 mm) (Sravanthi et al., 2014). *A. auriculiformis* leaves extracts (petroleum ether, chloroform, ethyl acetate, ethanol, and water) showed antifungal activities against three stains of fungus (*C. albicans*, *A. niger,* and *Cryptococcus* sp.) in agar disc diffusion assay. Result revealed the mean minimum inhibitory concentration of ethanol, ethyl acetate, chloroform, petroleum ether, and water extract against fungal stains were found to be 2, 2, 6, 5 and 1 mg/mL, respectively (Kumar et al., 2017). Acaciasides A and B (acylated bisglycoside saponins) isolated from the *A. auriculiformis* funicles has been shown to inhibit the *Curvularia lunata* and *Aspergillus ochraceus* fungal strains at concentration equal to or <300 μg/mL in another study (Mandal et al., 2005). Furthermore, greater antifungal activity of *A. auriculiformis* was observed when compared with *A. mangium* due to the presence of the higher levels of flavonoids, that is, 3,4',7,8-tetrahydroxyflavanone and teracacidin (Mihara et al., 2005). In another study, scent composition was established through GCMS and found to have antioxidant and antifungal effect (SamLing and Umaru, 2018).

4.3.12 Activity Against the Worms

A. auriculiformis, consisted of two compounds mainly Acaciasides A and B (triterpenoid saponins) with 4 mg/mL concentration, has been shown to kill 97% *Setaria cervi* microfilaria in 100 min and in 35 min kills 100% of adults using in vitro procedure. When administered 100 mg/kg p.o. in rats, implanted with *S. cervi* adults, after the 10 days of treatment during the first phase increased the microfilaria count in blood by 1.5-fold but reduced microfilaria density >80% following the third phase of treatment suggesting the very high level of physiological stress on adult worms and increased the microfilaria discharge and death (Ghosh et al., 1993). Further, the mixture (1:1) ethanol extract of *Centella asiatica* leaves and *A. auriculiformis* funicles, when orally administered (0.04 mg/g, bw/day) for 45 days to pariah dogs infected with *Dirofilaria immitis* showed the reduction/suppression of microfilaria density up to 86% reduction after 30 days and 68% reduction after 120 days (Sarkar et al., 1998). Similarly, in another study on ethanol extract of *A. auriculiformis* (150 mg/kg/day/po) to four pariah dogs infected with *D. immitis* naturally for 50 days showed the 98% and 99% reduction in microfilarial density on 45 days and 75 days of treatment (Chakraborty et al., 1995). Further, acetone and water extract of *A. auriculiformis* barks possess significant activity against the *Bactrocera cucurbitae* (Coquillett) and reduced the percentage of pupation, oviposition, emergence, and egg hatching in a dose-dependent manner (Kaur et al., 2010). In a study focussing on the anticestocidal activity of ethanol extract of *A. auriculiformis* in rats against the *Hymenolepis diminuta.* It was observed that the administration of ethanol extract of *A. auriculiformis* (300 mg/kg/day/po) to each rat for 20 days of the inoculation of single cysticercoid of *H. diminuta* resulted in the expulsion of adult worms within 5 days of the treatment (Ghosh et al. 1996). It has been reported that the two acylated triterpenoid bisglycosides, that is, Acaciasides A and B possess significant antihelmintic activity (Sinha Babu et al., 1997). Further, it has been observed that the addition of the Acaciasides obtained from *A. auriculiformis* to tetracycline antibiotic would further improve its efficacy against microfilariae of *D. immitis* (Datta et al., 2009).

4.3.13 Antilarvicidal Activity/Antimalarial Activity

Leaves extracts of *A. auriculiformis* also possess the larvicidal activity against vectors *Culex quinquefasciatus* and *Aedes albopictus* in dose-dependent manner. The LC_{50} values obtained for *A. albopictus* and *C. quinquefasciatus* were 6.1 and 4.2 μg/mL, respectively. The toxicity studies performed on *Gambusia affinis* using methanol extract of leaves suggested that larvicidal activity against malarial and Japanese encephalitis vectors, that is, *A. albopictus* and *C. quinquefasciatus* (Kumar et al., 2017). Further, variation in results was obtained with the fruits extracts of *A. auriculiformis* (100, 200, and 300 ppm) against third larval instars with highest mortality at 300 ppm. It was found that the mortality was higher in chloroform: methanol (1:1 v/v) extract (Barik et al., 2016). The ethyl acetate extracts of fruits of *A. auriculiformis* was investigated against several larval instars of JE vector *Culex vishnui.* In this connection crude extracts showed good activity against larval instars with highest mortality at 0.09% and 300 ppm concentration (Barik et al., 2019). The ethanol extract of *A. auriculiformis* leaves possess the antiplasmodial activity. It was shown that ethanol leaves extract (350–1050 mg/kg/day) showed suppressive activity in the 4-day test and in Rane's test. *A. auriculiformis* exhibited significant blood schizonticidal activity against the *Plasmodium berghei* infected mice suggesting the potential benefit in malaria (Okokon et al., 2010). Recently, larvicidal potentiality of crude and ethyl acetate extracts *A. auriculiformis* fruits showed good results against all the larval instars with highest mortality at 0.09% with 300 ppm of ethyl acetate extract. The mosquito larval death activity was due to presence of three compounds namely 1-methyl ether ester, Ethane 2-chloro-1,1-dimethoxy acetic acid, and [4-[1-[3,5-Dimethyl-4[(trimethylsilyl)oxy) phenyl]-1,3-dimethylbutyl)-2,6dimethylphenoxy) (trimethyl) silane (Barik et al., 2019).

4.3.14 Sperm Immobilizing and Spermicidal Activity

Sperm immobilizing activity of two partially triterpenoid saponins mixture isolated from *A. auriculiformis* through a modified Sander–Cramer test was observed at 0.35 mg/mL in physiological saline. Further, the abilities of the compound as a potential sperm immobilizing agent was compared with Triton X-100 resulted compound to be more potent (Pakrashi et al., 1991). Fraction enriched with Acaciaside B isolated from the *A. auriculiformis* seeds shown significant spermicidal activity without any sign of mutagenic effect and adverse impacts on lactobacilli growth. This result suggested its safe and effective utilization as a prospective future spermicide for the development of vaginal contraceptive formulation (Pal et al., 2009).

KEYWORDS

- ***Acacia auriculiformis***
- **leguminosae**
- **metabolites**
- **isolation**
- **characterization**

REFERENCES

Anderson, D. M. W. Chemotaxonomic Aspects of the Chemistry of Acacia Gum Exudates. *Kew Bull.* **1978,** *32* (3), 529–536.

Arora, S.; Kaur, K.; Kaur, S. Indian Medicinal Plants as a Reservoir of Protective Phyto-chemicals. *Teratogen. Carcinogen. Mutagen. Suppl.* **2003,** *1*, 295–300.

Asati, N.; Yadava, R. N. New Triterpenoid Saponin from *Acacia auriculiformis* Cunn. *Int. J. Pharmaceut. Res. Bio-Sci.* **2014,** *3* (5), 341–349.

Barik, M.; Rawani, A.; Chandra, G. Mosquito Larvicidal Activity of Solvent Extracts of Fruits of *Acacia auriculiformis* Against Japanese Encephalitis Vector *Culex vishnui* Group. *J. Mos. Res.* **2016,** *6* (13), 1–8.

Barik, M.; Rawani, A.; Laskar, S.; Chandra, G. Evaluation of Mosquito Larvicidal Activity of Fruit Extracts of *Acacia auriculiformis* Against the Japanese Encephalitis Vector *Culex vishnui*. *Nat. Prod. Res.* **2019,** *33* (11), 1682–1686.

Barry, K. M.; Mihara, R.; Davies, N. W.; Mitsunaga, T.; Mohammed, C. L. Polyphenols in *Acacia mangium* and *Acacia auriculiformis* Heartwood with Reference to Heart Rot Susceptibility. *J. Wood Sci.* **2005,** *51*, 615–621.

Chaki, S.; Ghosh, B.; Bandyopadyhay, S.; Mookerjee, M.; Das, S.; Dastidar, S. G. Detection of Various Phytochemical Compounds from Seeds of *Acacia auriculiformis* for Possibilities of Obtaining Potent Antimicrobial Agents. *Int. J. Biol. Pharma Res.* **2015,** *6*, 120–128.

Chakraborty, L.; Pandit, P.; Roy-Maulik, S. *Acacia auriculiformis*—A Natural Dye Used for Simultaneous Coloration and Functional Finishing on Textiles. *J. Clean. Prod.* **2020,** *245*, 118921.

Chakraborty, T.; Sinhababu, S. P.; Sukul, N. C. Antifilarial Effect of a Plant *Acacia auriculiformis* on Canine Dirofilariasis. *Trop. Med.* **1995,** *37* (1), 35–37.

Chew, Y. L.; Chan, E. W.; Tan, P. L.; Lim, Y. Y.; Stanslas, J.; Goh, J. K. Assessment of Phytochemical Content, Polyphenolic Composition, Antioxidant, and Antibacterial Activities of Leguminosae Medicinal Plants in Peninsular Malaysia. *BMC Complement. Altern. Med.* **2011,** *11* (1), 12.

Chowdhury, A. R.; Banerji, R.; Misra, G.; Nigam, S. K. Chemical Composition of Acacia Seeds. *J. Am. Oil Chem. Soc.* **1983,** *60* (11), 1893–1894.

Datta, S.; Maitra, S.; Gayen, P.; Babu, S. S. Improved Efficacy of Tetracycline by Acaciasides on *Dirofilaria immitis*. *Parasitol. Res.* **2009,** *105* (3), 697.

Dhawan, B. N.; Dubey, M. P.; Mehrotra, B. N.; Tandon, J. S. Screening of Indian Plants for Biological Activity: Part IX. *Indian J. Exp. Biol* **1980,** 18, 594–602.

Drewes, S. E.; Roux, D. G. A New Flavan-3, 4-diol from *Acacia auriculiformis* by Paper Ionophoresis. *Biochem. J.* **1966,** *98* (2), 493.

Garai, S.; Mahato, S. B. Isolation and Structure Elucidation of Three Triterpenoid Saponins from *Acacia auriculiformis*. *Phytochemistry* **1997,** *44* (1), 137–140.

Ghosh, M. A.; Babu, S. P.; Sukul, N. C.; Mahato, S. B. Antifilarial Effect of Two Triterpenoid Saponins Isolated from *Acacia auriculiformis*. *Indian J. Exp. Biol.* **1993,** *31* (7), 604–606.

Ghosh, N. K.; Babu, S. S.; Sukul, N. C.; Ito, A. Cestocidal Activity of *Acacia auriculiformis*. *J. Helminthol.* **1996,** *70* (2), 171–172.

Kaur, A.; Sohal, S. K.; Arora, S.; Kaur, H. Phenolic Rich Fractions from the Bark of *Acacia auriculiformis*. *Bull. Pure Appl. Sci. Zool.* **2014,** *32a* (1 and 2), 1–5.

Kaur, A.; Sohal, S. K.; Singh, R.; Arora, S. Development Inhibitory Effect of *Acacia auriculiformis* Extracts on *Bactrocera cucurbitae* (Coquillett) (Diptera: Tephritidae). *J. Biopest.* **2010,** *3* (2), 499–504.

Kaur, K.; Arora, S.; Hawthorne, M. E.; Kaur, S.; Kumar, S.; Mehta, R. G. A Correlative Study on Antimutagenic and Chemopreventive Activity of *Acacia auriculiformis* A. Cunn. and *Acacia nilotica* (L.) Willd. ex Del. *Drug Chem. Toxicol.* **2002,** *25* (1), 39–64.

Kirtikar, K. R.; Basu B. D. *Indian Medicinal Plants*; Lalit Mohan Basu, Allahabad, 1984; p. 2.

Kumar, S. D.; Srinivasan, P.; Rajalakshmi, M.; Raj, R. G.; Sathiyamurthy, K. Ecofriendly Larvicide Source from *Acacia auriculiformis* and Its Antimicrobial Activity Against Clinical Pathogens. *Int. J. Phytomed.* **2017,** *9* (1), 60–71.

Kushalapa, K. A. Performance of *Acacia auriculiformis* in India. In *Advances in Tropical Acacia Research. ACIAR Proc.* **1991,** *35*, 189–193.

Langeland, K. A.; Burks, K. C. *Identification and Biology of Non-Native Plants of Florida's Natural Areas*; University Press of Florida: Gainesville, FL, 1998.

Loganayaki, N.; Siddhuraju, P.; Manian, S. A Comparative Study on In Vitro Antioxidant Activity of the Legumes *Acacia auriculiformis* and *Acacia ferruginea* with a Conventional Legume *Cajanus cajan. CYTA J. Food* **2011,** *9* (1), 8–16.

Mahato, S. B. Saponins with antifilarial activity from *Acacia auriculiformis*. In: Waller GR, Yamasaki K (eds). *Saponins used in traditional and modern medicine.* New York, USA. Springer: **1996a,** pp. 173–184.

Mahato, S. B. Saponins with Antifilarial Activity from *Acacia auriculiformis*. *Adv. Exp. Med. Biol.* **1996b,** *404*, 173–184.

Mahato, S. B.; Pal, B. C.; Nandy, A. K. Structure Elucidation of Two Acylated Triterpenoid Bisglycosides from *Acacia auriculiformis* Cunn. *Tetrahedron* **1992,** *48* (32), 6717–6728.

Mahato, S. B.; Pal, B. C.; Price, K. R. Structure of Acaciaside, a Triterpenoid Trisaccharide from *Acacia auriculiformis*. *Phytochemistry* **1989,** *28* (1), 207–210.

Mandal, P.; Babu, S. S.; Mandal, N. C. Antimicrobial Activity of Saponins from *Acacia auriculiformis*. *Fitoterapia* **2005,** *76* (5), 462–465.

Mihara, R.; Barry, K. M.; Mohammed, C. L.; Mitsunaga, T. Comparison of Antifungal and Antioxidant Activities of *Acacia mangium* and *A. auriculiformis* Heartwood Extracts. *J. Chem. Ecol.* **2005,** *31* (4), 789–804.

Nakanishi, K.; Sasaki, S. I.; Goh, J.; Kakisawa, H.; Ohashi, M.; Goto, M. et al. Phytochemical Survey of Malaysian Plants Preliminary Chemical and Pharmacological Screening. *Chem. Pharm. Bull.* **1965,** *13* (7), 882–890.

Nalawade, P.; Mukherjee, P.; Kapoor, S. Biosynthesis, Characterization and Antibacterial Studies of Silver Nanoparticles Using Pods Extract of *Acacia auriculiformis*. *Spectrochim Acta A Mol. Biomol. Spectrosc.* **2014,** *129*, 121–124.

Nandi, B.; Roy, S.; Bhattacharya, S.; Sinha Babu, S. P. Free Radicals Mediated Membrane Damage by the Saponins Acaciaside A and Acaciaside B. *Phytother. Res.* **2004,** *18*, 191–194.

Okokon, J. E.; Jackson, O.; Opara, K. N.; Emmanuel, E. In Vivo Antimalarial Activity of Ethanolic Leaf Extract of *Acacia auriculiformis*. *Int. J. Drug Dev. Res.* **2010,** *2*, 482–487.

Orwa, C.; Mutua, A.; Kindt, R.; Jamnadass.; Simons, A. Agroforestry Database 4.0, 2009; pp 1–6.

Pakrashi, A.; Ray, H.; Pal, B. C.; Mahato, S. B. Sperm immobilizing Effect of Triterpene Saponins from *Acacia auriculiformis*. *Contraception* **1991,** *43* (5), 475–483.

Pal, D.; Chakraborty, P.; Ray, H. N.; Pal, B. C.; Mitra, D.; Kabir, S. N. Acaciaside B-Enriched Fraction of *Acacia auriculiformis* Is a Prospective Spermicide with No Mutagenic Property. *Reproduction* **2009,** *138* (3), 453–462.

Pennacchio, M.; Kemp, A. S.; Taylor, R. P.; Wickens, K. M.; Kienow, L. Interesting Biological Activities from Plants Traditionally Used by Native Australians. *J. Ethnopharmacol.* **2005,** *96* (3), 597–601.

Prakash, D.; Upadhyay, G.; Pushpangadan, P. Antioxidant Potential of Some Under-Utilized Fruits. *Indo-Global J. Pharm. Sci.* **2011b,** *1* (1), 25–32.

Prakash, D.; Upadhyay, G.; Pushpangadan, P.; Gupta, C. Antioxidant and Free Radical Scavenging Activities of Some Fruits. *J. Complement. Integr. Med.* **2011a,** *8* (1), 23.

Pullaiah, T. *Encyclopedia of World Medicinal Plants*, Vol. 1; Regency Publications: New Delhi, India, 2006, p. 20.

Rangra, N. K.; Samanta, S.; Pradhan, K. K. A Comprehensive Review on Phytopharmacological Investigations of *Acacia auriculiformis* A. Cunn. ex Benth. *Asian Pac. J. Trop. Biomed.* **2019a,** *9* (1), 1–11.

Rangra, N. K.; Samanta, S.; Pradhan, K. K. *In vivo* Antiinflammatory Potential of Leaf Extracts of *Acacia auriculiformis* Benth. *Indian J. Pharm. Sci.* **2019b,** *81* (4), 709–719.

Rao, A. S.; Shobha, K. L.; Shetty, M. S.; Pai, K. S. R. In Vitro Antibacterial and Antifungal Activities of Aqueous and Ethanolic Leaf Extracts of *Acacia auriculiformis*. *Asian J. Pharm. Clin. Res.* **2018,** *11* (12), 480–482.

Ray, B.; Ghosal, P. K.; Thakur, S.; Majumdar, S. G. Structural Studies of an Acidic Polysaccharide from the Seeds of *Acacia auriculiformis* A. Cunn. *Carbohydr. Res.* **1989,** *185* (1), 105–112.

Sahai, R.; Agarwal, S. K.; Rastogi, R. P. Auriculoside, a New Flavan Glycoside from *Acacia auriculiformis*. *Phytochemistry* **1980,** *19* (7), 1560–1562.

Samling, B.; Umaru, I. J. Phytochemical Screening, Antioxidant, Antifungal Potentials of *Acacia auriculiformis* Floral Scent Composition. *J. Anal. Pharm. Res.* **2018,** *7* (6), 646–650.

Sarkar, P.; Sinha Babu, S. P.; Sukul, N. C. Antifilarial Effect of a Combination of Botanicals from *Acacia auriculiformis* and *Centella asiatica* on Canine Dirofilariasis. *Pharm. Biol.* **1998,** *36* (2), 107–110.

Sastri, B. N. *The Wealth of India: A Dictionary of Indian Raw Materials and Industrial Products*; National Institute of Science Communication and Information Resources, CSIR: New Delhi, 1950.

Sathya, A.; Siddhuraju, P. Role of Phenolics as Antioxidants, Biomolecule Protectors and as anti-Diabetic Factors-Evaluation on Bark and Empty Pods of *Acacia auriculiformis*. *Asian Pac. J. Trop. Med.* **2012,** *5* (10), 757–765.

Sathya, A.; Siddhuraju, P. Effect of Indigenous Processing Methods on Phenolics and Antioxidant Potential of Underutilized Legumes *Acacia auriculiformis* and *Parkia roxburghii*. *J. Food Qual.* **2013a,** *36* (2), 98–112.

Sathya, A.; Siddhuraju, P. Protective Effect of Bark and Empty Pod Extracts from *Acacia auriculiformis* Against Paracetamol Intoxicated Liver Injury and Alloxan Induced Type II diabetes. *Food Chem. Toxicol.* **2013b,** *56*, 162–170.

Sharma, A.; Shetty, M.; Parida, A.; Adiga, S.; Kamath, S. Effect of Ethanolic Extract of *Acacia auriculiformis* Leaves on Learning and Memory in Rats. *Pharmacogn. Res.* **2014,** *6* (3), 246–250.

Sharma, N.; Singh, S.; Singh, S. K. Review on Phytopharmacological Properties of *Acacia auriculiformis* A. Cunn. ex. Benth. *Inventi Rapid: Planta Activa* **2015,** *2016* (1), 1–6.

Sharma, N.; Singh, S.; Singh, S. K. Pharmacognostical Standardization and Preliminary Phytochemical Investigations on *Acacia auriculiformis* A. Cunn. ex. Benth Stem Bark. *J. Med. Plants Stud.* **2017,** *5* (1), 398–402.

Singh, R.; Singh, S.; Kumar, S.; Arora, S. Evaluation of Antioxidant Potential of Ethyl Acetate Extract/Fractions of *Acacia auriculiformis* A. Cunn. *Food Chem. Toxicol.* **2007b,** *45* (7), 1216–1223.

Singh, R.; Singh, S.; Kumar, S.; Arora, S. Free Radical-Scavenging Activity of Acetone Extract/ Fractions of *Acacia auriculiformis* A. Cunn. *Food Chem.* **2007c,** *103* (4), 1403–1410.

Singh, R.; Singh, S.; Kumar, S.; Arora, S. Studies on Antioxidant Potential of Methanol Extract/Fractions of *Acacia auriculiformis* A. Cunn. *Food Chem.* **2007a,** *103* (2), 505–511.

Singh, S. U.; Sharma, N. I. Evaluation of Wound Healing Activity of *Acacia auriculiformis* A. Cunn. Stem Bark. *Asian J. Pharm. Clin. Res.* **2014,** *7* (2), 204–207.

Sinha Babu, S.; Sarkar, D.; Ghosh, N. K.; Saha, A.; Sukul, N. C.; Bhattacharya, S. H. Enhancement of Membrane Damage by Saponins Isolated from *Acacia auriculiformis*. *Jpn. J. Pharmacol.* **1997,** *75*, 451–454.

Sravanthi, S.; Santosh, C.; Mohan, M. M. Phytochemical Analysis, Antioxidant and Antimicrobial Activities of Ethanolic Extract of *Acacia auriculiformis*. *J. Environ. Appl. Biores.* **2014,** *2* (1), 1–4.

Uniyal, S. K.; Badoni, V.; Sati, O. P. A New Triterpenoidal Saponin from *Acacia auriculiformis*. *J. Nat. Prod.* **1992,** *55* (4), 500–502.

Urmi, K. F.; Chowdhury, S. Y.; Hossain, M. K.; Bhusal, P.; Hamid, K. Comparative Antioxidant Activity and Brine Shrimp Lethality Bioassay of Different Parts of the Plant-*Acacia auriculiformis*. *Int. J. Pharm. Sci. Res.* **2013,** *4* (2), 875–880.

CHAPTER 5

Phytochemistry and Pharmacology of Red Bead Seed Tree *Adenanthera pavonina* L.

DIGAMBAR N. MOKAT* and TAI D. KHARAT

Department of Botany, Savitribai Phule Pune University, Pune 411007, Maharashtra, India

Corresponding author E-mail: parimalsai@gmail.com

ABSTRACT

The leaves, stem bark and seeds of *Adenanthera pavonina* contain therapeutically active compounds which show anti-inflammatory, analgesic, antihypertensive, antifungal, anti-oxidant, anticancerous, hepatoprotective, renal protective, CNS depressant, anticonvulsant, anti-hyperlipidemic, antibacterial, anthelmintic and analgesic activities. The detailed account on part wise traditional uses, phytoconstituents, pharmacological activities, are given in the chapter.

5.1 INTRODUCTION

Adenanthera pavonina L., is commonly known as a red-bread tree, belongs to the family Fabaceae. The scientific name is derived from a combination of two Greek words aden, "a gland," and anthera, "anther." It is an important medicinal plant that is known as "*Saga*" in Malaysia, "*Raktakambal*" in India, Red-Bead in English. *A. pavonina* is a deciduous tree, which is 18–24 m tall, erect and is widely distributed in the Asian and African countries, whereas

Bioactives and Pharmacology of Legumes. T. Pullaiah, PhD (Ed.)

as an indigenous plant it is grown and cultivated mostly in the south-eastern region of Bangladesh (Mujahid et al., 2016).

The seeds of *A. pavonina* are bright red, hard, and heart shaped, these seeds are formed in curved pods, which are then released from the pods at maturity. The seeds of *A. pavonina* have medicinal importance in treating various human ailments including inflammation, arthritis, rheumatism, cholera, treatment of boils, blood disorders, convulsion, spasm, and indigestion (Adedapo et al., 2009; Kumar et al,. 2012). The leaves and bark of this plant are used by the tribal people for curing various ailments and diseases (Rastogi and Mehrotra, 1991; Partha and Chowdhury, 2015).

This plant is used as traditional herbal medicine against a variety of diseases. Decoction of the seeds was used in pulmonary infection and externally applied in chronic opthalmia. Raw seeds are poisonous, so seeds may require boiling to neutralize toxicity. The red and glossy seeds were used in making toys and for Jewellery, and in earlier days were used to weigh gold, silver, and diamonds, because they have a narrow range in weight for example four seeds are equal to about 1 g (Orwa et al., 2009). Leaves and barks of the plant are used as a remedy for chronic rheumatism, gout, haematuria, haematemesis, ulcer, and diarrhoea (Warrier, 2003).

A. pavonina has been used in traditional and folk medicine in Bangladesh, for the cure of diarrhea, asthma, gout, rheumatism, inflammations, ulcers, and tumor as well as a tonic (Arzumand et al., 2013).

5.2 BIOACTIVE PHYTO-CONSTITUENTS

The phytochemical constituents of *A. pavonina* are octacosanol, dulcitol, glucosides of β-sitosterol, and stigmasterol in leaves, the bark constitutes the major component as stigmasterol glucoside, whereas seeds consists of glycosides, saponins, and steroid. The seeds contain saponins, alkaloids, phenols and tannins, cardiac glycosides, and steroids (Khare, 2007).

Earlier phytochemical investigation showed that the leaves contain octacosanol, dulcitol, glucosides of β-sitosterol, flavones and stigmasterol, and alkaloid. The bark contains beside stigmasterol glycosides, butein, chalcone, dihydromyricetin, 2, 4-dihydrobenzoic acid, robinetin, and saponins. The wood contains robenetin, chalcone, butein, and the flavones. The seeds contain nonprotein amino acid, methylene glutamine, and traces of ethledine glutamic acid. It is reported to have a rich amount of flavanoids mainly gallic acid, terpenoids, tannins, sterols (beta-sitosterol, beta-sitosterol-3β –D-glucoside), triterpinoids (nonacosane and hentriacontane), and saponins (sapogenins) (Muhammad et al., 2005).

The extract of seeds contains many amino acids namely arginine, cystine, histidine, isoleucine, leucine, lysine, methionine, phenylalanine, tyrosine, valine, etc. The kernels contain pale yellow fat. The fatty acid presents are palmitic, stearic, arachidonic, lignoceric, and eicosenoic. The kernels also contain stigmasterol and its glycoside, dulcitol and a polysaccharide, oleanolic, and echinocystic acid. The dried powdered leaves of *A. pavonina* were successively extracted with petroleum ether, chloroform, and methanol. From the chloroform extract, the hydrocarbon nonacosane and hentriacontane, the triterpenoid squalene, and the long chain fatty acid ester palmitate have been isolated. The methanolic extract contains β-sitosterol-3β-D-glucoside (Shaiq et al., 2005).

Its bark contains the reducing sugar (1.01%) as glucose. The percentages of various amino acids present in the crude protein (5.25%) were found to be aspartic acid (0.10%), threonine (0.24%), serine (0.08%), glutamic acid (0.52%), glycine (0.09%), alanine (0.07%), valine (0.10%), methionine (0.13%), isoleucine (0.06%), tyrosine (0.27%), histidine (0.11%), lysine (0.88%), and arginine (0.25%). The fatty acid compositions were found to be lauric (5.23%), palmitic (38.16%), oleic acid (6.29%), and stearic acid (8.93%) (Kolhe and Chaudhari, 2019). The methanolic extracts of leaf and bark showed positive tests for carbohydrates, proteins, alkaloids, glycosides, saponins, flavonoids, steroids, tannins, etc. (Ghosh and Chowdhury, 2015).

This plant revealed the presence of robinetin, chalcone, flavonal ampelopsin, tannins, flavonoids, terpenoids, saponins, alkaloids, steroids, butin and stigmasterol glucosides, oleanolic acid, echinocystic acid, sapogenins, and many other bioactive phytoconstituents (Silva and Soysa, 2011). The phytochemical screening of the oil revealed the presence of saponin, alkaloid, and terpenoids. Proximate analysis indicated that seed oil of *A. pavonina* had an appreciable quantity of crude fiber 5.82%, protein 27.72%, and ash contents 2.51%. It also showed the organic matter, moisture content, pH, Refractive index, and carbohydrate, to be 97.49%, 13.34%, 5.96, 1.62, and 54.36%, respectively. The mineral analysis showed that the seed oil contained manganese (0.001 ± 0.0002 %) and iron (13.52 mg/kg), while sodium, potassium, nickel, cobalt, lead, chromium, and cadmium were present. The structure of the oil was clarified using physicochemical analyses and spectral data FT-IR, UV, 1H-, 13C-NMR, and mass spectral data. The seed oil could provide access to increase dietary preparation owed to its large amount of protein content (Ara et al., 2012; Ajani 2017).

5.3 PHARMACOLOGY

This plant has been reported to demonstrate anti-inflammatory, analgesic activities, antihypertensive effect, antifungal, antioxidant, anticancer, hepatoprotective, renal protective, CNS depressant, anticonvulsant, anti-hyperlipidemic, and antibacterial. *A. pavonina* ethanolic bark extract fractions showed antihyperlipidemic activity (Das et al., 2011). The medicinal properties are due to the presence of flavonoids, glycosides, saponins, and steroids. Seed extract has the potential to cause a blood pressure lowering effect, antidiarrheal, reducing the progression of diabetic nephropathy, and regulating blood glucose level (Kolhe and Chaudhari, 2019).

5.3.1 *Antimicrobial Activity*

The antibacterial and antifungal activity of solvent extract of *A. pavonina* on microbes isolated from dairy cattle rearing unit was studied. The results revealed that these extracts have antibacterial activity against *Salmonella enteritidis*, *Klebsiella pneumoniae*, *Bacillus subtilis*, *Staphylococcus aureus*, *Escherichia coli*, and *Pseudomonas aeruginosa* and hence these crude extracts can be used as anti-infection agents in dairy cattle rearing unit to avoid infections especially to calves, which are prone to infections at birth (Hussain et al., 2011).

The antifungal activity of peptides extracted from *A. pavonina* seeds was investigated. Peptides were extracted and fractionated by DEAE-Sepharose chromatography and inhibited the growth of the pathogenic fungi from nonretained D1 fraction. This fraction was fractionated by reversed-phase chromatography and resulting, 23 subfractions; all the fractions were separated by tricine-SDS-PAGE. H11 and H22 fractions strongly inhibited growth of *Candida albicans* and *Saccharomyces cerevisiae*. H11 fraction caused 100% death of *S. cerevisiae* in an antimicrobial activity study (Soares et al., 2015). *A. pavonina* extract consisting of phytochemicals flavonoids, terpenes, and tannins showed maximum activity against *Campylobacter jejuni* and inhibited growth at 62.5–125 μg ml^{-1}(Achara et al., 2012).

5.3.2 *Antioxidant and Antimalarial Activity*

A. pavonina leaves and bark are used by the tribal people for curing various ailments and diseases. Amount of total phenolics in leaf is 8.53 mg/g and in bark, it is 8.51 mg/g. DPPH scavenging activity of methanolic extracts of leaf and bark (200 mg/mL methanol) is 32.31% and 30.23% ascorbic acid equivalent, respectively (Ghosh and Chowdhury, 2015). Methanolic extract of *A. pavonina* leaves (MEAP) showed the presence of glycosides, alkaloids, carbohydrates, tannins, phenolic compounds, flavonoids, saponins, terpenoids, proteins, sterols, and resins. Scavenging activity was evaluated by DPPH free radical and nitric oxide anion scavenging assays. The IC_{50} values for the scavenging of DPPH free radical and nitric oxide anion were 425 and 352 μg/ml as compared with standard ascorbic acid 320 and 280 μg/ml, respectively. Total reducing power of the extract was found to be dose dependent. The authors concluded that this extract is a potent scavenger of ROS and can counteract oxidative damages by ROS (Mujahid et al., 2015).

The antimalarial and antioxidant activities of *A. pavonina* seed methanol extract were investigated in *Plasmodium berghei* infected mice. *P. berghei* was injected into Swiss albino mice intraperitoneally with an inoculum size of 1×10^7 on day zero (D0). The vehicle (1% DMSO), *A. pavonina* seed methanol extract at dose 100, 200, 400, 600, and 800 mg/kg or chloroquine 10 mg/kg were administered from D0 to D3. The crude extract, at a dose of 800 mg/kg showed an antimalarial activity (92.11%) greater than of chloroquine (88.73%). Blood samples were collected from these animals through cardiac puncture for biochemical activity and determined effect of *A. pavonina* seed methanol extract on the oxidative stress in infected mice. *In vitro* antioxidant activities of *A. pavonina* seed methanol extract were evaluated using the 1, 1-diphenyl-2-picrylhydrazyl (DPPH)-based activity. The extract had an $IC_{50} > 400$ μg/ml, which was significantly greater than the standard antioxidant drug as ascorbic acid ($1C_{50} = 1.20$ μg/ml). In the case of biochemical and *in vivo* assay, there was not any statistical difference in plasma malondialdehyde (MDA), total protein, and H_2O_2 levels in all the treated groups as compared to the parasite control group, there was a statistical important reduction in glutathione levels at 400 and 800 mg/kg doses as compared to the parasitized without treated control group. Authors concluded that *A. pavonina* seed methanol extract showed significant of antimalarial activity but no antioxidant effect over the treated mice (Adedapo et al., 2014).

5.3.3 Anticancer Activity

A. pavonina seed and leaves methanolic extracts showed effective antimicrobial and anticancer activity in contradiction of various pathogens and its effective against bone cancer cell line and acetone seed extracts not indicated biological activity (Chauhan et al., 2015). A seed of *A. pavonina* possesses good antimicrobial and anticancer activity (Ajani, 2017). Evaluation of *A. pavonina* bark extracts for antioxidant activity and cytotoxicity against cancer cell lines was carried out. Methanolic seed extracts similarly were also tested at 50% and 100% concentration. MTT assay was performed and IC_{50} values were analyzed with respect to the boiled and un-boiled crude seed extracts on HEK (normal) cell line and HepG2 (cancerous) cell line. Cytotoxic effect was predictable to be more effect in HepG2 as compared with HEK; but as per the IC_{50} values of MTT Assay, higher effect in HEK than HepG2 cell line. Cytotoxic activity was evident in cancer cell line (HepG2) and normal cell line (HEK). Boiled samples were safer than un-boiled samples as per the IC_{50} values of cancer (HepG2) and normal (HEK) cell lines (Nair, 2017; Sophy and Fleming, 2015).

A. pavonina methanolic extracts were appraised on Dalton's ascetic lymphoma at doses of 125 and 250 mg/kg/day and tumor was induced in mice through intraperitoneal injection of DAL cells. The antitumor result of the extract was estimated by *in vitro* cytotoxic assay, tumor volume, mean survival time, percentage Increase in Life Span, nonviable, and viable tumor cells count. The detections of this study showed that the methanolic extract of *A. pavonina* (MAP) possesses important antitumor activity (Kumar et al., 2012).

Lindamulage and Soysa (2016) stated that *A. pavonina* is presently used in the treatment of cancer patients. Lactate Dehydrogenase (LDH) release, (3-(4, 5-Dimethylthiazol-2-yl)-2 and 5-diphenyltetrazolium bromide) MTT in addition Sulforhodamine B (SRB) evaluations were carried out to investigate cytotoxicity and antiproliferative activity compared to the HEp-2 cells and 24 h post-treatment with the decoction. Results showed that the mean values (±SD) of EC_{50} were 77.06 (±8.80), 120.02 (±29.82), and 195.50 (±40.68) μg/ml for SRB, MTT, and LDH, respectively. These results are equal to the morphological changes detected in cells treated by the decoction. Induction of apoptosis was detected by fluorescence microscopy with ethidium bromide/acridine orange dye mix stain. Furthermore brine shrimp lethality assay showed an EC_{50} value at a more concentration (1.96 mg/mL) resulted that the decoction prepared by *A. pavonina* showed antiproliferative activity and induces apoptosis on the HEp-2 cancer cells however no toxicity was observed contrary to *Artemia salina*.

5.3.4 Anti-Inflammatory and Analgesic Activity

The *A. pavonina* seed contains an anti-inflammatory active principle, O-acetylethanolamine (Ara et al., 2009; Chauhan et al., 2015). Analgesic activity of butanol fraction of *A. pavonina* bark extract was investigated by Dash et al. (2017) using acetic acid-induced abdominal writhing reflex pain model. Butanol fraction of 200 mg/kg showed significant inhibition of abdominal constrictions or writhing formation up to 65.20% when compared with vehicle treated control group (73.79%).

The anti-inflammatory and analgesic effects of *A. pavonina* seeds extract were investigated (Olajide et al., 2004). *A. pavonina* seeds methanol extract (50–200 mg/kg) produced statistically significant inhibition of the carrageenan-induced paw oedema in the rat, and the acetic-acid-induced vascular permeability in mice. A total of 100 and 200 mg/kg at doses of pleurisy induced with carrageenan were also inhibited. A total of 50–200 mg/kg extract showed a dose-dependent and important analgesic activity in the acetic-induced writhing in mice. Acute toxicity studies discovered that the extract formed decreased motor activity. The LD_{50} value of the extract was found to be 1.36 g/kg. *A. pavonina* leaves ethanolic extracts were evaluated at 250 and 500 mg/kg doses for anti-inflammatory effects by using acute and chronic inflammatory models. The doses possessed inhibitory effects on the subacute study of cotton pellet-induced granuloma formation as well as in acute phase of inflammation as seen in carrageenan-induced hind paw edema. The anti-inflammatory assay was stimulated by the leaf extracts due to the influence of the active phyto-constituents such as β-sitosterol and stigmasterol (Mayuren and Ilavarasan, 2009).

5.3.5 *Antiviral Activity*

The *A. pavonina* seeds activity of sulfated polysaccharide against poliovirus type 1 in cell cultures was evaluated by using method of dimethylthiazolyl-diphenyltetrazolium bromide and assay *i.e.* plaque reduction. The sulfated polysaccharide caused antiviral effect in steps after virus entry into the cells with a low cytotoxicity (Godoi et al., 2014).

5.3.6 *CNS Depressant and Anticonvulsant Activity*

A. pavonina seed of methanolic extract (MESAP) was examined for CNS depressant effect at doses of 1 at 50–200 mg/kg. MESAP (50–200 mg/kg) protected mice against picrotoxin- and pentylenetetrazole-induced seizures and prolonged phenobarbitone-induced sleeping time dose dependently (Olukayode et al., 2009).

5.3.7 Antidiarrheal Activity

In castor oil-induced diarrhoea model on rat and *A. pavonina* bark methanol extract have significantly deceased the accumulative wet fecal mass; at the doses of 500 and 1000 mg/kg body weight found 17.91% and 34.32% deceased as compared with control. In acute toxicity investigated on mice and resulted the *A. pavonina* bark methanol extract LD_{50} value was 1453.44 mg/kg found (Arzumand et al., 2013). The *A. pavonina* seed aqueous extract was orally administered in to 3 groups of animals (6 per group) to investigate antidiarrheal activity against magnesium sulfate and castor oil- induced diarrhoea in rats. The effect of extract on gastrointestinal by using castor oil and charcoal-induced enter-pooling was evaluated by using Loperamide 3 mg/kg. Oral administration of *A. pavonina* seed aqueous extract at 50, 100, and 200 mg/kg doses, showed dose-dependent significant decrease in propulsive movement in castor oil-induced gastrointestinal transit by using charcoal meal in rats when compared Loperamide. These results show that *A. pavonina* seed aqueous extract shows significant antidiarrhoeal potential (Pandhare et al., 2017).

5.3.8 Hepatoprotective Activity

A. pavonina of 50% methanolic extract was used as hepatoprotective at doses of 100 and 200 mg/kg and silymarin 100 mg/kg was used as a standard drug for 28 days. The glutamate pyruvate transaminase (SGPT), serum levels of glutamic oxaloacetic transaminase (SGOT), lactate dehydrogenase (LDH), alkaline phosphatase (ALP), bilirubin, albumin, and total protein, were appraised along with activities of catalase, superoxide dismutase (SOD), thiobarbituric acid reactive substances (TBARS), glutathione. The *A. pavonina* leaves methanolic extract exhibited hepatoprotective effects against RIF and INH-induced hepatic damage in rats against standard drug of silymarin (Mujahid et al., 2013).

5.3.9 Hypoglycaemic and Antihyperglycaemic Activity

A. pavonina mature leaves hot water extract at different doses 500, 750, and 1000 mg/kg and 1 ml distilled water as negative control as well as 22.5 mg/kg of tolbutamide as positive control were orally administrated to normal hypoglycaemic rats and their fasting (n = 6 per group) and random (n = 6 per group) as well as postprandial (n = 6 per group) (after 5 ml/kg 50% oral glucose challenge). Serum glucose levels were determined at intervals for 4 h and resulted significant of hypoglycaemic effects in both fasted in all 3 doses, up to 2 h and fed only mid dose tested, up to 2 h rats respectively, and antihyperglycaemic effect on all doses, up to 3 h and 5 ml/kg 50% oral glucose challenge. Excluding for the production of watery stools the subchronic in 21 days administration of *A. pavonina* leaves extract was well tolerated with no hepatotoxic in terms of SGOT and SGPT levels and nephrotoxic in terms of serum creatinine as well as urea levels or neurotoxic in terms of response time of bar and bond tests effects. It was concluded that *A. pavonina* leaves hot water extract show harmless oral hypoglycaemic and antihyperglycaemic activities (Dissanayake et al., 2016).

Hyperglycaemia in rats was induced through intraperitoneal injection of fresh prepared alloxan monohydrate. Normal control rats (Group I), normal rats treated by aqueous and alcoholic extract of *A. pavonina* seeds (500 mg/kg) (Groups II and III), respectively, diabetic control rats (160 mg/dl) (Group IV) and diabetic rats treated by *A. pavonina* seeds aqueous and alcoholic extract (500 mg/kg) (Groups V and VI) for 10 days, respectively. After the experiment period rats were sacrificed by cervical beheading and blood samples collect and analyzed for biochemical parameters. Groups V and VI formed important reversal in hyperglycemic status and produced important helpful effects on lipid profile in diabetes rats *i.e.* reduction in total cholesterol, increasing HDL, LDL, and triglycerides significantly. Consequently, aqueous and alcoholic extract of *A. pavonina* seeds produced favourable changes in lipid profile in diabetic rats for better glycemic control (Krishnaveni et al., 2011). The renal protective effects of *A. pavonina* seed aqueous extract was investigated in STZ-induced diabetic rats and this extract given daily dose at 50, 100, and 200 mg/kg per day for 13 weeks. Blood glucose, serum parameters namely total protein, albumin, urea, creatinine, glycated haemoglobin (HbA1c), lipid profile as well as urine parameters *i.e.* albumin and urine protein were studied and kidney histopathology also done. In STZ-induced diabetic rats, severe hyperglycemia was developed, with marked increase in proteinuria and albuminuria after 13 weeks treatment. *A. pavonina* seed aqueous extract (dose) treatment significantly decreased proteinuria, lipid levels, albuminuria, and HbA1c deposition in diabetic rats. Therefore, *A. pavonina* seed aqueous extract has decreased growth of diabetic nephropathy in streptozotocin (STZ)-induced diabetic rats and useful in decreasing the progression of diabetic nephropathy (Pandhare and Sangameswaran, 2012).

The antidiabetic effect of galactomannans *A. pavonina* seeds extract in STZ-induced diabetic mice was investigated. The initial galactomannan yields from *A. pavonina* and extraction products composition were assessed. Different methods used such as thin layer chromatography, 1H and 13C nuclear magnetic resonance, Fourier transform infrared spectroscopy, and molecular weight by gel permeation chromatography and have been working to describe the extracted galactomannan. The mice were divided into four different groups namely normal and diabetic control, galactomannans *A. pavonina* seeds extract (1% and 2%) treated as well as standard drug treated. Daily treatment diabetic mice for 30 days and induced through STZ at only one dose of 120 mg/kg. Water and food intake, body weight, total cholesterol, fasting blood glucose, and triglycerides were measured and resulted galactomannans *A. pavonina* seeds extract showed mannose:galactose ratio (1.46:1), and molar weight high. Both galactomannans *A. pavonina* seeds extract enriched food reduced glycaemia, total cholesterol, and triacylglycerol. Galactomannans of *A. pavonina* seeds extract did not interfere on body weight and food intake although it improved water intake. In addition, the relative liver weight showed toxic galactomannan effects on the histopathological changes of the pancreas in STZ-induced diabetes. Therefore concluded that galactomannans of *A. pavonina* seeds extract is beneficial for avoiding and treating diabetes (Vieira et al., 2018).

5.3.10 Antinociceptive Activity

The antinociceptive activity of ethanol extract of leaves of *A. pavonina* was investigated by Moniruzzaman et al. (2015) using different nociceptive models induced thermally and chemically in mice including acetic acid-induced writhing, hot plate, and tail immersion test, and glutamate- and formalin-induced licking tests at the different doses of 50, 100, and 200 mg/kg of body weight. Furthermore to evaluate the possible mechanisms, participation of opioid system was confirmed by naloxone (2 mg/kg) as well as cyclic guanosine monophosphate (c-GMP) signaling pathway using methylene blue (MB; 20 mg/kg). The results showed that ethanol extract of leaves of *A. pavonina* produced an important and dose-dependent increase in the hot plate latency and tail extraction time. It also decreased the number of abdominal constrictions and paw lickings induced by glutamate and acetic acid. Ethanol extract of leaves of *A. pavonina* inhibited the nociceptive reactions in all phases of formalin test. Moreover, the reverse effects of naloxone showed the association of opioid receptors on the effort of ethanol extract of leaves of *A. pavonina* doing centrally. Furthermore, the enhancement of writhing inhibitory activity by MB proposes the possible participation of c-GMP pathway in ethanol extract of leaves of *A. pavonina* of mediated antinociception (Moniruzzaman et al., 2015).

5.3.11 Anthelmintic Activity

Dash et al. (2017) investigated the anthelmintic activity of the butanol fraction of the bark extract of *A. pavonina.* The butanol fraction of the bark extract demonstrated paralysis as well as death of worms in a comparable time as compared to piperazine citrate especially at higher concentration of 100 mg/ml.

5.3.12 Blood Cholesterol Lowering Effect

A. pavonina seed extract effect was evaluated on the blood cholesterol level of atherogenic diet rats. Adult 12 male Wistar rats, divided 3 groups each group of four animals and were treated orally with normal saline (control group), atherogenic diet (+ve control group) and 200 mg/kg of *A. pavonina* seed extract (treatment group) over 4-week period. The total body weight, total cholesterol (TC), triglyceride (TG), low-density lipoprotein (LDL), and high-density lipoprotein (HDL) as well as very low-density lipoprotein (VLDL) were estimated. Results of the investigation showed that *A. pavonina* seed extract was the potential to cause a blood cholesterol-lowering effect (Maruthappan and Shree, 2010).

KEYWORDS

- ***Adenanthera pavonina***
- **phytochemistry**
- **pharmacology**
- **anti-microbial activity**
- **medicinal uses**

REFERENCES

Achara, D.; Tim, T. C. P.; Nathanon, T. Antibacterial Activity of Three Medicinal Thai Plants Against *Campylobacter jejuni* and Other Foodborne Pathogens. *Nat. Prod. Res.* **2012,** *26* (4), 356–363.

Adedapo, A. D. A.; Olayinka, J. N.; Abiodun, O. O.; Oyagbemi, A. A.; Azeez, O.; Adedapo, A. A.; Adeyemi, A. A.; Moody, J. O. Evaluation of Antimalarial and Antioxidant Activities of the Methanol Seed Extract of *Adenanthera pavonina* (Linn.) in *Plasmodium berghei* Infected Mice. *Asian J. Med. Sci.* **2014,** *5* (4), 44–51.

Adedapo, A. D. A.; Osude, Y. O.; Adedapo, A. A. Blood Pressure Lowering Effect of *Adenanthera pavonina* Seed Extract on Normotensive Rats. *Rec. Nat. Prod.* **2009,** *3*, 82–89.

Ajani, O. O. Proximate Composition, Structural Characterization and Phytochemical Screening of the Seed Oil of *Adenanthera pavonina* Linn. *Rasayan J. Chem.* **2017,** *10* (3), 807–814.

Ara, A.; Arifuzzaman, M.; Ghosh, C. K. Anti-Inflammatory Activity of *Adenanthera pavonina* L., Fabaceae, in Experimental Animals. *Braz. J. Pharmacogn.* **2009,** *20* (6), 929–932.

Ara, A.; Hashem, A.; Muslima, T. Chemical Investigation of the Bark of *Adenanthera pavonina* Linn. *Int. J. Chem. Sci.* **2012,** *10* (1), 98–103.

Arzumand, A. R. A.; Saleh-E-In, M.; Ahmed, N. U.; Hashem, A.; Bachar, S. C. Anti-Diarrheal Activity and Acute Toxicity of Methanolic Bark Extract of *Adenanthera pavonina* Linn (Fabaceae) and Its Elemental Composition. *Turk. J. Pharm. Sci.* **2013,** *10* (2), 263–272.

Chauhan, R.; D'Souza, H. L.; Shabnam, R. S.; Jayanthi, A. Phytochemical and Cytotoxicity Analysis of Seeds and Leaves of *Adenanthera pavonina*. *Res. J. Pharm. Tech.* **2015,** *8* (2) 198–203.

Das, S.; Dash, S.; Sahoo, A. C.; Giri, R. K.; Sahoo, D. C.; Guru, P. Anti-Hyperlipidemic Activity of *Adenanthera pavonina* Linn. Ethanolic Bark Extract Fractions. *Nat. Pharm. Technol.* **2011,** *1*, 1–4.

Dash, S.; Kanungo, S. K.; Sahoo, A. C.; Barik, C.; Mishra, B. Phytochemical Evaluation, Anthelmintic and Analgesic Activities of Butanol Fraction of *Adenanthera pavonina* L. Bark Extract. *MIT Int. J. Pharm. Sci.* **2017,** *3* (1), 7–9.

Dissanayake, D. M. R. K.; Wijayabandarab, M. D. J.; Ratnasooriya, W. D. Hypoglycaemic and Antihyperglycaemic Activities of an Aqueous Leaf Extract of *Adenanthera pavonina* (Fabaceae) in Rats. *Int. J. Pharm. Res. Allied Sci.* **2016,** *5* (1), 34–39.

Ghosh, P.; Chowdhury, H. R. Pharmacognostic, Phytochemical and Antioxidant Studies of *Adenanthera pavonina* L. *Int. J. Pharmacogn. Phytochem. Res.* **2015,** *7* (1), 30–37.

Godoi, A. M.; Galhardi, L. C. F.; Lopes, N.; Rechenchoski, D. Z.; Almeida, R. R.; Ricardo, N. M. P.; Nozawa, C.; Linhares, R. E. C. Antiviral Activity of Sulfated Polysaccharide of *Adenanthera pavonina* Against Poliovirus in HEp-2 Cells. *Evid. Based Complem. Altern. Med.* **2014,** 1–6.

Hussain, A.; Rizvi, A.; Wahab, S.; Zareen, I.; Ansari, S.; Hussain, S. Antibacterial Screening of the Bark of *Adenanthera pavonina*. *Int. J. Biomed. Res.* **2011,** *2* (2), 110–122.

Khare, C. P. *Indian Medicinal Plants—An Illustrated Dictionary*; Springer-Verlag: Berlin, 2007.

Kolhe, R. C.; Chaudhari, R. Y. Ethnopharmacology, Phytochemistry and Pharmacology of *Adenanthera pavonina* L. (Mimosaceae). *Res. J. Pharmacol. Pharmacodyn.* **2019,** *11* (4), 140–146.

Krishnaveni, A.; Selvi, S.; Mohandass, S.; Antidiabetic, Hypolipidemic Activity of *Adenanthera pavonina* Seeds in Alloxan Induced Diabetic Rats. *J. Pharm. Res.* **2011,** *4* (5), 1440–1442.

Kumar, A. S. G.; Javvadi, R. K., Kumar, V. K.; Reddy, M. E.; Reddy, V. Y.; Harshvardhan, G.; Akbar, M. D. Effect of Methanolic Extract of *Adenanthera pavonina* Linn on Dalton'sascitic Lymphoma. *Indian J. Res. Pharm. Biotech.* **2012,** *1* (1), 138–141.

Lindamulage, I. K. S.; Soysa, P. Evaluation of Anticancer Properties of a Decoction Containing *Adenanthera pavonina* L. and *Thespesia populnea* L. *BMC Complement Altern Med.* **2016,** *16* (7), 1–8.

Maruthappan, V. G.; Shree, K. S. Blood Cholesterol Lowering Effect of *Adenanthera pavonina* Seed Extract on Atherogenic Diet Induced Hyperlipidemia Rats. *Int. J. Pharma. Sci. Res.* **2010,** *1* (7), 87–94.

Mayuren, C.; Ilavarasan, R. Anti-Inflammatory Activity of Ethanolic Leaf Extracts from *Adenanthera pavonina* (L.) in Rats. *J. Young Pharm.* **2009,** *1* (2), 125–128.

Moniruzzaman, M.; Khatun, A.; Imam, M. Z. Evaluation of Antinociceptive Activity of Ethanol Extract of Leaves of *Adenanthera pavonina. Evidence-Based Complem. Altern. Med.* **2015,** *2015*, Article ID 412497.

Muhammad, S. A.; Farman, A.; Iqbal, A.; Muhammad, K. P. Pavonin: A New Five Membered Lactone from *Adenanthera pavonina* Linn. (Mimosaceae). *Nat. Prod. Res.* **2005,** *9*, 37–40.

Mujahid, M.; Siddiqui, H. H.; Arshad Hussain, M. D.; Azizur Rahman, Mohd. Khushtar, Yasmeen Jahan. Phytochemical Analysis and Evaluation of Scavenging Activity of Methanolic Extract of *Adenanthera pavonina* Linn Leaves. *J. Drug Deliv. Ther.* **2015,** *5* (3), 55–61.

Mujahid, M.; Siddiqui, H. H.; Hussain, A.; Hussain, M. S. Hepatoprotective Effects of *Adenanthera pavonina* (Linn.) Against Anti-Tubercular Drugs-Induced Hepatotoxicity in Rats. *Pharmacogn. J.* **2013,** *5*, 286–290.

Mujahid, Md.; Ansari, V. A.; Sirbaiya, A. K.; Kumar, R.; Usmani, A. An Insight of Pharmacognostic and Phytopharmacology Study of *Adenanthera pavonina. J. Chem. Pharm. Res.* **2016,** *8* (2), 586–596.

Nair, A. Cytotoxic Effect of *Adenanthera pavonina* Seed Extracts on Cancer and Normal Cell Lines. *Conference: NCRM NICHE Inter-Disciplinary Conclave* (IDC), 2017. DOI: 10.13140/RG.2.2.25863.24484.

Olajide, A. O.; Echianu, C. A.; Adedapo, A. D.; Makinde, J. M. Anti-Inflammatory Studies on *Adenanthera pavonina* Seed Extract. *Inflammo, Pharmacol.* **2004,** *3*, 196– 202.

Olukayode, O. J.; Emmanuel, A. O.; Olajide, A. O.; Makinde, M. J. Anticonvulsant and Depressant Activities of the Seed Extracts of *Adnanthera parvonina. J. Nat. Prod.* **2009,** *2*, 74–80.

Orwa, C.; Mutua, A.; Kindt, R.; Jamnadass, R.; Anthony. Agroforestree Database: A Tree Reference and Selection Guide Version 4, 2009. http://www.worldagroforestry.org/sites/treedbs/treedatabases.asp.

Pandhare, R.; Balakrishnan. S.; Bangar. G.; Dighe, P.; Deshmukh, V. Antidiarrheal Potential of *Adenanthera pavonina* Linn Seed Aqueous Extract in Experimental Animals. *Intern. J. Chinese Med.* **2017,** *1* (4), 116–120.

Pandhare, R.; Sangameswaran, B. Extract of *Adenanthera pavonina* L. Seed Reduces Development of Diabetic Nephropathy in Streptozotocin-Induced Diabetic Rats. *Avicenna J. Phytomed.* **2012,** *2* (4), 233–242.

Partha, G.; Chowdhury, H. R. Pharmacognostic, Phytochemical and Antioxidant Studies of *Adenanthera pavonina* L. *Int. J. Pharmacogn. Phytochem. Res.* **2015,** *7* (1), 30–37.

Rastogi, R. P.; Mehrotra, B. N. *Compendium Indian Medicinal Plants*, Vol. 2; PID: New Delhi, 1991; p 23.

Sanjeewani, N. A.; Fernando, P. H. P.; Wickramarathne, D. B. M. Determination of Hypoglycemic Activity of the Leaves of *Adenanthera pavonina* (Madatiya) an Animal

Study. *Proc. Peradeniya University International Research Sessions*, Sri Lanka, 5th & 6th Nov **2015,** *19*, 189.

Shaiq, A. M.; Ahmed, F.; Azhar, I; Pervez, M. K. Pavinin: A New Five Membered Lactone from *Adenanthera pavonina* L. (Mimoaceac). *Nat. Prod. Res.* **2005,** *19* (1), 37–40.

Silva I. K.; Soysa, P. Evaluation of Phytochemical Composition and Antioxidant Capacity of a Decoction Containing *Adenanthera pavonina* L. and *Thespesia populnea* L. *Pharmacogn. Mag.* **2011,** *7* (27), 193–199.

Soares, J. R.; Carvalho, A. O.; Santos, I. S.; Machado, O. L. T.; Nascimento, V. V.; Vasconcelos, I. M.; Ferreira, A. T. S.; Perales, J. E. A.; Gomes, V. M. Antimicrobial Peptides from *Adenanthera pavonina* L. Seeds: Characterization And Antifungal Activity. *Protein Pept. Lett.* **2015,** *19* (5), *520* – 529.

Sophy, A. J. R.; Fleming, A. T. Evaluation of *Adenanthera pavonina* Bark Extracts for Anti-oxidant Activity and Cytotoxicity Against Cancer Cell Lines. *Int. J. Sci. Res.* **2015,** *4* (12), 48–51.

Vieira, I. G. P.; Mendes, F. N. P.; Silva, S. C.; Paim, R. T. T.; Silva, B. B.; Benjamin, S. R.; Tramontina Florean, E. O. P.; Guedes, M. I. F. Antidiabetic Effects of Galactomannans from *Adenanthera pavonina* L. in Streptozotocin-Induced Diabetic Mice. *Asian Pac. J. Trop. Med.* **2018,** *11* (2), 116–122.

Warrier, P. K. *Indian Medicinal Plants, a Compendium of 500 Species*; Orient Longman Pvt Ltd. **2003,** *4* (1), 58.

CHAPTER 6

Phytochemistry and Pharmacology of *Abrus precatorius* L.

AVINASH PATIL[1*], SONALI PATIL[2], and DARSHANA PATIL[3]

[1]*Department of Botany-Biotechnology, B. K. Birla College of Arts, Science & Commerce (Autonomous), Kalyan (West), Maharashtra, India*

[2]*Department of Bioanalytical Sciences, B. K. Birla College of Arts, Science & Commerce (Autonomous), Kalyan (West), Maharashtra, India*

[3]*Department of Botany, Smt. C.H.M. College, Ulhasnagar-03, Maharashtra, India*

[*]*Corresponding author. E-mail: dravinashpatil@rediffmail.com*

ABSTRACT

Abrus precatorius is said to have a wide range of therapeutic effects, including the treatment of ulcers, inflammation, throat sores, scratches and, wounds. It is also reported to have various pharmacological activities like anti-fungal, anti-bacterial, analgesic, anti-tumor, anti-inflammatory, anti-diabetic, anti-spasmodic, anti-serotonergic, anti-migraine etc. Presently it is regarded as a valuable source of many natural products for the development of both industrial products and medications against various ailments. The goal of this review is to update knowledge of the traditional uses, phytochemistry, and pharmacological actions of *Abrus precatorius*.

Bioactives and Pharmacology of Legumes. T. Pullaiah, PhD (Ed.)

6.1 INTRODUCTION

Abrus precatorius L., a member of the Fabaceae family is pantropical in distribution. Indian licorice, jequirity, crab-eye vine, wild liquorice, prayer beads, coral pea, and rosary pea are some of its common names. It is a flexible, deciduous, and short-branched climber. Leaves paripinnate, 5–10 cm long, leaflets 20–40, ligulate, oblong, and 1–1.6 cm long. Racemes have many white flowers, are densely packed, and are usually shorter than the leaves. White corolla. Pods are rectangular and turgid, measuring 2.5–3.7 cm in length. Seeds are tiny, oval, crimson, and have a black mark on the hilum. Pure white, black, and entirely red seeds have also been discovered. The flowering season lasts from June to August. It thrives in mountainous areas. During the flowering season in the autumn, the roots, stem, and leaves are harvested. Fresh or dried, they are used to treat a variety of ailments. Because the seeds are poisonous, they are only utilized on the outside of the body. In traditional medicinal systems, seeds of *A. precatorius* are used to promote fertility such as in Ayurveda and Unani (Bhakta and Das, 2020).

6.2 BIOACTIVES

Alkaloids (Ghosal and Dutta, 1971), steroids, and other triterpenoids (Gupta et al., 1969), protein, flavanoids (Amites et al., 2012), phenolic substance, amino acid, fixed oil (Arora, 2011), and the flavones abrectorin, luteolin, orientin, isoorientin, and desmethoxycentaureidin 7-0-rutinoside, (Bharadwaj et al., 1980). Carbohydrates-Galactose, arabinose, and xylose are also present in the aerial parts like leaves (Saxena and Sharma, 1999).

6.2.1 Leaves

A. precatorius leaves contain chemicals like abrine, choline, trigonelline (Bharadwaj et al., 1980), hypaphorine, glycyrrhizin, precatorine, (Akinloye and Adalumo, 1981), montanyl alcohol (Lefar et al., 1968), hemiphloin, abruslactone A, (Ragasa et al., 2013), abrusoside A (Choi et al., 1989b), abrusoside D, abrusoside C, abrusoside B (Choi et al., 1989a), arabinose, galactose, xylose (Karawya et al., 1981), inositol, pinitol, and D monomethyl ether (Ali and Malek, 1966).

6.2.2 Root

Abrol, precasine, abrasine, and precol (Khaleqe et al., 1966), Triterpenoids and saponins (Chang et al., 1983), glycyrrhizin (Karawya et al., 1981), and oleanolic acid are present in the roots of *A. precatorius*. Protein, abraline, abrussic-acid, abrusgenic-acid, abruslactone, abrusgenic-acid-methyl-esterabricin, campesterol, anthocyanins, delphinidin, calcium, trigonelline, cycloartenol, gallic-acid, hypaphorine (Bharadwaj, et al., 1980), N, Ndimethyl- tryptophan-metho-cation-methyl-ester,choline, P coumaroyl-galloyl glucodelphinidin, N, N dimethyl-tryptophan, pectin, phosphorus, gallic-acid, pentosans, picatorine, delphinidin, polygalacturonic-acids, precatorine, polysaccharide (Willaman and Li, 1970), isoflavonoids, quinones-abruquinones A, B, C, D, E, F, (Kuo et al., 1995), O, G, abrus-genic acid-methanol-solvate abruslactone a, (Chang et al., 1983), galactose, arabinose, and xylose (Ragasa et al., 2013) are present in the root. The roots of *A. precatorius* yielded a new 7,5-dihydroxy-6,49-dimethoxy isoflavone 7-O-b-D- galactopyranoside (Saxena and Sharma, 1999).

6.2.3 Seed

Several essential amino acids like serine, Abrusin-2′-0-apioside, Abrusin, hederagenin, sophoradiol, kaikasaponin III, sophoradiol-22-0-acetate, tryptophan (Desai et al., 1971), alanine (Glasby, 1991), alpha, amyrin, ursolic acid (Maiti et al., 1970), trimethyl (Kinjo et al., 1991), valine (Kinjo et al., 1991), and methyl ester were found in seeds. Seeds are poisonous and contain principle compound, abrine, abrin C (Wei et al., 1974), abrin A, abrin B (Lin et al., 1971), abrin ll, abrin abrus agglutinin APA-l, labrin lll, Abrus agglutinin APA-ll (Hegde et al., 1991), abrisapogenol, abrus-saponins I and II, arachidyl alcohol, β-amyrin, decan-1-ol, docos-13-enoic acid. docosan-1-ol, brassicasterol, docosane, N, heptacosan-1-ol, N, eicos-11-enoic acid, N, dodecan-1-ol, dotriacontane, eicosane, heneicosan-1-ol, N, N, octadeca-9,12-dienoic acid, N, lignoceric acid, elaidic alcohol, heneicosane, N, hexacosan-1-ol, heptadecan-1-ol, hexadec-9-enoic acid, N, nonacosane, hexacosane, hexadecan-1-ol, hexadecane, N, pentatriacontane, nonadecan-1-ol, octacosan-1-ol, octacosane, pentacosan-1-ol, octadecane, n, octanoic acid, pentacosane, N, pentadecan-1-ol (Bhaumik, 1987), squalene, abridin, abricin, (Zia-Ul-Haque et al., 1983), cycloartenol,abrulin, â-sitosterol campesterol, and cholesterol (Hameed et al., 1961).

Seeds include a novel triterpinoid saponin 3-O-D-glucopyranosyl-(12)-D-glucopyranosyl subprogenin D, as well as six recognized terpinoids (Xiao et al., 2011). *A. precatorius* seeds have been identified as C-glucosylscutelarein 6,7-dimethylether (abrusin) and its 2″-O-apioside. Both are novel natural compounds that contain flavone-c glycosides for the first time containing a trioxygenated A-ring (Markham et al., 1989).

Seed of *A. precatorius* also contain calcium, sodium, magnesium, potassium, mucilage phosphorous, manganese, zinc, copper, iron, and cellulose. Cystalline abrin contained 4–9% of neutral sugar in addition to 9-3 residues of glucosamine per mole of abrin (molecular weight 65,000). The neutral sugars consist of mannose, xylose, and fucose in ratios of 2.08:1.00:0.94 (Lin et al., 1971). Tetracos-15-enoic acid, tetracosane, tetracosan-1-ol, tetradecanoic acid, N, triacosan-1-ol, N, tetradecan-1-ol, tetratriacontane, N, tricosane, triacontane, tritriacontan-1-ol, N, tridecan-1-ol, tritriacontane, N, undecan-1-ol, anthocyanins (List and Horhammer, 1969–1979), arabinose, behenic acid, linolenic acid, arachidic acid, oleic acid (Begum, 1992), galactose, xylose (Karawya et al., 1981), palmitic acid: stearic acid (Begum), and rhamnose, N-N-dimethyl metho-cation (Desai et al., 1971) have been found in the seed of *A. precatorius.* Lectin (Maiti et al., 1970), dimethoxycentaureidin-7-0-rutinoside, abrectorin, flavonoids and anthocyanins-abrectorin, centaureidin,precatorins I, II, and III, luteolin, iso, orientin, demethoxy 7-O-beta-drutinoside, orientin, *A. precatorius* plant growth inhibitor (Roy et al., 1976), and xyloglucosyl-delophinidin have been extracted from the seeds.

6.3 PHARMACOLOGY

6.3.1 Anti-Inflammatory Activity

On inflammation caused by croton oil in a rat ear model, the anti-inflammatory effect of *A. precatorius* extract was examined. When *A. precatorius* root extract was combined with croton oil and administered to the rat ear for 6 hours, the inflammatory response was reduced by 2% compared with croton oil alone. This finding explains why the leaves of this plant are useful in the treatment of inflammation (Georgewill and Georgewill, 2009).

Anam (2001) used the croton oil ear model to test two triterpenoid saponins derived from the aerial portions of *A. precatorius* and their acetate derivatives for anti-inflammatory efficacy. All of the compounds had anti-inflammatory properties, but the acetates were more effective than the parent compounds at inhibiting inflammation.

Abruquinones A, D, B, and F, which were isolated from the roots of *A. precatorius*, were found to have potent anti-inflammatory and antiallergic properties. For the suppression of the release of both beta-glucuronidase and lysozyme from rat neutrophils and the release of both beta-glucuronidase and histamine from rat neutrophils, the IC50 values were <0.3 g/mL (Jadon et al., 1984).

6.3.2 Antiarthritic Activity and Analgesic Activity

The antiarthritic activity of *A. precatorius* was investigated in a rat model of croton oil-induced inflammation. Aqueous extracts of *A. precatorius* leaves were given orally at two different concentrations (200 and 400 mg/kg), and both extracts indicated a reduction in paw inflammation (Georgewill, 2010).

Another study looked at the effects of *A. precatorius* white and *A. precatorius* red seed extracts on Freund's complete adjuvant-induced arthritis in rats. The results showed that *A. precatorius* white significantly inhibited FCA-induced arthritis and increased paw withdrawal latency, indicating a protective effect against arthritis. However, in the case of *A. precatorius* red, inflammation was suppressed at a significant level at the late stage, indicating a protective effect against arthritis (Sudaroli and Chatterjee, 2007).

6.3.3 Neuroprotective Effect

In hypoxic neurotoxicity-induced rats, the neuroprotective effects of petroleum ether extract from aerial portions of *A. precatorius* at various concentrations (100 and 200 mg/kg) were studied. When compared with hypoxic rats, tested dosages of extract significantly promoted spatial behavior. At measured concentrations, the extract restored lowered levels of enzymes such as dopamine, glutamate, and acetylcholinesterase, demonstrating neuroprotective properties when taken orally (Premanand and Ganesh, 2010).

6.3.4 Anticonvulsant Effects

In metrazole-induced convulsions, an ethanol (70%) extract of fresh root of *A. precatorius* delivered intraperitoneally to mice of both sexes at varied dosage levels was considerably active, while the results were the opposite when tested in strychnine-induced convulsions (Adesina, 1982). When an ethanol/water (1:1) extract of the aerial portions of *A. precatorius* was administered at a dose of 500 mg/kg in electroshock-induced convulsions, no statistically significant change was found.

6.3.5 Neuromuscular Effects

Crude extracts from *A. precatorius* leaves were tested on isolated tissue such as the rectus abdominis, rat phrenic nerve-diaphragm muscle, and isolated tissue from young chicks. The ethanol extract of the leaves reduced acetylcholine-induced contractions in rectus abdominis and rat phrenic nerve-diaphragm muscle preparations. The outcomes were concentration dependent and reverifiable. When the extract was given intravenously into young chicks, it induced flaccid paralysis. Direct electrical stimulation of the rat diaphragm had no effect on the ethanol extract. In the presence of raised magnesium ions, reduced calcium ions, or reduced potassium ions, the inhibitory action of the ethanol extract on the rat phrenic nerve diaphragm preparation was amplified (Wambebe and Amosun, 1984).

6.3.6 Antidepressant Activity

Antidepressant efficacy was demonstrated in mice of both sexes after treatment with ethanol (70%) extract of fresh root of *A. precatorius* at various dose levels (Adesina, 1982).

6.3.7 Memory Enhancer Activity

In an Alzheimer's disease model, a hydroalcoholic root extract of *A. precatorius* was studied. The work involved identifying the activation of microglial cells (MGCs) in autoptic brain tissues using glycohistochemistry. *A. precatorius* agglutinin identifies MGC in the cerebral white matter, revealing rod-like cells that are particularly dense near oligodendroglial cells. Active component lectin from *A. precatorius* roots was used to histochemically identify MGC activation in autopic brain tissues from Alzheimer's disease patients (Zambenedetti et al., 1998).

6.3.8 Antiepileptic Activity

In a cross-sectional investigation, it was discovered that when *A. precatorius* leaves were boiled with water, they had antiepileptic action and could be used orally as three tablespoonfuls twice daily for the treatment of epilepsy (Moshi et al., 2005).

6.3.9 Nephroprotective Activity

Sohn et al. (2009) used aqueous leaves extract of *A. precatorius* to study the recovery impact on HEK 293 cells following cisplatin and acetaminophen-induced nephrotoxicity. The assay revealed that *A. precatorius* had the best recovery effect and could be utilized to prevent or treat kidney disease.

Following alcohol-induced kidney injury in adult male Sprague Dawley Wister rats, Ae et al. (2009) studied the renal protective effects of the seed extract of *A. precatorius*. Alcohol administration raised serum potassium and sodium levels, as well as creatinine and malondialdehyde levels, significantly alcohol-induced kidney damage was suppressed when the same doses of alcohol and *A. precatorius* seed extract were given at the same time. The result was linked to the seed extract's mitigation of alcohol-induced lipid peroxidation, according to measurements of malondialdehyde. It was revealed that *A. precatorius* seed extract could protect the kidney from alcohol-related damage.

6.3.10 Antiyeast Activity

The antiyeast activity of dried seeds of *A. precatorius* against *Cryptococcus neoformans* was tested on agar plate at a concentration of 1.0%, and the opposite results were observed using an ethanol/water (1:1) extract of the aerial parts (25.0 mcg/mL) on agar plate, which was inactive against *Candida albicans* and *C. neoformans* (Adelowotan et al., 2008).

6.3.11 Antimicrobial Activity

The antimicrobial effects of *A. precatorius* extracts from leaves, seed oil, and the stem were checked against some of the microorganisms *Staphylococcus aureus, Enterococcus faecalis, S. epidermidis, Bacillus subtilis, Streptococcus anginosus, Corynebacterium* spp., *Proteus mirabilis, Escherichia coli, Klebsiella pneumoniae, C. albicans*, and *Pseudomonas aeruginosa* using the agar well diffusion technique. The most sensitive organism was *S. aureus,* which had a MIC of 8 mcg/mL for the leaf extract. *E. faecalis, S. anginosus,* and some Gram-negative bacteria were resistant to extracts from the stem and seed oil, but not to *E. faecalis, S. anginosus,* and some Gram-negative bacteria (Adelowotan et al., 2008).

Another study used the agar well diffusion technique to test the anti-bacterial activity of chloroform, hexane, and methanolic crude extracts of *A. precatorius* seeds against clinical isolates of *E. coli, E. faecalis, Lactobacillus fermentum, Micrococcus luteus, Streptococcus thermophilus, K. pneumoniae, S. mutans, S. mitis,* and *P. aeruginosa* using the agar well diffusion technique. When compared with hexane and chloroform extracts, methanolic seed extracts demonstrated more effective antimicrobial activity (Bobbarala and Vadlapudi, 2009).

Various solvent fractions were used to investigate the antibacterial activity of *A. precatorius* root extracts. Antibiotic activity was found in aqueous and methanolic extracts against 13 Gram-positive and Gram-negative bacteria. In a disc diffusion assay, the extract's antibacterial activity was highest, with 56% inhibition of *S. aureus* compared with the standard antibiotic Ampicillin (Zore et al., 2007).

Desai and Sirsi (1966) explored *A. precatorius'* antibacterial properties with a concentration of 25 mg/mL, an ethanol/water (1:1) extract of the aerial portions proved inert against *B. subtilis, Salmonella typhi, Agrobacterium tumefaciens*, and *E. coli,* and *S. aureus.* Ether extract of seeds, was inhibiting the growth of *S. aureus*. The ethanol (95%) extract was effective against on *E. coli* as well as *S. aureus* on agar plate.

A. precatorius' antibacterial efficacy against five clinically relevant bacterial strains, namely *Pseudomonas pseudoalcaligenes* ATCC17440, *B. subtilis* ATCC6633, *Staphylococcus epidermidis* ATCC12228, *Salmonella typhimurium* ATCC23564, and *Proteus vulgaris* NCTC8313 was carried out by Parekh et al. (2005). Agar well diffusion and agar disc diffusion were used to analyze the aqueous and methanol extracts. The methanol extract was more effective against Gram-negative bacteria than the aqueous extract against Gram-positive bacteria. *B. subtilis, P. vulgaris,* and *S. epidermidis* were the most sensitive bacteria, whereas *S. typhimurium* and *P. vulgaris* were the most resistant bacteria (Parekh et al., 2005).

Using methanol as a solvent, the antifungal and antibacterial activity of powdered seed extracted materials of *A. precatorius* was studied.

The antibacterial activity was investigated using the Agar well diffusion method against *S. aureus* MTCC-902, *P. aeruginosa* MTCC1934, and *E. coli* MTCC-405. The study revealed that the extract was potent against *S. aureus* than *E. coli* and *P. aeruginosa*. Antifungal activity of extracts was discovered to suppress fungal growth on seeds. When compared with control seeds, extract-treated seeds had 100% germination and seed infection was significantly lower (10% DMSO). The antibacterial and antifungal action of both extracts may be due to the presence of saponins, flavonoids, and alkaloids (Kekuda et al., 2010).

Mobin et al. (2018) found antifungal activity against *Trichophyton mentagrophytes, Microsporum canis, Aspergillus niger* using an ethanol/water (1:1) aerial component extract at a concentration of 25 mcg/mL on an agar plate.

6.3.12 Antiviral Activity

Otake et al. (1995) reported antiviral activity of an ethanol/water (1:1) extract of *A. precatorius* aerial parts against Ranikhet virus and Vaccinia virus in cell culture at a concentration of 50 mcg/mL, and using the cell culture approach, water and methanol extracts of dried seeds of this plant were found to be ineffective against the virus HLTV-1.

6.3.13 Antidiabetic Activity

In alloxan diabetic rabbits, the antidiabetic effects of a chloroform–methanol extract of *A. precatorius* seed (50 mg/kg) were investigated. The chloroform–methanol extract reduced blood glucose levels by a percentage at different intervals, indicating that the chloroform–methanol extract of *A. precatorius* seed possesses antidiabetic effects (Dhawan et al., 1977).

The ethanolic *A. precatorius* extract was tested for antidiabetic activity in diabetic rats produced with streptozotocin and nictonamide. Fasting blood glucose levels were lowered to normal levels when fed as 100 and 200 mg/kg b.wt (Reddy et al., 2010).

6.3.14 Antioxidant Activity

An ethanol seed extract of *A. precatorius* was used to test antioxidant activities in vitro. Total phenolic compound was found to be 95 mg/g of extract calculated as gallic acid equivalent in ethanol seeds extract of *A. precatorius,* while total flavonoids compound was found to be 21 mg/g of extract calculated as rutin equivalent (BHT) (Pal et al., 2009).

6.3.15 Hepatoprotective Activity

In rats, Battu and Kumar (2009) tested the hepatoprotective efficacy of *A. precatorius* hydroalcoholic seed extract against paracetamol-induced liver injury. In rats, the plant extracts of 100 and 200 mg/kg were efficient in protecting the liver from paracetamol-induced damage. This was demonstrated by a considerable decrease in serum enzymes such as aspartate aminotransferase, alanine aminotransferase, bilirubin, and alkaline phosphatase.

6.3.16 Anticancer Activity

At a concentration of 2.0 μL/mL, a water extract of fresh seeds proved ineffective against mitogenic activity in human lymphocytes (Itokawa et al., 1990). In another study, the isolated chemical abrin from *A. precatorius* seeds showed anticancer activity in vitro and in vivo through inducing apoptosis (Desai et al., 1966).

A. precatorius seed showed a protein that has been demonstrated to have anticancer effect in rats and mice with Yoshida sarcoma (solid and ascites types). In vitro investigations have shown that the extract has a direct cytotoxic impact on tumor cells, resulting in vacuolation, karyolysis, cytoplasm disruption, and chromosomal aberrations (Subba Reddy and Sirsi, 1969).

Anbu et al. (2011) used an Ehrlich Ascitis Carcinoma (EAC) tumor model to test the antitumor potential of the petroleum ether extract of *A. precatorius* (PEEAP). PEEAP was given orally to the EAC-bearing mice, which enhanced their survival time. In EAC-bearing mice, the PEEAP restored the altered levels of haematological parameters in a dose-dependent manner. The results were equivalent to those obtained with the conventional medication 5-flurouracil (20 mg/kg body weight) in mice.

A. precatorius and nitrosodiethylamine have been shown to protect HepG2 cells from hepatocellular cancer in Swiss albino rats. On HepG2 cells, an aqueous/ethanol (50%) extract of *A. precatorius* demonstrated severe cytotoxic effects. When compared with the control group, the AP extract-treated group had lower mean and relative liver weights at doses of 100 and 200 mg/kg (Ghosh and Maiti, 2007).

When water extract (5 mcg/kg) and the protein fraction of seed extract (20 mcg/kg) were delivered intraperitoneally, it was found to be effective against sarcoma (Ramaswami et al., 2010).

6.3.17 Immunomodulating Activity

Immunomodulating activity and the effect of abrin on cellular immune responses in normal and tumor-bearing mice were documented. Abrin significantly increased natural killer cell activity in both the normal (49.8% cell lysis on day 9) and tumor-bearing groups (51.7% cell lysis on day 9), and NK cells were active earlier than the control. Complement-mediated cytotoxicity and Antibody-dependent cellular was significantly increased in the abrin-treated tumor-bearing group on the 9th day (44% cell lysis) and the 15th day (27.6% cell lysis), indicating that abrin has immunomodulatory properties (Kusumot et al., 1992).

For murine splenocyte proliferation, NK-cell activation, cytokine secretion, and thrombocyte proliferation, Abrus agglutinin could be a potential immunomodulator in both native and heat denatured (HDA) conditions. The conditioned media of adherent splenocytes caused by native agglutinin and HDA could activate nonadherent splenocytes and vice versa. Although heat denatured agglutinin was able to activate NK cells at a lower dosage than NA, the amount of NK cell activation was greater with NA. NA and HDA were also found to promote thymocyte proliferation (Ramnath et al., 2006).

Researchers discovered that abrin can enhance the host's humoral immune response. In normal mice, a nontoxic dose of abrin (1.25 mg/kg body weight) given 5 days in a row elicited specific humoral responses. Total leucocyte count, circulating antibody, spleen and thymus weights, bone marrow cellularity, antibody-forming cells, and alpha-esterase positive B cells all were increased in amount (Ramnath et al., 2002).

6.3.18 Activity on Blood Cells

Platelet aggregation was considerably suppressed by the abruquinones A, D, and B and isoflavoquinones isolated from *A. precatorius* (Kuo et al., 1995). Another study found that the red blood cells of human A, B, and O groups, as well as fifteen other animals, were resistant to the drug. The red blood cells of buffalo, ant (leafcutter), chicken, cat, dog, guinea pig, duckling, lamb, horse, and mice were all affected by water seeds extract (Misra et al.,

1966). On human lymphocytes, agglutinin activity was shown in cell culture with water seeds extract cell (2.0 microliters/mL). Methylene chloride and methanol fractions of the *A. precatorius* plant inhibited thrombin by 53% and 31%, respectively, demonstrating its antithrombin efficacy (Khan et al., 1966).

6.3.19 *Antifertility Activity*

In an experiment on male rats, ethanol extract of seeds delivered intragastrically at doses of 100.0 and 250.0 mg/kg for 60 days demonstrated antifertility action. The 20 females matched with the 10 males had no pregnancies (Rao, 1987). Female rats were given an ethanol (80%) extract of seeds orally and subcutaneously, and the results were opposite (Zia-Ul-Haque et al., 1983) and the same results were found with ethanol (95%), water extracts of seeds to mice. When female mice are given ethanol (95%), water, or petroleum ether extracts of leaves, the antifertility effect remains inactive (Samad et al., 1974). The number of pregnant women rats has decreased significantly (Bhaduri et al., 1968).

In albino rats, Sinha (1990) explored the effects *of A. precatorius* seed extract on post-testicular antifertility. In albino rats, oral treatment of a 50% ethanol extract of *A. precatorius* seeds (250 mg/kg) for 30 and 60 days caused reversible absolute infertility in males. The treatment's most notable side effect was a reduction in sperm motility in the cauda epididymis Such treatment may affect the oxidative/energy metabolism of the cauda epididymis.

Another study looked at sperm production using *A. precatorius* seed extract, as well as the DNA integrity of spermatozoa in adult male BALB/c mice. *A. precatorius* seed extract could be used as an antifertility drug or contraceptive, but that there is a danger of DNA damage in spermatozoa, which could result in teratogenic effects (Agarwal et al., 1970). Another study was cited as an example (Jahan et al., 2009).

6.3.20 *Abortifacient Activity*

From day 1 to day 10, an aqueous seed suspension *of A. precatorius* (125 mg/kg) caused 51% of rats to get pregnant. When the same dose was given from day 6 to day 15, the activity was reduced. The chloroform/methanol extract (50 mg) and water extract (125 mg/kg) of seeds given to rats yielded positive results (Das et al., 1987). In pregnant hamsters, oral administration of an ethanol extract of seeds at a dosage of 200 mg/kg was ineffective, while it was effective in pregnant rats (Sethi et al., 1990).

The embryotoxic effect of various *A. precatorius* extracts was investigated. At 200 mg/kg, ethanol seed extract was inactive in pregnant hamsters and female rats. At a dose of 150 mg/kg, petroleum ether extract had no effect on rats (Agarwal et al., 1970). Intragastrically given water extract of dried seeds to pregnant rats at a dose of 125 mg/kg proved ineffective (Prakash and Mathur, 1976).

6.3.21 Estrous Cycle Disruption Effect

A. precatorius seed extract delivered orally to female rats at various doses was inactive. Seed extract (1 mg/kg) in chloroform/methanol (2:1) was active when given subcutaneously to female rats at a dose of (Munsho et al., 1977). Another study found the same results at different doses (10, 5, and 2 gm/kg), however, 80%, 50%, and 25% of the rats had widespread leukocytic smears respectively (Samad et al., 1974). Ethanol (95%) dried seed extract applied orally to mice at a dose of 150 mg/kg had an antigonadotropin action (Prakesh et al., 1980).

6.3.22 Anti-Implantation Effect

When pregnant rats were given 50 mg/kg of body weight of chloroform/methanol seeds extract of *A. precatorius*, anti-implantation action was discovered (Maiti et al., 2009). When given orally to rats, ethanol (95%) root extract (100 mg/kg) produced significant results (Jadon et al., 1984), whereas when given orally to rats, ethanol (95%) seed extract produced significant results (Jadon et al., 1984) vice versa results were obtained in ethanol (95%) seed extract to rats and hamsters at a dose of 200.0 mg/kg (Sethi et al., 1990).

When pregnant rats were given 50 mg/kg of body weight of chloroform/methanol seeds extract of *A. precatorius*, anti-implantation action was also reported (Maiti et al., 2009). When given orally to rats, ethanol (95%) root extract (100 mg/kg) produced significant results, whereas when given orally to rats, ethanol (95%) seed extract produced significant results (Jadon et al., 1984).

6.3.23 Antispermatogenic Effect

When male rats were given ethanol seed extract (100 mg/kg) intragastrically for 60 days, the results were inconsequential (Samad et al., 1974). After 60 days, no significant histopathological alterations in the testes or sperm count were seen when rats were given dried seeds (250 mg/kg) 1:1 ratio of ethanol/water extract shown by Bhaduri et al. in 1968. When injected intramuscularly, the sterol component of dried seeds had excellent outcomes. The stoppage of spermatogenesis and a considerable reduction in the diameter of the seminiferous tubules were also observed in testicular lesions (Agarwal et al., 1970).

6.3.24 Cardiovascular Protective Activity

Abrus causes venous thrombosis (because to hemolysis, dehydration, and endothelial cell damage), pulmonary edema (related to endothelial damage and increased capillary permeability), hypertension, and arrhythmias in the cardiovascular system (due to electrolyte imbalance and through one or other associated mechanisms as noted in other pathologies) (Bhalla et al., 2015)

6.3.25 Antihelminthic Activity

In comparison to 0.25% DDT, *A. precatorius* seed extract (10.0%) had limited efficacy on *Musca domestica* (Hartzell and Wilcoxon, 1941). On *Culex quinquefasciatus*, acetone extracts of dried root and dried stem were inactive. Molgaard et al. (2001) studied the antihelminthic activity of extracts from 23 plant species in Zimbabwe and discovered that *A. precatorius* stem and root extracts were showing maximum effect. To establish molluscicidal efficacy, Abrin reduced protein, free amino acid, RNA, and DNA levels in *Lymnaea acuminata* nerve tissue, as well as inhibiting acetylcholinesterase, lactic dehydrogenase, and acid and alkali phosphatase activity (Fennell et al., 2004). The antihelminthic efficacy of an aqueous extract of *A. precatorius* stem and root was investigated. It was also found to have a deadly effect on cestodes. Root extracts (0.6 mg/mL) and stem extracts (1.5 mg/mL) were showing maximum effect against schistosomules (Singh and Singh, 1999).

6.3.26 *Antimalarial Activity*

Cytotoxicity and antiplasmodial activity were assessed in the evaluation of antimalarial activity, and *A. precatorius* extract had an IC_{50} value of <20 g/mL (Limmatvapirat et al., 2004).

6.3.27 *Smooth Muscle Stimulant Activity*

The seed extract of *A. precatorius* (0.2 mg/mL) in a methanol-water (1:1) chromatographic fraction was active on the guinea pig ileum and 0.5 mg/mL was active on the rat stomach (Menan et al., 2006). The ileum of guinea pigs was treated with seed oil (1.8 mcg/mL) and showed similar outcomes (Anam, 2001).

The same extract was found to have antispasmodic efficacy in rat uterus contractions caused by PGE, Ach, oxytocin, and epinephrine. The ethanol (95%) leaves extract was found to have a relaxing effect on the phrenic nerve diaphragm. The extract (4.0 mg/mL) was active on muscular stimulation at different concentrations, and 1.0 mg/mL was active on toad rectus abdominus muscle induced by Ach (Wambebe and Amosun, 1984). Other effects, such as reduction of intestinal motility, intestinal fluid retention, and antidiarrheal activity, were discovered to be active on the small intestines using the chromatographic fraction of dried seeds (10.0 mg/kg). In an intestinal motility model, the chromatographic fraction of dried seeds generated similar results (Nwodo and Botting, 1983).

6.3.28 *Antiallergic Activity*

Abruquinones A, B, F, and D, which were isolated from the leaves of *A. precatorius,* were found to have potent antiallergic properties. Inhibition of superoxide generation in rat neutrophils was <0.3 g/mL, and histamine inhibition in mast cells was <1 g/mL. Polymyxin B-induced hind paw oedema was reduced by abruquinone A in both normal and adrenalectomized mice. With these chemical ingredients, bradykinin, serotonin, and histamine, substance P-induced plasma extravasation in ear oedema was inhibited to a larger extent than with diphenhydramine and methysergide (Chukuo et al., 1995).

6.3.29 Wound Healing Activity

The presence of gums, mucilage, tannins, or phenolic compounds in the seeds resulted in early wound healing activity, which could be attributable to the presence of gums, tannins, mucilage, or phenolic compounds in the seeds. This adds to the growing body of data showing seed extracts and fractions are effective in vivo at preventing infection (Nwodo and Alumanah, 1991).

6.3.30 The Antiserotonergic Activity

When an albino rat was used in an in vitro investigation, petroleum ether extracts of *A. precatorius* showed smooth muscle contraction at various concentrations, with the reaction increasing as the dose increased; However, when ethyl acetate extract was compared with conventional sumatriptan at various doses, it only showed a baseline elevation (Alagesaboopathi and Sivakumar, 2008).

Choudhari et al. (2011) conducted pharmacological screenings of *A. precatorius* leaf extract on a frog fundus strip and found that it has smooth muscle contraction properties (Graded dosage response) when compared with a standard of sumatriptan. The antiserotonergic effect of ethyl acetate extract was suggested by this graded dosage response.

KEYWORDS

- **medicinal plant**
- ***Abrus precatorius***
- **pharmacological properties**
- **phytochemistry**

REFERENCES

Adelowotan, O.; Aibinu, I.; Aednipekun, E.; Odugbemi, T. The In-Vitro Antimicrobial and Antifungal Activity of *Abrus precatorius* (L) Fabaceae Extract on Some Clinical Pathogens. *Niger Postgrad. Med. J.* **2008,** *15* (1), 32–37.

Adesina, S. K. Studies on Some Plants Used as Anticonvulsants in Amerindian and African Traditional Medicine. *Fitoterapia* **1982,** *53*, 147–162.

Ae, L.; Bnrl, J.; Nf, N. Protective Effect of *Abrus precatorius* Seed Extract Following Alcohol Induced Renal Damage. *Eur. J. Sci. Res.* **2009,** *25* (3), 428–436.

Agarwal, S. S.; Ghatak, N.; Arora, R. B.; Bhardwaj M. M. Antifertility Activity of the Roots of *Abrus precatorius*. *Pharmacol. Res. Comm.* **1970,** *2*, 159–163.

Akinloye, B. A.; Adalumo, L. A. *Abrus precatorius* Leaves-a Source of Glycyrrhizin. *Niger J. Pharm.* **1981,** *12*, 405.

Alagesaboopathi, C.; Sivakumar, R. Studies on Cytotoxicity and Antitumor Screening of Red and White Forms of *Abrus precatorius* L. *Afr. J. Biotech.* **2008,** *7* (22), 3984–3988.

Ali, E.; Malek, A. Chemical Investigations on *Abrus precatorius* Linn. (Beng. Kunch). *Sci. Res. lll.* **1966,** *3*, 141–145.

Amites, G.; Karmakar, S.; Biswas, T. Pharmacognostical Studies and Chromatographic Evaluation of the Different Extracts of *Abrus precatorius* Linn. *Intern. J. Pharma. Res. Dev.* **2012,** *4* (3), 225–233.

Anam, E. M. Anti-Inflammatory Activity of Compounds Isolated from the Aerial Parts of *Abrus precatorius* (Fabaceae). *Phytomed.* **2001,** *8* (1), 24–27.

Anbu, J.; Velayuthan, R.; Sumithra, M.; Chowdary, S. B.; Kumar, S.; Kannadhasan, R.; Kumar, R. Anticancer Activity of Petroleum Ether Extract of *Abrus precatorius* on Ehrlich Ascitis Carcinoma in Mice. *Int. J. Pharm. Bio Sci.* **2011,** 2, 24–31.

Arora, R. Phytopharmacological Evaluation of Ethanolic Extract of the Seeds of *Abrus precatorius* L. *J. Pharmacol. Toxicol.* **2011,** *6* (6), 580–588.

Battu, G. R.; Kumar, B. M. Hepatoprotective Activity of *Abrus precatorius* Linn. Against Paracetamol Induced Hepatotoxicity in Rats. *Pharmacol. Online* **2009,** *3*, 366–375.

Begum S. Chemical Investigation of White Seeded Variety of *Abrus precatorius* Linn. *Pakistan J. Sci. Indus. Res.* **1992,** *35* (7/8), 270–271.

Bhaduri, B.; Ghose, C. R.; Bose, A. N.; Moza, B. K.; Basu, U. P. Anti-Fertility Activity of Some Medicinal Plants. *Indian J. Exp. Biol.* **1968,** 6, 252–253.

Bhakta, S.; Das, S. The Medicinal Values of *Abrus precatorius*: A Review Study. *J. Adv. Biotechnol. Exp. Ther.* **2020,** *3* (2), 84–91.

Bhalla, A.; Thirumalaikolundusubramanian, P.; Fung, J.; Cordero-Schmidt, G.; Soghoian, S.; Sikka, V.; Dhindsa, H.; Singh, S. Native Medicines and Cardiovascular Toxicity. *Heart Toxins*. **2015,** 175–202.

Bharadwaj, D. K.; Bisht, M. S.; Mehta, C. K. Flavonoids from *Abrus precatorius*. *Phytochemistry* **1980,** *19*, 2040–2041.

Bhaumik, H. L. Hydrocarbons, Fatty Acids, Triterpenoid and Sterols in the Seeds of *Abrus precatorius*. *Sci. Cult.* **1987,** *53* (1), 23–24.

Bobbarala, V.; Vadlapudi, V. *Abrus precatorius* L. Seed Extracts Antimicrobial Properties Against Clinically Important Bacteria. *Intern. J. Pharm.Tech. Res.* **2009,** *1* (4), 1115–1118.

Chang, H. M.; Chiang, T. C.; Mak, T. C. New Oleanene-Type Triterpenes from *Abrus precatorius* and X-Ray Crystal Structure of Abrusgenic Acid-Methanol 1:1 Solvate. *Planta Med.* **1983,** *49* (11), 165–169.

Choi, Y. H.; Hussain, R. A.; Pezzuto, J.M; Kinghorn A. D.; Morton J. F. Abrusosides A-D; Four Novel Sweet-Tasting Triterpene Glycosides from the Leaves of *Abrus precatorius*. *J. Nat. Prod.* **1989a,** *52* (5), 1118–1127.

Choi, Y. H.; Kinghorn, A. D.; Shi, X. B.; Zhang, H.; Teo, B. K. Abrusoside A: A New Type of Highly Sweet Triterpene Glycoside. *J. Chem. Soc. Chem. Commun.* **1989b,** *13*, 887–888.

Choudhari, A. B.; Nazim, S.; Khairnar, A. S.; Gomase, P. V.; Afsar, S. Evaluation of Antiserotonergic Activity of Ethyl Acetate Extract of *Abrus precatorius* Leaves. *J. Pharm. Res.* **2011,** *4* (3), 570–572.

Chukuo, S.; Chen, S. C.; Chen, L. H.; Wu, J. B.; Wang, J. P.; Teng, C. M. Potent Anti-Platelet; Anti-Inflammatory and Anti-Allergic Isoflavanquinones from the Roots of *Abrus precatorius*. *Plant Medica* **1995,** *61* (4), 307–312.

Das, P. C.; Sarkar, A. K.; Thakur, S. Studies on Animals of a Herbo-Mineral Compound for Long Acting Contraction. *Fitoterapia* **1987,** *58* (4), 257–261.

Desai, V. B.; Sirsi, M. Antimicrobial Activity of *Abrus precatorius*. *Indian J. Pharm.* **1966,** *28*, 164.

Desai, V. B.; Sirsi, M.; Shankarappa, M.; Kasturibai, A. R. Studies on the Toxicity of *Abrus precatorius* Effect of Aqueous Extracts of Seeds on Mitosis and Meiosis in Grasshopper (*Poecilocera picta*). *Indian J. Exp. Biol.* **1966,** *4*, 164–166.

Desai, V. B.; Sirsi, M.; Shankarappa, M.; Kasturibai, A. R. Chemical and Pharmacological Investigations on the Seeds of *Abrus precatorius* Linn. II. Effect of Seeds on Mitosis and Meiosis in Grasshopper; *Poecilocera picta* and Some Ciliates. *Indian J. Exp. Biol.* **1971,** *9* (3), 369–371.

Dhawan, B. N.; Patnaik, G. K.; Rastogi, R. P.; Singh, K. K.; Tandon J. S. Screening of Indian Plants for Biological Activity VI. *Indian J. Exp. Biol.* **1977,** *15*, 208–219.

Fennell, C. W.; Lindsey, K. L.; Mcgaw, L. J.; Sparg, S. G.; Stafford, G. I.; Elgorashi, E. E.; Grace, O. M.; Staden, J. V. Assessing African Medicinal Plants for Efficacy and Safety: Pharmacological Screening and Toxicology. *J. Ethnopharmacol.* **2004,** *94*, 205–217.

Georgewill, O. A. Anti-Arthritic Activity of *Abrus precatorius* in Albino Rats. *Internet J. Lab. Med.* **2010,** *4*, 1.

Georgewill, O. A.; Georgewill, U. O. Evaluation of the Anti-Inflammatory Activity of Extract of *Abrus precatorius*. *Eastern J. Med.* **2009,** *14*, 23–25.

Ghosal, S.; Dutta, S. K. Alkaloids of *Abrus precatorius*. *Phytochemistry* **1971,** *10*, 195–198.

Ghosh, D.; Maiti, T. K. Immunomodulatory and Anti-Tumor Activities of Native and Heat Denatured *Abrus precatorius* Agglutinin. *Immunobiology* **2007,** *212* (7), 589–599.

Glasby, J. H. *Dictionary of Plants Containing Secondary Metabolites*; Taylor and Francis: New York: 1991; p 488.

Gupta, N. C.; Singh, B.; Bhakuni, D. S. Steroids and Triterpenes from *Alangium lamarckii, Allamanda cathartica, Abrus precatorius* and *Holoptelea integrifolia. Phytochemistry* **1969,** *8*, 791–792.

Hameed, A. K.; Hasmi, M. A.; Khan, M. I. *Abrus precatorius*. I. Isolation and Toxic Properties of Abrulin, a Protein Fraction from the Seeds. *Pak. J. Sci. Ind. Res.* **1961,** *4*, 53–56.

Hartzell, A.; Wilcoxon, F. A Survey of Plant Products for Insecticidal Properties. *Contr. Boyce Thompson Inst.* **1941,** *12*, 127–141.

Hegde, R.; Maiti, T. K.; Podder, S. K. Purification and Characterization of Three Toxins and Two Agglutinins from *Abrus precatorius* Seed by Using Lactamyl-Sepharose Affinity Chromatography. *Anal. Biochem.* **1991,** *194* (1), 101–109.

Itokawa, H.; Hirayama, F.; Tsuruoka, S.; Mizuno, K.; Takeya, K.; Nitta A. Screening Test for Antitumor Activity of Crude Drugs (III): Studies on Antitumor Activity of Indonesian Medicinal Plants. *Shoyakugaku Zasshi* **1990,** *44* (1), 58–62.

Jadon, A.; Mathur, R.; Jiwaji, J. Effects of *Abrus precatorius* Linn. Seed Extract on Biochemical Constituents of Male Mice. *J. Jiwaji Univ.* **1984,** *9* (1), 100–103.

Jahan, S.; Rasool, S.; Khan, M. A.; Ahemad, M.; Zafar, M.; Arsahd, M.; Abbasi, A. M. Antifertility Effects of Ethanolic Seed Extract of *Abrus precatorius* L. on Sperm Production and DNA Integrity in Adult Male Mice. *J. Med. Plant Res.* **2009,** *3* (10), 809–814.

Karawya, M. S.; El Gengaihi, S.; Wassel, G.; Ibrahim, N. A. Carbohydrates of *Abrus precatorius*. *Fitoterapia* **1981,** *52*, 179–181.

Kekuda, T. R. P.; Vinayaka, K. S.; Soumya, K. V.; Ashwini, S. K.; Kiran, R. Antibacterial and Antifungal Activity of Methanolic Extract of *Abrus pulchellus* Wall and *Abrus precatorius* Linn—A Comparative Study. *Intern. J. Toxicol. Pharmacol. Res.* **2010,** *2* (1), 26–29.

Khaleqe, A.; Aminuddin, M.; Mulk, S. A. U. Investigations of *Abrus precatorius* L. Constituents of Dry Root. *Pak. C S I R Bull. Monogr.* **1966,** *3*, 203.

Khan, A. H.; Gul, B.; Rahman, M. A. The Interactions of the Erythrocytes of Various Species with Agglutinins of *Abrus precatorius*. *J. Immunol.* **1966,** *96* (3), 554–557.

Kinjo, J.; Matsumoto, K.; Inoue, M.; Takeshita, T.; Nohara, T. A New Sapogenol and Other Constituents in Abrin Semen; the Seeds of *Abrus precatorius* L. *Chem. Pharm. Bull.* **1991,** *39* (1), 116–119.

Kuo, S. C.; Chen, S. C.; Chen, L. H.; Wu, J. B.; Wang, J. P.; Teng, C. M. Potent Antiplatelet; Anti-Inflammatory and Antiallergic Isoflavanquinones from the Roots of *Abrus precatorius*. *Planta Med.* **1995,** *61*, 307–312.

Kusumot, I. T.; Shimada, I.; Kakiuchi, N.; Hattori, M.; Namba, T.; Supriyatna, S. Inhibitory Effect of Indonesian Plant Extracts on Reverse Transcriptase of an RNA Tumour Virus. *Phytother. Res.* **1992,** *6* (5), 241–244.

Lefar, M. S.; Firestone, D.; Coleman, E. C.; Brown, N.; Shaw, D. W. Lipids from the Seeds of *Abrus precatorius*. *J. Pharm. Sci.* **1968,** *57*, 1442–1444.

Limmatvapirat, C.; Sirisopanaporn, S.; Kittakoop, P. Antitubercular and Antiplasmodial Constituents of *Abrus precatorius*. *Planta Med.* **2004,** *70*, 276–278.

Lin, J. Y.; Cheng, Y. C.; Liu, K.; Tung, T. C. Carbohydrate in Abrin. *Toxicon.* **1971,** *9*, 333–360.

List, P. H.; Horhammer, L. *Hager's handbuch der pharmazeutischen praxis*, Vols. 2–6; Springer-Verlag: Berlin, 1969–1979.

Maiti, P. C.; Mukherjea, S.; Chatterjee, A. Chemical Examination of Seeds of *Abrus precatorius*. *J. Indian Acad. Forensic Sci.* **1970,** *9*, 64–68.

Maiti, T. K.; Bhutia, S. K.; Mallick, S. K. In Vitro Immunostimulatory Properties of Abrus Lectins Derived Peptides in Tumor Bearing Mice. *Phytomedicine* **2009,** *16*, 776–778.

Markham, K. R.; Wallace, J. W.; Babu, Y.N; Krishnamurty, V.; Rao, M. G. 8-C-Glucosylscutellarein 6, 7-Dimethyl Ether and Its 2″-O-apioside from *Abrus precatorius*. *Phytochemistry* **1989,** *28* (1), 299–301.

Menan, H.; Banzouzi, J. T.; Hocquette, A.; Pelissier, Y.; Blache, Y.; Kone, M.; Mallie, M.; Assi, L. A.; Valentin, A. Antiplasmodial Activity and Cytotoxicity of Plants Used in West African Traditional Medicine for the Treatment of Malaria. *J. Ethnopharmacol.* **2006,** *105*, 131–136.

Misra, D. S.; Sharma, R. P.; Soni, B. K. Toxic and Hemagglutinating Properties of *Abrus precatorius*. *Indian J. Exp. Biol.* **1966,** *4* (3), 161–163.

Mobin, L.; Sayeed, S. A.; Ali, R.; Saeed, S. G.; Ahmed, R. Antibacterial and Antifungal Activities of the Polyphenolic Fractions Isolated from the Seed Coat of *Abrus precatorius* and *Caesalpinia crista*. *Nat. Prod. Res.* **2018,** *32* (23), 2835–2839.

Molgaard, P.; Nielsen, S. B.; Rasmussen, D. E.; Drummond, R.; Makaza, N.; Andreassen, J. Anthelmintic Screening of Zimbabwean Plants Traditionally Used Against Schistosomiasis. *J Ethnopharmacol.* **2001,** *74*, 257–264.

Moshi, M. J.; Kagashe, GA. B.; Mbwambo, Z. H. Plants Used to Treat Epilepsy by Tanzanian Traditional Healers. *J. Ethnopharmcol.* **2005,** *97*, 327–336.

Munsho, S. R.; Shetye, T. A.; Nair, R. K. Antifertility Activity of Three Indigenous Plant Preparations. *Planta Med.* **1977,** *31* (1), 73–75.

Nwodo, O. F. C.; Alumanah, E. O. Studies on *Abrus precatorius* Seeds II Antidiarrheal Activity. *J. Ethnopharmacol.* **1991,** *31* (3), 395–398.

Nwodo, O. F. C.; Botting, J. H. Uterotonic Activity of Extracts of the Seeds of *Abrus precatorius*. *Planta Med.* **1983,** *47* (4), 230–233.

Otake, T.; Mori, H.; Morimoto, M.; Ueba, N.; Sutardio, S.; Kusumoto, I.; Hattori, M.; Namba, T. Screening of Indonesian Plant Extracts for Antihuman Immunodeficiency Virus-Type *1* (HIV-1) Activity. *Phytother. Res.* **1995,** *9* (1), 6–10.

Pal, R. S.; Ariharasivakumar, G.; Kundlik G.; Upadhyay A. In-Vitro Antioxidative Activity of Phenolic and Flavonoid Compounds Extracted from Seeds of *Abrus precatorius. Int. J. Pharm. Pharm. Sci.* **2009,** *1* (2), 136–140.

Parekh, J.; Jadeja, D.; Chanda, S. Efficacy of Aqueous and Methanol Extracts of Some Medicinal Plants for Potential Antibacterial Activity. *Turk. J. Biol.* **2005,** *29*, 203–210.

Prakash, A. O.; Mathur, R. Screening of Indian Plants for Antifertility Activity. *Indian J. Exp. Biol.* **1976,** *14*, 623–626.

Prakesh, A. O.; Gupta, R. B.; Mathur, R. Effect of Oral Doses of *Abrus precatorius* Linn. Seeds on the Oestrus Cycle; Body Weight; Uterine Weight and Cellular Structures of Uterus in Albino Rats. *Probes* **1980,** *19*, 286–292.

Premanand, R.; Ganesh, T. Neuroprotective Effects of *Abrus precatorius* Linn. Aerial Extract on Hypoxic Neurotoxicity Induced Rats. *Int. J. Chem. Pharma. Sci.* **2010,** *1* (1), 9–15.

Ragasa, C. Y.; Lorena, G. S.; Mandia, E. H.; Raga, D. D.; Shen, C. C. Chemical Constituents of *Abrus precatorius. Am. J. Essent. Oils Nat.Prod.* **2013,** *1* (2), 7–10.

Ramaswami, K.; Chandana, V. R.; Palpu, P.; Sunil, P. T.; Gaddam, D. R. Exploring the Protective Effects of *Abrus precatorius* in HepG2 and N-Nitrosodiethylamine-Induced Hepatocellular Carcinoma in Swiss Albino Rats. *Iran J. Pharm. Sci. Spring.* **2010,** *6* (2), 99–114.

Ramnath, V.; Kuttan, G.; Kuttan, R. Immunopotentiating Activity of Abrin, a Lectin from *Abrus precatorius* L. *Indian J. Exp. Bio.* **2002,** *40* (8), 910–913.

Ramnath, V.; Kuttan, G.; Kuttan, R. Effect of Abrin on Cell-Mediated Immune Responses in Mice. *Immunopharmacol. Immunotoxicol.* **2006,** *28* (2), 259–268.

Rao, M. V. Antifertility Effects of Alcoholic Seed Extracts of *Abrus precatorius* Linn. in Male Albino Rats. *Acta Eur. Fertil.* **1987,** *18* (3), 217–220.

Reddy, M.; Theja, D.; Bhagvan, R.; Reddy, P.; Gupta, A.; Raidu, S.; Rao, A. Antidiabetic Activity of the Seeds of *Abrus precatorius* in Streptozotocin & Nicotinamide Induced Diabetic Rats. *Pharmacologyonline* **2010,** *1*, 701–705.

Roy, J.; Som, S.; Sen, A. Isolation, Purification, and Some Properties of a Lectin and Abrin from *Abrus precatorius* Linn. *Arch. Biochem. Biophys.* **1976,** *174*, 359–361.

Samad, F.; Mukhtar, A.; Jan, Z. A.; Khan, Z. U. Effect of Alcohol Extract of Ratti Seeds (*Abrus precatorius*) on the Reproduction of Female Rats. *J. Math Sci.* **1974,** *12*, 157.

Saxena, V. K.; Sharma, D. N. A New Isoflavone from the Roots of *Abrus precatorius. Fitoterapia* **1999,** *70*, 328–329.

Sethi, N.; Nath, D.; Singh, R. K. Teratological Aspects of *Abrus precatorius* Seeds in Rats. *Fitoterapia* **1990,** *61* (1), 61–63.

Singh, S.; Singh, D. K. Effect of Molluscicidal Components of *Abrus precatorius*; *Argemone mexicana* and *Nerium indicum* on Certain Biochemical Parameters of *Lymnaea acuminata*. *Phytother Res.* **1999,** *13* (3), 210–213.

Sinha, R. Post-Testicular Anti-Fertility Effects of *Abrus precatorius* Seed Extract in Albino Rats. *J. Ethnopharmacol.* **1990,** *28* (2), 173–181.

Sohn, S. H.; Lee, H.; Nam, J. Y.; Kim, S. H.; Jung, H. J.; Kim, Y.; Shin, M.; Hong, M.; Bae, H. Screening of Herbal Medicines for the Recovery of Cisplatin Induced Nephrotoxicity. *Env. Tox. Pharmacol.* **2009,** *28*, 206–212.

Subba, Reddy, V. V.; Sirsi, M. Effect of *Abrus precatorius* L. on Experimental Tumors. *Cancer Res.* **1969,** *29*, 1447–1451.

Sudaroli, M.; Chatterjee, T. K. Evaluation of Red and White Seed Extracts of *Abrus precatorius* Linn. Against Freund's Complete Adjuvant Induced Arthritis in Rats. *J. Med. Plants Res.* **2007,** *1* (4), 086–094.

Tripathi, S.; Maiti, T. K. Immunomodulatory Role of Native and Heat Denatured Agglutinin from *Abrus precatorius*. *Intern. J. Biochem. Cell Biol.* **2005,** *37*, 451–462.

Wambebe, C.; Amosun, S. L. Some neuromuscular Effects of the Crude Extracts of the Leaves of *Abrus precatorius*. *J. Ethnopharmacol.* **1984,** *11* (1), 49–58.

Wei, C. H.; Hartman, F. C.; Pfuderer, P.; Yang, W. K. Purification and Characterization of Two Major Toxic Proteins from Seeds of *Abrus precatorius*. *J. Biol. Chem.* **1974,** *249*, 3061–3067.

Willaman, J. J.; Li, H. L. *Alkaloid-Bearing Plants and Their Contained Alkaloids*; Agricultural Research Service; U. S. Department of Agriculture: USA, 1970; pp 1–286.

Xiao, Z. H.; Wang, F. Z.; Sun, A. J.; Li, C. R.; Huang, C. G.; Zhang, S. A New Triterpenoid Saponin from *Abrus precatorius* Linn. *Molecules* **2011,** *17*, 295–302.

Zambenedetti, P.; Giordano, R.; Zatta, P. Histochemical Localization of Glycoconjugates on Microglial Cells in Alzheimer's Disease Brain Samples by Using *Abrus precatorius*, *Maackia amurensis, Momordica charantia,* and *Sambucus nigra* lectins. *Exp. Neurolog.* **1998,** *153* (1), 167–171.

Zia-Ul-Haque, A.; Qazi, M. H.; Hamdard, M. E. Studies on the antifertility properties of active components isolated from the seeds of *Abrus precatorius* Linn I. *Pakistan J. Zool.* **1983,** *15* (2), 129–139.

Zore, G. B.; Awad, V.; Thakre, A.D; Halde, U. K.; Meshram, N. S.; Surwase, B. S.; Karuppayii, S. M. Activity-Directed-Fractionation and Isolation of Four Antibacterial Compounds from *Abrus precatorius* L. Roots. *Nat. Prod. Res.* **2007,** *21* (9), 838–845.

CHAPTER 7

Biomolecules and Pharmacology of *Alhagi maurorum* Medik (Family: Fabaceae)

N. V. L. SIRISHA MULUKURI[1*] and PASALA PRAVEEN KUMAR[2]

[1]*Department of Natural Products, Nitte College of Pharmaceutical Sciences, Bangalore, Karnataka, India*

[2]*Department of Pharmacology, Santhiram College of Pharmacy, Kurnool, Andhra Pradesh, India*

[*]*Corresponding author. E-mail: sirishamulukuri056@gmail.com*

ABSTRACT

Local communities have identified the potential role of *Alhagi maurorum* for its numerous medicinal properties for the treatment of various ailments like diaphoretic, diuretic, purgative, treatment of piles, warts, migraine, and rheumatism etc. Many bioactives like tannins, alkaloids, saponins, flavonoids are present in the plant. The current chapter deals with the Phytochemical and Biological investigation of *Alhagi maurorum*. Bioactivities include hepatoprotective, wound healing, antireheumatic, antiviral, antibacterial, antifungal, antidepressant and analgesic activities.

7.1 INTRODUCTION

Alhagi maurorum is an under shrub, 60–90 cm tall, hairless or velvet-hairy. Simple leaves with dimensions of 1.0–2.5 cm length, 3–8 mm width, and

Bioactives and Pharmacology of Legumes. T. Pullaiah, PhD (Ed.)

obovate or elliptic-oblong shape are the characteristic features with a stalk of 2 mm long. Flowers of this shrub are borne in lateral and the fruit is 2.0–3.5 cm long, 2–3 mm broad, and is 1–9-seeded. *A. maurorum* is widely distributed in Afghanistan, Africa, India, Iran, Iraq, Pakistan, Palestine, Russia, Syria, and Turkey. Its flowering mainly occurs from April to September (Tackholm, 1974; Boulos, 2000).

Common names of *A. maurorum* include Camel Thorn; Caspian manna (English); Persian Mannaplant (Persian); Javasau (Gujarati); Javasa and Oont-jhari (Hindi); Nelingala (Kannada); Durlav (Nepali); Manna (Oriya); Dhanvayasa, Dhanvayasaka, and Duralabha (Sanskrit); Thurnjbin and Turanchipin (Tamil); Tella-giniyachettu (Telugu); and Bikh jawansa, Jawasa, and Turanjabeen (Urdu).

Traditionally, *A. maurorum* was used in folk medicine against various diseases like asthma, rheumatic-infections, diseases associated with urinary tract infections, and gastritis (Boulos, 1983; Archad et al., 2002). *A. maurorum* is also popular in folk medicine to treat glandular tumors, nasal polyps, and bile duct-related complications (James, 2017).

7.2 BIOACTIVES

A. maurorum is enriched with several phytochemicals like alkaloids, glycosides, flavonoids, saponins, steroids, tannins, and anthraquinone (Rahman et al., 2011; Samejo et al., 2012). The flavonoids isolated from *A. maurorum* include tamarixtin 3-O-dirhamnoside, isorhamnetin 3-O-glucosylneo-hesperidoside, isorhamnetine 3-O-robinobioside, lupeol, quercetin 3-O-rhamnoside, kampferol 3-O-galactoside, quercetin 3, 7-diglycoside, isorhamnetin 3-rutinoside, daidzein 7, 4-dihydroxyisoflavone, calycisin 3-hydroxyformononetin, isorhamnetin, tamarxtin aglycones, isorhamnetin-3-O-[-alpha-l-rhamnopyranosyl-(1→3)]-beta-D-gluco-pyranoside, 3'-O-methylorobol, and quercetin 3-O-beta-d-glucopyranoside (Al-Snafi, 2015). Several other constituents like triglyceride, aliphatic ester, aliphatic ketone (Mudawy et al., 2007), thiophene derivatives (Marashdah et al., 2006, 2007; Marashdah, 2014), and oleanane-type triterpene glycosides (e.g., lupeol) were isolated from *A. maurorum* (Hameda et al., 2012) roots. Interestingly, the exudates of the *A. maurorum* showed the presence of different types of monomeric units like galactose and uronic acids (Goncharov et al., 2001). *A. maurorum* leaves contain significant number of compounds like 4-hexyl-2,5-dihydro-2,5-dioxo-3-furanacetic acid; 2-nonadecanone; 9- octylheptadecane; 13-tetradecen-1-ol acetate; enuciferol; octadecane; hexadecanoic acid methyl ester; eicosane; docosane; tricosane;

tetracosane; 1,21- docosadiene; pentacosane; squalene; and octacosane. *A. maurorum* stem contains various oily compounds like β damascenone; E-geranyl acetone; actinidiolide; 2-(1,3-butadienyl)-1,3,5-trimethylbenzene; isopropyl myristate; drimenol; octadecane; 6,10,14-trimethyl-2- pentadecanone; nonadecane; farnesyl acetate; isopropyl palmitate; E-15-heptadecenal; neophytadiene; tricosane; tetracosane; squalene; nonacosane; hentriacontane; oxygenated monoterpenes; triterpene hydrocarbons; oxygenated sesquiterpenes; terpene-related compounds; and hydrocarbons, along with different acid derivatives (Samejo et al., 2012). Fourteen flavonoids, including four aglycones (1–4) and 10 glycosides (5–14), were isolated from the alcoholic extract of root by (Al-Saleem et al., 2019). As for the first time, flavonoid glycosides (6–14) were reported (Figure 7.1).

7.3 PHARMACOLOGICAL ACTIVITIES

7.3.1 Hepato-Nephroprotective Activity

Al-Saleem et al. (2019) evaluated the hepatoprotective activity of flavonoid fractions of root against CCl_4-induced hepatotoxicity rats. They also exhibited antioxidant and antiproliferative activities and exhibited weak cytotoxic activity.

Khalifa et al. (2020) reported that *A. maurorum* aqueous extract prevented norfloxacin-induced hepato-nephrotoxicity in rats. Extract-treated rats ameliorated norfloxacin-induced elevation in tissue malondialdehyde reduction of activities of catalase, glutathione peroxidase, and superoxide dismutase, and reduced glutathione concentration. Extract-treated rats' serum alkaline phosphatase, aminotransferases, urea, creatinine, and uric acid levels were reduced and increased RBC count, Hb concentration, PCV, leucocyte, and monocyte counts compared with the norfloxacin-treated rats.

7.3.2 Effect on Infertility

Ahmed (2019) investigated *A. maurorum* leaves' aqueous extract against CCl_4-induced oxidative stress on spermatogenesis, glutathione, superoxide dismutase, and malondialdehyde levels in adult rats. The author concluded that at the dose of 300 mg/kg b.wt., *A. maurorum* had a positive effect on spermatogenesis and increased glutathione and superoxide dismutase and reduced malondialdehyde levels in toxic rats.

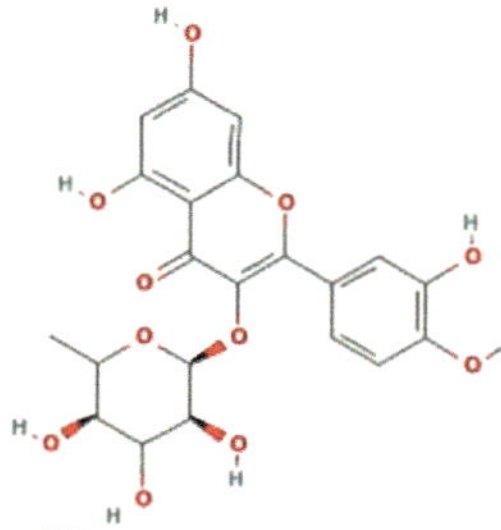

[1]

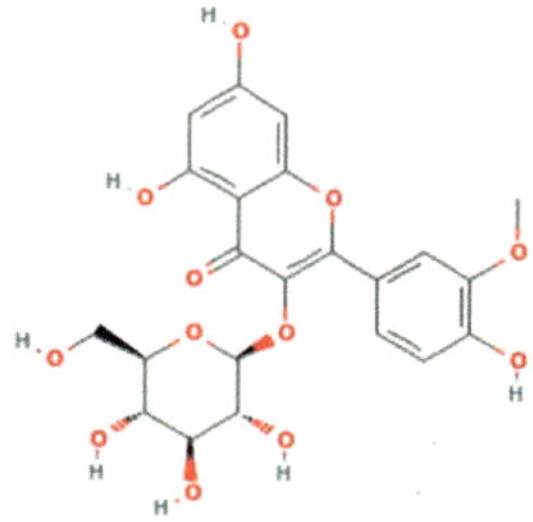

[2]

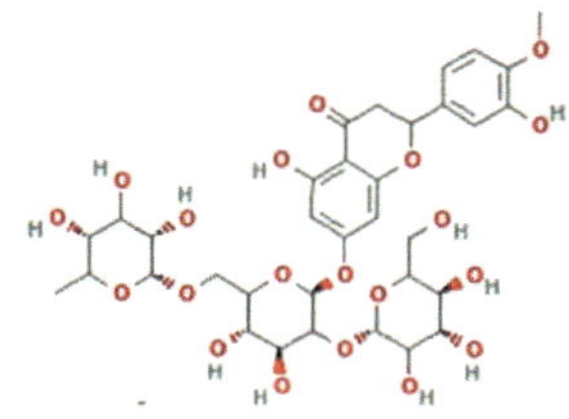

[3]

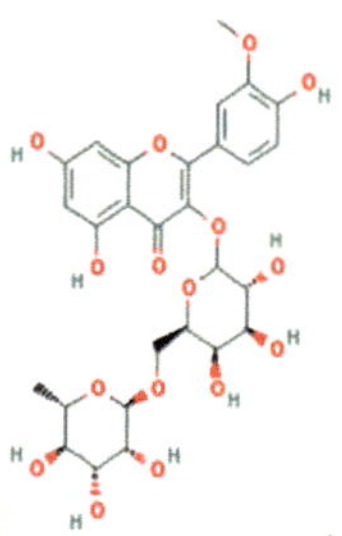

[4]

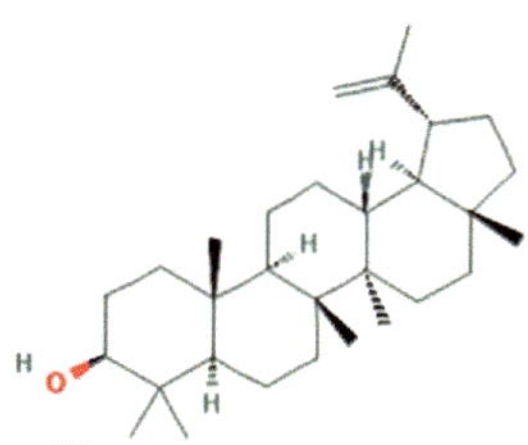

[5]

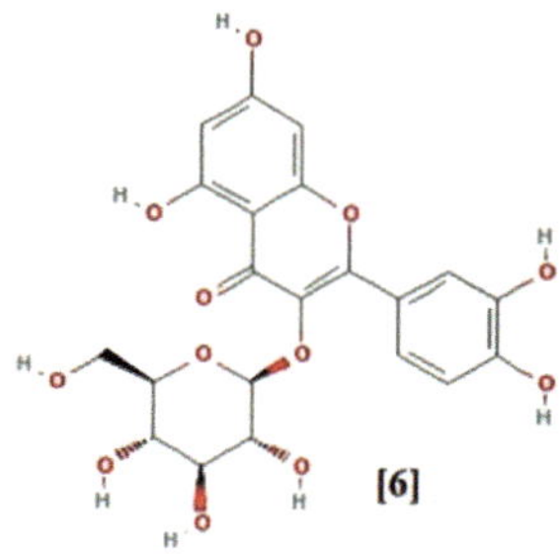

[6]

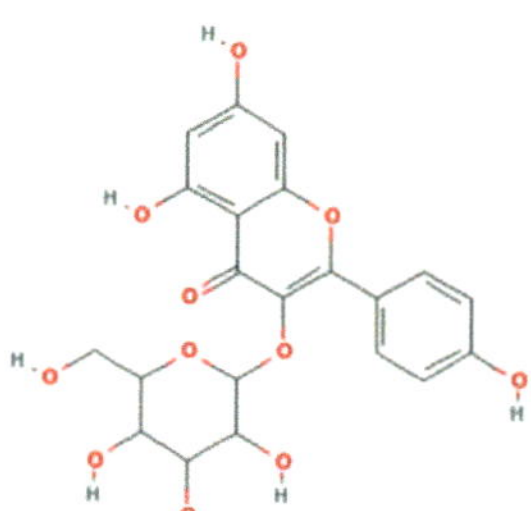

[7]

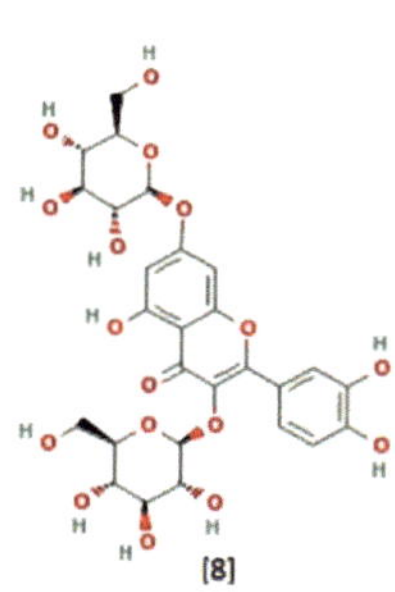

[8]

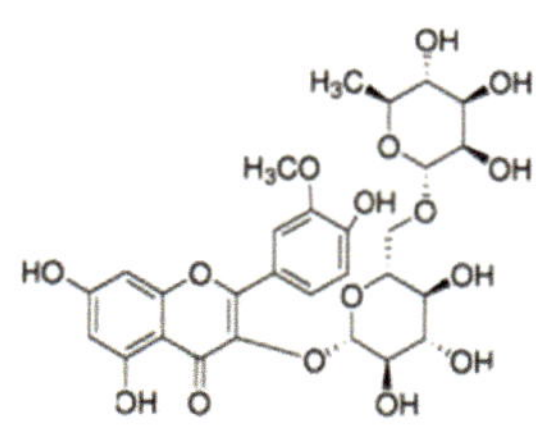

[9]

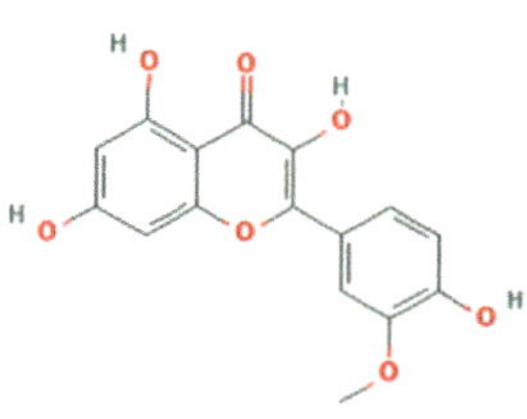

[10]

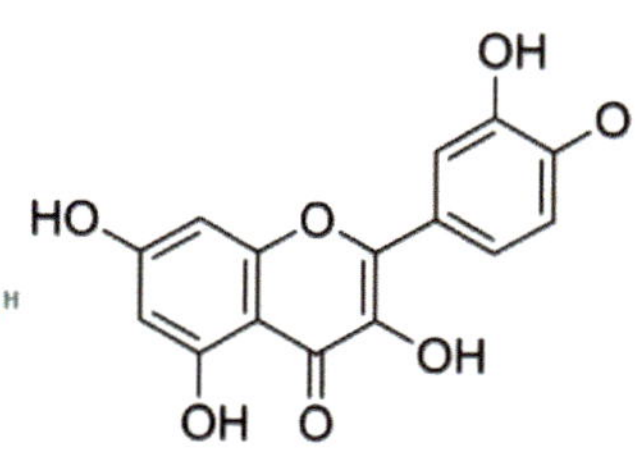

[11]

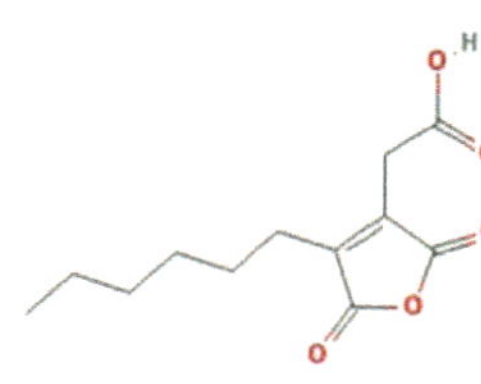

[12]

FIGURE 7.1 *(Continued)*

[13]

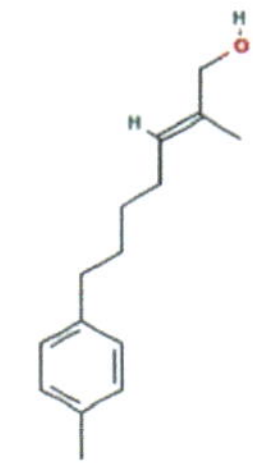

[14]

[15]

[16]

[17]

[18]

[19]

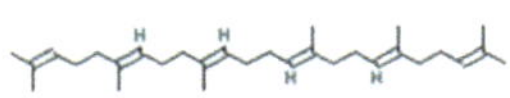

[20]

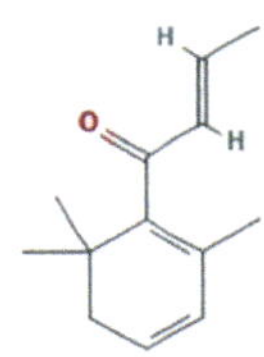

[21]

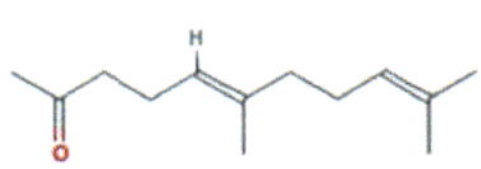

[22]

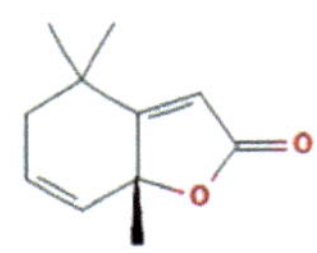

[23]

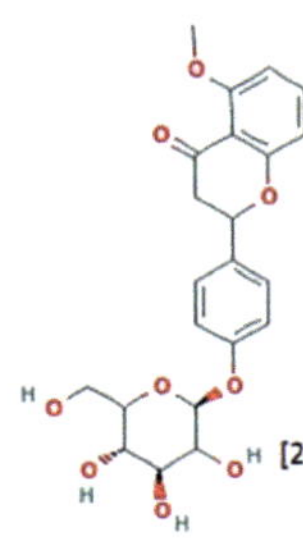

[24]

[25]

[26]

[27]

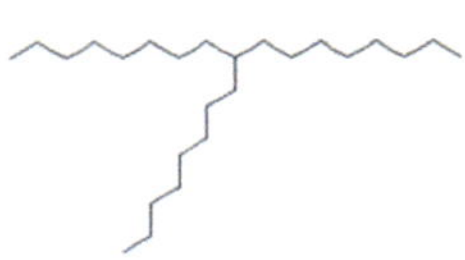

[28]

[29]

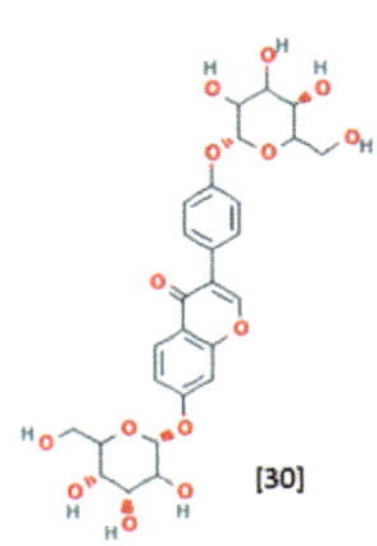

[30]

FIGURE 7.1 *(Continued)*

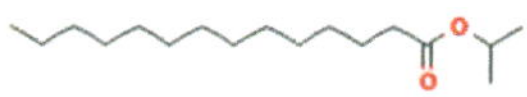
[31]

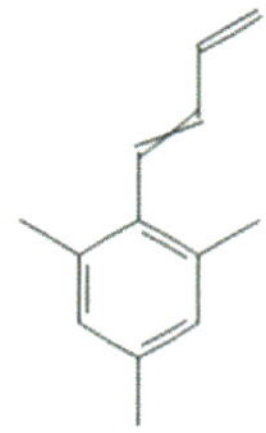
[32]

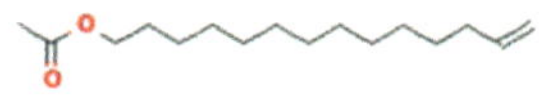
[33]

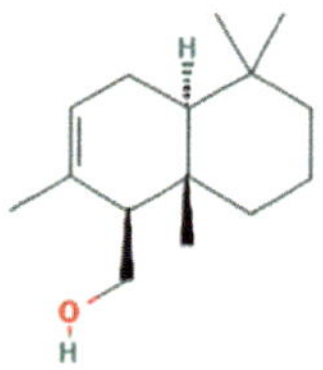
[34]

[35]

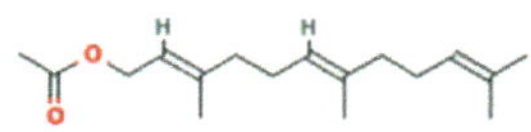
[36]

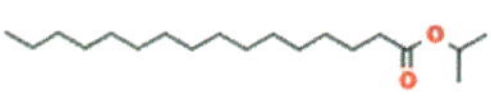
[37]

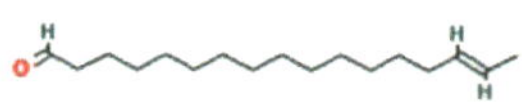
[38]

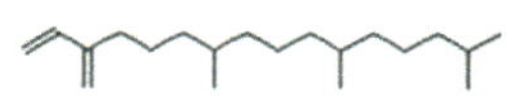
[39]

[40]

[41]

[42]

[43]

FIGURE 7.1 Structures of (1) tamarixtin 3-O-dirhamnoside, (2) isorhamnetin, 3-O-glucoside, (3) alhagidin, (4) isorhamnetine 3-O-robinobioside, (5) lupeol, (6) quercetin 3-O-rhamnoside, (7) kampferol 3-Ogalactoside, (8) quercetin 3, 7-diglycoside, (9) isorhamnetin 3- rutinoside, (10) isorhamnetin, (11) tamarxtin aglycones, (12) 4-hexyl-2,5-dihydro-2,5- dioxo-3-furanacetic acid, (13) 2-nonadecanone, (14) Enuciferol, (15) hexadecanoic acid methyl ester, (16) eicosane, (17) docosane, (18) tricosane, (19) tetracosane, (20) squalene, (21) β damascenone, (22) E-geranyl acetone, (23) actinidiolide, (24) alhagitin, (25) octacosane, (26) octadecane, (27) pentacosane, (28) 9- octylheptadecane, (29) 1,21- docosadiene, (30) daidzein 7, 4–dihydroxyisoflavone, (31) isopropyl myristate, (32) 2-(1,3- butadienyl)-1,3,5-trimethylbenzene, (33) 13-tetradecen-1-ol acetate, (34) drimenol, (35) 6,10,14-trimethyl-2- pentadecanone, (36) farnesyl acetate, (37) isopropyl palmitate, (38) E-15-heptadecenal, (39) neophytadiene, (40) tricosane, (41) Tetracosane, (42) nonacosane, and (43) hentriacontane.

7.3.3 Wound Healing Activity

Pseudomonas aeruginosa alginate and *A. maurorum* aqueous extract combined gel showed wound-healing properties due to decrease in the time for re-epithelialisation and increase in wound-contraction percentage, after 21 days treatment on Wistar rats (Pourali and Yahyaei, 2019).

7.3.4 Anti-Rheumatic Activity

A. maurorum aqueous extract has the power of reducing TNF-α, IL-1β, and COX-2 gene expressions and decreased production rate of NO and PGE2 in chondrocytes of monocyte/macrophages; hence, it is used as medicine for the treatment of rheumatic diseases traditionally (Maghsoodi et al., 2019).

7.3.5 Phytoacaricidal Activity

Mostafa et al. (2019) investigated the activity of extracts of *A. maurorum* aerial parts against *Panonychus citri*. Methanol extract has reduced the viability (58.16%) due to 4H-pyran-4-one,2,3-dihydro-3,5-dihydroxy-6-methyl (3.66%), 2-Furan-carboxaldehyde, 5-(hydroxymethyl) (3.33%), Bis(2-ethylhexyl) phthalatehalic acid, and bis(2-ethylhexyl)ester (5.58%) in methanol observed in gas chromatography/mass spectrometry (GC-MS) analysis, which are responsible for insecticidal activity (Mostafa et al., 2019).

7.3.6 Antibacterial and Antiheamolytic Activity

Methanol extracts of *A. maurorum* aerial parts (23 mg/mL) showed its significant activity against Gram positive bacteria *Bacillus cereus, Clostridium perfringens*, and *Staphylococcus aureus* and Gram negative organisms such as *Klebsiella oxytoca* and *K. pneumoniae* (Rahman et al., 2011). Hexane extracts of *A. maurorum* fresh aerial parts (6 mg/mL) showed zone of inhibition against Gram positive bacteria such as *Listeria ivanovii, L. monocytogenes, Bacillus cereus*, and *Staphylococcus aureus* and Gram negative organisms such as *E. coli, Yersinia enterocolitica, Klebsiella oxytoca*, and *K. pneumoniae* (Rahman et al., 2011).

Hussein et al. (2019) investigated the antibacterial and antihaemolytic efficiency of *A. maurorum,* native to Iraq. Zone of inhibition of *A. maurorum* against *Staphylococcus aureus* was 12.66 mm, while its effect against *Pseudomonas aeruginosa* generated an 8.33 mm ZOI and did not have any toxic effects on human red blood cells.

7.3.7 Antifungal Activity

The flavonoids from *A. maurorum* are mainly responsible for antifungal activity against *Alternaria alternata, Candida albicans, Cladosporium cladosporioides, Cryptococcus neoformans*, and *Trichophyton mentagrophytes*. Methanolic extract of *A. maurorum* (23 mg/mL) exhibited antifungal activity against *Aspergillus flavus, Alternaria alternata, Fusarium oxysporum, F. solani, Bipolaris oryzae, Chetomium*, and *Mucor*, with a percentage of growth inhibition of 33.4, 89.4, 89.3, 94.6, 91.7, 59.0, and 94.1% (Abd-Ellatif et al., 2001). Butanol and ethyl acetate extracts at a dose of 2 mg/disc has showed significant antifungal activity against the fungal specie (*C. albicans*) by disc diffusion susceptibility assay (Bakht et al., 2014).

7.3.8 Antidepressant Activity

A. maurorum extracts have been studied by Abdulbary et al. (2020) for its antidepressant effect using tail suspension test and open field test. The results of this study showed a high significant antidepressant effect of methanolic extract of *A. maurorum* in 2000 μg/ kg in contrast to both the positive and negative control (imipramine drug and normal saline).

7.3.9 Analgesic Activity, Anti-inflammatory Activity, and Anti-ulcer Activity

At the dose 200 and 400 mg/kg of *A. maurorum*, aerial parts aqueous methanolic extract (AMAM) exhibited analgesic effect in hot plate method. Anti-inflammatory effect using hind paw edema rats, activity was confirmed by a significant dose-dependent decrease hind paw edema thickness compared with the control group AMAM showed significant control gastric in indomethacin induced ulcer rats. At the doses of 200 and 400 mg/kg, AMAM showed decreased ulcer score 1.25 ± 0.045 and 0.25 ± 0.004 (Khalifa et al., 2018).

7.3.10 Antiviral Activity

A. maurorum aerial parts methanolic extract showed antiviral activity against foot mouth disease. MTT assay has revealed that methanolic extract in a concentration of 3 mg/mL or less has no cytotoxic impact, and the 50% cell cytotoxicity (CC_{50}) was about 30 μg/mL (Shakiba et al., 2017).

7.3.11 Antihyperglycemic Effect

A. maurorum aerial parts ethanolic extract (300 mg/kg) lowered the blood glucose levels, and the SOD, GPx, and GST level increased in STZ diabetic group of rats (Sheweita et al., 2016).

7.3.12 Antiplatelet Activity

A. maurorum root on oxidative protein/lipid damages in human blood platelets and human plasma after treatment with hydrogen peroxide was investigated in vitro, along with blood platelet activation corresponding to thrombin-induced arachidonic acid pathway (Olas et al., 2015). Study results showed that *A. maurorum* extract (50 mg/mL) exhibited 37.9 ± 14.3 of inhibition of platelet adhesion (%).

7.3.13 Antiproliferative Activity

Loizzo et al. (2014) investigated the antiproliferative and antioxidant activities *A. maurorum* extracts. Highest radical scavenging activity against ABTS radical with an IC_{50} value of 4.7 μg/mL was observed with Diethyl ether (DE) extract, along with reduced iron transition with a *A. maurorum (*69.6 μM Fe(II)/g) that is compared with standard butylated hydroxytoluene 63.2 μM Fe(II)/g. Strong inhibitory action (IC_{50}: 2.7 μg/mL) on C32 melanoma cell lines was also observed with *A. maurorum* DE.

7.3.14 Antidiarrhoeal Activity

A. maurorum extract exhibited a significant antidiarrhoeal activity against castor oil-induced diarrhoea due to calcium channel blocking effect (Atta and Mouneir, 2004).

KEYWORDS

- ***Alhagi maurorum***
- **bioactives**
- **pharmacology**
- **antimicrobial**
- **antirheumatic**

REFERENCES

Abd-Ellatif, S.; Abdel Rahman, S. M.; Deraz, S. F. Promising Antifungal Effect of Some Folkloric Medicinal Plants Collected from El-Hammam Habitat, Egypt Against Dangerous Pathogenic and Toxinogenic Fungi. *ARPN J. Agric. Biol. Sci.* **2001,** *6* (9), 25–32.

Abdulbary, M.; Al-Hilaly, I. A.; Ghani, S. Z.; Fhadil A. Detection of Antidepressant Effect of Alcoholic Extract of *Alhagi maurorum* Medik on Mice. *Plant Arch.* **2020,** *20* (1), 2727–2730.

Ahmed, M. A. Protective Effect of Aqueous Extract of *Alhagi maurorum* in Spermatogenesis and Antioxidant Status of Adult Rats Exposed to Carbon Tetrachloride. *Iraqi J. Vet. Sci.* **2019,** *33* (1), 1–7.

Al-Saleem, M. S.; Al-Wahaib, L. H.; Abdel-Mageed, W. M.; Gouda, Y. G.; Sayed, H. M. Antioxidant Flavonoids from *Alhagi maurorum* with Hepatoprotective Effect. *Pharmacogn. Mag.* **2019,** *15* (65), 592–599.

Al-Snafi, A. E. *Alhagi maurorum* as a Potential Medicinal Herb: An Overview. *Intern. J. Pharm. Rev. Res.* **2015,** *5* (2), 130–136.

Archad, M.; Akbar, G.; Rashid S. Wealth of Medicinal Plants of Cholistan Desert, Pakistan. Conventional Strategies. *Hamdard Medicus* **2002,** *45*, 25–34.

Atta, A. H.; Mouneir, S. M. Antidiarrhoeal Activity of Some Egyptian Medicinal Plant Extracts. *J. Ethnopharmacol.* **2004,** *92* (2–3), 303–309.

Bakht, J.; Gohar, N.; Shafi, M. In Vitro Antibacterial and Antifungal Activity of Different Solvent Extracted Samples of *Alhagi maurorum*. *Pak. J. Pharma. Sci.* **2014,** *27*, 1955–1961.

Boulos, L. *Medicinal Plants of North Africa*; Reference Publications Inc.: Cairo, Egypt, 1983.

Boulos, L. *Flora of Egypt*, Vol. II; Al-Hadara Publishing: Cairo, Egypt, 2000.

Goncharov, M. Y.; Yakovlev, G. P.; Vitovskaya, G. A. Composition of Polysaccharides from Above-Ground Part of *Alhagi maurorum* Medic. *Rastitel'nye Resursy* **2001,** *37*, 60–63.

Hameda, A.; Perronec, A.; Mahalel, U.; Oleszek, W.; Stochmal, A.; Piacente, S. Oleanane Glycosides from the Roots of *Alhagi maurorum*. *Phytochem. Lett.* **2012,** *5* (4), 782–787.

Hussein, N. N.; Marzoog, T. R.; Al-Niaame, A. E. The Antibacterial, Antiheamolytic, and Antioxidant Activities of *Laurus nobilis* and *Alhagi maurorum* Native to Iraq. *Baghdad Sci. J.*. **2019,** *16* (3 Suppl), 707–712.

James, A. D. *Alhagi maurorum* (Fabaceae). Dr. Duke's Phytochemical and Ethnobotanical Databases, 2017.

Khalifa, H. A.; Elattar, A. M.; Eleiwa, N. Z. Possible Analgesic, Anti-Inflammatory and Anti-Ulcerogenic Effects of *Alhagi maurorum* Methanolic Extract in Rats and Mice. *Slovenian Vet. Res.* **2018,** *55*, 375–382.

Khalifa, H. A.; Shalaby, S. I.; Abdelaziz AS. *Alhagi maurorum* Aqueous Extract Protects Against Norfloxacin-Induced Hepato-Nephrotoxicity in Rats. *Chinese Herbal Med* **2020,** *12* (2), 156–162.

Loizzo, M. R.; Rashed, K.; Said, A.; Bonesi, M.; Menichini, F.; Tundis, R. Antiproliferative and Antioxidant Properties of *Alhagi maurorum Boiss* (Leguminosae) Aerial Parts. *Ind. Crops Prod* **2014,** *53*, 289–295.

Maghsoodi, H.; Hagalahyari, S.; Narouie A. Reduced TNF-α, IL-1β and COX-2 Expressions and Decreased Production Rate of PGE2 and NO in Articular Chondrocytes of Monocyte/Macrophage Using the Aqueous Extract *Alhagi maurorum* L. *Scientific J. Ilam Univ. Medical Sci.* **2019,** *27* (2), 135–147.

Marashdah, M. S.; Mudawi, B. M.; Al-Hazimi, H. M.; Abdallah, M. A. New Triglyceride and New Aliphatic Ester from the Roots of *Alhagi maurorum* Medik. *J. Saudi Chem. Soc*, **2006,** *10* (2), 367–372.

Marashdah, M. S.; Mudawi, B. M.; Al-Hazimi, H. M.; Abdallah, M. A. New Aliphatic Ketone and New Aliphatic Ester from the Roots of *Alhagi maurorum* Medik. *J. Saudi Chem. Soc.* **2007,** *10* (3), 509–514.

Marashdah, S. New Natural Compound for the Enlargement of the Ureter. *Arabian J. Chem.* **2014,** *7*, 381–383.

Mostafa, R. M.; Essawy, H. S.; Baz, M. M. Potency of *Alhagi maurorum* Plant Extracts as Phytoacaricidal Against *Panonychus citri* (Acari: Tetranychidae). *Egyptian Acad. J. Biol. Sci. H. Botany.* **2019,** *10* (2), 41–53.

Mudawy, B. M.; Al-Hazimi, H. M.; Abdullah, M. A.; Marashdah, M. S. New Aliphatic Ester and New Thiophene Derivative from the Roots of *Alhagi maurorum* Medik. *J. Saudi Chem. Soc.* **2007,** *11* (1), 87–92.

Olas, B.; Hamed, A. I.; Oleszek, W.; Stochmal, A. Comparison of Biological Activity of Phenolic Fraction from Roots of *Alhagi maurorum* with Properties of Commercial Phenolic Extracts and Resveratrol. *Platelets.* **2015,** *26* (8), 788–794.

Pourali, P.; Yahyaei, B. Wound Healing Property of a Gel Prepared by the Combination of *Pseudomonas aeruginosa* Alginate and *Alhagi maurorum* Aqueous Extract in Rats. *Dermatologic Ther.* **2019,** *32* (1), e12779.

Rahman, S. M. A.; Abd-Ellatif, S. A.; Deraz, S. F.; Khalil, A. A. Antibacterial Activity of Some Wild Medicinal Plants Collected from Western Mediterranean Coast, Egypt: Natural Alternatives for Infectious Disease Treatment. *Afr. J. Biotechnol.* **2011,** *10*, 10733–10743.

Samejo, M. Q.; Memon, S.; Bhanger, M. I.; Khan, K. M. Chemical Composition of Essential Oils from *Alhagi maurorum*. *Chem. Nat. Compd.* **2012,** *48*, 898–900.

Sheweita, S. A.; Mashaly, S.; Newairy, A. A.; Abdou, H. M.; Eweda, S. M. Changes in Oxidative Stress and Antioxidant Enzyme Activities in Streptozotocin-Induced Diabetes Mellitus in Rats: Role of *Alhagi maurorum* Extracts. *Oxidat. Med. Cell. Longevity* **2016,** Article ID 5264064.

Shakiba, Y.; Rezatofighi, S. E.; Nejad, S. M. S.; Ardakani, M. R. Antiviral Activity of *Alhagi maurorum* Medik's Methanolic Extract on Foot and Mouth Disease Virus (FMDV) in Cell Cultures. *Jundishapur J. Nat. Pharm. Prod.* **2017,** Doi: 10.17795/jjnpp-30641

Tackholm, V. *Students Flora of Egypt*, 2nd ed.; Cairo University Publications: Beirut, 1974.

CHAPTER 8

A Systematic Review of *Astragalus membranaceus* (Fisch.) Bge. (Fabaceae)

PASALA PRAVEEN KUMAR[1*] and SILVIA NETALA[2]

[1]*Department of Pharmacology, Santhiram College of Pharmacy, Nandyal, Kurnool, Andhra Pradesh, India*

[2]*Department of Pharmacognosy, Shri Vishnu College of Pharmacy, Bhimavaram, West Godavari, Andhra Pradesh, India*

[*]*Corresponding author. E-mail: praveenpharmaco@gmail.com*

ABSTRACT

Astragalus membranaceus is a member of Fabaceae family. Nearly 200 compounds were isolated from this plant which consists of polysaccharides, saponins and flavonoids; this include 142 kinds of cycloartane-type saponins and 19 kinds of oleanane-type saponins, 63 flavonoids comprising of isoflavones, isoflavans, pterocarpans, flavonols, flavones, and flavonones. Out of 14 polysaccharides isolated from *A. membranaceus* 13 of them have β-D-(1→3)-galactan moieties branched with β-D-(1→6)-galactooligosaccharide side chains. Astragalus-based treatments have demonstrated significant amelioration of the toxicity induced by other concurrently administered orthodox drugs such as immunosuppressants and cancer chemotherapeutics. In this chapter we summarize the bioactives and their pharmacological activities *A. membranaceus*.

8.1 INTRODUCTION

Astragalus membranaceus (Fisch.) Bunge (Syn. *Astragalus propinquus*) is a member of Fabaceae family and distributed widely in Asia, Europe,

Bioactives and Pharmacology of Legumes. T. Pullaiah, PhD (Ed.)

and North America. *A. membranaceus* has been used commonly in many herbal formulations in traditional Chinese medicine for the treatment of a wide variety of diseases and body disorders or promoted as life-prolonging. It is appraised for its potential to reinforce the principal energy of the body, which we know as the immune system, together with the metabolic, respiratory, and eliminative functions. This fact is being increasingly justified by pharmacological studies showing different activities (Anon, 2003; Ryu et al., 2008; Chan et al., 2009; Ma et al., 2011; Nalbantsoy et al., 2012).

8.2 BIOACTIVES

Fifteen saponins were isolated from root ethanolic extract of *A. membranaceus* and named as astragaloside I (1), isoastragaloside II (2), astragaloside II (3), agroastragaloside I (4), cyclogaleginoside B (5), cycloaraloside A (6), brachyoside B (7), agroastragaloside II (8), astragaloside III (9), astragaloside IV (10), astramembranoside A (11), astramembranoside B (12), cylocanthoside E (13), cyclounifolioside B (14), and azukisaponin V methyl ester (15) by spectroscopic methods. Ten compounds 1–3, 5–7, 9–11, and 14 have cycloastragenol as an aglycon, and four compounds 4, 8, 12, and 13 have cyclocanthogenin as an aglycon. Previously unreported cycloartane-type saponins such as agroastragalosides I (4) and II (8) and cycloastragenol 3-O-β-D-xyloside (5) are produced from the hairy roots of *A. membranaceus*, together with the known saponins. The saponins (4, 5, and 8) are reported for the first time from the intact plant. Although the presence of the oleanane-type triterpene saponin, azukisaponin V methyl ester (15), in *Astragalus* plants has been demonstrated by others, this is the first report of the azukisaponin V methyl ester (15) from the *Astragalus* plants (Kim et al., 2008a; Kumar et al., 2018) (Figure 8.1).

From the ethanolic extract of *A. membranaceus* root, eleven flavonoid derivatives and a lignin were isolated and identified as liquiritigenin (1), daidzein (2), formononetin (3), sophorophenolone (4), calycosin (5), methylnissolin (6), isomucronulatol (7), isomucronulatol 7-O-glucoside (8), methylnissolin 3-O-glucoside (9), calycosin 7-O-glucoside (10), (+)-syringaresinol O-β-D-glucoside (11), and isomucronulatol 7,2'-di-O-glucoside (12) by spectroscopic methods (Lee et al., 2008) (Figure 8.2).

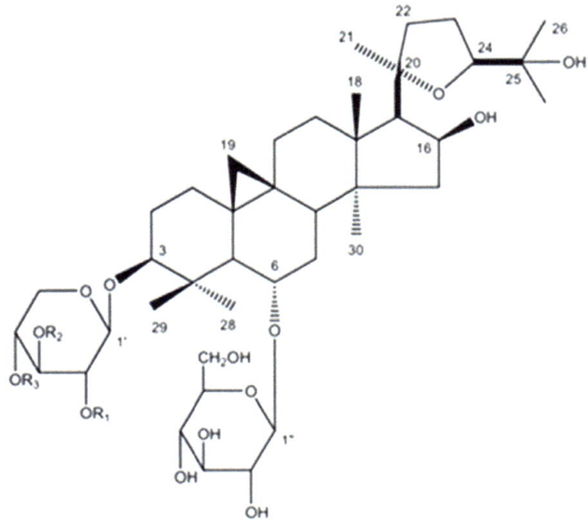

1	$R_1 = R_2 = Ac$ $R_3 = H$	astragaloside I
2	$R_1 = R_3 = H$ $R_2 = Ac$	isoastragaloside II
3	$R_1 = Ac$ $R_2 = R_3 = H$	astragaloside II
10	$R_1 = R_2 = R_3 = H$	astragaloside IV

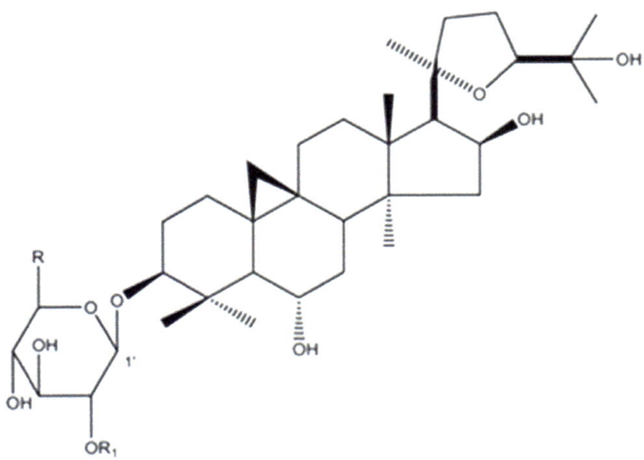

5	$R = R_1 = H$	cyclogaleginoside B
6	$R = CH_2OH$ $R_1 = H$	cycloaraloside A
9	$R_1 = Glc$ $R = H$	astragaloside III
14	$R = CH_2OH$ $R_1 = Glc$	cyclounifolioside B

FIGURE 8.1 *(Continued)*

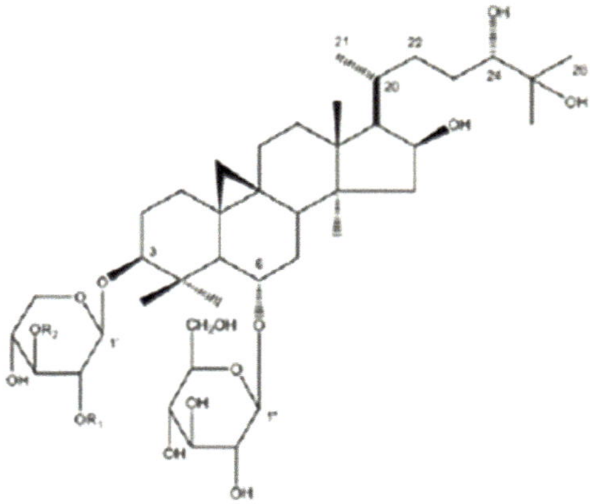

4	$R_1 = R_2 = Ac$	agroastragaloside I
8	$R_1 = Ac$ $R_2 = H$	agroastragaloside II
13	$R_1 = R_2 = H$	cylocanthoside E

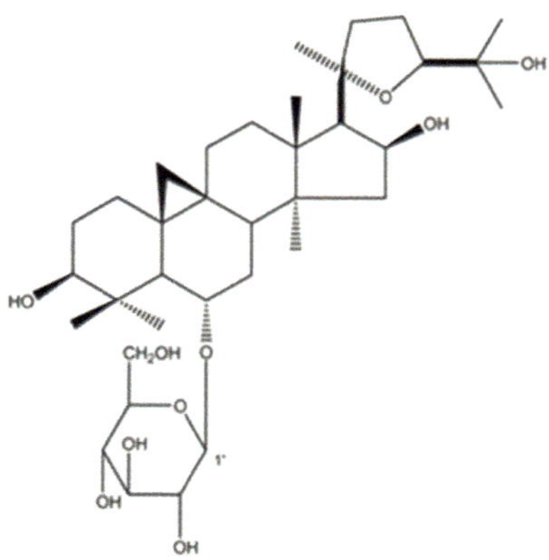

7 brachyoside B

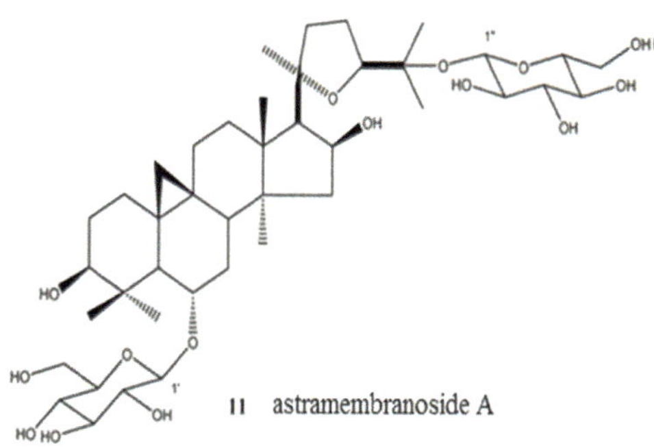

11 astramembranoside A

FIGURE 8.1 *(Continued)*

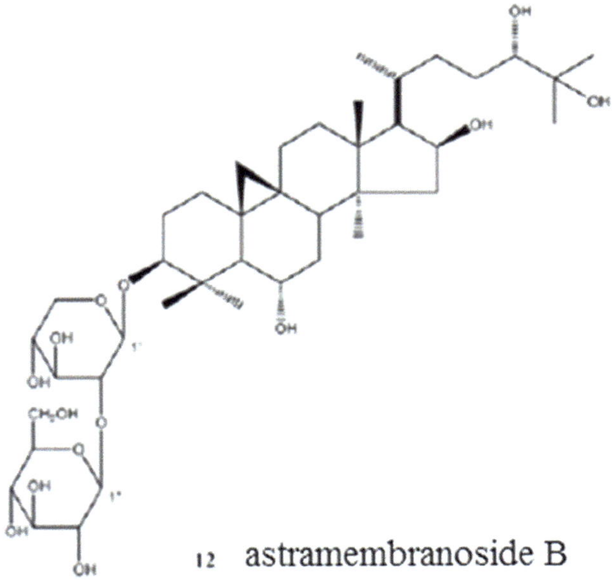

12 astramembranoside B

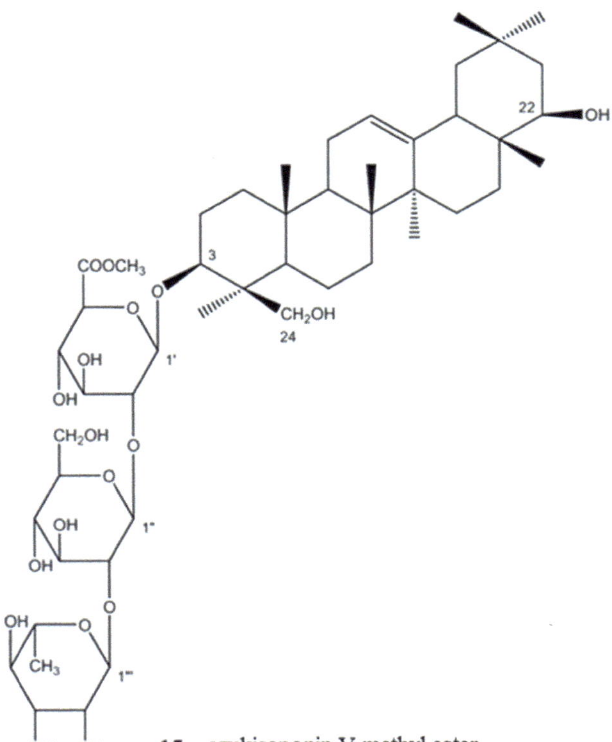

15 azukisaponin V methyl ester

FIGURE 8.1 Saponins from *Astragalus membranaceus* (compounds 1–15) (Kim et al., 2008b).

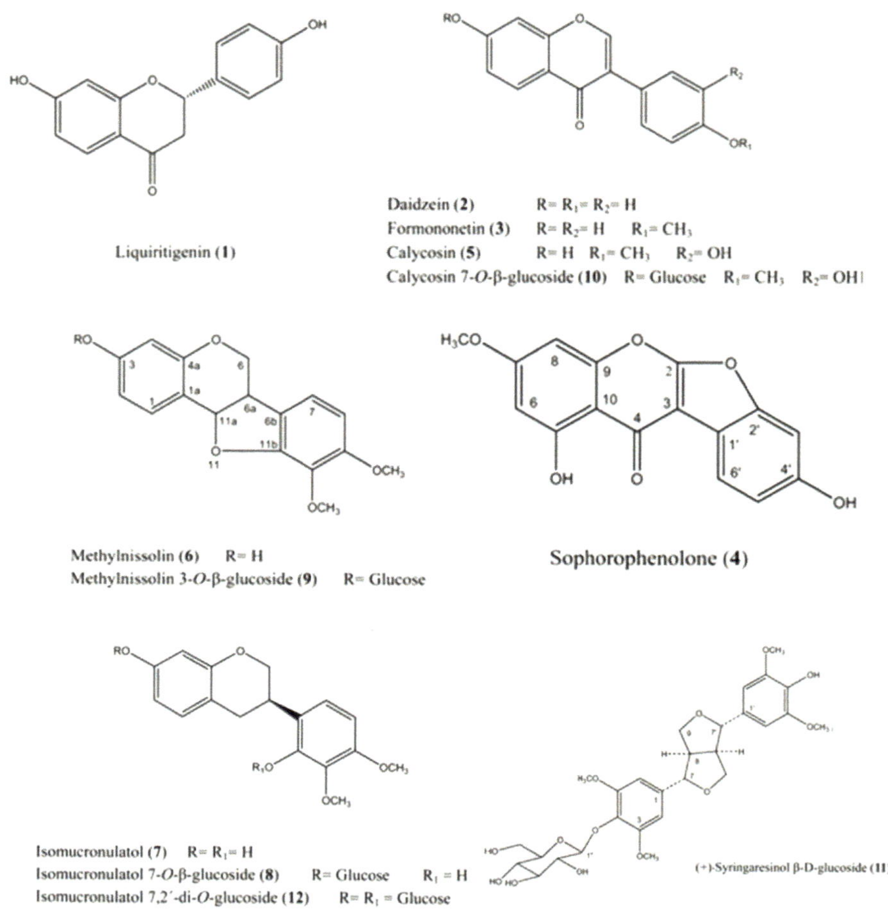

FIGURE 8.2 Flavonoid derivatives and a lignin from *A. membranaceus* (compounds 1–12) (Lee et al., 2008).

Kaempferol, isorhamnetin, rhamnocitrin, kumatakenin, rhamnocitrin-3-glucoside, and quercetin-3-glucoside are the flavones isolated from *A. membranaceus* (Jeong et al., 2018). From the dried sprouts of *A. membranaceus* tryptophan, linoleic acid and adenine were isolated (Jeong et al., 2018). Cycloartane-type triterpenoid, Astramembranosides B, was isolated from the roots of *A. membranaceus* (Kim et al., 2008b). Saponins, mongholicoside A and mongholicoside B, were isolated from aerial parts of *A. membranaceus* var. *mongholicus* (Yu et al., 2007). Secocycloartane triterpenoidal saponins, Huangqiyenin E & F, Huangqiyegenin III & IV, and trideacetylhuangqiyegenin III were isolated from leaves (Kuang et al., 2009). Flavonoids isolated from *Astragalus* root are calycosin-7-O-β-D-glucopyranoside, calycosin, formononetin, isomucronulatol, licoagroside D, vesticarpan, and pendulone (Li et al., 2014), and aulfuretin from leaves (Kavtaradze et al., 2011).

Chemical analysis conducted by Royal Botanic Gardens, Kew, United Kingdom, confirmed the presence of simple phenolics (caffeic, *p*-coumaric acids), various flavonoids, isoflavones, and saponins (astragalosides) in *A. membranaceus* extracts prepared by superheated water extraction method from the material grown by hydroponic or soil growing conditions in a glasshouse (Curnow et al., 2016).

Several types of organic compounds, namely, *Astragalus* polysaccharides, saponins, flavonoids, isoflavonoids, sterols, amino acids, and volatile oils have been isolated from *A. membranaceus* (Agyemang et al., 2013). Polysaccharides, *Astragalans* I, II, and III are obtained by hot water extraction. Astragalans II and III were *α*-(1,4)-glucans. By water extraction and alcohol precipitation technique, APS I and APS II were isolated. It was found by structural and content analyses that APS I consisted of arabinose and glucose in the ratio of 1:3.45 and ASP II consisted of rhamnose, arabinose, and glucose in the ratio of 1:6.25:17.86. AMem-P, AH-1, and APSID3, which are acidic polysaccharides, have also been isolated. AMem-P is a complex acidic polysaccharide consisting mainly of hexuronic acid and has terminal and *β*-1,5-linked-arabinofuranose, terminal and *β*-1,3-, *β*-1,4, *β*-1,6-linked, 3,6-branched-D-galactose, and 2,4-branched-L-rhamnose residue groups attached. AH-2, AE, AEF-1, AEF-2, and astroglucans A, B, and C are the other *Astragalus* polysaccharides. Various *Astragalus* saponins including isoastragalosides III and IV, astramembrannin II, cyclogaleginoside B, cycloaraloside A, brachyoside B, cyclocanthoside E, cyclounifolioside B, and astramembranosides A and B have also been isolated. Kaempferol, isorhamnetin, rhamnocitrin, kumatakenin and rhamnocitrin-3-glucoside, and quercetin-3-glucoside are the flavones. Formononetin, ononin, calycosin, calycosin-7-O-*β*-D-glucoside-6″-O-malonate, 3′-methoxy-5′-hydroxy-isoflavone-7-O-*β*-D-glucoside, and (3R)-2′,3′-dihydroxy-4′,7-dimethoxyisoflavone are the isoflavones. The isoflavanones compounds include 2′-hydroxy-3′,4′-dimethoxyisoflavone-7-O-*β*-D-glucopyranoside, 2′-hydroxy-3′,4′,7-trimethoxyisoflavone, 2′,7-dihydroxy-3′,4′,7-trimethoxyisoflavone, 3′,4′-dimethoxyisoflavone-7-O-*β*-D-glucoside, 8,2′-dihydroxy-4′,7-dimethoxyisoflavone, and 2′,3′,7-trihydroxy-4′-methoxyisoflavone. The isolated pterocarpans include 3,9,10-trimethoxypterocarpan, (6aR,11aR)-10-hydroxy-3,9-dimethoxypterocarpan, and 9,10-dimethoxypterocarpan-7-O-*β*-D-glucopyranoside (Agyemang et al., 2013).

Thirty-one compounds, including four phenolic acids, nineteen flavonoids, three isoflavones, two pterocarpans, and three saponins were identified using UPLC-QTOF-MS in the fraction obtained from ethanolic extract on AB-8 macroporous resin column from flowers of *A. membranaceus*. Phenolic acids include vanillic acid, protochatechuic acid, caffeic acid, and ferulic acid. Flavonoids include quercetin, rutin, hyperoside, isoquercitrin, quercetin-O-hexoside-pentoside, quercetin-O-rutinoside-hexoside, quercetin-O-malonyl-hexoside, isorhamnetin, isorhamnetin-O-hexoside, isorhamnetin-3-O-rutinoside, isorhamnetin-3-O--D-glucoside, isorhamnetin-O-hexoside-hexoside derivative, rhamnocitrin, astragalin, kaempferol, rhamnocitrin-O-hexoside, rhamnocitrin-O-malonyl-glucoside-rhamnoside, and rhamnocitrin-O-malonyl-hexoside. Isoflavones include calycosin-7-O--D-glucoside, calycosin, and pratensein-7-O--D-glucoside. Pterocarpans include (-)-methylinissolin 3-O--D-glucoside and 9, 10-diMP-3-O-acetyl-glucoside. Saponins include isoastragaloside II, malonylastragaloside II isomer, and soyasaponin I (Li et al., 2019). 1-hydroxy-5-methylolbenzol-2-O-beta-D-glucoside and (3R, 4R)-4,7-hydroxy-2',3'-dimethoxyisoflavane-4'-O-beta-D-glucoside were isolated from *A. membranaceus* var. *mongholicus* (Wang et al., 2014). *A. membranaceus* root gave eight flavonoids including formononetin, formononetin-7-O-β-D-glucoside, calycosin, calycosin-7-O-β-Dglucoside, 3'-hydroxy-7,4'-dimethoxyisoflavone, 10'-hydroxy-3,9'-dimethoxypeterocarpan, 8,3'-dihydroxy-3,9'-dimethoxyisoflavone, and 2'-hydroxy-3',4'-dimethoxyisoflavan-7-O-β-D-glucoside (Jiang et al., 2014).

8.3 PHARMACOLOGICAL ACTIVITIES

Pharmacological activities of *A. membranaceus* have been reviewed by Fu et al. (2014) and Zhang et al. (2020).

8.3.1 Immunomodulatory Activity

Huang et al. (2019) isolated the *A. membranaceus* waste protein from *A. membranaceus* and evaluated the immunomodulatory activity using cell model experiments. Study results showed that increased expression of inflammatory cytokines (IL-6, IL-12 p40, IL-1β, IL-1α), chemokines (CXCL1, CCL3), and enhancement of phagocytosis was observed in *A.*

membranaceus waste protein. The study concluded that *A. membranaceus* waste could be a natural immunomodulator.

8.3.2 Neuroprotective Activity

Costa et al. (2019) reported that Astragaloside IV, the main active compound present in *A. membranaceus* root attenuates Parkinson's disease, cognitive impairments (Li et al., 2017), and autoimmune encephalomyelitis (He et al., 2013) and exhibited anti-inflammatory and antioxidant activity.

8.3.3 Cardioprotective Effect

Sun et al. (2019) identified the effect of Astragalus polysaccharide (APS), the bioactive ingredient of *A. membranaceus* root and proved anti-apoptotic mechanism APS on diabetic cardiomyopathy rats (DCM). Study results showed improved heart function and attenuated myocardial apoptosis in APS-treated DCM rat. Jia et al. (2014) reported that the cardioprotective activity of AS-IV *via* activation of the PI3K/Akt pathway in doxorubicin induced cardiotoxicity.

8.3.4 Anticancer

Astragaloside IV exerted beneficial anticancer effects against various tumor types based on immune-enhancement activity, inhibition of tumor growth, and inhibition of tumor migration and invasion (Zhang et al., 2014). AS-IV had certain protective and adjuvant therapeutic effects on lung-related diseases and lung cancer, inhibited migration ability, cell viability, and invasiveness of Hepato carcinoma cells (Li et al., 2018) and proved anti-metastasis activity on gastric cancer cells *via* inhibition of PI3K/Akt/NF-κB pathway. AS-IV inhibits cell proliferation of CRC cells by downregulating the expression of B7-H3 (Zhu et al., 2018).

8.3.5 Effect on Diabetic Complications

Lu et al. (2015) showed that astragaloside IV effect on diabetic nephropathy (DN) rats was induced by streptozotocin injection. Protective effect was confirmed by significantly drop of blood glucose levels, urine micro-albumin levels, integrin β1, ILK, and α-actinin-4 in AS-IV-treated rats.

Ding et al. (2014) found that reduction of aldose reductase activity, ERK1/2 phosphorylation and NF-kB, apoptosis in AS-IV treatment in diabetic retinopathy db/db mice. Chen et al. (2018) showed that AS-IV protect against free fatty acids-induced proximal renal tubular epithelial cells injury in diabetes by reducing ROS, Bax expression, and caspase 3, and increasing Bcl-2 expression and phosphorylated Nrf2.

8.3.6 Cardioprotective Effect

Zheng et al. (2018) evaluated the cardioprotective mechanisms of AS-IV on myocardial ischemia/reperfusion (I/R) injury. AS-IV significantly reduces myocardial infarct size and left ventricular ejection fraction, increases shortening fraction, significantly decrease cardiac enzymes and cardiac troponin, and increases decline degree in ST-segment. AS-IV reported protective effect against 3,4-Benzopyrene-induced abdominal aortic aneurysm, which is due to the abrogation of oxidative stress and the improvement of phosphorylation of PI3K/AKT (Wang et al., 2018). Dai et al. (2014) investigated the effect of anti-hypertrophic actions of APS on cardiac hypertrophy induced by isoprenaline. Results showed dose-dependent suppression of intracellular free calcium, upregulated expression of calcineurin, translocation of nuclear factor of activated T cells, cytoplasmic 3 into nucleus, and activation of calmodulin kinase II.

8.3.7 Toxicity of AS-IV

Xuying et al. (2010) investigated the effect of AS-IV (0.25, 0.5, and 1.0 mg/kg/day i.v. for 4 weeks) on reproductive system in female rats. Results showed that the AS-IV had no maternal toxicity at 0.25–1.0 mg/kg in rats, inhibitory effect on female fertility, devoid of early embryonic developmental toxicity. Maternal AS-IV exposure at 1.0 mg/kg/day led to a significant delay in time for pup fur development, eye opening, and cliff parry reflex compared to the control group, but had no effect on pup cognition.

8.3.8 Antifatigue Effect

Yeh et al. (2014) evaluated *A. membranaceus* effect on ergogenic and antifatigue functions on male ICR strain mice. Antifatigue function was evaluated by forelimb grip strength, exhaustive swimming time, and estimation of serum lactate, ammonia, glucose, and creatine kinase after 15-min swimming exercise. *A. membranaceus* supplementation enhanced exercise performance and had antifatigue effects in mice.

8.3.9 Anti-ulcerative Colitis

Zhao et al. (2014) investigated the prophylactic and curative effects of crude polysaccharides (QHPS) prepared from *Lycium barbarum* and *Astragalus membranaceus* at a ratio of 2:3, treatment included QHPS 0.18 g/kg for 14 days or prior to acetic acid-induced ulcerative colitis in rats. Protective effect was confirmed by weight loss reduction and control diarrhea and decrease colonic mucosal damage in colitis rats.

8.3.10 Antiarthritic Effect

Wang and Chen (2014) evaluated astragaloside IV (AST) (100 mg/kg/day) anti-arthritic effect on adjuvant-induced arthritis (AIA). Anti-arthritic activity of AST was confirmed by inhibited IL-1β, TNFα, NO production, reduced splenocyte proliferation, and reduced swelling, which induced and protected against IL-1β-induced damage of cartilage proteoglycan synthesis and chondrocyte proliferation.

8.3.11 Anxiolytic and Anti-Inflammatory Effects

Oh et al. (2014) investigated antistress, anxiolytic, and anti-inflammatory effects of *A. membranaceus* root ethanolic extract (50, 200, and 500 mg/kg) in immobilization stress model. *A. membranaceus* extract increased time spent on open arms and open arm entries in the elevated plus maze test and also reduced corticosterone, IL-6, and TNF-α levels. *A. membranaceus* may ameliorate immobilized stress-induced anxiety and inflammation.

Two active compounds, namely, 1-hydroxy-5-methylolbenzol-2-O-b-Dglucoside (1) and (3R, 4R)-4,7-hydroxy-20, 30 -dimethoxyisoflavane-40 -O-b-D-glucoside (2) were isolated and identified from the EtOAc crude extract of *A. membranaceus* (40, 60, and 100 mg/kg p.o) and evaluated anti-inflammatory activity on egg albumin-induced inflammation in rats. Compound 1 had higher and significant suppression of inflammation in a dose-dependent manner anti-inflammatory potential than acetylsalicylic acid (ASA) (Wang and Wang, 2014).

8.3.12 Antioxidant Effect

Zuo et al. (2012) reported that *A. membranaceus* root powder (5–10 g/kg for 10 weeks) improved the antioxidant status of AMP, the serum, and egg yolk with supplementation of AMP in a dose-dependent manner in laying hens.

8.3.13 Effect on Allergic Dermatitis

Kim et al. (2013) examined *A. membranaceus* effect on atopic dermatitis on skin lesions in Chloro-2, 4-dinitrobenzene (DNCB)-induced allergic dermatitis in BALB/c mice. Topical *A. membranaceus* markedly improved the allergic dermatitis skin lesions in DNCB-induced mice by suppressing the expression of Th2 cytokines (IL-4, -5, -6, and -13) and significantly decreasing the TNF-α level, restored nuclear factor-κB (NF-κB) expression, and reduced the serum IgE level. *A. membranaceus* may be an alternative or complementary therapeutic option for treating patients with allergic dermatitis.

8.3.14 Effect on Reproductive Toxicity

Kim et al. (2012) investigated the effect of *A. membranaceus* on cyclophosphamide (CP)-induced reproductive toxicity in mouse. *A. membranaceus* treatment increased diminished relative testes weight, and sperm count and motility in mice treated with CP. *A. membranaceus* has beneficial influences due to ameliorate relative testes weight, sperm parameters, and cAMP-responsive element modulator expression against CP-induced reproductive toxicity.

8.3.15 Anti-asthmatic Effects

Jin et al. (2013) determined the effect of *A. membranaceus* (10 g/kg/day, for 28 days) on ovalbumin-induced airway hyperreactivity and mucus hypersecretion in allergic mice. *A. membranaceus* treatment significantly decreased inflammatory infiltration and mucus secretion in the lung tissues, decreased IFN-, IL-5, and IL-13 by 28-day injection. Study also showed a 28-day *A. membranaceus* injection markedly suppressed increased airway hyperreactivity in allergic mice.

8.3.16 Antimicrobial Activity

Kopparapu et al. (2011) investigated the antifungal activity of chitinase (100μg/ml), isolated from *A. membranaceus* seed (Fisch.), against *Trichoderma viride, Botrytis cinerea, Fusarium oxysporum*, and *Fusarium solani.* Chitinase showed a potent inhibitory action on fungal species, which would certainly be significant for its utilization as an antifungal agent. Shang et al. (2011) investigated the anti-adenovirus effect of astragaloside IV on HAdV-3 (human adenovirus type 3) in A549 cell. Astragaloside IV possessed anti-HAdV-3 capabilities by inhibiting HAdV-3 replication and HAdV-3-induced apoptosis. Astragalus polysaccharide (APS) inhibited the replication of porcine circovirus type 2 by reducing oxidative stress and activation of the NF-κB signaling pathway (Xue et al., 2015) and inhibited astrocytes proliferation infected by the herpes simplex virus (Shi et al., 2014). APS exerted dose-dependent bacteriostatic effect on *Streptococcus, Escherichia coli*, and *Staphylococcus aureus* (Yang et al., 2010).

KEYWORDS

- **anti-inflammatory**
- **anticancer**
- ***Astragalus membranaceus***
- **flavonoids**
- **polysaccharides**
- **saponins.**

REFERENCES

Agyemang, K.; Han, L.; Liu, E.; Zhang, Y.; Wang, T.; Gao, X. 2013. Recent Advances in *Astragalus membranaceus* Anti-Diabetic Research: Pharmacological Effects of Its Phytochemical Constituents. *Evid. Based Complement Altern. Med.* **2013,** *2013*, 1–9.

Anon. *Astragalus membranaceus*. Monograph. *Altern Med Rev.* **2003,** *8*, 72–77.

Chan, J. Y.; Lam, F.; Leung, P.; Che, C.; Fung, K. Antihyperglycemic and Antioxidative Effects of an Herbal Formulation of Radix Astragali, Radix Codonopsis and Cortex Lycii in a Mouse Model of Type 2 Diabetes Mellitus. *Phytother Res*. **2009,** *23*, 658–665.

Chen, Q. Q.; Su, Y.; Ju, Y. H.; Ma, K. K.; Li, W. Z.; Li, W. P. Astragalosides IV Protected the Renal Tubular Epithelial Cells from Free Fatty Acids-Induced Injury by Reducing Oxidative Stress and Apoptosis. *Biomed. Pharmacother.* **2018,** *108*, 679–686.

Costa, I. M.; Lima, F. O.; Fernandes, L. C.; Norrara, B.; Neta, F. I.; Alves, R. D.; Queiroz, D. B. Astragaloside IV Supplementation Promotes a Neuroprotective Effect in Experimental Models of Neurological Disorders: A Systematic Review. *Curr. Neuropharmacol.* **2019,** *17* (7), 648–665.

Curnow, A.; Owen, S. J. An Evaluation of Root Phytochemicals Derived from *Althaea officinalis* (Marsh Mallow) and *Astragalus membranaceus* as Potential Natural Components of UV Protecting Dermatological Formulations. *Oxid. Med. Cell. Longev*. **2016,** *2016.*

Dai, H.; Jia, G.; Liu, X.; Liu, Z.; Wang, H. Astragalus Polysaccharide Inhibits Isoprenaline-Induced Cardiac Hypertrophy via Suppressing $Ca2^{+}$Mediated Calcineurin/NFATc3 and CaMKII Signaling Cascades. *Environ. Toxicol. Pharmacol.* **2014,** *38* (1), 263–271.

Ding, Y. Z.; Yuan, S. T.; Liu, X. Y.; Mao, P. A.; Zhao, C.; Huang, Q. Protective Effects of Astragaloside IV on db/db Mice with Diabetic Retinopathy. *PLoS One*. **2014,** *9* (11) e112207.

Fu, J.; Wang, Z.; Huang, L.; Zheng, S.; Wang, D.; Chen, S.; Yang, S. Review of the Botanical Characteristics, Phytochemistry, and Pharmacology of *Astragalus membranaceus* (Huangqi). *Phytother. Res*. **2014,** *28* (9), 1275–1283.

He, Y.; Du, M.; Gao, Y.; Liu, H.; Wang, H.; Wu, X.; Wang, Z. Astragaloside IV Attenuates Experimental Autoimmune Encephalomyelitis of Mice by Counteracting Oxidative Stress at Multiple Levels. *PloS One* **2013,** *8* (10), e76495.

Huang, H.; Luo, S. H.; Huang, D. C.; Cheng, S. J.; Cao, C. J.; Chen, G. T. Immunomodulatory Activities of Proteins from *Astragalus membranaceus* Waste. *J. Sci. Food Agric.* **2019,** *99* (8), 4174–4181.

Jeong, S. Y.; Moon, M. Y.; Ryu, E. K.; Lee, J. S.; Cha, J. Identification of the Phytochemical Compounds and Their Type I Procollagen Induction in *Astragalus membranaceus* Sprouts Grown Under Different Light Conditions. *J Appl. Pharm. Sci*. **2018,** *8* (06), 1–7.

Jia, Y.; Zuo, D.; Li, Z.; Liu, H.; Dai, Z.; Cai, J.; Wu, Y. Astragaloside IV Inhibits Doxorubicin-Induced Cardiomyocyte Apoptosis Mediated by Mitochondrial Apoptotic Pathway via Activating the PI3K/Akt Pathway. *Chem. Pharm. Bull.* **2014,** *62* (1), 45–53.

Jiang, Z. P.; Wei, G. M.; Li, Z.; Liu, Y. H.; Wang, Z. D. Flavonoids from the Roots of *Astragalus membranaceus* (Fisch.) Bge. Prevent Development of Diabetic Nephropathy In Vitro. *Lat. Am. J. Pharm.***2014,** *33* (2), 339–43.

Jin, H.; Luo, Q.; Zheng, Y.; Nurahmat, M.; Wu, J.; Li, B.; Dong, J. CD4+ CD25+ Foxp3+ T Cells Contribute to the Antiasthmatic Effects of *Astragalus membranaceus* Extract in a Rat Model of Asthma. *Int. Immunopharmacol.* **2013,** *15* (1), 42–49.

Kavtaradze, N. S.; Alaniya, M. D.; Mshvildadze, V. D.; Skhirtladze, A. V.; Lavoie, S.; Pichette, A. Flavonoids from *Astragalus microcephalus*. *Chem. Nat. Compd.* **2011,** *46*, 971–973.

Kim, J. H.; Kim, M. H.; Yang, G.; Huh, Y.; Kim, S. H.; Yang, W. M. Effects of Topical Application of *Astragalus membranaceus* on Allergic Dermatitis. *Immunopharmacol. Immunotoxicol.* **2013,** *35* (1), 151–156.

Kim, J. S.; Yean, M. H.; Lee, E. J.; Jung, H. S.; Lee, J. Y.; Kim, Y. J.; Kang, S. S. Two New Cycloartane Saponins from the Roots of *Astragalus membranaceus*. *Chem. Pharm. Bull.* **2008a,** *56*, 105–108.

Kim, J. S.; Yean, M. H.; Lee, S. Y.; Kang, S. S. Phytochemical Studies on Astragalus Root (1)-Saponins. *Nat. Prod.* **2008b,** *14* (1), 1–10.

Kim, W.; Kim, S. H.; Park, S. K.; Chang, M. S. *Astragalus membranaceus* Ameliorates Reproductive Toxicity Induced by Cyclophosphamide in Male Mice. *Phytother Res.* **2012,** *26* (9), 1418–1421.

Kopparapu, N. K.; Liu, Z.; Fei, F.; Yan, Q.; Jiang, Z. Purification and Characterization of a Chitinase (sAMC) with Antifungal Activity from Seeds *of A. membranaceus*. *Process Biochem.* **2011,** *46* (6), 1370–1374.

Kuang, H.; Okada, Y.; Yang, B.; Tian, Z.; Okuyama, T. Secocycloartane Triterpenoidal Saponins from the Leaves of *Astragalus membranaceus* Bunge. *Helv. Chim. Acta.* **2009,** *92*, 950–958.

Kumar, S.; Sephuhle, N.; Bouic, P. J.; Rosenkranz, B. HPLC/LC-MS Guided Phytochemical and in Vitro Screening of *Astragalus membranaceus* (Fabaceae), and Prediction of Possible Interactions with CYP2B6. *J. Herb. Med.* **2018,** *14* (10), 35–47.

Lee, E. J.; Yean, M. H.; Jung, H. S.; Kim, J. S.; Kang, S. S. Phytochemical Studies on Astragalus Root (2); Flavonoids and a Lignan. *Nat. Prod.* **2008,** *14* (2), 131–137.

Li, M.; Li, H.; Fang, F.; Deng, X.; Ma, S. Astragaloside IV Attenuates Cognitive Impairments Induced by Transient Cerebral Ischemia and Reperfusion in Mice via Anti-Inflammatory Mechanisms. *Neurosci. Lett.* **2017,** *639*, 114–119.

Li, X.; Qu, L.; Dong, Y.; Han, L.; Liu, E.; Fang, S.; Zhang, Y.; Wang, T. A Review of Recent Research Progress on the *Astragalus* genus. *Molecules*. **2014,** *19*, 18850–18880.

Li, Y. L.; Ye, Y.; Chen, H. Y. Astragaloside IV Inhibits Cell Migration and Viability of Hepatocellular Carcinoma Cells Via Suppressing Long Noncoding RNA ATB. *Biomed. Pharmacother.* **2018,** *99*, 134–141.

Li, Y.; Guo, S.; Zhu, Y.; Yan, H.; Qian, D. W.; Wang, H. Q.; Yu, J. Q.; Duan, J. A. Flowers of *Astragalus membranaceus* var. *mongholicus* as a Novel High Potential By-Product: Phytochemical Characterization and Antioxidant Activity. *Molecules* **2019,** *24*, 434–451.

Lu, W. S.; Li, S.; Guo, W. W.; Chen, L. L.; Li, Y. S. Effects of Astragaloside IV on Diabetic Nephropathy in Rats. *Genet. Mol. Res.* **2015,** *14* (2), 5427–5434.

Ma, J.; Qiao, Z.; Xiang, X. Aqueous Extract of *Astragalus mongholicus* Ameliorates High Cholesterol Diet Induced Oxidative Injury in Experimental Rat Models. *J. Med. Plant Res.* **2011,** *5*, 855–858.

Nalbantsoy, A.; Nesil, T.; Yilmaz-Dilsiz, O.; Aksu, G.; Khan, S.; Bedi, E. Evaluation of the Immunomodulatory Properties in Mice and In Vitro Anti-Inflammatory Activity of Cycloartane Type Saponins from *Astragalus* Species. *J. Ethnopharmacol.* **2012,** *139*, 574–581.

Oh, H-A., Hyuck J-C., Kim, N. J.; Kim, D-H. Anti-Stress Effect of Astragaloside IV in Immobilized Mice. *J. Ethnopharmacol.* **2014,** *153*, 928–932.

Ryu, M.; Kim, E. H.; Chun, M.; Kang, S.; Shim, B.; Yu, Y. B. Astragali Radix Elicits Anti-Inflammation via Activation of MKP-1, Concomitant with Attenuation of p38 and Erk. *J Ethnopharmacol.* **2008,** *115*, 184–193.

Shang, L.; Qu, Z.; Sun, L.; Wang, Y.; Liu, F.; Wang, S.; Jiang, F. Astragaloside IV Inhibits Adenovirus Replication and Apoptosis in A549 Cells In Vitro. *J. Pharm. Pharmacol.* **2011,** *63* (5), 688–694.

Shi L.; Yin F.; Xin X.; Mao S.; Hu P.; Zhao C. Astragalus Polysaccharide Protects Astrocytes from Being Infected by HSV-1 Through TLR3/NF-κB Signaling Pathway. *Evidence-Based Complement. Altern. Med.* **2014,** 1–6.

Sun, S.; Yang, S.; An, N.; Wang, G.; Xu, Q.; Liu, J.; Mao, Y. Astragalus Polysaccharides Inhibits Cardiomyocyte Apoptosis During Diabetic Cardiomyopathy via the Endoplasmic Reticulum Stress Pathway. *J. Ethnopharmacol.* **2019,** *238*, 111857.

Wang, B.; Chen, M. Z. Astragaloside IV Possesses Antiarthritic Effect by Preventing Interleukin 1β-Induced Joint Inflammation and Cartilage Damage. *Arch. Pharm. Res.* **2014,** *37* (6), 793–802.

Wang, J. N.; Zhou, Y. Y.; Wu, S. Z.; Huang, K. Y.; Thapa, S.; Tao, L. Y. Astragaloside IV Attenuated 3,4-Benzopyrene-Induced Abdominal Aortic Aneurysm by Ameliorating Macrophage-Mediated Inflammation. *Front. Physiol.* **2018,** *9*, 496.

Wang, Q. H.; Wang, X. L. Anti-Inflammatory Effects and Structure Elucidation of Two New Compounds from *Astragalus membranaceus* (Fisch) Bge. var. *mongholicus* (Bge) Hsiao. *J. Mol. Struct.* **2014,** *1074*, 284–288.

Xue H. X.; Gan F.; Zhang Z. Q.; Hu J. F.; Chen X. X.; Huang K. H. Astragalus Polysaccharides Inhibits PCV2 Replication by Inhibiting Oxidative Stress and Blocking NF-κB Pathway. *Int. J. Biol. Macromol.* **2015,** *81*, 22–30.

Xuying, W.; Jiangbo, Z.; Yuping, Z.; Xili, M.; Yiwen, Z.; Tianbao, Z.; Weidong, Z. Effect of Astragaloside IV on the General and Peripartum Reproductive Toxicity in Sprague-Dawley Rats. *Int. J. Toxicol.* **2010,** *29* (5), 505–516.

Yang Y. A.; Yang L. F.; Yang L. Y.; Chu X. H.; Bi B. L.; Sun Y. K. Inhibitory Effect of Astragalus Polysaccharide on Main Pathogenic Bacteria of Dairy Bovine Mastitis In Vitro. *Prog. Vet. Med.* **2010,** *31*, 47–49.

Yeh, T. S.; Chuang, H. L.; Huang, W. C.; Chen, Y. M.; Huang, C. C.; Hsu, M. C. *Astragalus membranaceus* Improves Exercise Performance and Ameliorates Exercise-Induced Fatigue in Trained Mice. *Molecules.* **2014,** *19* (3), 2793–2807.

Yu, Q. T.; Li, P.; Bi, Z. M.; Luo, J.; Gao, X. D. Two New Saponins from the Aerial Part of *Astragalus membranaceus* var. *mongholicus. Chin. Chem. Lett.* **2007,** *18*, 554–556.

Zhang, A. l.; Zheng, Y. H.; Que, Z. J.; Zhang, L. L.; Lin, S. C.; Le, V. Astragaloside IV Inhibits Progression of Lung Cancer by Mediating Immune Function of Tregs and CTLs by Interfering with IDO. *J. Cancer Res. Clin.* **2014,** *140* (11), 1883–1890.

Zhang, J.; Wu, C.; Gao, L.; Du, G.; Qin, X. Astragaloside IV Derived from *Astragalus membranaceus*: A Research Review on the Pharmacological Effects. *Adv. Pharmacol.* **2020,** *87*, 89–112.

Zhao, L.,; Wu, H.; Zhao, A.; Lu, H.; Sun, W.; Ma, C.; Jia, W. The In Vivo and In Vitro Study of Polysaccharides from a Two-Herb Formula on Ulcerative Colitis and Potential Mechanism of Action. *J. Ethnopharmacol.* **2014,** *153* (1), 151–159.

Zheng, Q.; Zhu, J. Z.; Bao, X. Y.; Zhu, P. C.; Tong, Q.; Huang, Y. Y. A Preclinical Systematic Review and Meta-Analysis of Astragaloside IV for Myocardial Ischemia/Reperfusion Injury. *Front. Physiol.* **2018,** *9*, 795.

Zhu, J.; Wen, K. Astragaloside IV Inhibits TGF-β1-Induced Epithelial-Mesenchymal Transition Through Inhibition of the PI3K/Akt/NF-κB Pathway in Gastric Cancer Cells. *Phytother. Res.* **2018,** *32* (7), 1289–1296.

Zuo, Z. Y.; Yang, W. R.; Wang, Y.; Yang, Z. B.; Jiang, S. Z..; Zhang, G. G. Effects of *Astragalus membranaceus* on Laying Performance and Antioxidant Status of Laying Hens. *J. Appl. Poultry Res.* **2012,** *21* (2), 243–250.

CHAPTER 9

Chemical Composition and Biological Properties of *Bauhinia purpurea* L. Benth. (Family: Fabaceae)

MEGHNA ADHVARYU[1*] and BHASKER VAKHARIA[2]

[1]*Department of Botany, Government Science College, Vankal, Mangrol, Surat, Gujarat*

[2]*Founder & Head, R&D, Bhuma Research in Ayurvedic & Herbal Medicine (BRAHM), 1-Pathik Duplex, Rani Laxmibai Marg, Opp Navyug College, Surat, Gujarat*

[*]*Corresponding author. E-mail: meghna.adhvaryu@gmail.com*

ABSTRACT

Bauhinia purpurea, commonly known as purple Orchid tree, Camel foot tree, is an exotic tropical tree distributed worldwide in tropical and subtropical zones. It is medicinally important plant, used in several Ayurvedic formulations for various therapeutic uses. Various phytochemical studies show presence of several important bioactive compounds such as flavonoids, phenolic acids, steroids, lectins, statins, tannins, coumarins etc. Its pharmacological actions describe antibacterial, antifungal, antioxidant, cerebro-protective, anti-inflammatory, antiarthritic, analgesic, nephroprotective, hepatoprotective, and anticancer properties.

9.1 INTRODUCTION

Medicinal plants are used as remedies for different ailments since centuries. India is rich in the Ayurvedic system of medicine for treating different

Bioactives and Pharmacology of Legumes. T. Pullaiah, PhD (Ed.)

diseases with the help of different formulations that include medicinal plants in addition to other nonplant natural products. *Bauhinia purpurea* is one of the important medicinal plants, used in several formulations for various therapeutic uses. It is a small to medium size, ornamental tree, found throughout the Indian subcontinent and Myanmar; up to 1300 m height from sea level in the sub-Himalayan range. The bark of this tree is grayish brown, nearly smooth. Leaves measure 0.75–1.5 cm long with cleft apex, bilobed, very minutely pubescent beneath, inflorescence—raceme panicle, flowers in September–November. Fruits are leguminous. Pod size ranges from 15 to 25 cm × 1.5–2 cm. Seeds 12–15 per pod, suborbicular, flattened, 1.3 cm, dark brown, and smooth. Many active secondary metabolites and their activities attracted researchers to work with this plant in different pharmaceutical areas.

9.2 BIOACTIVES

Phytochemical studies carried out by many workers revealed dihydrodibenzoxepins, flavonoids, phenolic acids, steroids, lectins, statins, tannins, coumarins, etc., which exhibit various biological activities. Kuo and Yeh (1997) isolated 22 compounds from heartwood of *B. purpurea* that includes six steroids, two triterpenoids, three fatty alcohol, acid and ester, two glycerols, five flavonoids, two phenols, one chromone, and one sugar. Twenty-two components are: six steroids [β-sitosteryl hexadecanoate and octadecanoate (1), β-Sitostcrol (2), stigmasterol (3), 3 β-hydroxystigmast-5-en-7-one (4), 6′(β-sitosteryl-3-*O*-β-glucopyranosidyl) hexadecanoate and tetraeicosanoate (new compound isolated) (5), and β-sitosteryl-3-*O*-β-gJucopyranosidc (6)], two triterpenes [lupenone (7) and lupeol (8)], three fatty ester, alcohol, and acid [methyl hexadecanoate and octadecenoate (9), I-tetracosanol, 1-hexacosanol and I-octacosanol (10), and hexadecanoic acid (11)], two glycerols [2,3-dihydroxypropyl hexadecanoate and octadecenoate (12), and glycerol (13)], five flavonoids [eriodictyol (14), kaempferol (15), 3-methoxy-5,7,3′,4′-tctrahydroxyflavone (16), quercetin (17), and taxifolin (18). One chromone [5,7-dihydroxychromone (19)], two phenols [vanillic acid (20) and Isotachioside (21) and one sugar [bornesitol (22)].

Pettit et al. (2006) isolated four new components from the ethanolic root extract of *B. purpurea* which are four Bauhiniastatin (1–4) and their eluted structures are shown below.

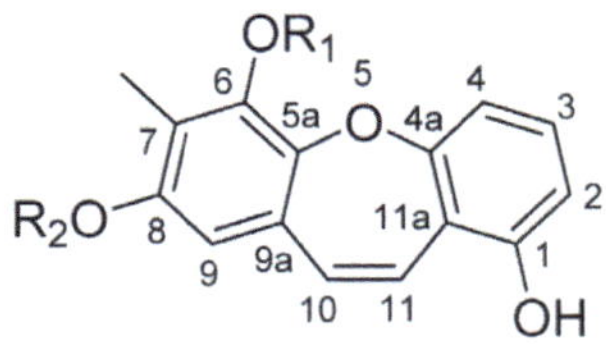

1, $R_1 = H, R_2 = CH_3$

5, $R_1 = CH_3, R_2 = H$

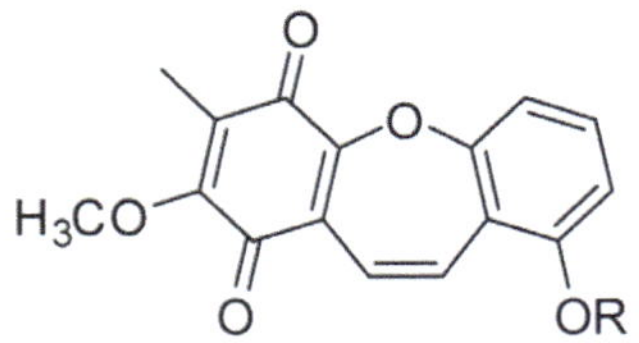

2a, R = H

b, R = Ac

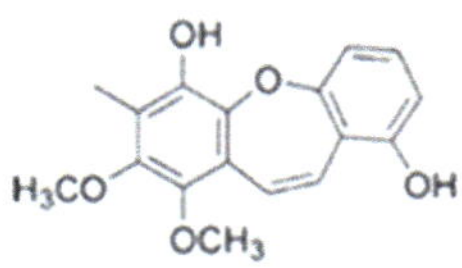

3

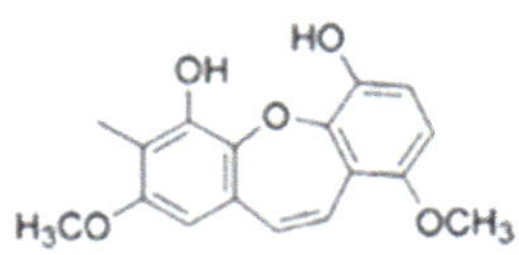

4

Bauhiniastatins (1–4)

Boonphong et al. (2007) isolated 11 new compounds from the roots of *B. purpurea*. Their eluted structures are given below, including eight dihydrodibenzoxepins (1–8), a dihydrobenzofuran (9), a novel spirochromane-2,1′-hexenedione (10), and a new bibenzyl (11).

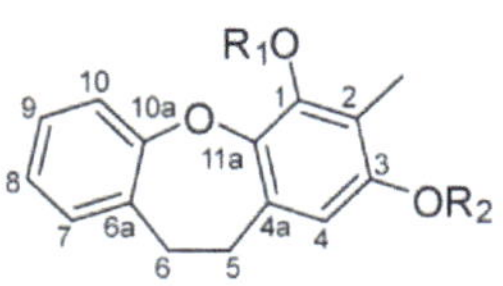

1, $R_1 = H, R_2 = Me$

2, $R_1 = Me, R_2 = H$

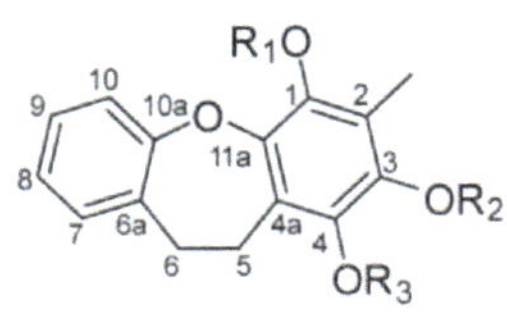

1, $R_1 = Me, R_2 = Me, R_3 = H$

2, $R_1 = Me, R_2 = H, R_3 = Me$

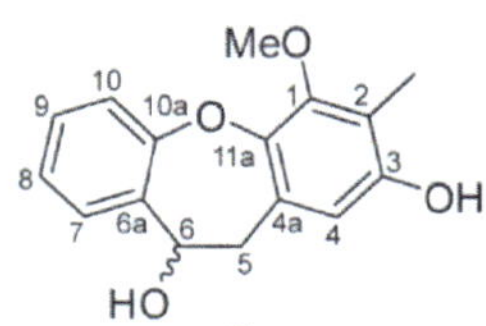

5

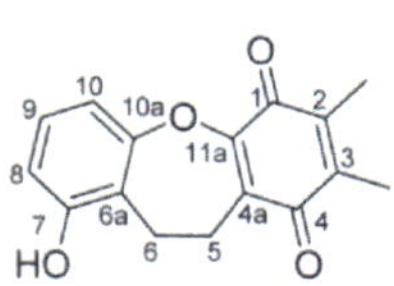

6

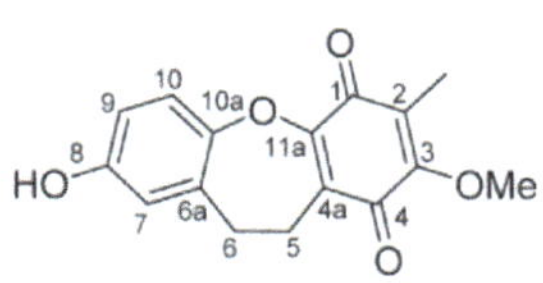

7

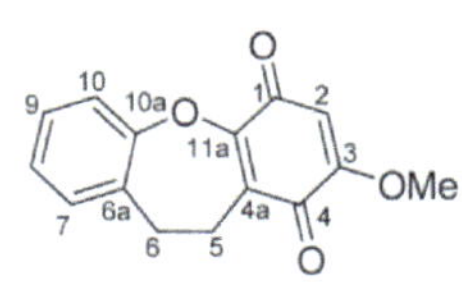

8

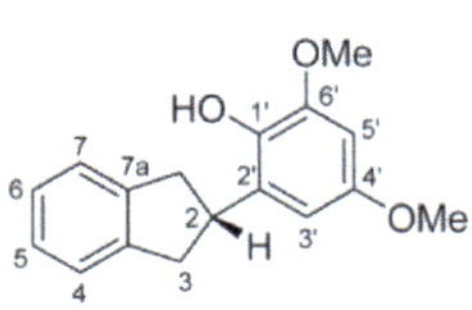

9

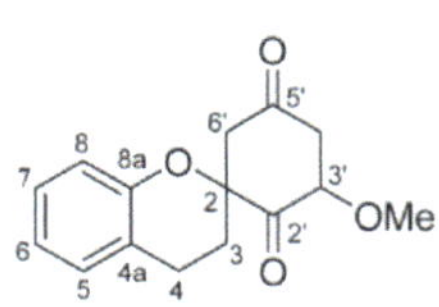

10

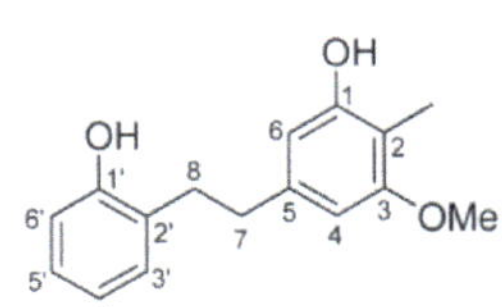

11

Eleven new secondary metabolites were identified and extracted from the roots of *B. purpurea.*

Ni et al. (2018) isolated new phenylpropanoid glucoside, which they gave the name purpuroside from the bark of *B. purpurea.* They also isolated seven other compounds similar to other workers, namely, wilfordonol A-known nor-sesquiterpene, dehydrovomifoliol, roseoside, 3-*O*-α-ionol-β-D-glucopyranoside, phenyl-*O*-β-D-glucopyranoside, methyl-2-*O*-β-D-glucopyranosyl benzoate, and benzyl-*O*-α-rhamnopyranosyl (1-6)-β-D-glucopyranoside.

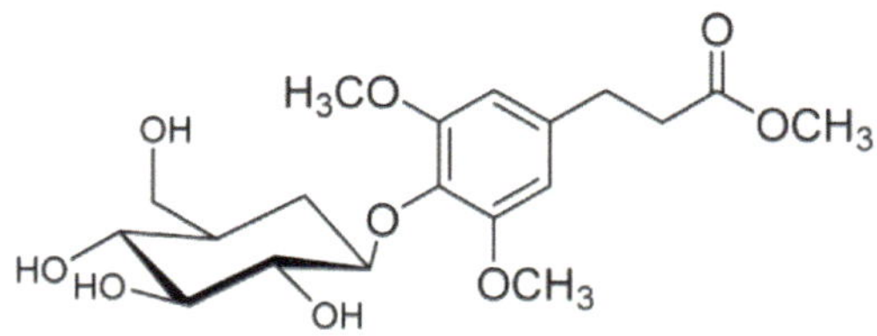

Purpuroside

Negi et al. (2012) isolated 11 compounds from *B. purpurea* leaf extract and studied by GC– MS analysis. The compounds identified in *B. purpurea* leaf extract were lupeol, stigmasterol, lanosterol, ergosterol, beta-tocopherol, Hexadecanoic acid, methyl ester, Hexadecanoic acid, 9,12-octadecadienoic acid (Z,Z) methyl, 9,12,15-octadecatrienoic acid, methyl, phytol, ergost-5-en-3-ol, acetate, vitamin E acetate, stigmast-5-en-3-ol, lupeol, and lanosterol.

9.3 PHARMACOLOGY

9.3.1 *Antimicrobial, Antifungal Activity*

Antibacterial activity was done by minimum inhibitory concentration (MIC) using the disk diffusion method for hexane, acetone, and aqueous leaf extracts of *B. purpurea* extract. In total, 1500 mcg/mL concentration was shown to have inhibitory activity against *Staphylococcus aureus* and *Bacillus subtilis* (Gram-positive) while no inhibition was observed against Gram-negative *Salmonella typhi, Escherichia coli,* and *Pseudomonas aeruginosa* or against fungus *Candida albicans* (Negi et al., 2012). *S. aureus* and *Lactobacillus acidophilus* are responsible for dental caries and conventional treatment by repeated courses of antibiotics may cause damage to normal flora of buccal cavity with no long-term benefit of treatment. *B. purpurea* saponins were found to have strong antimicrobial activity on *Streptococcus mutans* (MTCC

890), *S. aureus*, and *L. acidophilus* at MIC concentration ranges from 20 to 29.6 mcg/mL, and MBC from 36.8 to 48.00 mcg/mL. Thus, *B. purpurea* is one of the potential candidates from the plant kingdom, that can be utilized for reducing the incidence of caries, periodontal problems, and may help to sustain a healthy oral environment (Jyothi and Seshagiri, 2012). According to Boonphong et al. (2007), isolated compounds 1, 4, 9, 15, and 18 exhibited antifungal activity IC_{50} 49.6–130.1 μM concentration. Among Bauhiniastatins, Bauhinoxepin J (compound 8) showed potent antimycobacterial activity with MIC 24.4 μM concentration. While compounds 4 and 7 inhibited COX-2 anti-inflammatory marker showing IC_{50} 6.9 & 10.1 μM, respectively.

9.3.2 Cerebroprotective and Antioxidant Activity

Cerebrovascular strokes have been found to be the second leading cause of death in the world. The in vivo pharmacological studies with flower extract of *B. purpurea* indicated a significant reduction in cerebral infarct size in dose-dependent manner, when ischemia was induced by 30-min bilateral carotid artery occlusion in succession to reperfusion for 4 h in Wistar rats. The antioxidant parameters, namely, SOD, and catalase levels were significantly augmented, and MDA end product of lipid peroxidation significantly reduced in magnitude by *B. purpurea* flower extract (Annapurna et al., 2017).

9.3.3 Antiarthritic, Anti-inflammatory, and Analgesic activity

In vivo studies conducted on Wistar rats and albino mice showed analgesic and anti-inflammatory properties of EtOH extract of *B. purpurea* bark. Two different doses of ethanolic extract were administered intraperitoneally to evaluate the protective effect against acetic acid induced abdominal constriction (writhing) model for analgesic activity and Eddy's hot plate method for the anti-inflammatory activity. Diclofenac sodium for anti-inflammatory and acetyl salicylic acid (aspirin) for analgesic activity were taken as positive controls. Ethanolic bark extract of *B. purpurea* at 100 mg/kg dose was highly effective in reducing 77% of foot edema and effective in pain relief which was equivalent to positive controls aspirin and diclofenac (Shreedhara et al., 2009).

Another in vivo study by Sunil (2019) on antiarthritic activity of bark extract of *B. purpurea* confirmed the folklore use of bark in arthritis. The study demonstrated 50, 100, and 200 mg/kg of bark extract activity on arthritis, induced by λ-carrageenan, which was injected into the left hind paw of Wistar albino rats in subplantar region. Results showed a significant decrease in paw edema in dose-dependent manner. Also, a significant decrease in oxidative stress indicators and circulating pro-inflammatory cytokines like tumor necrosis factor-alpha (from 45.91 to 37.44), interleukin-1 (IL-1) β (from 18.24 to 16.06), and IL-6 (from 69.77 to 58.44), and a marked increase in anti-inflammatory cytokine level (IL-10) from 8.07 to 12.07 was observed.

9.3.4 Wound Healing and Antioxidant Activity

Healing of wounds occurs naturally by repairing of skin and surrounding soft tissue injuries. This process involves inflammatory responses initially, later followed by increased collagen synthesis below the dermis and at last epithelialization to cover the wound. Ananth et al. (2010) carried out a study to evaluate the effect of methanolic and chloroform leaf extract of *B. purpurea* on wound healing in Sprague Dawley rats model of four kinds of injuries namely excision of 500 mm^2 of skin area, incision of 6 mm over skin, burn by 80°C heated wax and dead space wound by implanting foreign body in subcutaneous plane in the ventral region. Rats treated with high doses of MeOH and chloroform extracts achieved significantly earlier epithelization and wound contraction (50%) in comparison to control. A significant enhancement was found in breaking strength in incision wound model with methanolic and chloroform extracts compared to ointment, *Aloe vera* and carbopol treated group. In the dead space wound model, methanolic and $CHCl_3$ extract treatment (100 and 500 mg/kg) per oral route, resulted in a significant enhancement in the breaking strength, and markers of healing like dry tissue weight and hydroxyproline content of the granulation tissue in comparison to that observed in controls. Among the extracts, methanolic extract exhibited superior activity compared to the chloroform extract.

9.3.5 Antiulcer Activity

Under normal physiological conditions, local production of prostaglandins helps to stimulate and maintain the mucus and bicarbonate synthesis that protects the gastric mucosa. Disruption in these defense mechanisms will lead to the formation of gastric or duodenal ulcers. It is evident that antiulcer agents should work either to enhance mucosal protection or to decrease the acidity level. The in vivo study by Hisam et al. (2012) showed the antiulcer potential of leaf extract of *B. purpurea*. Here male Sprague–Dawley rats were used for acute toxicity study and antiulcerogenic activity. No sign of toxicity was seen in the acute toxicity study carried out by giving 5000 mg/kg dose. Rats were given ethanol and indomethacin to induced gastric ulcers. Histopathology of the stomach and mucus content of gastric wall was the parameters to determine the comparative potential of antiulcer activity of *B. purpurea* with omeprazole as the positive control. In total, 100, 500, and 1000 mg/kg chloroform extract of *B. purpurea* to three experimental groups and 30 mg/kg omeprazole to the positive control group were administered orally to check antiulcerogenic activity. The extract showed its marked antiulcer activity in a dose-dependent manner for the ethanol-induced model compared to the positive control while a weaker protective potential in the Indomethacin-induced model. It also significantly increased the gastric content pH as well as gastric wall mucus production while significantly reducing the total volume and acidity of the gastric juices in the pylorus ligation group.

9.3.6 Nephroprotective Activity

Antibiotics are used to treat susceptible bacterial infections that may affect various organs and systems of body. Aminoglycosides are commonly used for Gram-negative as well as staphylococcal infections in hospitalized patients. Prolonged use of these agents can cause nephrotoxicity in the form of glomerular congestion, epithelial shedding, vascular congestion with inflammatory cells, and lastly necrosis of the kidney cells. A study undertaken by Lakshmi et al. (2009) revealed nephroprotective effect of *B. purpurea* ethanolic extract of leaves and unriped pods in Wistar rats against nephrotoxicity induced by intraperitoneal injection of gentamicin 100 mg/kg once a day for 8 days. Simultaneous administration of ethanolic extract of leaves and unriped pods at a dose of 300 mg/kg daily was done by oral route. Damage was measured by serum creatinine, serum uric acid, blood urea nitrogen, and serum urea as indicators for kidney damage. Results suggested a significantly decreased level of kidney damage markers. Histopathological studies were also done to check the physical damage at the tissue level; protection was seen at the tissue level by lesser glomerular congestion. Out of two ethanolic extracts, unriped pod extract showed significantly superior activity compared to leaf ethanolic extract.

Another in vivo study was carried out by Rana et al. (2016) on nephrotoxicity induced by drug cisplatin. The study was carried out on albino rats of either sex; they were given 200 and 400 mg/kg/day per oral ethanolic extract of *B. purpurea* bark and unripe pods, respectively, for 9 days. On the sixth day, cisplatin was administered 6 mg/kg intraperitoneally as a single dose. Blood urea, S creatinine, Cr–Cl (creatinine clearance), and urine glucose were taken as direct parameters for the evaluation of kidney function. The resultant impact on body weight, kidney weight, kidney volume, and urine volume were also measured. On the 10th day, histopathology for tissue texture study and malonyl dialdehyde (MDA) test as lipid peroxidation markers and for antioxidant status catalase and reduced GSH, was measured in homogenized kidney tissue. The finding indicated *B. purpurea* unripe pod extract at 400 mg/kg reduced the blood urea, serum creatinine, and urine glucose by 29.22, 26.32, and 19.90%, respectively, and increased the creatinine clearance by 12.74%. The bark extract at 400 mg/kg reduced the blood urea, serum creatinine, and urine glucose by 28.29, 24.21, and 19.56%, respectively, and increased creatinine clearance by 12.42%. Also, protection against lipid peroxidation and increase in the activity of catalase, and the level of GSH against cisplatin-induced oxidative stress in kidney were significant. The significant beneficial effects of *B. purpurea* as evidenced by biochemical and histological findings in cisplatin toxicity might be due to the free radical scavenging effect of the extracts (Rana et al., 2016).

9.3.7 Hepatoprotective and Antioxidant Activity

Yahya et al. (2031) studied the potential of *B. purpurea* leaf extract against paracetamol (PCM) induced hepatic damage. Rats were administered daily 50, 250, and 500 mg/kg of methanolic extract of *B. purpurea* leaves (MEBP), 200 mg/kg silymarin as positive control, and 10% DMSO as a negative control for 7 days. Liver enzymes and histopathology were tested as parameters for this study. MEBP (500 mg/kg) exhibited a significant decrease in ALT and AST enzyme levels compared to only PCM-treated rats and were comparable to positive control group. The extract was also subjected to antioxidant study using the 2, 2-diphenyl-1-picrylhydrazyl (DPPH) radical scavenging assay. The extract also exhibited antioxidant activity comparable to ascorbic acid. Antioxidant activity and hepatoprotective activity could be attributed to high phenolic content present in leaves of *B. purpurea* (Yahya et al., 2013).

9.3.8 Anticancer Activity

Antiproliferative activity of N-acetyl galactosamine specific plant lectins from seeds of *B. purpurea* (BPL) was studied against MCF-7 (breast cancer cell lines) using MTT, lactate dehydrogenase (LDH) leakage assay, reactive oxygen species, and caspase-3 assays by flow cytometry for cytotoxic effect as well as cell cycle analysis. The agglutinin BPL 446 μg/mL (2.2 μM) showed remarkable concentration-dependent antiproliferative potential on MCF-7; binding of the lectin to the cell surface inducing apoptosis by S and G2 phase arrest and activation of caspase-3 leading to cell death (Agarwal et al., 2020).

9.4 AYURVEDIC PROPERTIES

B. purpurea is one of the important plants in ayurvedic system (Patra et al., 2019). It is known as *Kachanara*, the most commonly used parts are the leaves, fruit, stem, bark, and flowers. According to different *Acharya's*, the plant parts used in ayurvedic preparations are as follows:

- Raj Nighntu—leaves, flowers, and root bark.
- Dhanvantari Nighntu—stem, bark, and flowers.
- Madanpal Nighantu—flowers.
- Kaiydev Nighntu—flowers.

The largely accepted action of the plant by major *Acharya's* is *Kaphapittaghna.* The properties of *B. purpurea* in Ayurvedic terminology have been said to be *kashaya or katu Rasa, katu Vipaka, Sheeta Veerya* and *laghu, ruksa, Gunas.* In *Ayurveda*, bark and flower have different significance and individual properties are shown in the table given below:

Parts of plant	**Properties**	**Effects**
Stem–Bark	Rasa—(Taste)	*Kasaya*
	Guna—(Quality)	*Ruksa, Laghu*
	Veerya—(Strength)	*Sheeta*
	Vipaka—(Digestive effect)	*Katu*
	Karma—(Action)	*Kapha-pittahara, Grahni, Dipana*
	Prabhava—(Impression)	*Gandamalanashana*
Flower	Rasa—(Taste)	*Madhura, Kasaya*
	Guna—(Quality)	*Laghu, Ruksha*
	Veerya—(Strength)	*Sheeta*
	Vipaka—(Digestive effect)	*Madhur*
	Karma—(Action)	*Vrana Shodhana-Vrana ropana*
	Prabhava—(Impression)	*Gandmalanashana,*

Famous formulations given by different *Acharya's* in *Ayurveda* are as follows:

5:1 *Kachanar guggulu*—(Sharndhar Samhita, Gandamala–Apachi Rogadhikar, 100, P-206).

1:8 *Usirasava*—(Bhaisajyaratnavali, Raktapittaadhikara, 137–141, P63).

Also, it is formulated in different forms like *Kanchanar arishta (Asavaandarista), Kanchanaradrava (Bhasma), Kanchanara Guggulu, Kanchana gutika, Kanchanaradi kvath, Gandamala kandan rasa.*

9.5 OTHER ACTIVITIES

9.5.1 *B. purpurea* Lectins Predictor of Gastric Cancer

Takuro et al. (2015) conducted a lectins microarray study in gastric cancer patients and found 24 lectins behaved differently among the studied 60. Thirteen showed increased expression and 11 showed decreased expression in cancer tissue in comparison to normal epithelium. The lectins isolated from *B. purpurea* (BPL–Galβ1-3GalNAc) exhibited decreased expression in patients who suffered from gastric cancer and recurrence, and were identified as an independent predictive factor in addition to lymph node metastasis by multivariate analysis. Thus, a molecular marker of herbal origin lectin (BPL) has been discovered to be a predictor of gastric cancer recurrence after gastrectomy.

KEYWORDS

- ***Bauhinia purpurea***
- **anti-inflammatory**
- **anti-oxidant**
- **hepatoprotective**
- **nephroprotective**
- **anti-arthritic**

REFERENCES

Agrawal, S. B.; Gupta, N.; Bhagywant, S. S.; Gakwad, S. M. Anticancer Activity of Lectins from *Bauhinia purpurea* and *Wisteria floribunda* on Breast Cancer MCF-7 cell lines. *Protein Pept. Lett.* **2020,** *27* (9), 870–877. DOI: 10.2174/0929866527666200408143614

Annapurna, A.; Vishala, T. C.; Bitra, V. R.; Rapaka, D.; Shaik, A. Cerebroprotective Actions of *Triticum aestivum* Linn Powder and *Bauhinia purpurea* Flower Powder in Surgically Induced Cerebral Infraction in Rats. *Pharmacogn. Mag.* **2017,** *13* (4), S737–S741.

Ananth, K. V.; Asad, M.; Prem, K. N.; Asdaq, S. M.B.; Rao, G. S. Evaluation of Wound Healing Potential of *Bauhinia purpurea* Leaf Extracts in Rats. *Indian J. Pharm. Sci.* **2010,** *72* (1), 122–127. DOI: 10.4103/0250–474X.62250.

Boonphong, S.; Puangsombat, P.; Baramee, A.; Mahidol C.; Somsak, R.; Prasat, K. Bioactive Compounds from *Bauhinia purpurea* Possessing Antimalarial, Antimycobacterial, Antifungal, Anti-Inflammatory and Cytotoxic Activities. *J. Nat. Prod.* **2007,** *70* (5), 795–801. https://doi.org/10.1021/np070010e

Hisam, E. E.A.; Zakaria, Z. A.; Mohtaruddin, N.; Rofiee, M. S.; Hamid, H. A.; Othman, F. Antiulcer Activity of the Chloroform Extract of *Bauhinia purpurea* Leaf. *Pharma. Biol.* **2012,** *50* (12), 1498–1507. DOI: 10.3109/13880209.2012.685945

Jyothi, S. K.; Seshagiri, M. In Vitro Activity of Saponins of *Bauhinia purpurea, Madhuca longifolia, Celastrus paniculatus* and *Semecarpus anacardium* on Selected Oral Pathogens. *J. Dentistry, Tehran Univ. Med. Sci.* **2012,** *9* (4), 216–223.

Kuo, Y.-H.; Yeh, M.-H. Chemical Constituents of Heartwood of *Bauhinia purpurea. J. Chinese Chem. Soc.* **1997,** *44*, 379–383.

Lakshmi, B. V. S.; Neelima, N.; Kasthuri, N.; Umarani, V.; Sudhakar, M. Protective Effect of *Bauhinia purpurea* on Gentamicin-Induced Nephrotoxicity in Rats. *Indian J. Pharm. Sci.* **2009,** *71* (5), 551–554. DOI: 10.4103/0250–474X.58196.

Negi, B.; Dave, B.; Agarwal, Y. Evaluation of Antimicrobial Activity of *Bauhinia purpurea* Leaves Under In Vitro Conditions. *Indian J. Microbiol.* **2012,** *52* (3), 360–365. DOI 0.1007/s12088–012–0264–0

Ni, L.; Huang, W.; Shi, Y.; Wang, H.; Qiu, Y.; Xu, H. Chemical Constituents from the Bark of *Bauhinia purpurea* and Their NO Inhibitory Activities. *Nat. Prod. Res.* **2018,** *34* (17), 2424–2429. DOI: 10.1080/14786419.2018.1543672

Patra, G. S.; Gupta, R. Joshi, P. K.; Rout, O. P. *Bauhinia purpurea* (*Kanchanara*): A Critical Review on the Medicinal Plant Used in Ayurveda. *World J. Pharma. Res.* **2019,** *8* (4), 497–512. DOI: 10.20959/wjpr20194–14514.

Pettit, G.; Atsushi, N.; Iwamoto, C.; Yoshihide, U.; Takeshi, Y.; Hirofumi, O.; Gordon, C. Antineoplastic Agents. 551. Isolation and Structures of Bauhiniastatins 1–4 from *Bauhinia purpurea. J. Nat. Prod.* **2006,** *69*, 323–327.

Rana, M. A.; Khan, R. A.; Nasiruddin, M.; Khan, A. A. Amelioration of Cisplatin-Induced Nephrotoxicity by Ethanolic Extract of *Bauhinia purpurea*: An In Vivo Study in Rats. *Saudi J Kidney Dis Transpl.* **2016,** *27* (1), 41–48.

Shreedhara, C.; Vaidya, V.; Vagdevi, H.; Latha, K.; Muralikrishna, K.; Krupanidhi, A. Screening of *Bauhinia purpurea* Linn. for Analgesic and Anti-Inflammatory Activities. *Indian J. Pharmacol.* **2009,** *41* (2), 75–79. DOI: 10.4103/0253–7613.51345

Sunil, K.; Rohit, K.; Gupta, Y. K.; Surender, S. In Vivo Anti-Arthritic Activity of *Bauhinia purpurea* Linn. Bark Extract. *Indian J. Pharmacol.* **2019,** *51* (1), 25–30. DOI: 10.4103/ijp.IJP_107_16: 10.4103.

Takuro, F.; Tsuyoshi, E.; Kentaro, N.; Tsutomu, D.; Hidefumi, S.; Norio, S.; Seigo, K.; Masafumi, I. Decreased Expression of *Bauhinia purpurea* Lectin Is a Predictor of Gastric Cancer Recurrence. *Surg Today.* **2015,** *45* (10), 1299–306. DOI: 10.1007/s00595–015–1127–1

Yahya, F.; Mamat, S. S.; Kamarolzaman, M. F. F.; Seyedan, A. A.; Jakius, K. F.; Mahmood, N. D.; Shahril, M. S.; Suhaili, Z.; Mohtarrudin, N.; Susanti, D.; Somchit, M. N.; Teh, L. K.; Salleh, M.Z; Zakaria, Z. A. Hepatoprotective Activity of Methanolic Extract of *Bauhinia purpurea* Leaves Against Paracetamol-Induced Hepatic Damage in Rats. *Evidence-Based Complem. Altern. Med.* **2013,** (Sp. Issue), 1–10. http://dx.doi.org/10.1155/2013/636580

CHAPTER 10

Medicinal Properties of the *Bowdichia virgilioides* Kunth: An Evidence-Based Review

MARIA DANIELMA DOS SANTOS REIS*,
JAMYLLE NUNES DE SOUZA FERRO, FELIPE LIMA PORTO,
RAFAEL VRIJDAGS CALADO, TAYHANA PRISCILA MEDEIROS SOUZA,
and EMILIANO BARRETO

Laboratory of Cell Biology, Federal University of Alagoas, Brazil

**Corresponding author. E-mail: danielma.reis@icbs.ufal.br*

ABSTRACT

Bowdichia virgilioides Kunth (Fabaceae) is a medicinal plant from South America popularly known as "sucupira-preta". Parts of this plant are used in traditional medicine to treat various diseases, especially inflammation and diabetes. Phytochemical analysis of extracts obtained from *B. virgilioides* showed a wide variety of bioactive compounds, such as anthocyanins, alkaloids, terpenoids, essential oils, condensed tannins, saponins and steroids. *In vivo* and *in vitro* studies have corroborated the popular use of this plant, in addition to elucidating the action of some of its isolated compounds.

10.1 INTRODUCTION

Bowdichia virgilioides Kunth (Fabaceae) is a medium-sized native tree from tropical forests of South America, spreading from Venezuela to Northeast Brazil (Juck et al., 2006). It is popularly known in Brazil as sebipìra, sebupìra, sicopira, paricarana, sucupira, sucupira-preta, and sucupira-do-cerrado

Bioactives and Pharmacology of Legumes. T. Pullaiah, PhD (Ed.)

(Smiderle and Schwengber, 2011; Breitbach et al. 2013; Ribeiro et al., 2017). The tree can reach about 20 m tall, and its pod of 4–7.5 cm carry 6–10 reddish seeds (5–6 mm long with weight of 0.02 g) (Rodrigues et al., 2009; Smiderle and Schwengber, 2011).

The bark, leaves, and seeds from *B. virgilioides* are used in the traditional medicine for the treatment of disease and symptoms, such as pain, arthritic tumors, weakness by the syphilitic virus, hydrops, impingements, skin diseases, inflammation of the genitourinary system, menstrual cramps, parasitic and viral infections, osteoporosis, wound healing, and diabetes (de Albuquerque et al., 2007; Brandão et al., 2008; Breitbach et al. 2013; de Medeiros et al., 2013; Ribeiro et al., 2017).

In addition to its traditional use, sucupira also is used for forest restoration and furniture industry (Albuquerque et al., 2007). It is also an option for ornamentation with beautiful lilac flowers.

10.2 BIOACTIVES

Phytochemical analysis of the extracts from seeds, the bark of stem, and root showed that *B. virgilioides* Kunth contains a wide variety of bioactive compounds, such as anthocyanins, alkaloids, terpenoids, essential oils, isoflavonoids, benzofuran, condensed tannins, saponins, and steroids (Machado, 1936; Torrenegra et al., 1989; Arriaga et al., 1998; Velozo et al., 1999a; Melo et al., 2001; Silva et al., 2016; Beserra et al., 2018).

From the seed extract of the *B. virgilioides* was isolated 10 diterpernoids, one of them being a new pimarane-type diterpenoid named sucupiol (Kubo et al., 2016). Moreover, 17 new furanocassane-type diterpenoids named sucupiranins, A–L (Endo et al., 2017) and M–Q (Endo et al., 2019), were recently been identified from methanolic extract of the seeds. Also, the phytochemical characterization from the essential oil of *B. virgilioides* seeds revealed the presence of sesquiterpene hydrocarbons (82.5%) and oxygen-containing sequiterpenes (13.2%), with β-caryophyllene representing the main chemical compound (44.1% of the oil) (Rodrigues et al., 2009).

The stem barks are a rich source of terpenoids sitosterol, stigmasterol, lupeol acetate, and the 2,3-dihydrobenzofuran bowdenol (Melo et al., 2001). Moreover, alkaloids as the acosmine, ormosanine, podopetaline, bowdichin, and homoormosanine were identified in methanolic extract from *B. virgilioides* barks (Torrenegra et al., 1985, 1989; Barbosa-Filho et al., 2004). The compounds homopiptanthine and homo-18-epiormosanine and the cathepsin inhibitors as *p*-coumaric acid derivatives and 8-methoxycoumestrol were found in the extracts from stem barks (Silva et al., 2019). As from the organic crude extracts, in cyclohexane and methanolic extracts from barks were identified the isoflavonoids medicarpin, maackiain, 7,8,4′-trimethoxyisoflavone, and 7,8,4′-trimethoxyisoflavanone (Juck et al., 2006; Bezerra-Silva et al., 2015). Phytochemical screening of aqueous extract of *B. virgilioides* stem bark identified the presence of other biologically active compounds, such as flavonoids, phenols, terpenes, tannins, and saponins (Assis et al., 2018).

The isoflavone and flavone derivatives odoratin, afromosin, wistin, cladrastin, cladrastin 7-*O*-β-D-glucoside, fujikinetin, fujikinin, isohemiphloin, odoratin 7-*O*-β-D-glucopyranoside (Velozo et al., 1999a, 1999b), 3,6-dimethoxy-6″,6″-dimethylchromene-(7,8,2″,3) flavone, and 3,5,6-trimethoxyfuran-(7,8,2″,3″)-flavone (Arriaga et al., 2000) were isolated from *B. virgilioides* root extracts.

10.3 PHARMACOLOGY

10.3.1 Anti-Inflammatory Activity

In traditional medicine, different extracts of leaves, bark, and roots of *B. virgilioides* are frequently used for the treatment of several kinds of human diseases. Biological properties of this species were confirmed by distinct preclinical studies. It has been demonstrated that oral administration of the ethanolic extract from *B. virgilioides* stem bark attenuated the inflammatory response by suppressing the plasma leakage, leukocyte infiltration, and granuloma formation (Barros et al., 2010; Thomazzi et al., 2010). It is important to point out that in the specific inflammatory response, such as allergic reactions, the extract of *B. virgilioides* stem bark also inhibited the oedema formation, leukocyte influx, gene expression, and synthesis of pro-inflammatory mediators, a phenomenon that may be associated with suppression of histamine release by tissue mast cells (Silva et al., 2016).

10.3.2 Antinociceptive Activity

The ethanolic extract of the stem bark of *B. virgilioides* showed an antinociceptive effect in acetic acid-induced writhing test and in the second phase of the formalin test in mice (Barros et al., 2010). Corroborating with these findings, the oral administration of the aqueous extract of the stem bark and leaves also inhibited the acetic acid effect and reduced the formalin effect in the second phase (Silva et al., 2010; Thomazzi et al., 2010).

10.3.3 Antimicrobial Activity

The essential oil from leaves from *B. virgilioides* showed relevant bactericidal activity by reducing the growth of *Micrococcus luteus* and antifungal activity against *Candida albicans*, *Candida guilliermondii*, *Candida stellatoidea*, and *Trichophyton rubrum* (Almeida et al., 2006). The essential oil derived from seeds showed relevant antibacterial effect against *Bacillus subtilis*, *Bacillus bulgaricus*, *Enterococcus faecalis*, and *Staphylococcus aureus* (Rodrigues et al., 2009). Besides the essential oils, organic extracts from *B. virgilioides* barks also showed antifungal activity against *Aspergillus fumigatus*, *Trichophyton mentagrophytes*, and *Aspergillus niger* (Silva-Júnior et al., 2009). Other authors have reported the bactericidal effects of the aqueous extract from *B. virgilioides* stem barks against the standard strain of *S. aureus*, methicillin-resistant *S. aureus* (MRSA), *S. epidermidis*, and *S. saprophyticus* (Agra et al., 2013; Assis et al, 2018).

10.3.4 Wound Healing Properties

In the study performed by Agra et al. (2013), the topical treatment once a day for nine consecutive days with the aqueous extract of the *B. virgilioides* stem bark accelerated the noninfected and infected with *S. aureus* wound closure. Furthermore, the histopathological analysis showed that after treatment with this extract occurred an increase in collagen type I deposition associated with an augment in the fibroblast count in the scar tissue. It is worth mentioning that these findings may be associated with the pharmacological effects of the pentacyclic triterpene lupeol, which was found in the extract of the *B. virgilioides* stem bark. The lupeol is a phytoconstituent with a wide range of pharmacological activities against various diseases. Beserra et al. (2018) reported that lupeol extracted from stem barks of *B. virgilioides* enhanced keratinocytes migration and improved dermal fibroblasts functions in vitro.

10.3.5 Antioxidant Efficacy

It was shown that oral administration of the hydroethanolic extract from the inner bark of *B. virgilioides* for 25 days reduced markers of oxidative stress and muscle damage in resistance-trained rats (dos Santos et al., 2014).

10.3.6 Anxiolytic Activity

The aqueous extract from the stem bark of *B. virgilioides* orally administered was effective in inducing anxiolytic effects in mice with single acute (1 h) and subchronic treatment (7 days), without affecting the spontaneous locomotor activity (Vieira et al., 2013).

10.3.7 Antihyperglycemic Effect

Orally administrated *B. virgilioides* root methanolic extract promoted an acute antihyperglycemic effect in streptozotocin-induced diabetic rats (Silva et al., 2015).

10.3.8 Antimalarial Activity

The antimalarial activity of *B. virgilioides* extracts and fractions was performed in vitro against chloroquine-sensitive and resistant *Plasmodium falciparum* strains cultured in human erythrocytes. It was shown that hydroethanolic, total alkaloid fraction, and the alkaloid ormosanine were able to inhibit 100% (IC_{50} = 1 μg/mL) and 89% (IC_{50} = 5 μg/mL) of the parasitaemia in vitro, respectively (Bravo et al., 2002). Similarly, the ethanolic extract from *B. virgilioides* stem bark at 1 μg/mL reduced in 50% of chloroquine-sensitive and resistant *P. falciparum* parasitaemia. The extract also inhibited 51% of the *P. berghei* growth in infected mice (Deharo et al., 2001).

10.3.9 Antineoplastic Activity

The diterpenoids sucupiranin B and sucupiranin K isolated from the methanolic extract from the seeds of *B. virgilioides* showed cytotoxic activity against human cervical cancer cells (HeLa) and human promyelocytic leukaemia cells (HL-60) with IC_{50} of 9.8 and 32.6 μM, respectively (Endo et al., 2017).

10.3.10 Larvicidal Activity

The ethanolic (200 μg/mL) and the cyclohexane extract (50 μg/mL) of the heartwood of *B. virgilioides* were effective against *Aedes aegypti* larvae (Bezerra-Silva et al., 2015).

10.3.11 Immunomodulatory Activities

The ethanolic extracts from leaves and stem augmented the production of tumor necrosis factor-α (TNF-α) by the human monocyte cell line THP-1 stimulated with lipopolysaccharide (LPS) (Henriques et al., 2016). Oral administration of the aqueous extract from *B. virgilioides* stem bark decreased thymus weight and cellularity and increased the laminin deposition in the thymus microenvironment. Especially, at a low dose (20 mg/kg), the *B. virgilioides* extract reduced thymocytes motility on laminin substrates, B lymphocytes proliferation, antibody production, and enhanced the interleukin (IL)-10 secretion (dos Santos Reis et al., 2020).

The isoflavonoid odoratin-7-*O*-β-D-glucopyranoside isolated from the methanolic extract of the roots was able to inhibit mitogen-induced B- and T-lymphocyte proliferation (Velozo et al., 1999b). The furanocassane-type diterpenoids, sucupiranins J (IC_{50} = 30.6 μM), P (IC_{50} 44.0 ± 7.6 μM) and Q (IC_{50} 54.3 μM) isolated from *B. virgilioides* seeds inhibited LPS-induced nitric oxide production in RAW 264.7 cells (Endo et al., 2017, 2019).

10.3.12 Antileishmanial and Anticonvulsant Activity

The ethanolic and hexane extracts from the leaves of *B. virgilioides* did not show in vitro antileishmanial activity (IC_{50} > 200 μg/mL) (Ribeiro et al., 2014) and were not effective in inhibiting the seizures induced by electroshock and pentylenetetrazole (Quintans-Júnior et al., 2002).

KEYWORDS

- ***Bowdichia virgilioides***
- **traditional medicine**
- **bioactive compounds**
- **pharmacology**
- **plant extract**
- **Fabaceae**

REFERENCES

Agra, I. K.; Pires, L. L.; Carvalho, P. S.; Silva-Filho, E. A.; Smaniotto, S.; Barreto, E. Evaluation of Wound Healing and Antimicrobial Properties of Aqueous Extract from *Bowdichia virgilioides* Stem Barks in Mice. *Anais da Academia Brasileira de Ciências* **2013,** *85* (3), 945–954.

Albuquerque, K. S.; Guimarães, R. M.; Almeida, Í. F. D.; Clemente, A. D. C. S. Métodos para a superação da dormência em sementes de sucupira-preta (*Bowdichia virgilioides* Kunth.). *Ciência e Agrotecnologia* **2007,** *31* (6), 1716–1721.

Almeida, J. R. G. D. S., Silva-Filho, R. N. D., Nunes, X. P., Dias, C. D. S., Pereira, F. D. O., Lima, E. D. O. Antimicrobial Activity of the Essential Oil of *Bowdichia virgilioides* Kunt. *Revista Brasileira de Farmacognosia* **2006,** *16*, 638–641.

Arriaga, A. M. C.; Gomes, G. A.; Braz-Filho, R. Constituents of *Bowdichia virgilioides*. *Fitoterapia* **2000,** *71* (2), 211–212.

Arriaga, Â. M. C.; Machado, M. I. L.; Gomes, G. A.; Craveiro, A. A. Volatile Constituents from Roots of *Bowdichia virgilioides* Kunth. *J. Essential Oil Res.* **1998,** *10* (2), 205–206.

Assis, I. B.; da Rocha, E. M. M.; Guimarães, D. S. M.; do Nascimento Pereira, G. A.; Pereira, F. P.; Ferreira, J. M. S.; Barreto, E. O. Evaluation of Biological Activity, Toxicity, and Phytochemical Content of *Bowdichia virgilioides* (Fabaceae) aqueous extract. *Pharmacogn. Mag.* **2018,** *14* (57), 403–408.

Barbosa-Filho, J. M.; Guedes Da Silva Almeida, J. R.; De Oliveira Costa, V. C.; Leitão Da-Cunha, E. V.; Da Silva, M. S.; Braz-Filho, R. Bowdichine, a New Diaza-Adamantane Alkaloid from *Bowdichia virgilioides*. *J. Asian Nat. Prod. Res.* **2004,** *6* (1), 11–17.

Barros, W. M.; Rao, V. S.; Silva, R. M.; Lima, J.; Martins, D. T. Anti-Inflammatory Effect of the Ethanolic Extract from *Bowdichia virgilioides* HBK Stem Bark. *Anais da Academia Brasileira de Ciências* **2010,** *82* (3), 609–616.

Beserra, F. P.; Xue, M.; Maia, G. L. D. A.; Rozza, A. L.; Pellizzon, C. H.; Jackson, C. J. Lupeol, a Pentacyclic Triterpene, Promotes Migration, Wound Closure, and Contractile Effect In Vitro: Possible Involvement of PI3K/Akt and p38/ERK/MAPK Pathways. *Molecules* **2018,** *23* (11), 2819.

Brandão, M. G.; Zanetti, N. N.; Oliveira, P.; Grael, C. F.; Santos, A. C.; Monte-Mór, R. L. Brazilian Medicinal Plants Described by 19th Century European Naturalists and in the Official Pharmacopoeia. *J. Ethnopharmacol.* **2008,** *120* (2), 141–148.

Bravo, J. A.; Lavaud, C.; Bourdy, G.; Deharo, E.; Giménez, A.; Sauvain, M. Antimalarial Activity of Ormosanine and Homoormosanine from *Bowdichia virgiloides*. *Revista Boliviana de Quimica* **2002,** *19* (1), 12–17.

Breitbach, U. B.; Niehues, M.; Lopes, N. P.; Faria, J. E.; Brandão, M. G. Amazonian Brazilian Medicinal Plants Described by CFP von Martius in the 19th Century. *J. Ethnopharmacol.* **2013,** *147* (1), 180–189.

de Albuquerque, U. P.; Monteiro, J. M.; Ramos, M. A.; de Amorim, E. L. C. Medicinal and Magic Plants from a Public Market in Northeastern Brazil. *J. Ethnopharmacol.* **2007,** *110* (1), 76–91.

de Medeiros, P. M.; Ladio, A. H.; Albuquerque, U. P. Patterns of Medicinal Plant Use by Inhabitants of Brazilian Urban and Rural Areas: A Macroscale Investigation Based on Available Literature. *J. Ethnopharmacol.* **2013,** *150* (2), 729–746.

dos Santos Reis, M. D.; Vieira, L. F. A.; Brandão, A. R. A.; Barreto, E.; Smaniotto, S. The Effects of Aqueous Extract of *Bowdichia virgilioides* Kunth in the Immune Functions of Thymocytes and B-Lymphocytes. *J. Med. Plants Res.* **2020,** *14* (6), 247–257.

dos Santos, J. L.; Dantas, R. E. A.; Lima, C. A.; de Araújo, S. S.; de Almeida, E. C. V.; Marçal, A. C.; dos Santos Estevam, C. Protective Effect of a Hydroethanolic Extract from *Bowdichia virgilioides* on Muscular Damage and Oxidative Stress Caused by Strenuous Resistance Training in Rats. *J. Intern. Soc. Sports Nutr.* **2014,** *11* (1), 1–10.

Endo, Y.; Kasahara, T.; Asakura, K.; Mori, A.; Funasaki, M.; Akisue, G.; Etoh, T.; Harada, K.; Fukuyama, Y.; Matsuzaki, K.; Ishibashi, M.; Ohsakia, A. Sucupiranins M–Q. Five New Furanocassane-Type Diterpenoids from the Seeds of *Bowdichia virgilioides*. *Tetrahedron* **2019,** *75* (37), 130511.

Endo, Y.; Kasahara, T.; Harada, K.; Kubo, M.; Etoh, T.; Ishibashi, M.; Ishiyama, A.; Iwatsuki, M.; Otoguro, K.; Ōmura, S.; Akisue, G.; Hirano, T.; Kagechika, H.; Fukuyama, Y.; Ohsaki, A. Sucupiranins A–L, Furanocassane Diterpenoids from the Seeds of *Bowdichia virgilioides*. *J. Nat. Prod.* **2017,** *80* (12), 3120–3127.

Henriques, B. O.; Corrêa, O.; Azevedo, E. P. C.; Pádua, R. M.; Oliveira, V. L. S. D.; Oliveira, T. H. C.; Boff, D.; Dias, A. C. F.; de Souza, D. G.; Amaral, F. A.; Teixeira, M. M.; Castilho, R. O.; Braga, F. C. In Vitro TNF-Inhibitory Activity of Brazilian Plants and Anti-Inflammatory Effect of *Stryphnodendron adstringens* in an Acute Arthritis Model. *Evid.-Based Compl. Altern. Med.* **2016,** Article ID 9872598.

Juck, D. B.; de Rezende, L. C.; David, J. P.; de Queiroz, L. P.; David, J. M. Two New Isoflavonoids from *Bowdichia virgilioides*. *Nat. Prod. Res.* **2006,** *20* (1), 27–30.

Kubo, M.; Uemura, M.; Harada, K.; Ohsaki, A.; Fukuyama, Y. A New Pimarane-Type Diterpenoid from the Seeds of *Bowdichia virgilioides*. *Nat. Prod. Commun.* **2016,** *11* (7), 1934578X1601100713.

Machado, A. Oil (Resin) of "sucupira" (*Bowdichia virgilioides* Seeds). *Rev. Assoc. Bras. Farm.* **1936,** *17*, 117–118.

Melo, F. N.; Navarro, V. R.; Silva, M. S. D.; da-Cunha, E. V.; Barbosa-Filho, J. M.; Braz-Filho, R. Bowdenol, a New *2*, 3-Dihydrobenzofuran Constituent from *Bowdichia virgilioides*. *Nat. Prod. Lett.* **2001,** *15* (4), 261–266.

Quintans-Junior, L. J.; Almeida, R. N.; Falcão, A. C. G. M.; Agra, M. F.; Sousa, M. F. V.; Barbosa-Filho, J. M. Avaliação da Atividade anticonvulsivante de plantas do Nordeste Brasileiro. *Acta Farmacêutica Bonaerense* **2002,** *21* (3), 179–184.

Ribeiro, R. V.; Bieski, I. G. C.; Balogun, S. O.; de Oliveira Martins, D. T. Ethnobotanical Study of Medicinal Plants Used by Ribeirinhos in the North Araguaia Microregion, Mato Grosso, Brazil. *J. Ethnopharmacol.* **2017,** *205*, 69–102.

Ribeiro, T. G.; Chávez-Fumagalli, M. A.; Valadares, D. G.; Franca, J. R.; Lage, P. S.; Duarte, M. C.; Andrade, P. H.; Martins, V. T.; Costa, L. E.; Arruda, A. L.; Faraco, A. A.; Coelho, E. A.; Castilho, R. O. Antileishmanial Activity and Cytotoxicity of Brazilian Plants. *Exp. Parasitol.* **2014,** *143*, 60–68.

Rodrigues, M. O.; Alves, P. B.; Nogueira, P. C.; Machado, S. M.; Moraes, V. R.; Ribeiro, A. D. S.; Silva, E. S.; Feitosa, J. G. R. Volatile Constituents and Antibacterial Activity from Seeds of *Bowdichia virgilioides* Kunt. *J. Essential Oil Res.* **2009,** *21* (3), 286–288.

Silva, A. C. M.; dos Santos, M. P.; de França, S. A.; da Silva, V. C.; da Silva, L. E.; de Figueiredo U. S.; Dall'Oglio, E. L.; de Sousa Júnior, P. T.; Lopes, C. F.; Baviera, A. M.; Kawashita, N. H. Acute and Subchronic Antihyperglycemic Activities of *Bowdichia virgilioides* Roots in Non-Diabetic and Diabetic Rats. *J. Intercult. Ethnopharmacol.* **2015,** *4* (1), 57–63.

Silva, J. P.; Rodarte, R. S.; Calheiros, A. S.; Souza, C. Z.; Amendoeira, F. C.; Martins, M. A.; Silva, P. M. R.; Frutuoso, V. S.; Barreto, E. Antinociceptive Activity of Aqueous Extract of *Bowdichia virgilioides* in Mice. *J. Med. Food* **2010,** *13* (2), 348–351.

Silva, J. P.; Ferro, J. N.; Filho, H. S.; Silva, L. A.; Souza, T. P.; Matos, H. D.; Carvalho, V. D.; Rodarte, R. S.; Barreto, E. Aqueous Extract of *Bowdichia virgilioides* Stem Bark Inhibition of Allergic Inflammation in Mice. *J. Med. Plants Res.* **2016,** *10*, 575–584.

Silva, T. L.; Fernandes, J. B.; Silva, M.; Consolaro, H. N.; de Sousa, L. R.; Vieira, P. C. New Cathepsin V Inhibitor from Stems of *Bowdichia virgilioides*. *Revista Brasileira de Farmacognosia* **2019,** *29* (4), 491–494.

Silva-Júnior, I. E.; Cechinel Filho, V.; Zacchino, S. A.; Lima, J. C. D. S.; Martins, D. T. D. O. Antimicrobial Screening of Some Medicinal Plants from Mato Grosso Cerrado. *Revista Brasileira de Farmacognosia* **2009,** *19* (1B), 242–248.

Smiderle, O. J.; Schwengber, L. A. M. Superação da dormência em sementes de paricarana (*Bowdichia virgilioides* Kunth.). *Revista Brasileira de Sementes* **2011,** *33* (3), 407–414.

Souza, R. K. D.; da Silva, M. A. P.; de Menezes, I. R. A.; Ribeiro, D. A.; Bezerra, L. R.; de Almeida Souza, M. M. Ethnopharmacology of Medicinal Plants of Carrasco, Northeastern Brazil. *J. Ethnopharmacol.* **2014,** *157*, 99–104.

Thomazzi, S. M.; Silva, C. B.; Silveira, D. C. R.; Vasconcellos, C. L. C.; Lira, A. F.; Cambui, E. V. F.; Estevam, C. S.; Antoniolli, A. R. Antinociceptive and Anti-Inflammatory Activities of *Bowdichia virgilioides* (sucupira). *J. Ethnopharmacol.* **2010,** *127* (2), 451–456.

Torrenegra, R. G., Escarria, R. S., Bauereiß, P., Achenbach, H. Homoormosanine, The Major Alkaloid of the Bark from *Bowdichia virgiloides*. *Planta medica* **1985,** *51* (3), 276–277.

Torrenegra, R.; Bauereiß, P.; Achenbach, H. Homoormosanine-Type Alkaloids from *Bowdichia virgiloides*. *Phytochemistry* **1989,** *28* (8), 2219–2221.

Velozo, L. S. M.; da Silva, B. P.; da Silva, E. M. B.; Parente, D. J. Constituents from the Roots of *Bowdichia virgilioides*. *Fitoterapia* **1999a,** *70* (5), 532–535.

Velozo, L. S.; da Silva, B. P.; Bernardo, R. R.; Parente, J. P. Odoratin 7-O-β-D-glucopyranoside from *Bowdichia virgilioides*. *Phytochemistry* **1999b,** *52* (8), 1473–1477.

Vieira, L. F. A.; Reis, M. D. S.; Brandão, A. R. A.; Viana, I. M. M.; da Silva, J. P.; Barreto, E.; Smaniotto, S. Anxiolytic-Like Effect of the Extract from *Bowdichia virgilioides* in Mice. *Revista Brasileira de Farmacognosia* **2013,** *23* (4), 680–686.

CHAPTER 11

Phytoconstituents and Biological Activities of *Caesalpinia mimosoides* Lam.

PRADEEP BHAT[1], HARSHA V. HEGDE[1], SAVALIRAM G. GHANE[2], and SANTOSHKUMAR JAYAGOUDAR[3*]

[1]*ICMR-National Institute of Traditional Medicine, Nehru Nagar, Belagavi, Karnataka 590010, India*

[2]*Plant Physiology Laboratory, Department of Botany, Shivaji University, Kolhapur, Maharashtra 416004, India*

[3]*Department of Botany, G. S. S. College & Rani Channamma University P. G. Centre, Belagavi, Karnataka 590006, India*

[*]*Corresponding author. E-mail: santoshjayagoudar@gmail.com*

ABSTRACT

Caesalpinia mimosides Lam. is a prickly scandent shrub belongs to the family Fabaceae, native to Southeast Asia. It is traditionally used by the healers of India and Thailand to treat epilepsy, boils, wounds, ulcers and arthritis. Immature leaves are used as a vegetable, appetizer, carminative and to get rid of giddiness. It displayed diverse phytochemical constituents and well-documented biological activities. Mimosol A–D, caesmimosins A-F, taepeenin A, nortaepeenin A, taepeenin L, tetracosylcaffeate, resveratrol, bergenin, pterocarpol, quercetin, gallic acid, ethyl gallate, eucomin, intricatinol, 8-methoxybonducellin, quercetin, quercetin-3-O-glucoside, sulfuretin, isoliquiritigenin, butein, are the major bioactive compounds

Bioactives and Pharmacology of Legumes. T. Pullaiah, PhD (Ed.)

isolated from different parts of *C. mimosoides*. The plant is demonstrated for both *in vitro* and *in vivo* biological activities. Different extracts from shoot tips, stems, leaves, fruits and the whole plant showed significant antimicrobial potentiality against various microbial pathogens that are responsible for emerging infectious diseases. The other *in vitro* biological activities exhibited by the extracts include antioxidant, anticancer, anticholinesterase inhibition, anti-inflammatory, neuritogenic and neuroprotective activities. The results related to *in vivo* pharmacological activities were carefully reviewed and summarized. It was found that petroleum ether, chloroform, ethyl acetate, ethanol and aqueous extracts as well as Ayurvedic preparations (Panchanga kwatha and Panchanga kashaya) exhibited wide range of anti-inflammatory, wound healing, antiarthric and analgesic activities.

11.1 INTRODUCTION

Caesalpinia mimosides Lam. belongs to the family Fabaceae. Synonyms of this species include *Caesalpinia simora* Buch. Ham. ex Roxb.; *Caesalpinia resupinata* Roxb.; *Caesalpinia horrida* Herv. Madr. ex Wall.; *Caesalpinia armata* Graham; *Biancaea mimosoides* (Lam.). Mimisa thorns, Kanjigina kudi, Kenchigi balli, Pananjimullu, Pansi, and Puliyindam are some of its common names.

A very prickly scandent shrub, sometimes erect; branches clothed with brown tomentum; young parts glandular bristly; prickles erect and recurved, and leaves 30–60 cm long, pinnae 18–20 pairs, leaflets sensitive, opposite, subsessile, membranous, 10–20 pairs, oblong; flowers yellow, 2–2.5 cm in simple, axillary and terminal racemes; pods 4–5 cm long, bristly, obliquely ovoid-oblong; seeds 1–4, mottled. The nativity of *C. mimosoides* is Southeast Asia and distributed in India, southwest China, Bangladesh, Sri Lanka, Laos, Vietnam, Myanmar, and Thailand.

The tribal communities of Northern and Northeastern Thailand traditionally consume tender shoots and immature leaves of the plant as a vegetable, appetizer, carminative, and to get rid of giddiness (Chanwitheesuk et al., 2007; Tangsaengvit et al., 2013; Manasa et al., 2014). However, in many southern states of India, different parts of the plant are frequently used in the treatment of several diseases. The *Mullu Kuruma* tribe of Kerala state utilizes the plant leaves in the treatment of epilepsy (Silja et al., 2008). Traditional practitioners of Uttara Kannada district in Karnataka state claimed that the oral consumption of juice extracted from tender shoots cures all kinds of skin disorders and it plays a major role as a blood purifier. Further, they use the paste of tender shoots in the topical application for suppuration and rupture of boils, followed by curing the wounds (Bhat et al., 2016). The traditional healers of Udupi district of Karnataka state use the root paste of plants in the treatment of wounds, ulcers, and arthritis (Rekha, 2011).

11.2 BIOACTIVES

Preliminary phytochemical screening of the extracts of stem, leaves, and tender shoots of the plant *C. mimosoides* showed the presence of phenolics, flavonoids, tannins, alkaloids, glycosides, saponins, phytosterols, steroids, resins, anthraquinones, proteins, reducing sugars, and ascorbic acid (Chanwitheesuk et al., 2005; Supriya et al., 2013; Manasa et al., 2014; Bhat et al., 2016; Viji and Wilson, 2017).

Bioassay-guided fractionation and isolation of dichloromethane and acetone extracts of roots reported four diterpenes: mimosol A–D, a dimer mimosol E, and two dibenzo [b,d] furans, such as mimosol F and G (Yodsaoue et al., 2010). Further, they have isolated the compounds taepeenin A; nortaepeenin A; taepeenin L; (E)-7-hydroxy-3-(4-methoxybenzyl) chroman-4-one; (E)-7,8-dihydroxy-3-(4-methoxybenzyl) chroman-4-one; (E)-7-hydroxy-8-methoxy-3-(4-methoxybenzyl) chroman-4-one; tetracosylcaffeate; resveratrol; bergenin; and pterocarpol from the active subfractions. Tangsaengvit et al. (2013) have isolated the compound quercetin from ethyl acetate extract of young shoots and leaves through flash column chromatography technique. GC–MS analysis of ethanol extract from tender shoots revealed gallic acid and ethyl gallate as major compounds with the concentrations of 59.04 and 56.88 μg/mg, respectively (Bhat et al., 2016). Gallic acid was also isolated by Chanwitheesuk et al. (2007) and Rattanata et al. (2016) from shoot, leaves, and whole plant through Sephadex LH-20 and silica gel column chromatography methods. Bi et al. (2017) isolated six new cassane type diterpenes, namely, caesmimosins A–F from the ethyl acetate extract of fruits and further structurally elucidated. He et al. (2017) isolated 17 compounds from the ethyl acetate extract of twigs and structurally elucidated the compounds through ^{1}H and ^{13}C NMR spectroscopy. The identified compounds were mainly 10-methoxyprotosappanin B, 10-methoxyisoprotosappanin B, eucomin, intricatinol, 8-methoxybonducellin, quercetin, quercetin-3-*O*-glucoside, sulfuretin, isoliquiritigenin, butein, bergenin,

bergenin 11-*O*-I-ferulate, sitoindoside I, friedelin, coniferaldehyde, gallic acid, and ethyl gallate (Figure 11.1).

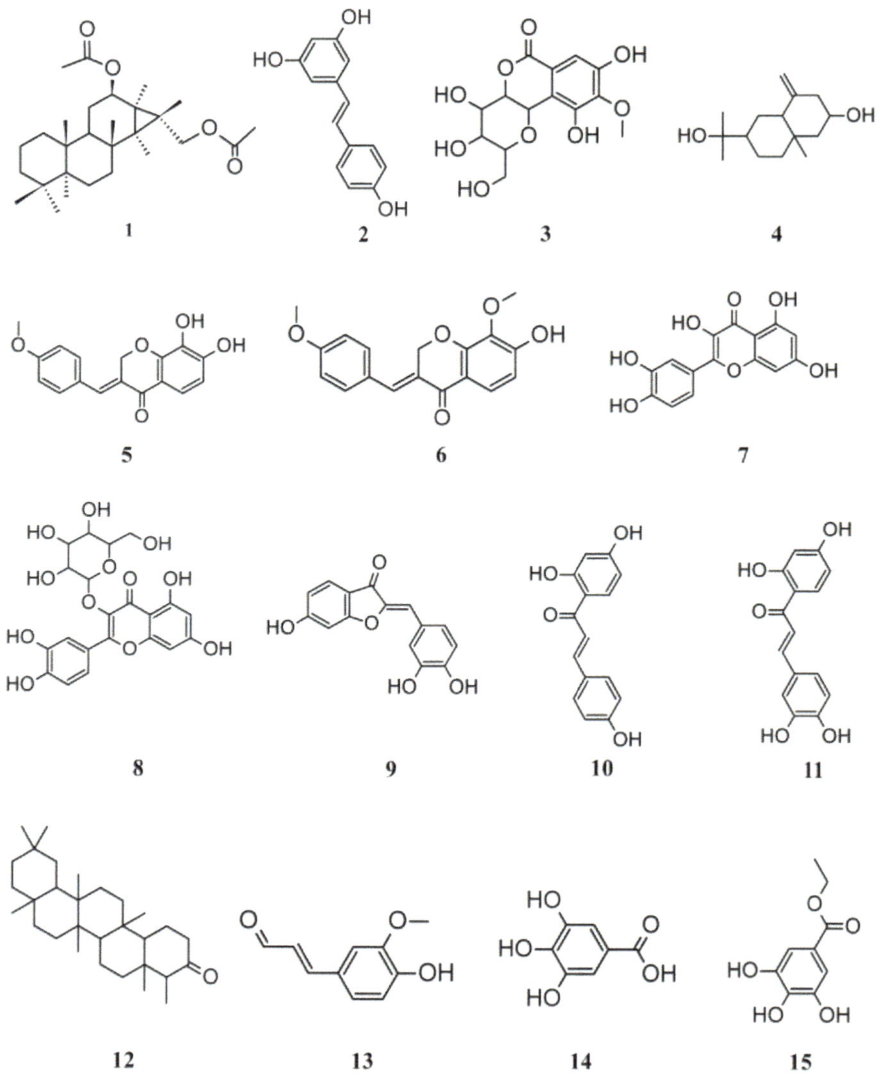

FIGURE 11.1 Structures of Mimosol-D (1), resveratrol (2), bergenin (3), pterocarpol (4), intricatinol (5), 8-Methoxybonducellin (6), quercetin (7), Quercetin-3-*O*-glucoside (8), sulfuretin (9), isoliquiritigenin (10), butein (11), friedelin (12), coniferaldehyde (13), gallic acid (14), and ethyl gallate (15).

11.3 PHARMACOLOGY

11.3.1 Antimicrobial Activity

Antimicrobial activity of hexane, chloroform, acetone, ethanol, as well as aqueous extracts of tender shoots was analyzed against different bacterial and fungal skin pathogens through the serial tube dilution method. Ethanol extract was the most effective against *Staphylococcus aureus* (0.104 ± 0.026 mg/mL), *Micrococcus flavus* (0.208 ± 0.052 mg/mL), *Micrococcus luteus* (0.026 ± 0.007 mg/mL), *Pseudomonas aeruginosa* (0.130 ± 0.026 mg/mL), *Proteus vulgaris* (0.260 ± 0.052 mg/mL), and *Salmonella typhimurium* (0.469 ± 0.156 mg/mL) when compared to the standard streptomycin. Moreover, it also showed promising antifungal activity against *Trichophyton mentagrophytes* (0.208 ± 0.052 mg/mL), *Microsporum gypseum* (0.039 ± 0.020 mg/mL), and *Malassezia furfur* (0.469 ± 0.156 mg/mL) as compared to standard voriconazole (Bhat et al., 2016). Chanwitheesuk et al. (2007) have carried out antimicrobial activity of chloroform, acetone, ethanol, and aqueous extracts of the whole plant. Wherein, the aqueous extract showed potent activity in terms of the highest zone of inhibition against *Vibrio cholerae* (29.75 ± 0.71 mm), *Staphylococcus epidermidis* (35.25 ± 0.87 mm), *M. gypseum* (42.50 ± 0.43 mm), and *Trichophyton rubrum* (40.50 ± 0.50 mm). Alcoholic extract of shoot tips, leaves, and fruits also showed potent antimicrobial activities (Daduang et al., 2011; Manasa et al., 2014). The gallic acid isolated from *C. mimosoides* was evaluated for its antibacterial potential against *Plesiomonas shigelloides* and *Salmonella* spp. (serogroup B) with MIC values 110 and 120 mM, respectively (Rattanata et al., 2016). Ethyl acetate and methanol extracts of stem and leaves exhibited significant growth inhibition against several pathogens (Supriya et al., 2013; Viji and Wilson 2017). Silver nanoparticles synthesized using aqueous extract of the shoot tip exhibited strong antibacterial activity against food-borne pathogens (*Escherichia coli* and *Listeria monocytogenes*) with MIC and MBC values of 8/16 and 16/64 μg/mL, respectively (Shankar et al., 2016).

11.3.2 Antioxidant Activity

Chanwitheesuk et al. (2005) have studied the antioxidant activity of methanol extract through estimation of carotene and linoleic acid coupled oxidation. The extract showed a moderate antioxidant index (5.43 ± 0.61) (Chanwitheesuk et al., 2005). Methanol extract of root scavenged the radicals more effectively (44.99 μg/mL), followed by leaf (0.59 μg/mL), flower (3.41 μg/mL), and root (7.76 μg/mL) (Manasa et al., 2014). The ethanol extract of aerial parts was investigated for its *in vivo* antioxidant activity by conducting various assays such as catalase, glutathione, glutathione reductase, superoxide dismutase, and glutathione peroxidase, and it possessed a significant antioxidant activity (Ranjith et al., 2014). The antioxidant potential of ethanol extract of tender shoots was analyzed through DPPH, lipid peroxidise, nitric oxide, and total antioxidant activities. The extract showed effective DPPH and nitric oxide scavenging activities (IC_{50} 19.25 and 132.40 μg/g, respectively) (Bhat et al., 2016). Antioxidant and antiaging properties of ethyl acetate, hexane, and methanol extracts of leaves and young twigs have been carried out by Rangsinth et al. (2019). It was found that methanol extract exhibited the highest antioxidant activities against DPPH and ABTS free radicals with EC_{50} of 8.20 ± 0.29 and 5.16 ± 0.98 μg/mL, respectively. Enhancement in the nuclear localization of DAF-16 transcription factor and SOD-3 gene expressions demonstrated its antioxidant activity that mediated through the DAF16/FOXO pathway. Tangsaengvit et al. (2013) tested antioxidant activity of ethyl acetate extract of leaves and young sprouts and noted strong DPPH activity (IC_{50} 4.83 ± 0.03 and 3.18 ± 0.07 μg/mL, respectively). Supriya et al. (2013) also proved the antioxidant potentiality of hexane, ethyl acetate, chloroform, methanol, and aqueous extracts of the stem.

11.3.3 Anticancer Activity

Ethanol extract of shoot tips has been tested against human epidermal carcinoma oral cavity cell line (KB) and showed the inhibitory effect with IC_{50} value of 27.17 ± 5.44 μg/mL (Daduang et al., 2011). Phenolic rich fraction (fraction 21) of shoot and leaf methanolic extracts were studied on cell proliferation and apoptotic induction in human cervical carcinoma cells, such as HeLa, SiHa, and C33A (Palasap et al., 2014), and they showed a marked suppression of cell lines with IC_{50} of 10, 20, and 12.36 μg/mL, respectively. Rattanata et al. (2016) carried out anticancer potential of gallic acid isolated from ethyl acetate extract of shoots and leaves against cholangiocarcinoma cell lines (M213 and M214) and found that it significantly inhibited the proliferation of both the cell lines by inducing apoptosis (IC_{50} value of 120 and 124 μM, respectively). Further, Bi et al. (2017) isolated six new cassane diterpenes (caesmimosins A–F) from ethyl acetate extract of fruits and evaluated anticancer potential against HL-60 (acute leukemia), SMMC-7721 (liver cancer), A-549 (lung cancer), MCF-7 (mammary cancer), and SW-480 (colon cancer) human tumor cell lines. The extract failed to exhibit effective anticancer activity against the targeted cells, with IC_{50} more than 40 μM.

11.3.4 Anti-Inflammatory Activity

Anti-inflammatory property of petroleum ether, ethyl acetate, and chloroform extracts of aerial parts on the carrageenan paw edema model was studied (Lakshmi and Bobby, 2015). Ethyl acetate extract (250 μg/mL) showed higher inhibition of proteinase and protein denaturation (IC_{50} 75.04 and 95.21 μg/mL, respectively). In addition, ethanol extract (at 200 and 400 mg/kg) also significantly reduced the paw edema. Yodsaoue et al. (2010) carried out bioassay-guided isolation and characterization of dichloromethane and acetone extracts. Further, compounds were evaluated for anti-inflammatory activity against TNF-α and lipopolysaccharide induced nitric oxide productions in RAW264.7 cells. Among all, the compound mimosol-D significantly inhibited the expressions of TNF-α and nitric oxide in the cells (IC_{50} 6.5 and 3.0 μM, respectively).

11.3.5 Wound Healing Activity

The wound healing potential of ethanol and aqueous extracts was analyzed through excision and incision wound parameters. A significant higher percentage of excision wound closure was found in ethanol and water extract treated groups at 5% w/w dose (99.60 ± 0.23 and 99.89 ± 0.10%, respectively), followed by the positive control (Providine-AM) treated group (98.79 ± 0.96%). Similarly, the water extract treated group showed significantly higher tensile strength in incised wound on the 15th day (928.5 ± 21.42 g), followed by ethanol extract and positive control treated groups (838.83 ± 29.55 and 778.0 ± 9.18 g). Meanwhile, hydroxyproline contents of water and ethanol extracts were significantly higher (91.55 ± 0.27 and 68.44 ± 0.59 μg/100 mg, respectively) compared to the standard Providine-AM group (54.54 ± 0.47 μg/100 mg (Bhat et al., 2016).

11.3.6 Antiarthritic Activity

An Ayurvedic herbal formulation "Panchanga kwatha" prepared from *C. mimosoides* was examined for antiarthritic activity using Wistar rat model. An evident suppression in paw volume was found in the test and standard group on the 17th, 19th, and 21st days (49.18, 55.74, and 54.5%, respectively) (Rekha, 2011).

11.3.7 Analgesic Activity

"Panchanga Kashaya" (an Ayurvedic formulation) has been tested for analgesic potential in male Wistar rats. The responses of the test drug at 150 and 120 min were highly significant (4.250 ± 0.243 and 4.517 ± 0.293 mg/kg, respectively) when compared to the same interval in standard group (3.667 ± 0.153 and 3.299 ± 0.484 mg/kg, respectively). It denotes the test drug "Panchangakashaya" prepared from the plant exhibit an appreciable analgesic activity (Rekha, 2011).

11.3.8 Anticholinesterase (AChE) Inhibition Activity

Active compounds (gallic acid and quercetin) from ethyl acetate extract of young sprouts possessed significant anticholinesterase activity in both cellular and purified AChE enzymes (IC_{50} 2.97 ± 0.17 and 12.73 ± 0.56 μg/mL for gallic acid; 36.60 ± 2.78 and 56.84 ± 2.64 μg/mL for quercetin) (Tangsaengvit et al., 2013).

11.3.9 Neuritogenic and Neuroprotective Activity

Tangsaengvit et al. (2013) isolated quercetin from ethyl acetate extract of young shoots. The compound was further assessed for the formation and maturation of neurite outgrowths in P19-derived neurons. Quercetin (1 nM) was treated with P19-derived neurons grown in a precoated poly-L-lysine-6-well plate for 24 h. Morphological features of 30 neuronal cells were investigated using phase-contrast microscopy. A significant increase in the number of neurites and neurite length was found in geldanamycin (positive control) treatment (2.03 ± 1.30 and 3.23 ± 1.55 μM), followed by quercetin (3.23 ± 1.55 and 139.00 ± 108.80 μM), respectively. The compound protected the neurons from oxidative stress and the cell death was also triggered by serum deprivation. The pretreatment of neuronal cells with quercetin (1 nM) was found effective with significant increase in survival rate (128.01% and 88.29%) in both the cultures with and without serum supplementations, respectively (Tangsaengvit et al., 2013).

KEYWORDS

- **phytoconstituents**
- **mimosol**
- **caesmimosins**
- **antiinflammatory**
- **wound healing**
- **analgesic activity**

REFERENCES

Bhat, P. B.; Hegde, S.; Upadhya, V.; Hegde, G. R.; Habbu, P. V.; Mulgund, G. S. Evaluation of Wound Healing Property of *Caesalpinia mimosoides* Lam. *J. Ethnopharmacol.* **2016,** *193*, 712–724.

Bi, D.; Xi, G.; Liang, X.; Li, Y.; Zhang, L.; Wang, L. New Cassane Diterpenes from the Fruits of *Caesalpinia mimosoides* Lam. *Phytochem. Lett.* **2017,** *21*, 283–286.

Chanwitheesuk, A.; Teerawutgulrag, A.; Kilburn, J. D.; Rakariyatham, N. Antimicrobial Gallic Acid from *Caesalpinia mimosoides* Lamk. *Food Chem.* **2007,** *100* (3), 1044–1048.

Chanwitheesuk, A.; Teerawutgulrag, A.; Rakariyatham, N. Screening of Antioxidant Activity and Antioxidant Compounds of Some Edible Plants of Thailand. *Food Chem.* **2005,** *92* (3), 491–497.

Daduang, J.; Vichitphan, S.; Daduang, S.; Hongsprabhas, P.; Boonsiri, P. High Phenolics and Antioxidants of Some Tropical Vegetables Related to Antibacterial and Anticancer Activities. *Afr. J. Pharm. Pharmacol.* **2011,** *5* (5), 608–615.

He, D.; Ma, R.; Li, Y.; Yang, X.; Wang, L. Chemical Constituents of the Twigs of *Caesalpinia mimosoides. Chem. Nat. Compd.* **2017,** *53* (3), 545–547.

Lakshmi, V. V.; Bobby, S. P. Anti-Inflammatory Activity of the Aerial Parts Extracts of *Caesalpinia mimosoides* Lamk. *Int. J. Pharm. Sci. Res.* **2015,** *6* (12), 5154–5158.

Manasa, M.; Vivek, M. N.; Kambar, Y.; Ramesh Kumar, K. A.; Kekuda, P. T. R. Mineral Content, Antimicrobial and Radical Scavenging Potential of *Caesalpinia mimosoides* Lamk. (Caesalpiniaceae). *World J. Pharm. Res.* **2014,** *3* (4), 1047–1063.

Palasap, A.; Limpaiboon, T.; Boonsiri, P. The Cytotoxic Effect of Phytophenolics form *Caesalpinia mimosoides* Lamk. on Cervical Carcinoma Cell Lines Through Apoptotic Pathway. *Asian Pac. J. Cancer Prev.* **2014,** *15*, 449–454.

Rangsinth, P.; Prasansuklab, A.; Duangjan, C.; Gu, X.; Meemon, K.; Wink, M.; Tencomnao, T. Leaf Extract of *Caesalpinia mimosoides* Enhances Oxidative Stress Resistance and Prolongs Lifespan in *Caenorhabditis elegans*. *BMC Complement Altern. Med.* **2019,** *19* (164). https://doi.org/10.1186/s12906-019-2578-5.

Ranjith, P. B.; Babu, H.; Ganesan, V.; Mathew, M. *In-Vivo* Antioxidant Activity of *Caesalpinia mimosoides* Lamk. *Res. J. Pharm. Biol. Chem. Sci.* **2014,** *5* (6), 1116–1120.

Rattanata, N.; Klaynongsruang, S.; Daduang, S.; Tavichakorntrakool, R.; Limpaiboon, T.; Lekphrom, R.; Boonsiri, P.; Daduang, J. Inhibitory Effects of Gallic Acid Isolated from *Caesalpinia mimosoides* Lamk on Cholangio Carcinoma Cell Lines and Food Borne Pathogenic Bacteria. *Asian Pac. J. Cancer Prev.* **2016,** *17* (3), 1341–1345.

Rekha, A. S. Pharmacognostical and Experimental Study on *Caesalpinia mimosoides* Lam.—A Folk Plant. Ph.D. Thesis, Rajiv Gandhi University of Health Sciences, Bangalore, Karnataka; 2011.

Shankar, S.; Tanomrod, N.; Rawdkuen, S.; Rhim, J. W. Preparation of Pectin/Silver Nanoparticles Composite Films with UV-Light Barrier and Properties. *Int. J. Biol. Macromol.* **2016,** *92*, 842–849.

Supriya, M. S.; Nagamani, J. E.; Komal Kumar, J. Cytoprotective and Antimicrobial Activities of Stem Extract of *Caesalpinia mimosoides*. *Int. J. Biol. Pharma. Res.* **2013,** *4* (1), 18–25.

Tangsaengvit, N.; Kitphati, W.; Tadtong, S.; Bunyapraphatsara, N.; Nukoolkarn, V. Neurite Outgrowth and Neuroprotective Effects of Quercetin from *Caesalpinia mimosoides* Lamk on Cultured P19-Derived Neurons. *Evid. Based Complement. Altern. Med.* **2013,** 838051. doi: 10.1155/2013/838051.

Viji, M. O.; Wilson, N. Studies on Phytochemical, Antimicrobial and Cytotoxic Properties of *Caesalpinia mimosoides* Leaf Extracts. *VISTAS A Multidisc. Res. J.* **2017,** *6* (1), 111–120.

Yodsaoue, O.; Karalai, C.; Ponglimanont, C.; Tewtrakul, S.; Chantrapromma, S. Potential Anti-Inflammatory Diterpenoids from the Roots of *Caesalpinia mimosoides* Lamk. *Phytochemistry* **2010,** *71*, 1756–1764.

CHAPTER 12

A Review on Phytochemistry and Pharmacology of *Cassia fistula* L.

LEPAKSHI M. D. BHAKSHU[1*], K. VENKATA RATNAM[2], and R.R. VENKATA RAJU[3]

[1]*Department of Botany, PVKN Government College (A), Chittoor, Andhra Pradesh 517002, India*

[2]*Department of Botany, Rayalaseema University, Kurnool, Kerala 518007, India*

[3]*Department of Botany, Sri Krishnadeveraya University, Ananthapuramu, Andhra Pradesh 515003, India*

[*]*Corresponding author. E-mail: lmbakshu@pvkngcchittoor.ac.in*

ABSTRACT

Golden shower or Indian laburnum is scientifically well known as *Cassia fistula* L. and belongs to the Family Fabaceae, subfamily Caesalpinoidae. It has pronounced medicinal properties and rich in bioactive metabolites with diversified biological activities. *C. fistula* is an ornamental as well as an important component in the forest ecosystem which blooms in summer and gives an aesthetic view. All parts of the plant are medicinally useful in the treatment of fevers, heart diseases, biliousness, hemorrhages, ulcers, wounds, piles, fistula, skin diseases, in traditional medicine. The plant is rich in flavonoids such as rhein, phenolics (fistulic acid), lignans, glycosides, anthracene derivatives, catechins etc., obtained from various parts. These phytochemicals possess pharmacological effects such as hepatoprotective, anti-inflammatory, antioxidant, antipyretic, antitussive activity, antiulcer, central nervous system activity, hypo-lipidaemia, larvicidal/ovicidal, laxative, leukotriene inhibition,

Bioactives and Pharmacology of Legumes. T. Pullaiah, PhD (Ed.)

anticancer, antidiabetic, antiepileptic, antifeedant, antifertility, urease inhibition, wound healing, antimicrobial, antifungal antibacterial, and antiviral activities. The present review has focused on the discussion of bio-actives of the *C. fistula* and the bio-effects on the medicinal claims.

12.1 INTRODUCTION

Cassia fistula L. (Family: Fabaceae/Caesalpiniaceae), is a well-known medicinal tree species besides an ornamental plant for its glorious appearance during the flowering conditions. It is widely distributed in the Amazon and Sri Lanka, and its nativity traced in India chiefly and spread to various countries including China, Mauritius, Bangladesh, East Africa, West Indies, Mexico, and South Africa (Anonymous, 2007; Bhalerao and Kelkar, 2012; Danish et al., 2011; Pawar et al., 2017; Kumar et al., 2017; Nyeem et al., 2017; Vijayrao et al., 2018; Hakim et al., 2019). Common names for this plant include Indian laburnum, Golden shower (English), Amalthas, Pudding-pipe tree, Purging cassia in different regions, and Rela in Telugu.

It is a potential medicinal plant and is used as an analgesic and antipyretic (Mazumder et al., 1998), cardioprotective (Chaminda et al., 2001) anti-inflammatory, and hypoglycaemic. It plays an important role in health wellness and proved as a potential both in vivo and in vitro studies (Rahmani, 2015). The phytochemical constituents were scrupulously reviewed and their relation to the biological activities were analyzed with special reference to flavonoids (Zhao et al., 2016). It has been proven scientifically for the therapeutic usage and its implication in the health management through its antioxidant, anti-inflammatory, antidiabetic, and other biological activities (Rahmani, 2015).

12.2 BIOACTIVES (PHYOTOCHEMICALS)

Presence of various secondary metabolites of *Cassia* genus has been reviewed and noted the rich diversity of flavonoids and polyphenols (Deshpande and Sanjivani, 2013; Zhao et al., 2016; Hakim et al., 2019) and discussed here some of the important constituents.

Barthakur et al. (1995) reported that the pods serve as a resource for minerals such as iron and manganese comparably greater than that of oranges, apple, peach, apricot in addition to the presence of glutamic acid, aspartic acid and lysine, and lysine in the pulp. Even the seeds showed similar amino acids. Misra et al. (1996, 1997) isolated triacontane, 2-hentriacontanone,

5-nonatetracontanone, β-sitosterol, and 16-hentriacontanone from the nonpolar portion of the fruits.

Sayeed et al. (1999) analyzed and reported monoglycerides and triglycerides from the seeds oil. The neutral lipids were accounted for over 89.80% of the total weight of the lipid and containing the saturated as well as unsaturated fatty acids especially the linoleic acid (42.42%), oleic acid (29.62%), stearic acid (14.33%), and palmitic acid (11.41%) besides, caprylic acid and myristic acid.

Nonpolar compounds including 5-nonatetracontanone, 2-hentricontanone, tricontane, 16-hentriacontanol, and β-sitosterol showing antibacterial activity have been isolated in *C. fistula* pods by Misra et al. (1996). The phytoconstituents such as proanthocyanidins containing flavan-3-ol (epiafzelechin and epicatechin) with an unusual 2S configuration isolated including flavan-3-ols and proanthocyanidins like catechin, epicatechin, procyanidin B-2, and epiafzelechin have been reported by Kashiwada et al. (1996) from pods. Misra et al. (1997) isolated a new diterpene, 3 beta-hydroxy-17-norpimar-8(9)-en-15-one. Meenarani (1998) characterized an anthraquinone derivative-3-formyl-1-hydroxy-8-methoxyanthraquinone-1.

Yueh-Hsiung et al. (2002) reported 5-(2-hydroxyphenoxymethyl) furfural, (22S)-7-hydroxy-5-hydroxymethyl-2-(22-hydroxypropyl) chromone, 5 hydroxy methylfurfural, benzyl 2-hydroxy-3,6-dimethoxybenzoate and benzyl 2β-*O*-D-glucopyranosyl-3,6-dimethoxybenzoate, together with four known compounds, (22 *S*)-7-hydroxy-2-(22-hydroxypropyl)-5-methylchromone and two oxyanthraquinones, chrysophanol and chrysophanein from the pods. Ali et al. (2003) isolated three lectins CSL-1, CSL-2, and CSL-3 from the *C. fistula* seeds and found that the sugar moity as galactose (CSL-1) and galactose-glucose (CSL-2) and galactose-mannose in CSL-3. Yadava and Verma (2003) isolated a new bioactive flavone glycoside from defatted seeds of *C. fistula,* that is, 5, 3′, 4′-trihydroxy-6-methoxy-7-*O*-alpha -L-rhamnopyranosyl-(1 -> 2)-*O*-β-D-galactopyranoside.

Agarwal and Paridhavi (2005), Nadkarni (2009) reported that fruit pulp consists of gum, gluten, sugar, gum, gluten, dying substance, astringents and soluble carbohydrates proteins, and amino acids such as arginine, leucine, methionine, phenylalanine, tryptophan, aspartic, and glutamic acids and a new dimeric proanthocyanidin CFI isolated along with (–) epiafzelechin, (+) catechin, kaempferol, dihydrokaempferol and 1, 8-dihydroxy-3-methylanthraquinone. Rastogi and Mehrotra (2004), Anonymous (2007) reported that the seed oil contains cyclopropenoid fatty acids, such as vernolic, malvalic, and sterculic acids. Anonymous (2007) reported that the pods of *C. fistula* contain 5-nonatetracontanone, 2-hentriacontanone. Two new aliphatic compounds, namely *cis*-heptacosanyl-5-hydroxypentadec-2-enoate 1 and octacosan-5, 8-diol 2 reported from *C. fistula* leaves (Singh et al., 2005). Agrawal and Paridhavi (2005) and Khare (2007) reported that pulp of the pod showed anthraquinone glycosides, sennosides A and B, rhein and its glucoside, barbaloin, aloin, formic acid, butyric acid, and their ethyl-esters and oxalic acid, tannins, and pectins. Sartorelli et al. (2009) characterized the active isoflavone biochanin A from the dichloromethane fraction of pods.

Twenty-seven compounds from the stem bark of *C. fistula* such as 1-hexacosanol, 1-octacosanol, palmitic acid, stearic acid, oleic acid, linoleic acid, heptacosyl eicosanate, glyceryl-1-tetraeicosanoate; three sterols namely β-sitosterol, stigmasterol, β-sitosteryl-3-*O* –D-glucopyranoside (long-chain hydrocarbons), lupeol (triterpene); chrysophanol; emodin, physcion, citreorosein, rhein, rhein-methyl-ester, ziganein, 1,4,5-trihydroxyanthraquinone (anthraquinones); isoscopoletin, scopoletin (coumarins); 2,5-dimethyl-7-hydroxychromone, 2,5-dimethyl-7-methoxychromone (chromones); isovanillic acid, vanillic acid, and 2,4-dihydroxybenzaldehyde (aromatic compounds), were isolated and identified by Lee et al. (2001). The phytochemical and biological studies of *Cassia* species that remain a potential source for new natural pharmacologically active components (Hakim et al., 2019). Isolation of phytochemicals such as Fistucacidin (3,4,7,8,4)-pentahydroxyflavan from heart wood, Oxyanthraquinone, dihydroxyanthraquinone from the bark, (-) epicatechin, (-) epiafzelechin, (−) epiafzelechin-3-oglucoside, procyanidin B2, bi/triflavonoids such as rhein, rhein-glucoside, sennoside-A/B, chrysophanol, physcion from the leaves, whereas kaempferol, leucopelargonidin tetramer [with free glycol unit], rhein, fistulin and the alkaloids or triterpenes identified from flowers, rhein, volatile oil, waxy, and resinous derivatives from the fruit-pulp, fistulic acid, 3-formyl-1-hydroxy-8-methoxy anthaquinone, 3B-hydroxy17-norpimar-8(9)-en-15-one from the pods, chrysophanol from the seeds, rhamnetin-3-O-gentiobioside from the roots. Whereas proanthocyanidins, certain flavonoids from the various parts were reported in *C. fistula* (Bahorun et al., 2004). Air-dried powdered leaves of *C. fistula* were extracted using methanol. From this extract, two compounds-(2′S)-7-hydroxy-5-hydroxymethyl-2-(2′-hydroxypropyl) chromone and benzyl-2-*O*-*β*-D-glucopyranosyl-3,6-dimethoxybenzoate were isolated (Nagpal et al., 2011). The fruit pulp decoction and its hydrolyzed solutions possess anthraquinone aglycones, which are effective against skin infective microbes especially on fungal strains. Moreover these extracts also demonstrated its stable effectiveness in different storage periods (Chewchinda et al., 2013). The qualitative phytochemical analysis and identification of groups of secondary metabolites were performed with leaves, flowers by Bahorun et al. (2005) and Selvi et al., (2014).

The fruit pulp of *C. fistula* analyzed by GC-MS revealed the presence of the 1,2,3,3a,4,5,6,7octahydro-1,4-dimethyl l–7-(1-methylethenyl), α-acorenol, spiro [5.5] undec-8-en-1-one, 16-nitrobicyclo [10.4.0] hexadecane-1-ol-13-one, 1a,2,5,5a,6,9,10,10a-octahydro-5,5a-dihydro, 1-acetyl-20á-hydroxy-16-methylene, 2,4,6-decatrienoic acid, 2-(3-acetoxy-4,4,14-trimethylandrost-8-en-17-yl), 2-nonanone, eicosanoic acid, 3-benzyloxy-2-fluoro-ß-hydroxy, 4-(2-propenyl), 4α-hydroxy-4-nitroperhydronaphthalen-1-one, azulene, benzyl benzoate, carda-4,20(22)-dienolide,3-[(6-deoxy-3-O-methyl-α-L-mannopyranosyl, caryophyllene, *cis*-13-eicosenoic acid, dasycarpidan -1-methanol, acetate (ester), D-glucose, 6-*O*-α-D-galactopyranosyl, eugenol, imidazole,2-amino-5-[(2-carboxy)vinyl], isoaromadendrene epoxide, N-[(3∞,5ß,12∞)-3,12-dihydroxy-24-oxocholan-24-yl]. n-isobutyl-(2E,4Z,8Z, 10E)-dodecatetra-enamide, oxacyclododecan-2-one, phenethylamine, phenylmethyl ester, phenol, propanoic acid, ß-copaene, strychane, tetraacetyl-d-xylonic nitrile, vitamin E, and glycine (Khadim et al., 2016). Further analyzed through FT-IR spectrum revealed the presence of different groups of secondary metabolites such as aliphatic fluoro compounds, alcohols, alkenes, carboxylic acids, ethers, esters, nitro compounds besides phenolics (Kadhim et al., 2016). The phytochemical analysis of the different parts of *C. fistula* (bark, seeds, pods, and leaves) revealed the presence of secondary metabolites, that is, alkaloids, tannins, anthraquinones, flavonoids, steroids, terpenoids, and glycosides besides providing the extractive values (Ishita and Punita, 2018).

The phytochemicals such as syringaresinol (lignan), quercetin-O-hexoside, apigenin-6,8-di-C-glycoside, kaempferol-rhamnosyl-xyloside, apigenin-6-C-pentoside-8-C-hexoside (isomer 1,2,3), apigenin-C-hexoside-O-pentoside, myricetin hexoside, proanthocyanidin B, a dimer (epi-catechin) (flavonoid), and 3,4-di-*O*-caffeoylquinic acid coumaric acid derivative (phenolic acid) were identified from the leaves (Castro-Lopez et al., 2018). The active ingredient in the herbal laxatives may be the presence of rhein used as phytochemical marker has been highlighted for quality analysis, extracted from fruit pulp through ultrasonic assisted extraction method and characterized by Yingngam et al. (2019). Cassioate (D, E, and F) were isolated and characterized as bioactive secondary metabolites against acetyl cholinesterase in bioassays as well as docking analysis (Aftab et al., 2020). The carbohydrate characterization in view of prebiotics in fruits were estimated and characterized by Dawood et al. (2021).

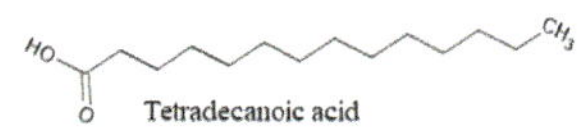

Tetradecanoic acid

Pentadecanoic acid, 13-methyl – methyl ester

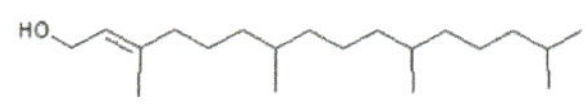

Phytol

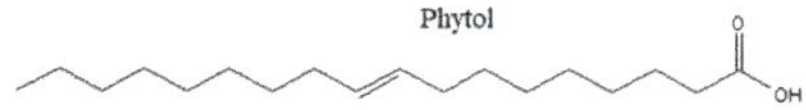

Octadec-9-enoic acid

Z-E-2Methyl, 3,13-Octadecadien-1-ol

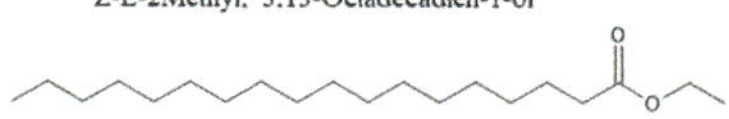

Isopropyl Stearate

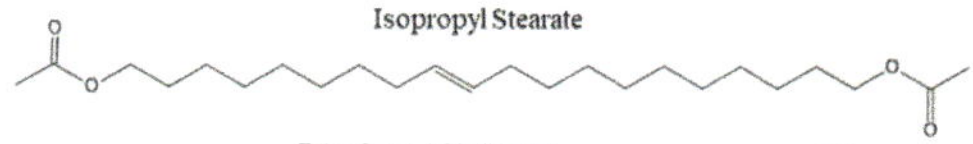

Ecios-9-ene-1,20-diacetate

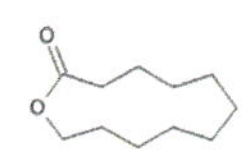

Oxycyclo dodecan-2-one

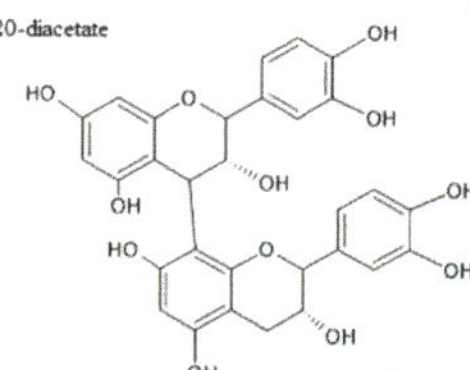

Procyanodin-B2

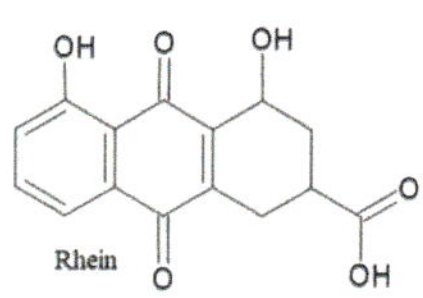

Rhein

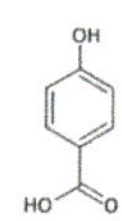

4-Hydroxy Benzoic acid

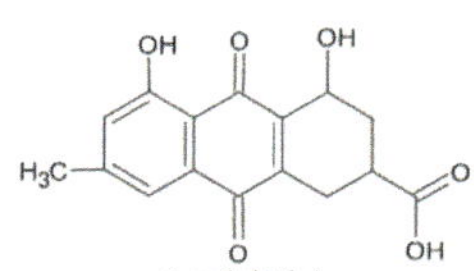

6-methyl Rhein

Eugenol

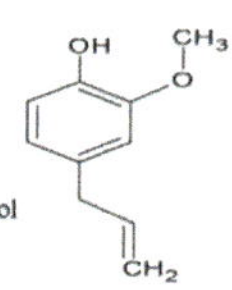

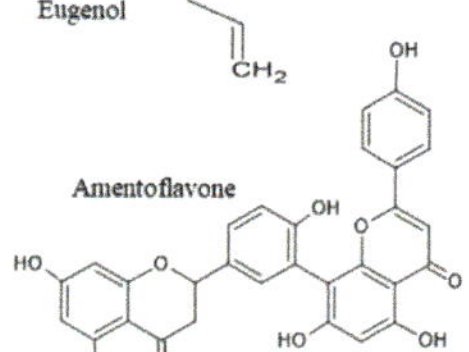

Amentoflavone

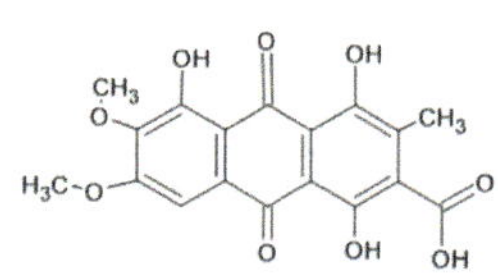

Fistulic acid

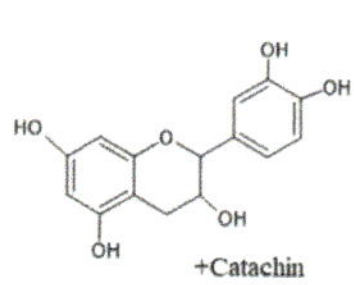

+Catachin

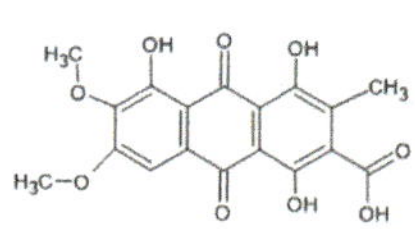

1-3-dihydroxy-2-methyl-5,6-dimethoxy anthraquinone

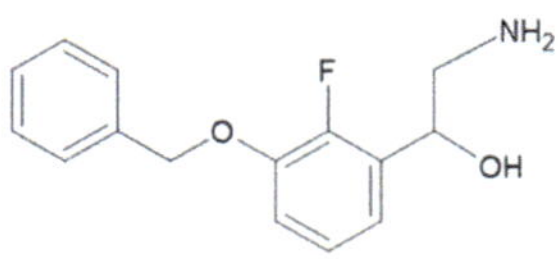

Phenethylamine,3-benzyloxy-2-flouro-β-hydroxy

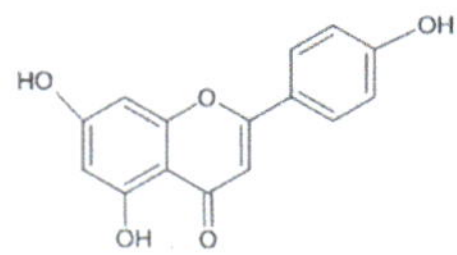

Kaempferol

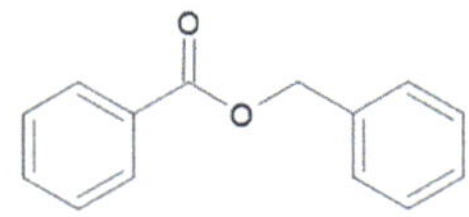

Benzyl benzoate

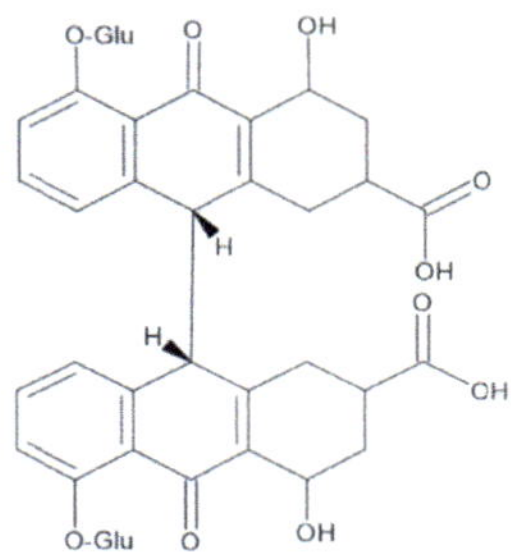

Sennoside-A/B

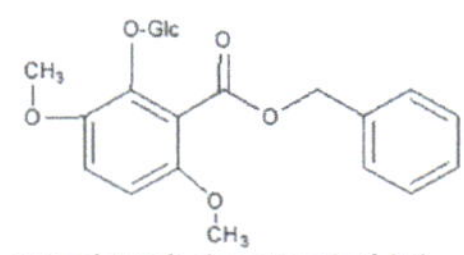

Benzyl-2-O-b-glucopyranosyl-3-6-dimethoxy benzoate

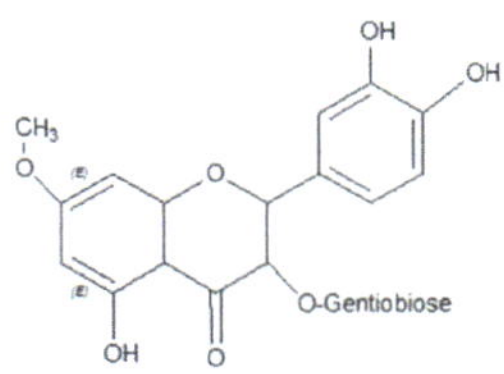

Rhametin 3-O-Gentiobiose

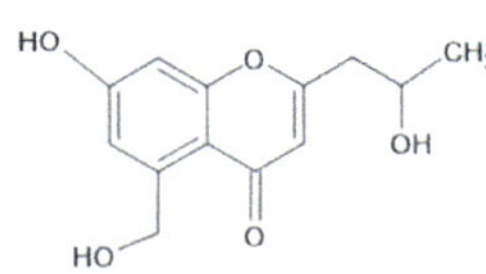

(2'S)-7-hydroxy-5-hydroxymethyl-2-(2'-hydroxypropyl) chromone

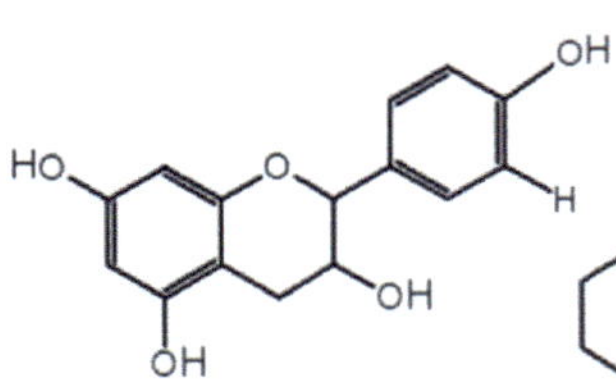

Epiafzelchin

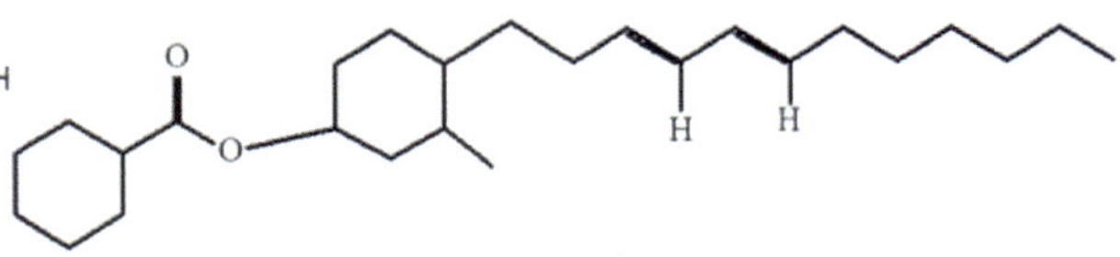

Cassioate D

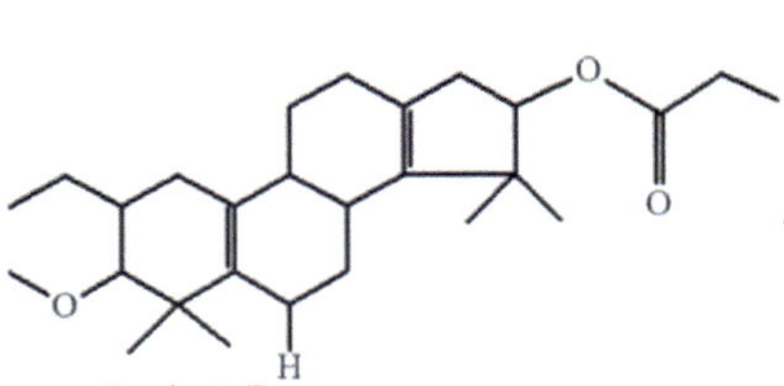

Cassioate E

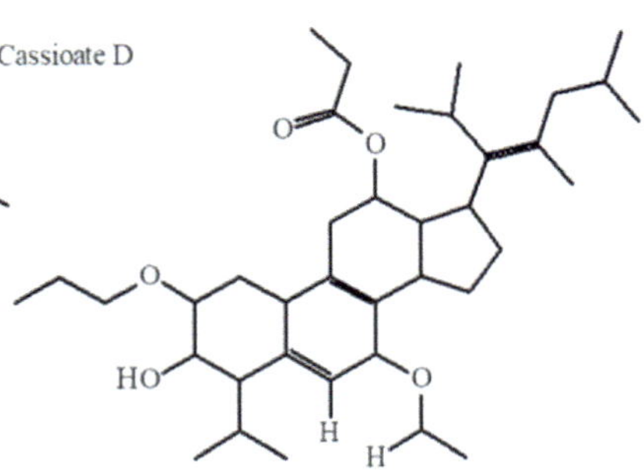

Cassioate F

12.3 PHARMACOLOGICAL ACTIVITIES

All parts of *C. fistula* such as leaves, bark, fruit, flowers, and seeds were effective in treating several diseases and established since ancient times. With the advancement in the scientific research there are many discoveries to find the leads from this plant to prove the claims like hepatoprotective, anti-inflammatory, antioxidant, antipyretic, antitussive activity, antiulcer, central nervous system activity, hypo-lipidaemia, larvicidal or ovicidal, laxative, leukotriene inhibition, anticancer, antidiabetic, antiepileptic, antifeedant, antifertility, urease inhibition, wound healing, antimicrobial, antifungal antibacterial, and antiviral activities (Gupta et al., 2000; Rizvi et al., 2009; Neelam et al., 2011; Rajagopal et al., 2013; Sanoria et al., 2020).

12.3.1 Antiviral Activity

The extracts obtained from the pods and leaves of *C. fistula* were tested and found effective on the infectious bovine rhinotracheitis (IBR) virus. Antiviral activity of *C. fistula* was assayed by using different concentrations of nontoxic doses on "IBR virus" in the "MDBK cell-line" system. Cytopathic effect was observed by microscopic examination and confirmed by using MTT dye uptake assay. The study revealed that the pod extract's significant effect depends on the extract dose, while the leaf extracts did not affect the proliferation of IBR virus (Sharma et al., 2010; 2021).

12.3.2 Antibacterial and Antifungal Activity

Patel and Patel (1956) have tested the antibacterial potential of *C. fistula* pulp and seed. The dealcoholized extract of the pulp showed greater effect on gram-positive bacteria than the aqueous extract, while in the case of Gram-negative bacteria both kind of extracts showed similar effect. The pattern of effect in the case of de-alcoholised seed extract on Gram-positive and Gram-negative bacteria was not different. However aqueous extract of seed was only effective against *Corynebacterium diphtheriae* and *Salmonella typhii.* Furthermore, it was observed that the action of dealcoholized extract of pulp was equivalent to about 2.5 units of penicillin sodium in the case of *Micrococci*, about 10 units for *Bacillus megaterium* and *Corynebacterium diphtheriae* and about 200 units of dihydrostreptomycine for *S. typhi* and *Escherichia coli.* The action of dealcoholized extract of seeds was equivalent to 0.5–1.5 unit of penicillin sodium against all gram-positive organisms and 100 unit of dihydro-streptomycin in case of *E. coli* and *S. typhi*. The aqueous extract of pulp showed an activity corresponding to 2.5 unit of penicillin sodium for *M. pyogenes*, 1.5 units for *M. citreus*, and *Corynebacterium diphtheriae*, 5–10 unit for *B. megaterium,* and more than 200 units of dihydro-streptomycin sulfate on *E. coli* and *S. typhi*. Aqueous, diethyl ether, ethyl acetate, dichloromethane, and methanol extract of leaves are reported to have significant zone of inhibition against *E. coli, Klebsiella aerogenes, Pemphigus vulgaris* and *Pemphigus aeruginosa* bacteria in concentration-dependent manner. The extract obtained from the dichloro-methane most effective on growth of *E. coli* as well as *P. vulgaris,* and methanolic extract on *P. aeruginosa,* while the water extract also exhibited significant antibacterial effect on the tested strains (Perumal et al., 1998).

The antibacterial effect of the organic solvent extracts of seeds, stem bark, and leaves against *Bacillus subtilis, B. megaterium, Staphylococcus haemolyticus, Streptococcus aureus, Streptococcus luteus, Streptococcus sonnei, E. coli, Klebsiella* sp., *Shigella boydii, Shigella flexneri, Shigella dysenteriae, S. typhi*, and *P. aeruginosa* revealed that the stem bark and pod were more active against all tested bacteria. Petroleum ether and ethyl acetate extracts of leaves showed activity against the tested bacteria but methanol extract was active only against on *B. megaterium, S. haemolyticus, S. typhi,* and *P. aeruginosa* (Abbas et al., 2004).

Samy et al. (1998), Phongpaichit et al. (2004), and Prashanth et al. (2006) reported antibacterial and antifungal activities through primary screening. The organic solvent extracts hexane, chloroform, ethyl acetate, methanol, and water extracts of the flowers significantly affected growth Gram-positive organisms with minimum inhibitory concentrations (MIC) ranged 0.078–2.5 mg/mL in the in vitro model, whereas *P. aeruginosa* is affected in Gram-negative strains. Among the tested Gram-negative bacteria, only *P. aeruginosa* was susceptible to the tested extracts. The ethyl acetate extract yielded, 4-hydroxy benzoic acid hydrate (showed antidermatophytic activity) demonstrated the antimycotic effect on *Trichophyton mentagrophytes* and *Epidermophyton floccosum* with MIC 0.5 mg/mL (Duraipandiyan and Ignacimuthu 2007). The antibacterial activity of flavone glycoside 5,3′,4′-trihydroxy-6-methoxy-7-*O*-alpha-L-rhamno-pyranosyl-(1->2)-*O*-beta-D-galacto-pyranoside characterized from *C. fistula* was effective moderately against bacterial and fungal strains, moreover the isolated compound was found significant on *A. niger* and *Fusarium oxysporum* (Yadava and Verma, 2003). The alcohol extract of *C. fistula* leaves exhibited antibacterial effect against *S. aureus* and *P. aeruginosa* and formulated ointment was topically applied on the infected wound and found effective on experimental rats (Senthilkumar et al., 2006).

The methanol extract of *C. fistula* seeds was effective on the tested micro-organism such as *Micrococcus* spp., *Salmonella* spp., *Staphylococcus* spp., *B. subtilis*, *B. thuringiensis* and *E. coli* with the MICs were determined in the range of 1.5–50.00 mg/mL in the in vitro studies (Lachumy et al., 2010). The protease inhibitor such as fistulin purified from the leaves demonstrated effective inhibition on the strains of *B. sublitis, S. aureus*, *E. coli,* and *K. pneumoniae* with dual role (Arulpandi and Sangeetha, 2012). The organic solvent extracts of pulp of pods showed effective inhibition on the broad species of microbes especially on *S. aureus, Streptococcus pyogenes, E. coli, P. aeruginosa,* and the mycotic strains such as *Aspergillus niger, Aspergillus clavatus* and *Candida albicans* in dose-dependent manner (Bhalodia et al., 2012).

The antifungal activity of decoction extracts of *C. fistula* pod pulp stored under varied storage settings and tested on the pathogenic fungal species such as *Trichophyton mentagrophytes* DMST 19735, *T. mentagrophytes* and *Microsporum gypseum* (dermatophytic fungi) was very poor with high MIC, while the hydrolyzed extracts were significantly effective as the standard compound rhein (Chewchinda et al., 2013). Senthil Kumar et al. (2014) demonstrated that the leaf alcoholic extract demonstrated in vivo effect when applied topically on the rat skin wounds, which are infected with *P. aeruginosa* and *S. aureus* strains.

Seyyednejad et al. (2014) reported that the effect of the extracts obtained from *C. fistula* flowers with methanol and ethanol on the growth of bacterial strains such as *S. epidermidis, S. aureus, Bacillus cereus, E. coli, P. aeruginosa, S. typhi, K. pneumoniae,* and *P. mirabilis* was studied, as well as the minimum inhibitory and bactericidal concentrations (MIC and MBC) and the activity has been attributed to the phenolic compounds in the extracts.

The extracts of *C. fistula* pods and leaves were analyzed with qualitative phytochemical studies and HPLC analysis. The effect of the extracts demonstrated significant effect on the tested antimicrobial strains such as *E. coli, S. aureus,* and *C. albicans*, *Pasteurella multocida*. Especially the leaf extracts were found significant effect against *C. albicans*. The microbiocidal property has been credited for the quercetin dihydrate, kaempferol in the extracts (Laxmi et al., 2015a). The leaf methanolic extract of *C. fistula,* reported for antifungal effect against *A. terreus* and antibacterial property on the eleven pathogenic bacterial strains (Kadhim et al., 2016). The inhibition of urease enzyme has been attributed to the leaf extracts antimicrobial effect on intestinal pathogens and the compound Mome-inositol, a derived sugar (80.83%) serves as a potent medicine for gastrointestinal infections (Sikri et al., 2019).

12.3.3 Effect on the Eukaryotic Parasites

12.3.3.1 Antiprotozoan Effect

The isoflavone, biochanin A demonstrates 50% effective concentration (EC_{50}) of 18.96 μg/mL against protozoan such as *Leishmania chagasi* and also effective on *Trypanosoma cruzi*, with EC_{50} of 18.32 μg/mL and a 2.4-fold effectiveness than standard benznidazole (Sartorelli et al., 2009).

12.3.4 Effect on Candidiosis

The fruit pulp and seed extracts showed growth inhibition of *C. albicans, Candida tropicalis,* and *Candida glabrata* and reduced ergosterol content in cell membrane of *C. albicans*, C. *tropicalis,* and *C. glabrata* using by fluconazole as standard (Irshad et al., 2011). The seed extracts demonstrated anticandidal effect using in vivo model with complete inhibition at 6.25 mg/mL (Jyothi et al., 2012). Seed oil of *C. fistula* was investigated for antifungal activities against the strains of *Candida* species. The seed oil analyzed through GC-MS and revealed the presence of beta-sitosterol, stigmasterol, ergosterol, betulinic acid, lupeol, fucosterol, alpha-amyrin, and friedelin. Tested oil inhibited by > or = 63.8% ergosterol bio-synthesis in *Candida* cell wall and fluconazole as standard (Irshad et al., 2013).

The antimycotic effect of extracts of leaves, bark, and seeds of C. *fistula* on the *Candida* strains (fluconazole resistant), *C. albicans*, *C. krusei*, *C. kefyr*, *C. glabrata*, *Candida parapsilosis,* and *C. tropicalis,* which are found in HIV patients. The ethanol extract of seed exhibited the high inhibition effect. Among the tested strains, *C. krusei* and *C. parapsilosis* were the most sensitive and *C. kefyr* was the resistant. The docking studies gallic acid of the seed, revealed that the excellent binding with lanosterol 14-α demethylase (Sony et al., 2018). Chaerunisa et al. (2020) demonstrated the alcoholic extract of stem bark on the resistant strains such as *S. dysenteriae* and *S. typhosa,* which is significant in in vitro as well as in vivo mouse models in the given dose.

12.3.5 Anthelmintic Activity

The pulp and seed extracts of *C. fistula* were found effective wormicidal property on the *Pheretima posthuma* checked with the standard Piparazine citrate, more over the seed extracts are significantly affected the worms (Irshad et al., 2010). The similar effect has been reported with the extracts of stem bark and pods and found ethanol extract as a potent inhibitor and suggested its usage. Ethanol extract of young bark and aqueous old bark extracts showed very good anthelmintic activity with respect to paralysis and death time, as compared to control and standard Albendazole (2 mg/mL) and piperazine citrate (1.5 mg/mL). Hence, extracts of bark and pod might be applied against the chronic infection caused by parasitic worms (Satpute et al., 2017).

12.3.6 Antipyretic Activity

Shade dried leaf ethanol extract of *C. fistula* effectively controlled the enhanced temperature in TAB vaccinated Wistar male albino rat effectively within 60 min with the treatment of 250 or 500 mg /kg BW whare as significantly reduced the temperature in 750 mg/kg BW treated animals in addition to anti-inflammatory effect, can be utilized effectively as a antipyretic medicament (Gobianand et al., 2010). Similarly the methanol extract of pods also demonstrated the effective controlling of induced pyrexia in rats (Patel et al., 1965; Singh et al., 2012).

12.3.7 Antioxidant Activity

The radical scavenging effect and lipid peroxidation effects of extracts of *C. fistula* among Sri Lankan plants were reported by Munasinghe et al. (2001). The antioxidant capacity of ethanol extracts (90%) of leaves, and methanol extracts (90%) of flowers, stem bark and pulp of *C. fistula* were investigated for the antioxidant activity and reported that stem bark extracts showed potent antioxidant capacity, which was correlated with poly phenolic content of the extracts. Similar type of results with stem bark extracts was reported by Siddhuraju et al. (2002).

The Trolox equivalent antioxidant capacity and ferric-reducing antioxidant power assays in the in vitro assays with different developmental stages of reproductive organs of *C. fistula* were strongly correlated with total phenols in all organs studied, and with proanthocyanidins. The extracts of the reproductive parts showed higher antioxidant potential than the vegetative organs. The extracts were rich in phenolic and flavonoids contents (Luximon-Ramma et al., 2002).

Antioxidant effects of water extract of *C. fistula* flowers (ACF) were assessed in alloxan-induced diabetic rats. It revealed that the extracts significantly enhanced lower levels of antioxidant enzymes in alloxan-induced animal models (Manonmani et al., 2005).

The antioxidant efficiency of EtOH extract of the leaves and methanolic extracts of the bark, pulp and flowers of *C. fistula* by using DPPH (1,1-Dipheny-2-picryl-hydrazyl) scavenging effect and subsequently investigated in alloxan-induced diabetic rats and reported significant reduction in peroxidation products and a prominent increase in the activities of the antioxidant enzymes such as superoxide dismutase, catalase, glutathione peroxidase, glutathione reductase an glutathione (Anonymous 2007). The extracts exhibited antioxidant activity in DPPH nitric oxide and hydroxyl radical-induced in vitro assay methods (Anonymous, 2008). The quantitative estimation of phenolics, tannins, and antioxidant activity of *C. fistula* extracts of leaf, stem, bark, and root collected at different stages were analyzed and discussed their antioxidant effects using in vitro studies (Lai and Liew, 2013).

Qualitative and quantitative phytochemical analysis and in vitro antioxidant capacity of the hydroalcoholic extract of fruit pulp revealed the presence of different types of secondary and good amount of phenolic components. The extract showed very feeble DPPH quenching activity (Bhalodia et al., 2013). Further many scientists reported that the extracts of *C. fistula* for in vitro or in vivo antioxidant studies using the radicals such as DPPH, hydroxyl radical, lipid peroxidation and generation of radicals in kidney and liver tissue homogenates, lipid peroxides test using ferric thiosynate method using Ascorbic acid, BHT or certain phenolic compounds as standards. The results indicated that different parts of *C. fistula* extracts have the capacity to act as radical scavenger and correlated with the presence of various phenolics or flavonoids in the tested samples (Akinyede and Amoo, 2009; Arya et al., 2011; Noorhajati et al., 2012; Jose and Reddy, 2013; Wankhade et al., 2014).

The total phenolic compounds, carotenoids, tocopherol and fatty oils and DPPH radical scavenging properties reported by Wati and Khabiruddin (2017). The antioxidant effect of aerial parts *C. fistula* was performed by the in vitro DPPH and CUPRAC assays and the tested extracts showed close antioxidant effect to that of BHA standard in CUPRAC assay and a relatively similar effect to that of EGCG and vitamin C in DPPH assay (Thabit et al., 2018).

12.3.8 Antidiabetic Activity

The antihyperglycemic effect of *C. fistula* leaves extracts was assessed in STZ-induced diabetic rat models. The extracts considerably decreased the levels of blood glucose, glycosylated haemoglobin and enhanced the status of total haemoglobin, plasma insulin, and liver glycogen as well as corrected the abnormalities in the activities of carbohydrate metabolizing enzymes in diabetic rats (Vasudevan et al., 2008).

Oral administration of crude ethanolic and ethyl acetate fraction obtained from *C. fistula* stem bark was investigated in alloxan-induced rat models. It revealed that the extract and fraction exhibited notable reduction in blood glucose levels and restored lipids in blood to normal levels. Glibenclamide was used as standard drug (Malpani et al., 2010). Antidiabetic efficiency of methanol extract and catechin isolated from methanol extract by bioassay-guided fractionation method was evaluated in STZ-challenged rat models. Catechin at tested concentrations markedly increased tissue glycogen and altered glycogen synthase, glucokinase, glycogen phosphorylase, and glucose-6-phosphate to normal levels. It further enhanced the protein expression levels of GLUT4 (Daisy et al., 2010). The antihyperglyceic effect of ETOH extracts of Golden-shower flowers was demonstrated in rats and the activity has been attributed to presence of phytoconstituents, which protects from oxidative damage in the STZ-induced diabetes (Jeyanthi, 2012).

Catachin, one of the important polyphenol of Golden-shower, demonstrated effective on glucose tolerance in male Wistar albino rats has been coincided with the molecular drug docking with structure-property interactions and also been demonstrated the potential agonist to PPARγ and insulin receptor, which clearly indicates its antihyper glyceamic effect (Pitchai and Manikkam, 2012). Similarly, the treatment with the gold nanoparticles of the flower extracts exhibited improvement in the diabetic cascade when tested with streptozotocin-induced rats, compared to the extracts, has been found beneficial in diabetes (Daisy and Saipriya 2012).

The ethanolic extract of *C. fistula* pod was evaluated for antidiabetic activities in streptozotocin-induced diabetic male Wistar rats. The diabetic rats were treated orally with the extracts at three different concentrations for a period of 2 months. Glibenclamide was used as standard drug control (5 mg/kg b. wt./day) in treated rats. The extracts showed significant reduction in the blood glucose and increased in the body weight and glycogen levels in liver tissues as compared to diabetic control rats, they also showed improvement of oral glucose levels in diabetic rats (Ashraf et al., 2012).

Methanolic extracts of *C. fistula* (bark and leaves) were administered to diabetic rats (induced by streptozotocin-nictotinamide) at 250 and 500 mg/kg doses for 21 days. The effectively reduced streptozotocin-nictotinamide-induced increased levels of glycosylated haemoglobin, reduced levels of plasma insulin. Histopathological studies revealed that the extracts strongly reduced chronic diabetic conditions in selected tissues (Einstein et al., 2013).

Three modes of extracts, that is, ethanol, water-soluble ethanol fraction, and petroleum ether of *C. fistula* showed significant antihyperglycemic activity at the given doses in alloxan-induced diabetic rats. In conclusion, the water-soluble fraction of ethanol extract was found to be exhibited strong activity than ethanol and glibenclamide (Jarald et al., 2013). Further, anti-diabetic study was conducted using streptozotocin antidiabetic model with nontoxic concentration of the extract (toxic above 2000 mg/kg b.wt.). The results confirmed that the tested extract has significant antidiabetic activity (Akhila and Aleykutty, 2015).

The solvent extracts of *C. fistula* roots analyzed for their antidiabetic activity using in vitro studies such as alpha-amylase inhibition and glucose diffusion assays. The results suggested that the ethanolic extract was significant in inhibiting the activity of alpha amylase with the value of 1200 µg/mL, glucose diffusion assay proved that the ethanolic extract has efficient in its antidiabetic potential (Balraj et al., 2016).

Jangir and Jain (2017) evaluated antidiabetic activities of hydro-alcoholic (70%) extract prepared from *C. fistula* pod in streptozotocin-induced diabetic male rats were administered orally with extract at three different doses (100, 250, and 500 mg/kg b. wt./day). After 2 months experimental schedule, the extract showed potent antidiabetic activity by reducing STZ-induced higher levels of diabetic profile.

12.3.9 Anti-Inflammatory Activity

The aqueous fraction of *C. fistula* exhibited significant inhibition of blood glucose levels in diabetic rats and correlated with the folklore uses, which is not coincided with *Cajanus cajan* extracts in the study (Esposito et al., 1991). The anti-inflammatory activity of the methanol and water extracts of *C. fistula* bark was assayed in Wistar albino rats. Both the extracts exhibited significant anti-inflammatory effects in carrageenan-induced oedema and cotton pellet-induced granuloma (Navanath et al., 2009). Anwikar and Bhitre (2010) studied on the *Solanum xanthocarpum* and *C. fistula* for the anti-inflammatory activity both alone and in combination. Both the extracts showed maximum anti-inflammatory activity at 500 mg/kg dose. Among the different dose combinations of both the extracts at 1:1 ratio showed maximum inhibition as >80% inhibition. The leaf extract of *C. fistula* exhibited significant antilipoxygenase (5-lipoxygenase (5-LOX, 12-lipoxygenase (12-LOX) using in vitro assay proved its effectiveness with a 50% inhibitory concentration (IC_{50}) of 6.23 mg/mL obtained for the 5-LOX assay and an IC_{50} of 3.22 mg/mL (Gopi and Manju, 2018).

Anti-inflammatory efficiency of rhein was evaluated in rat and mice models using croton oil/carrageenan-induced oedema of ear and hind paw, cotton pellet/acetic acid-induced granuloma, and vascular permeability models at different concentrations. The component exhibited notable anti-inflammatory effect in a concentration-dependent manner in both rat and mice models in the tested models. Further, it enhanced antioxidant enzymes such as SOD, CAT, and GSH-px and attenuated higher levels of inflammatory markers like IL-6, TNF-α, IL-1β, MDA, and VEGF. Western blot analysis revealed that the component diminished inflammatory markers, that is, COX-2 and iNOS, and enhanced levels of anti-inflammatory proteins such as HO-1, Nrf2, PPAR, and HSP-72 (Antonisamy et al., 2019).

12.3.10 Oxidative Stress and Diabetic Conditions

The alcoholic extracts of *C. fistula* stem bark were evaluated for antidiabetic effect and renal failure in animal models and antioxidant capacity in in vitro models. The extract significantly reduced glucose concentrations in blood samples and lipid markers in serum. Further, the extract strongly inhibited free radicals generated through in vitro antioxidant models (Agnihotri and Singh, 2013).

12.3.11 Hepatoprotective Activity

Hepatoprotective activity of methanol extract prepared from fruits of *C. fistula* was assessed in in vitro models in bovine polymorphonuclear leukocytes and bovine brain phospholipid liposomes. The extract strongly inhibited 5-lipoxygenase enzyme in bovine polymorphonuclear leukocytes and lipid peroxidation in bovine brain phospholipid liposomes (Kumar and Muller, 1998). The aqueous extract of *C. fistula* bark was evaluated for hepatoprotective activity against CCl_4 (carbon tetrachloride)-induced liver damage at two doses (250 and 500 mg/kg) and evaluated the biochemical markers of serum, protein levels in liver. They concluded that the extract showed significant protective effect against CCl_4-induced liver damage (Bhakta et al., 1999).

Hepatoprotective activity of *C. fistula* leaves n-heptane extract was assessed against paracetamol-induced hepatotoxicity in rat models. Oral administration of the extract at 400 mg/kg b.wt. significantly lowered the serum biochemical markers such as SGPT, SGOT and ALP (Bhakta et al., 2001). In another experiment aqueous extract of leaves and bark of *C. fistula* significantly reduced CCl_4-induced elevated levels of plasma markers (Wasu et al., 2009). While Chaudhari et al. (2009) assesses the hepatoprotective effect of methanolic extract of *C. fistula* at two doses (200 and 400 mg/kg). The extract significantly reversed back the altered level of biochemical markers to the near normal levels in the dose-dependent manner (Chaudhari et al., 2009). Fruit extracts (hydro-alcoholic) of *C. fistula* along with bromobenzene (induces liver injury) improved the biochemical parameters including glutamyl transpeptidase and normalised to native levels in mice has found effective hepatoprotective medicament (Kalantari et al., 2011).

Hepatoprotective effect of *C. fistula* leaf ethanolic extract was assessed in diethyl-nitrosamine-induced liver injury in rats for a period of 1 month. The extract significantly altered elevated levels of biochemical markers. It further enhanced antioxidant enzyme levels in treated animal liver tissues (Pradeep et al., 2007). In another study, hepatoprotective activity of ethyl acetate fraction from *C. fistula* leaves was evaluated in thioacetamide-induced liver damage. Results showed that the fraction significantly inhibited serum glutamic oxaloacetic transaminase and alkaline phosphatase. It further downregulated the expression levels of p-Akt, p-PI3K and p-mTOR. Chromatographic analysis of the fraction revealed the presence of phenolic components such as epicatechin, chlorogenic acid and catechin (Kaur et al., 2019).

12.3.12 Hypocholesterolaemic Effect

The hypocholesterolaemic effect of *C. fistula* (roasted fruit decoction) was investigated using hypercholesterolaemic male albino rats, which showed lower levels of total lipids in blood and liver tissues. Reduced levels of total cholesterol concentration in blood, kidney, spleen, and brain tissues were observed in extract treated animal group (El-Saadany, 1991).

12.3.13 Nephroprotective Role

The hydro-alcoholic extract of *C. fistula* fruit showed protective effect on the kidney against bromobenzene-induced toxicity. In the tested mice, the blood samples were evaluated and assessed blood urea nitrogen and creatinine levels, and kidneys in addition to histological examination found that bromobenzene induced significant nephrotoxicity reflected by an increase in levels of blood urea nitrogen and creatinine and these were prevented by the test extract indicating its significant protective role against nephrotoxicity (Heibatullah et al., 2015).

12.3.14 Neuroprotective Effects

The neuroprotective effects of a hydroalcoholic extract of the aerial parts of *C. fistula* in the *Caenorhabditis elegans* (a nematode) model was evaluated. The extract significantly reduced death rate of nematodes and accumulation of free radicals (Thabit et al., 2018).

12.3.15 Anticancer Activity

Anticancer effects of *C. fistula* seeds methanolic extract were investigated on the growth of Ehrlich ascites carcinoma and on the life span of mice with tumor. The extract significantly reduced tumor volume in mice and enhanced life span of mice. Further, the extract showed notable improvement in blood parameters such as bone marrow cell count, haemoglobin, and RBC count (Gupta et al., 2000). The extracts from flower and the isolated compound, rhein (flavonoid) showed anticancer effect on COLO 320 D revealed its potentiality (Duraipandiyan et al., 2012).

The anticancer efficiency of seeds and pulp extracts of fruits was assessed in SiHa and MCF-7 cell lines. The results revealed that the tested extracts significantly enhanced the activities of 3,7,9, and 10 types of caspases, upregulated p53 and Bax genes and down regulated Bcl-2 gene (Irshad et al., 2014). Al-fatlawi et al. (2014) studied anticancer potential of rhein in SiHa, HepG2, and MCF-7 cell lines. The synthesized AgNPs showed effective cytotoxic potential against MCF7 and the inhibitory concentration (IC_{50}) at 7.19 μg/mL. The apoptotic effects of the AgNPs were also confirmed by AO/EB staining. The investigation presents preliminary evidence that biosynthesized AgNPs have potential role as novel anticancer medicaments (Remya et al., 2015).

The whole plant extract was tested on the Human prostate cancer cell line and found significant inhibition and also induced the activities of caspase-3, -7, -9, and -10; these enzymes are crucial for apoptosis and they are quantified and compared with untreated control. The activity of these enzymes was increased from twofold to fivefold. The genomic DNA fragmentation was observed in extract treated cancer cells. Authors concluded that the methanol extract of *C. fistula* having anticancer agents and they showed anticancer property in MTT assay confirmed by acridine orange test. The extract inhibited the cell growth and induced the cell death by modulating caspase enzymes and cleaving genomic DNA (Kulkarni et al., 2015). The flavonoid namely Amentoflavone from the leaves exhibited moderate proliferation of HepG2 (hepatocarcinoma) cell line besides antioxidant properity (Srividhya et al., 2017).

The efficacy of anthraquinone derivatives was proved to the cytotoxicity on carcinomic human alveolar basal epithelial cell (A549), human leukaemia (NB4), human breast adenocarcinoma (MCF7), human neuroblastoma cell (SHSY5Y) and human prostate cancer cell from the stems of *C. fistula* (Zhou et al., 2017). Further the anticancer studies on the osteosarcoma (MG-63), neuroblastoma (IMR-32), and prostate adenocarcinoma (PC-3) demonstrated the apoptosis and antiproliferation effectively and yielded epiafzelchin from the leaves of golden-shower besides antioxidant activity (Kaur et al., 2020).

12.3.16 Immunomodulatory Effect

Ali et al. (2008) and Jadhav (2014) demonstrated the immunomodulatory effect of *C. fistula* in rats. The enhancement of humoral immune response demonstrated that the water extract of leaves and pods showed the stimulation of B-lymphocytes, T-helper cells, and macrophages subsets attributed to the presence of Quercetin dihydrate and kaempferol indicating its immunomodulatory activity (Laxmi et al., 2015b).

12.3.17 Toxicity Studies

The fruit ethanol extract of golden-shower studied for the toxic effects and found as nontoxic on the Wistar male rats at the different oral doses up to 5 g/kg bw for 14 days. The histopathological and biochemical studies supported the no detrimental effect (Abid and Mahmood, 2019). Jyothi et al. (2011) reported that the seed extracts are nontoxic and safe at the dose of 5 g/kg in the Wistar rats for 28 days. The bark extracts were also demonstrated and found nontoxic with the LD_{50} values in Wistar male (14.52 g/kg bw) and female (16.14 g/kg bw) rats in histopathology or biochemical parameters studied by Chaerunisa et al. (2018) including hepatoprotective and antioxidant effects.

12.3.18 Antifertility Effect

The aqueous extract of seed when given orally (100 or 200 mg/kg bw) in mated female rats in and found prevention of probability of pregnancy prevented by 57.14 or 71.43% whereas it is 100% when 500 mg/kg bw (Yadav and Jain, 1999). The pet.ether extract of seeds when fed to fertile female rats (albino Wistar) in 100–500 mg/kg bw daily showed decline effect of fertility index, implantation, or live foetus rate for 15 days (Yadav and Jain, 2009).

12.3.19 Wound Healing Property

The healing of wound has been an impact of methanol extract in the form of an ointment was demonstrated in wound models such as excision and inclusion in rats, and the ointment at two different doses exhibited efficient wound healing (Bhakta et al., 1998a). When a hydrocarbon ointment containing 10% w/w *C. fistula* foliage extract was administered topically to albino rat in infected wounds, it enhanced the healing effect (Senthilkumar et al., 2006). The fruit extract prepared as an emulsion in almonds oil, showed prominent inhibition of *P. vulgaris* erosions and the effect was attributed to the occurrence of secondary metabolites such as lupeol, rhein, and other flavonoids. The study demonstrated antioxidant, wound-healing, antibacterial, antifungal effects in addition to anti-inflammatory profile (Atarzadeh et al., 2017, 2018).

12.3.20 Antiaging and Tyrosinase Inhibition Studies

The butanol extract of flowers of golden shower exhibited effective anti-tyrosinase activity in addition to inhibition of collagenase/gelatinase might be enhancing the aging found beneficial in the development of antiaging cosmetic preparations for the delaying of hyperpigmentation and wrinkle of skin (Limtrakul et al., 2016).

12.3.21 Antitussive Activity

The methanol extract of leaves of golden shower were effective on the sulfur dioxide-induced cough model in mice in different doses as compared with standard (codeine) might be used as an antitussive agent (Bhakta et al., 1998b).

12.3.22 Laxative Activity

Sun dried (SD) and nonsun dried (NSD) fruit pulp were tested for purgative effect in rats and acute toxicity in mice. SD and NSD aqueous suspensions were given orally 60 min before the experiment in rats, and SD shortly before the toxicity trial in mice. SD and NSD at 1.0 g/kg doses both increased the number of defecations and faecal production 4 h after treatment. Both SD and NSD-treated rats demonstrated an increase in intestinal intraluminal fluid (ILF) accumulation and motility, although the SD group's ILF accumulation was less pronounced than the NSD group's. SD's stimulatory impact on ILF accumulation and intestinal motility might be attributed to its primary influence on nitric oxide production, since only L-NAME, a NOS inhibitor, inhibited both ILF accumulation and intestinal motility (Agarwal et al., 2012). The emulsion of extracts in mineral oil demonstrated its effectiveness in pediatric constipation and found improved defecation (Mozaffarpur et al., 2012).

12.3.23 Insecticidal Activity

The larvicidal effect of the fraction of flower extract on *Aedes albopictus, Culex tritaeniorhynchus,* and *Anopheles subpictus* demonstrated as potential in the controlling larvae (Govindarajan, 2001), in addition the leaf extracts demonstrated as effective ovicidal property (Govindarajan et al., 2008; Mehmood et al., 2014).

12.3.24 Antileishmaniasis Activity

The fruit extracts were demonstrated as effective for the leishmaniasis, as compared to the Glucantime (Abu et al., 1999) and further fractionation yielded clerosterol that is demonstrated as nontoxic and more effective than the Pentamidine (Sartorelli et al., 2007). The effective growth inhibition on the *Leishmania donovani* along with the intra-macrophagic amastigotes during the study of antileishmanial activity of *C. fistula* leaf methanol extract and proved in simulation models of molecular docking studies by Tabrez et al. (2021).

KEYWORDS

- ***Cassia fistula***
- **golden shower**
- **Indian laburnaum**
- **anti-inflammatory**
- **fistulin**
- **rhein**
- **cassiaotes**
- **chrysophanol**

REFERENCES

Abbas, A. M.; Abu-Sayeed, M.; Bhuiyan, M. S. A.; Sohel, F. I.; Sarmina, Y. Antimicrobial Screening of *Cassia fistula* and *Mesua ferrea*. *J. Med. Sci.* **2004,** *4*, 24–29.

Abid, R.; Mahmood, R. Acute and Sub-Acute Oral Toxicity of Ethanol Extract of *Cassia fistula* fruit in Male Rats. *Avicenna J. Phytomed.* **2019,** *9*, 117–125.

Abu, S. M.; Abbas, A. M.; Astaq, M. K. G. R. M.; Rahman, M. S. Studies on the Characterization and Glyceride Composition of *Cassia fistula* Seed Oil. *Bangladesh J. Sci. Indust. Res.* **1999,** *34*, 144–148.

Aftab, Z.; Bushra, K. H.; Khan, D. F.; Khan, A. Ullah, H.; Shahnaz; Khan, S. Three New Cholinesterase Inhibitory Cassioates from *Cassia fistula*. *Pharm Chem J.* **2020,** *53*, 1069–1075. https://doi.org/10.1007/s11094-020-02124-y.

Agarwal, S. S.; Paridhavi, M. *Clinically Useful Herbal Drugs*; Ahuja Publishing House, 2005; pp 281–282.

Agnihotri, A.; Singh, V. Effect of *Tamarindus indica* Linn. and *Cassia fistula* Linn. Stem Bark Extracts on Oxidative Stress and Diabetic Conditions. *Acta. Pol. Pharm.* **2013,** *70* (6), 1011–1019.

Agrawal, K.; Shivani, G.; Gautam, M. K.; Joshi, V. K.; Goel, R. K. Studies on Laxative Effect of Extract of Dried Fruit Pulp of *Cassia fistula*. *J. Natural Remedies* **2012,** *12* (2), 119–128.

Akhila, S.; Aleykutty, N. A. Antidiabetic Activity Studies on *Cassia fistula* Fruits. *Adv. J. Pharm. Life Sci. Res.* **2015,** *3*, 1–8.

Akinyede, A. I.; Amoo, I. A. Chemical and Functional Properties of Full Fat and Defatted *Cassia fistula* Seed Flours. *Pak. J. Nutr.* **2009,** *8* (6), 765–769.

Al-fatlawi, A. A.; Al-fatlawi, A. A.; Zafaryab, M.; Irshad, M.; Ahmad, I.; Kazim, Z. et al. Rhein Induced Cell Death and Apoptosis Through Caspase Dependent and Associated with Modulation of p53, bcl-2/bax Ratio in Human Cell Lines. *Int. J. Pharm. Pharm. Sci.* **2014,** *6*, 515–519.

Ali, M. A.; Sayeed, M. A.; Nurul, A. Antibacterial Activity and Cytotoxicity of Three Lectins Purified from *Cassia fistula* Linn. Seeds. *J. Medical Sci.* **2003,** *3* (3), 240–244.

Ali, N. H.; Kazmi, S. U.; Faizi, S. Modulation of Humoral Immunity by *Cassia fistula* and Amoxy-Cassia. *Pak. J. Pharm. Sci.* **2008,** *21* (1), 21–23.

Anonymous. *Australian Biological Resources Study. Flora of Australia Online*; Australian Biological Resources Study: Canberra, 2008.

Anonymous. *The Wealth of India*, First Supplement Series (Raw Materials), National Institute of Science Communication and Information Resources. CSIR, **2007,** *3*, 340–342.

Antonisamy, P.; Agastian, P.; Kang, C. W.; Kim, N. S.; Kim, J. H. Anti-Inflammatory Activity of Rhein Isolated from the Flowers of *Cassia fistula* L. and Possible Underlying Mechanisms. *Saudi J. Biol. Sci.* **2019,** *26* (1), 96–104.

Anwikar, S.; Bhitre, M. Study of the Synergistic Anti-Inflammatory Activity of *Solanum xanthocarpum* Schrad and Wendl and *Cassia fistula* Linn. *Int. J. Ayu. Res.* **2010,** *1* (3), 167–171.

Arulpandi, I.; Sangeetha, R. Antibacterial Activity of Fistulin: A Protease Inhibitor Purified from the Leaves of *Cassia fistula*. *ISRN Pharm.* **2012,** 584073.

Arya, V.; Bhardwaj, A.; Sharma, V. Pharmacology of Some Antioxidant Plants from District Kangra Himachal Pradesh, a Review. *Intern. J. Curr. Pharma. Res.* **2011,** *3* (2), 26–31.

Ashraf, A. M.; Sagar, H. A.; Sultana, M. C.; Azad, A. K.; Begum, K.; Wahed, M. Antihyperglycemic and Analgesic Activities of Ethanolic Extract of *Cassia fistula* (L.) Stem Bark. *Int. J. Pharm. Sci. Res.* **2012,** *3*, 416–423.

Atarzadeh, F.; Kamalinejad, M.; Amin, G. et al. Topical Application of *Cassia fistula* L. Fruit Gel in Management of Cutaneous Lesions of *Pemphigus vulgaris*: A Double-Blind, Placebo-Controlled Clinical Trial. *Avicenna J. Phytomed.* **2018,** *8* (6), 543–551.

Atarzadeh, F.; Kamalinejad, M.; Dastgheib, L.; Amin, G.; Jaladat, A. M.; Nimrouzi, M. *Cassia fistula*: A Remedy from Traditional Persian Medicine for Treatment of Cutaneous Lesions of *Pemphigus vulgaris*. *Avicenna J. Phytomed.* **2017,** *7* (2), 107–115.

Bahorun, T.; Neergheen, V. S.; Aruoma, O. L. Phytochemical Constituents of *Cassia fistula*. *Afr. J. Food Agric. Nutr. Dev.* **2004,** *4*, 1530–1540.

Bahorun, T.; Vidushi, S. N.; Okezie, I. A. Phytochemical Constituents of *Cassia fistula*. *African J. Biotechnol.* **2005,** 4 (13), 1530–1540.

Balraj, S.; Indumathy, R.; Jayshree, N.; Abirami, M. S. Evaluation of In Vitro Anti-Diabetic Activity of Various Root Extract of *Cassia fistula* L. *Imperial J. Interdisc. Res.* **2016,** *2* (6), 758- 761.

Barthakur, N. N.; Arnold, N. P.; Alli, I. The Indian Laburnum (*Cassia fistula* L.) Fruit: An Analysis of Its Chemical Constituents. *Plant Foods Human Nutr.* **1995,** *47* (1), 55–62.

Bhakta, T.; Banerjee, S.; Mandal, S. C.; Maity, T. K.; Saha, B. P.; Pal, M. Hepatoprotective Activity of *Cassia fistula* Leaf Extract. *Phytomedicine* **2001,** *8* (3), 220–224.

Bhakta, T.; Mukherjee, P. K.; Mukherjee, K.; Banerjee.; S.; Subhash.; C. M.; Maity, T. K.; Pal, M.; Saha, B. P. Evaluation of Hepatoprotective Activity of *Cassia fistula* Leaf Extract. *J. Ethnopharmacol.* **1999,** *66* (3), 277–282.

Bhakta, T.; Mukherjee, P. K.; Mukherjee, K.; Pal, M.; Saha, B. P. Studies on In Vivo Wound Healing Activity of *Cassia fistula* Linn. Leaves (Leguminosae) in Rats. *Nat. Prod. Sci.* **1998a,** *4*, 84–87.

Bhakta, T.; Mukherjee, P. K.; Pal, M.; Saha, B. P.; Pal, M. Studies on Antitussive Activity of *Cassia fistula* (Leguminosae) Leaf Extract. *Pharmaceut. Biol.* **1998b,** *36*, 140–143.

Bhalerao, S.; Kelkar, T. Traditional Medicinal Uses, Phytochemical Profile and Pharmacological Activities of *Cassia fistula* Linn. *Biological Sci.* **2012,** *1*, 79–84.

Bhalodia, N. R.; Nariya, P. B.; Acharya, R. N.; Shukla, V. J. In Vitro Antibacterial and Antifungal Activities of *Cassia fistula* Linn. Fruit Pulp Extracts. *Ayu* **2012,** *33*, 123.

Bhalodia, N. R.; Nariya, P. B.; Acharya, R. N.; Shukla, V. J. In Vitro Antioxidant Activity of Hydro Alcoholic Extract from the Fruit Pulp of *Cassia fistula* Linn. *Ayu.* **2013,** *34* (2), 209–214.

Castro-Lopez, C.; Rojas, R.; Martínez-Avila, G. C. G. Screening of the *Cassia fistula* Phytochemical Constituents by UPLC-ESI-QTOF-MS. *Clin. Oncol.* **2018,** *3*, 1477.

Chaerunisa, A. Y. Ramadhani, F. N.; Nurani, T. D.; Najihudin, A.; Susilawati, Y.; Subarnas, A. Hepatoprotective and Antioxidant Activity of the Ethanol Extract of *Cassia fistula* L. Barks. *J. Pharm. Sci. Res.* **2018,** *10* (6), 1415–1417.

Chaerunisa, A. Y.; Susilawati, Y.; Muhaimin, M.; Milanda, T.; Hendriani, R.; Subarnas, A. Antibacterial Activity and Subchronic Toxicity of *Cassia fistula* L. Barks in Rats. *Toxicol. Rep.* **2020,** *7*, 649–657.

Chaminda, T.; Munasinghe, T. C.; Seneviratne, C. K.; Thabrew, M. I.; Abeysekera, A. M. Antiradical and Antilipoperoxidative Effects of Some Plant Extracts Used by Sri Lankan Traditional Medical Practitioners for Cardio-Protection. *Phytother. Res.* **2001,** *15*, 519–523.

Chaudhari, N. B.; Chittam, K. P.; Patil, V. R. Hepatoprotective Activity of *Cassia fistula* Seeds Against Paracetamol Induced Hepatic Injury in Rats. *Arch. Pharm. Sci. Res.* **2009,** *1* (2), 218–221.

Chewchinda, S.; Wuthi-udomLert, M.; Gritsanapan, W. HPLC Quantitative Analysis of Rhein and Antidermatophytic Activity of *Cassia fistula* Pod Pulp Extracts of Various Storage Conditions. *Biomed. Res. Int.* **2013,** 821295.

Daisy, P.; Balasubramanian, K.; Rajalakshmi, M.; Eliza, J.; Selvaraj, J. Insulin Mimetic Impact of Catechin Isolated from *Cassia fistula* on the Glucose Oxidation and Molecular Mechanisms of Glucose Uptake on Streptozotocin-Induced Diabetic Wistar Rats. *Phytomedicine* **2010,** *17* (1), 28–36.

Daisy, P.; Saipriya, K. Biochemical analysis of *Cassia fistula* Aqueous Extract and Phytochemically Synthesized Gold Nanoparticles as Hypoglycemic Treatment for Diabetes Mellitus. *Intern. J. Nanomed.* **2012,** *7*, 1189.

Danish, M.; Singh, P.; Mishra, G.; Srivastava, S.; Jha, K.; Khosa, R. *Cassia fistula* Linn (Amulthus) -An Important Medicinal Plant: A Review of Its Traditional Uses, Phytochemistry and Pharmacological Properties. *J. Nat. Prod. Plant Resour.* **2011,** *1*, 101–118.

Dawood, H. D.; Darwish, M. S.; Asmaa, A. El-Awady, A. A.; Mohamed, A. H.; Ahmed, A. Z.; Tahe, M. A. Chemical Characterization of *Cassia fistula* Polysaccharide (CFP) and Its Potential Application as a Prebiotic in Synbiotic Preparation. *RSC Adv.* **2021,** *11*, 13329–13340. DOI: 10.1039-D1RA00380A.

Deshpande, H. A.; Sanjivani, R. B. Recent Advances in the Phytochemistry of Some Medicinally Important *Cassia* Species: A Review. *Int. J. Pharm. Med. Bio. Sci.* **2013,** *2* (3), 60–78.

Duraipandiyan, V.; Baskar, A. A.; Ignacimuthu, S.; Muthukumar, C.; Al-Harbi, N. A. Anticancer Activity of Rhein Isolated from *Cassia fistula* L. Flower. *Asian Pac. J. Trop. Dis.* **2012,** *2,* S517–523.

Duraipandiyan, V.; Ignacimuthu, S. Antibacterial and Antifungal Activity of *Cassia fistula* L.: An Ethnomedicinal Plant. *J. Ethnopharmacol.* **2007,** *112* (3), 590–594.

Einstein, J. W.; Rais, M. M.; Mohd, A. M. Comparative Evaluation of the Antidiabetic Effects of Different Parts of *Cassia fistula* Linn, a Southeast Asian Plant. *J. Chem.* **2013,** 714063, 1–10.

El-Saadany, S. S.; El-Massry, R. A.; Labib, S. M.; Sitohy, M. Z. The Biochemical Role and Hypocholesterolaemic Potential of the Legume *Cassia fistula* in Hypercholesterolaemic Rats. *Nahrung* **1991,** *35,* 807–815.

Esposito, A. M.; Diaz, A.; De, G. I.; De, T. R.; Gupta, M. P. Evaluation of Traditional Medicine: Effects of *Cajanus cajan* L. and of *Cassia fistula* L. on Carbohydrate Metabolism in Mice. *Rev. Med. Panama.* **1991,** *16,* 39–45.

Gobianand, K.; Vivekababdan, P.; Pradeep, K.; Mohan, C. V.; Karthikeyan, S. Anti-Inflammatory and Antipyretic Activities of Indian Medicinal Plant *Cassia fistula* Linn. (Golden Shower) in Wister Albino Rats. *Int. J. Pharmacol.* **2010,** *6,* 719–725.

Gopi, A.; Manju, J. Evaluation of anti-lipoxygenase activity of *Cassia fistula* Linn leaves using in vitro methods. *Intern. J. Basic & Clin. Pharmacol.* **2018,** *7* (9), 1678–1682.

Govindarajan, M. Larvicidal Activity of *Cassia fistula* Flower Against *Culex tritaeniorhynchus* Giles, *Aedes albopictus* Skuse and *Anopheles subpictus* Grassi (Diptera: Culicidae). *J. Pure Appl. Zool.* **2001,** *1,* 117–121.

Govindarajan, M.; Jebanesan, A.; Pushpanathan, T. Larvicidal and Ovicidal Activity of *Cassia fistula* Linn. Leaf Extract Against Filarial and Malarial Vector Mosquitoes. *Parasitol. Res.* **2008,** *102,* 289–292.

Gupta, M.; Mazumder, U. K.; Rath, N.; Mukhopadhyay, D. K. Antitumour Activity of Methanolic Extract of *Cassia fistula* L. Seed Against Ehrlich Ascites Carcinoma. *J. Ethnopharmacol.* **2000,** *72,* 151–156.

Hakim, F. A.; Haidy, A. G.; Radwan, R. A.; Ayoub, N.; Mohamed, E. S. Chemical Constituents and Biological Activities of *Cassia* Genus: Review. *Arch. Pharmaceut. Sci. Ain Shams Univ.* **2019,** *3* (2), 195–227.

Heibatullah, K.; Mohammadtaha, J.; Amir, J.; Masood, M.; Abobakr, S.; Bela, J.; Arpad, T.; Rudolf, G. Protective Effect of *Cassia fistula* Fruit Extract Against Bromobenzene-Induced Liver Injury in Mice. *Human Exp. Toxicol.* **2015,** *30* (8), 1039–1044.

Irshad, M.; Ahmad, A.; Zafaryab, M. et al. Composition of *Cassia fistula* Oil and Its Antifungal Activity by Disrupting Ergosterol Biosynthesis. *Nat. Prod. Comm.* **2013,** *8* (2), 261–264.

Irshad, M.; Mehdi, S. J.; Al-Fatlawi, A. A.; Zafaryah, M.; Ali, A.; Ahmad, I. Phytochemical Composition of *Cassia fistula* Fruit Extracts and Its Anti-Cancer Activity Against Human Cancer Cell Lines. *J. Biol. Active Prod. Nat.* **2014,** *4,* 158–170.

Irshad, M.; Singh, M.; Rizvi, M. A. Assessment of Anthelmintic Activity of *Cassia fistula*. *Middle East J. Sci. Res.* **2010,** *5* (5), 346–349.

Irshad, S. S.; Manzoor, N.; Khan, L. A.; Rizvi, M. M. Anticandidal Activity of *Cassia fistula* and Its Effect on Ergosterol Biosynthesis. *Pharm. Biol.* **2011,** *49* (7), 727–733.

Ishita, G.; Punita, S. Investigation of Phytoconstituents from Leaves, Seeds, Bark and Pods of *Cassia fistula* Plant. *J. Chem. Chem. Sci.* **2018,** *8* (5), 886–890.

Jadhav, S. N. Evaluation of Immunomodulatory Activity of *Cassia fistula*. *Int. J. Pharma. Chem. Sci.* **2014,** *3,* 291–293.

Jangir, R. N.; Jain, G. C. Evaluation of Antidiabetic Activity of Hydroalcoholic Extract of *Cassia fistula* Linn. Pod in Streptozotocin-Induced Diabetic Rats. *Pharmacogn. J.* **2017,** *9* (5), 599–606.

Jarald, E. E.; Joshi, S. B.; Jain, D. C.; Edwin, S. Biochemical Evaluation of the Hypoglycemic Effects of Extract and Fraction of *Cassia fistula* Linn. in Alloxan-Induced Diabetic Rats. *Indian J. Pharm. Sci.* **2013,** *75* (4), 427–34.

Jeyanthi, K. A. Beneficial Effect of *Cassia fistula* (L.) Flower Extract on Antioxidant Defence in Streptozotocin Induced Diabetic Rats. *Int. J. Pharm. Pharm. Sci.* **2012,** *4*, 274–276.

Jose, B.; Reddy, L. J. Evaluation of Antibacterial and DPPH Radical Scavenging Activities of the Leaf Extracts of *Cassia fistula* Linn. from South India. *Open Access Sci. Rep.* **2013,** *2* (8), 2–4.

Jothy, S. L.; Zakaria, Z.; Chen, Y.; Lau, Y. L.; Latha, L. Y.; Sasidharan, S. Acute Oral Toxicity of Methanolic Seed Extract of *Cassia fistula* in Mice. *Molecules* **2011,** *16* (6), 5268–5282.

Jothy, S. L.; Zakariah, Z.; Chen, Y.; Sasidharan, S. In Vitro, In Situ and In Vivo Studies on the Anticandidal Activity of *Cassia fistula* Seed Extract. *Molecules* **2012,** *17* (6), 6997–7009.

Kadhim, M. J.; Mohammed, **G. J.;** Hameed, I. H. In Vitro Antibacterial, Antifungal and Phytochemical Analysis of Methanolic Extract of Fruit *Cassia fistula. Oriental J. Chem.* **2016,** *32* (3), 1329–1346.

Kalantari, H.; Jalali, M.; Jalali, A. et al. Protective Effect of *Cassia fistula* Fruit Extract Against Bromobenzene-Induced Liver Injury in Mice. *Hum. Exp. Toxicol.* **2011,** *30* (8), 1039–1044.

Kashiwada, Y.; Toshika, K.; Chen, R.; Nonaka, G.; Nishioka, I. Tannins and Related Compounds. XCIII. Occurrence of Enantiomeric Proanthocyanidins in the Leguminosae Plants, *Cassia fistula* L. and *Cassia javanica* L. *Chem. Pharm. Bull.* **1996,** *38*, 888–893.

Kaur, S.; Kumar, A.; Thakur, S.; Kumar, K.; Sharma, R.; Sharma, A.; Singh, P.; Sharma, U.; Kumar, S.; Landi, M.; Brestic, M.; Kaur, S. Antioxidant, Antiproliferative and Apoptosis-Inducing Efficacy of Fractions from *Cassia fistula* L. Leaves. *Antioxidants* **2020,** *9* (9), 173. DOI: 10.3390/antiox9020173.

Kaur, S.; Sharma, D.; Singh, A. P.; Kaur, S. Amelioration of Hepatic Function, Oxidative Stress, and Histopathologic Damages by *Cassia fistula* L. Fraction in Thioacetamide-Induced Liver Toxicity. *Environ. Sci. Pollut. Res. Int.* **2019,** *26* (29), 29930–29945.

Khare, C. P. *Indian Medicinal Plants*; Springer, 2007; p 128.

Kulkarni, A.; Govindappa, M.; Ramachandra, Y. L.; Koka, P. GC-MS Analysis of Methanol Extract of *Cassia fistula* and Its In Vitro Anticancer Activity on Human Prostate Cancer Cell Line. *Indo Am. J. Pharm. Res.* **2015,** *5* (2), 937–944.

Kumar, K. A.; Satish, S.; Sayeed, I.; Hedge, K. Therapeutic Uses of *Cassia fistula*: Review. *Intern. J. Pharma Chem. Res.* **2017,** *3* (1), 38–43.

Kumar, K. C.; Muller, K. G. Inhibition of Leukotriene Biosynthesis and Lipid Peroxidation in Biological Models By the Extract of *Cassia fistula. Phytother. Res.* **1998,** *12* (7), 526–528.

Lachumy, S. J.; Zuraini, Z.; Sasidharan, S. Antimicrobial Activity and Toxicity of Methanol Extract of *Cassia fistula* Seeds. *Res. J. Pharm. Biol. Chem. Sci.* **2010,** *1*, 391–388.

Lai, T. K.; Liew, K. C. Total Phenolics Total Tannins and Antioxidant Activity of *Cassia fistula* L. Extracts of Bark, Stem, Leaf and Root Under Different Age Classes. *Asian J. Pharm. Res. Health Care* **2013,** *5*, 52–57.

Laxmi, V.; Bhatia, A. K.; Goel, A.; Wahi, N.; Sharma, A. Phytochemicals Screening and Analysis Using HPLC to Determine the Antimicrobial Efficacy of *Cassia fistula* Extract. *Adv. Biores.* **2015a,** *6* (3), 1–7.

Laxmi, V.; Wahi, N.; Sharma, A.; Goel, A.; Bhatia, Ashok. Investigating the Immunomodulatory Effect of *Cassia fistula* on Albino Rats. *Adv. Pharma. Ethnomed.* **2015b,** *3*, 1–5.

Lee, C. K.; Lee, P. H.; Kuo, Y. H. The Chemical Constituents from the Aril of *Cassia fistula*, L. *J. Chinese Chem. Soc.* **2001,** *48*, 1053–1058.

Limtrakul, P.; Yodkeeree, S.; Thippraphan, P.; Punfa, W.; Srisomboon, J. Anti-Aging and Tyrosinase Inhibition Effects of *Cassia fistula* Flower Butanolic Extract. *BMC Complement Altern. Med.* **2016,** *16* (1), 497.

Luximon-Ramma, A.; Bahorun, T.; Soobrattee, M. A.; Aruoma, O. I. Antioxidant Activities of Phenolic, Proanthocyanidins, and Flavonoid Components in Extracts of *Cassia fistula. J. Agric. Food Chem.* **2002,** *50*, 5042–5047.

Malpani, S. N.; Manjunath, K. P.; Sholapur, H.; Savadi, R. V.; Akki, K. S.; Darade, S. S. Antidiabetic Activity of *Cassia fistula* Linn. Bark in Alloxan Induced Diabetic Rats. *Intern. J. Pharmaceut. Sci. Res.* **2010,** *2* (1), 382–385.

Manonmani, G.; Bhavapriya, V.; Kalpana, S.; Govindasamy, S.; Apparanantham, T. Antioxidant Activity of *Cassia fistula* (Linn.) Flowers in Alloxan Induced Diabetic Rats. *J. Ethnopharmacol.* **2005,** *97* (1), 39–42.

Mazumder, U. K.; Gupta, M.; Rath, N. CNS Activities of *Cassia fistula* in Mice. *Phytother. Res.* **1998,** *12*, 520–522.

Meenarani, M.; Kalidhar, S. B. A New Anthraquinone Derivative from *Cassia fistula* Linn. Pods. *Indian J. Chem.* (Section B, Organic Including Medicinal), **1998,** *37* (12), 1314–1315.

Mehmood, S.; Lateef, M.; Omer, M. O.; Anjum, A. A.; Rashid, M. I.; Shehzad, W. Adulticidal and Larvicidal Activity of *Cassia fistula* and *Piper nigrum* Against Malaria Vector. *Sci. Int.* (Lahore), **2014,** *26* (1), 331–334.

Misra, T. N.; Singh, R. S.; Pandey, H. S.; Singh, B. K. A New Diterpene from *Cassia fistula* Pods *Fitoterapia* **1997,** *68* (4), 375–376.

Misra, T. N.;Singh, R. S.; Pandey, H. S.; Pandey, R. P. Chemical Constituents of Hexane Fraction of *Cassia fistula* Pods. *Fitoterapia* **1996,** *67* (2), 173–174.

Mozaffarpur, S. A.; Naseri, M.; Esmaeilidooki, M. R.; Kamalinejad, M.; Bijani, A. The Effect of *Cassia fistula* Emulsion on Paediatric Functional Constipation in Comparison with Mineral Oil: A Randomized, Clinical Trial. *Daru: J. Faculty Pharm. Tehran Univ. Medical Sci.* **2012,** *20* (1), 83.

Munasinghe, J. T. C.; Seneviratne, C. K.; Thabrew, M. I.; Abeysekera, A. M. Antiradical and Anti-Lipoperoxidative Effects of Some Plant Extracts Used by Sri Lankan Traditional Medical Practitioners for Cardioprotection. *Phytother. Res.* **2001,** *15* (6), 519–523.

Nadkarni, K. M. *Indian Materia Medica*; Popular Prakashan: Bombay **2009,** *1*, 285–286.

Nagpal, M. A.; Nagpal, N.; Rahar, S.; Shah, G.; Swami, G.; Kapoor, R. Phytochemical Investigation of Methanolic Extract of *Cassia fistula* Leaves. *J. Pharmacog.* **2011,** *3* (26), 61–69.

Navanath, M. S.; Naikwade, N.; Mule, S.; Krishna, P. P. Evaluation of Anti-Inflammatory Activity of *Cassia fistula* and *Ficus benghalensis*. *J. Pharm. Res*. **2009,** *2*, 1304–1306.

Neelam, C.; Ranjan, B.; Komal, S.; Nootan, C. Review on *Cassia fistula*. *Int. J. Res. Ayurveda Pharm.* **2011,** *2*, 426–430.

Noorhajati H; Tanjung, M.; Aminah, N. S.; Suwandi, J. S. A. Antioxidant Activities of Extracts of Trengguli Stem Bark (*Cassia fistula* L.). *Intern. J. Basic & Appl. Sci.* **2012,** *12* (4), 85–89.

Nyeem, M. A. B.; Haque, M. S.; Hoque, M. A.; Islam, M. M.; Shafiqul, I. *Cassia fistula* (Bundaralati) Linn: Phytochemical and Pharmacological Studies: A Review. *Intern. J. Adv. Sci. Res.* **2017,** (1), 25–30.

Patel, D.; Karbhari, D.; Gulati, D.; Gokhale, D.; Antipyretic and Analgesic Activities of *Aconitum spicatum* and *Cassia fistula*. *Pharma. Biol.* **1965,** *157*, 22–27.

Patel, R. P.; Patel, K. C. Antibacterial Activity of *Cassia fistula*. *Indian J. Pharm.* **1956,** *18* (4), 107–110.

Pawar, A. V.; Patil, S. J.; Killedar, S. G. Uses of *Cassia fistula* Linn. as a Medicinal Plant. *Int. J. Adv. Res. Dev*. **2017,** *2* (3), 85–91.

Perumal, R.; Samy, S. Iggnacimuthu, S.; Sen, A. Screening of Medicinal Plants Antibacterial Properties. *J. Ethnopharmacol.* **1998,** *62*, 173–182.

Phongpaichit, S.; Pujenjob, N.; Rukachaisirkul, V.; Ongsakul, M. Antifungal Activity from Leaf Extracts of *Cassia alata* L. *Cassia fistula* L. and *Cassia tora* L. *Songklanakarin J. Sci. Tech.* **2004,** *26*, 741–748.

Pitchai, D.; Manikkam, R. Hypoglycemic and Insulin Mimetic Impact of Catechin Isolated from *Cassia fistula*: A Substantiate in Silico Approach Through Docking Analysis. *Med. Chem. Res.* **2012,** *21*, 2238–2250.

Pradeep, K. Mohan, C. V.; Gobianand, K.; Karthikeyan, S. Effect of *Cassia fistula* Linn. Leaf Extract on Diethylnitrosamine Induced Hepatic Injury in Rats. *Chem. Biol. Interact.* **2007,** *167* (1), 12–18.

Prashanth, K. V.; Chauhan, N. S. Padh, H.; Rajani, M. Search for Antibacterial Antifungal Agents from Selected Indian Medicinal Plants. *J. Ethnopharmacol.* **2006,** *107*, 182–188.

Rahmani, A. H. *Cassia fistula* Linn: Potential Candidate in the Health Management. *Pharmacogn. Res*. **2015,** *7* (3), 217–224.

Rajagopal, P.; Premaletha, K.; Kiron, S.; Sreejith, K. Phytochemical and Pharmacological Review on *Cassia fistula* Linn. *Int. J. Pharm. Chem. Biol. Sci.* **2013,** *3*, 672–679.

Remya, R. R.; Rajasree, S. R. R.; Aranganathan, L.; Suman, T. Y. An Investigation on the Cytotoxic Effect of Bioactive AgNPs Synthesized Using *Cassia fistula* Flower Extract on Breast Cancer Cell MCF-7. *Biotechnol. Rep.* **2015,** *8*, 110–115.

Rizvi, M. M. A.; Irshad, M.; Hassadi, G. E.; Younis, S. B. Bio-Efficacies of *Cassia fistula*: An Indian labrum. *Afr. J. Pharm. Pharmacol.* **2009,** *3*, 287–292.

Samy, P. R.; Ignacimuthu, S.; Sen, A. Screening of 34 Medicinal Plants for Antibacterial Properties. *J. Ethnopharmacol.* **1998,** *62*, 173–182.

Sanoria, S; Qadrie, Z. L.; Gautam, S. P.; Barwal, A. *Cassia fistula*: Botany, Phytochemistry and Pharmacological Leverages—A Review. *Intern. J. Pharm. Pharma. Sci.* **2020,** *12* (6), 1–7.

Sartorelli, P.; Carvalho, C. S.; Reimao, J. Q.; Ferreira, M. J. P.; Tempone, A. G. Isolation of Antileishmanial Sterol from the Fruits of *Cassia fistula* Using Bio-Guided Fractionation. *Phytother. Res.* **2007,** *21* (7), 644–647.

Sartorelli, P.; Carvalho, C. S.; Reimao, J. Q.; Ferreira, M. J.; Tempone, A. G. Antiparasitic Activity of Biochanin A, an Isolated Isoflavone from Fruits of *Cassia fistula* (Leguminosae). *Parasitol. Res.* **2009,** *104*, 311–314.

Satpute, S. M.; Bhamburdekar, S. B.; Gaikwad, D. K. Anthelmintic Potential of Pods and Stem Bark Extracts of *Cassia fistula* L. *Intern. Res. J. Pharm.* **2017,** *8,* 70–72.

Sayeed, M. A.; Ali, A. M.; Astaq Mohal Khan, G. R. M.; Rahman, M. S. Studies on the Characterization and Glyceride Composition of *Cassia fistula* Seed Oil. *Bangladesh J. Sci. Indust*. **1999,** *34*, 144–148.

Selvi, V.; Isaivani, I.; R. Gomathi, R.; Karpagam, S. Phytochemical Screening of Crude Extract of Leaf and Flowers of *Cassia fistula* and *Cassia auriculata*. *Mintage J. Pharma. Medical Sci.* **2014,** *3* (4), 7–8.

Senthilkumar, M.; Sripriya, R.; Vijaya, R. H. Sehgal, P. K. Wound Healing Potential of *Cassia fistula* on an Infected Albino Rat Model. *J. Surg. Res.* **2006,** *131*, 283–289.

Seyyednejad, S. M.; Motamedi, H.; Vafei, M.; Bakhtiari, A. The Antibacterial Activity of *Cassia fistula* Organic Extracts. *Jundishapur J. Microbiol.* **2014,** *7* (1), e8921.

Sharma, A.; Kumar, A.; Jaitak, V. Pharmacological and Chemical Potential of *Cassia fistula* L- a Critical Review. *J. Herbal Med.* **2021,** *26*, 100407.

Sharma, A.; Laxmi, V.; Goel, A.; Sharma, V.; Bhatia, A. K. Anti-Viral Activity of *Cassia fistula* Against IBR Virus. *J. Immunol. Immunopathol.* **2010,** *12* (2), 114–119.

Siddhuraju, P.; Mohan, P. S.; Becker, K. Studies on the Antioxidant Activity of Indian Laburnum (*Cassia fistula* L.): A Preliminary Assessment of Crude Extracts from Stem Bark, Leaves, Flowers and Fruit Pulp. *Food Chem.* **2002,** *79* (1), 61–67.

Sikri, N.; Dhanda, S.; Dalal, S. Kinetics of Urease Inhibition by Different Fractions of *Cassia fistula. S. Afr. J. Bot.* **2019,** *120*, 274–279.

Singh, M. P.; Singh, A.; Alam, G.; Patel, R.; Datt, N. Antipyretic Activity of *Cassia fistula* Linn. Pods. *J. Pharm. Res.* **2012,** *5*, 2593–2594.

Singh, R. S.; Singh, H.; Pandey, H. S.; Pandey, R. P.; Sheela, S. Two New Aliphatic Compounds from *Cassia fistula* L. *Indian J. Chem.* **2005,** *44B*, 2372–2374.

Sony, P.; Kalyani, M.; Jeyakumari, D.; Kannan, I.; Sukumar, R. G. In Vitro Antifungal Activity of *Cassia fistula* Extracts Against Fluconazole Resistant Strains of *Candida* Species from HIV Patients. *J. Mycol. Med.* **2018,** *28* (1), 193–200.

Srividhya, M.; Hridya, H.; Shanthi, V.; Ramanathan, K. Bioactive Amento Flavone Isolated from *Cassia fistula* L. Leaves exhibits therapeutic efficacy. *3 Biotech.* **2017,** 7 (1), 33. DOI: 10.1007/s13205–017–0599–7.

Tabrez, S.; Rahman, F.; Ali, R.; Alouffi, S. A.; Alshehri, M. B.; Alshammari, F. A.; Alaidarous, M. A.; Banawas, S.; Dukhyil, A. B.; Rub, A. Assessment of the Antileishmanial Potential of *Cassia fistula* Leaf Extract. *ACS Omega* **2021,** *6*, 2318–2327.

Thabit, S.; Handoussa, H.; Roxo, M.; Elsayed, N. S.; Cestari, D. A. D.; Wink, M. Evaluation of Antioxidant and Neuroprotective Activities of *Cassia fistula* (L.) Using the *Caenorhabditis elegans* Model. *Peer J.* **2018,** *6*, e5159.

Vasudevan, K.; Manoharan, S.; Panjamurthy, K.; Vellaichamy, L.; Chellammal, A. Evaluation of Antihyperglycemic Effect of *Cassia fistula* (Linn.) Leaves in Streptozotocin Induced Diabetic Rats. *Electron. J. Pharmacol. Ther.* **2008,** *1*, 57–60.

Vijayrao, N. J.; Sanjay, G.; Kiran, R. *Chemical Studies in Cassia fistula.* Linn. *Bahava Study.* LAP LAMBERT Academic Publishing, 2018, 1–98.

Wankhade, T. R.; Ramrao, B. P.; Gavane, A. N.; Sadashiv, A. J. A Review Article on *Cassia fistula* Linn. *Intern. Ayurvedic Medical J.* **2014,** *2* (4), 614–617.

Wasu, S. J.; Muley, B. P. Hepatoprotective Effect of *Cassia fistula* Linn. *Ethnobotanical Leaflets* **2009,** *13*, 910–916.

Wati, M. S.; Khabiruddin, M. Phytochemical Properties of Seeds and Oils of *Cassia fistula* (Two Locations). *Oriental J. Chem.* **2017,** *33* (4), 1969–1975.

Yadav, R.; Jain, G. C. Antifertility Effect and Hormonal Profile of Petroleum Ether Extract of Seeds of *Cassia fistula* in Female Rats. *Int. J. Pharm. Tech. Res.* **2009,** *1* (3), 438–444.

Yadav, R.; Jain, G. C. Antifertility Effect of Aqueous Extract of Seeds of *Cassia fistula* in Female Rats. *Adv. Contracept.* **1999,** *15*, 293–301.

Yadava, R. N.; Verma, V. A New Biologically Active Flavone Glycoside from the Seeds of *Cassia fistula* (Linn.) *J. Asian Nat. Prod. Res.* **2003,** *5*, 57–61.

Yingngam, B. Zhao, H.; Baolin, B.; Pongprom, N.; Brantner, A. Optimization of Ultrasonic-Assisted Extraction and Purification of Rhein from *Cassia fistula* Pod Pulp. *Molecules* **2019,** *24* (10), 2013.

Yueh-Hsiung, K.; Ping-Hung, L.; Yung-Shun, W. Four New Compounds from the Seeds of *Cassia fistula. J. Nat. Prod.* **2002,** *65*, 1165–1167.

Zhao, Y.; Zhao, K.; Jiang, K. et al. A Review of Flavonoids from *Cassia* Species and Their Biological Activity. *Curr. Pharm. Biotechnol.* **2016,** *17* (13), 1134–1146.

Zhou, M.; Xing, H. H.; Yang, Y.; Wang, Y. D.; Zhou, K.; Dong, W.; Li, G. P.; Hu, W. Y.; Liu, Q.; Li, X. M.; Hu, Q. F. Three New Anthraquinones from the Twigs of *Cassia fistula* and Their Bioactivities. *J. Asian Nat.* **2017,** *19* (11), 1073–1078.

CHAPTER 13

Bioactives and Pharmacology of *Derris scandens* (Roxb.) Benth.

JITENDRA R. PATIL[1], SAVALIRAM G. GHANE[2], and GANESH C. NIKALJE[1*]

[1]*Department of Botany, Seva Sadan's R. K. Talreja College of Arts, Science and Commerce, Affiliated to University of Mumbai, Ulhasnagar, Maharashtra 421003, India*

[2]*Department of Botany, Shivaji University, Kolhapur, Maharashtra 416004, India*

[*]*Corresponding author. E-mail: ganeshnikalje7@gmail.com/ g.nikalje@ssrkt.edu.in*

ABSTRACT

Derris scandens Benth., a member of family Fabaceae, is commonly known as Jewel vine, "Gonj" in Hindi and "Thao-wan–priang" in Thailand. It is a spreading, evergreen, woody vine, or climbing shrub. The dried stem is used in muscular ache as well as in arthritis symptoms. It is used as folk medicine in Thailand as antidysenteric, antitussive, diuretic, expectorant, and against cachexia and muscular pain. It is also used traditionally in India and Sri Lanka as a fish poison and as an antidote against snake venom, for the treatment of bleeding of wounds, and stomach infections. The phytochemical studies of *D. scandens* revealed the presence of different flavanoids, predominantly isoflavones, and coumarins like robustic acid, warangalone (scandenone), chandalone, nallanin, lonchocarpic acid (Chandanin), scandenin, lonchocarpenin, scandinone, osajin, and 3-aryl coumarins. Along with

Bioactives and Pharmacology of Legumes. T. Pullaiah, PhD (Ed.)

flavonoids, presence of pterocarpene derivatives, terpenes, steroids, benzoic acid derivatives and benzyl acid derivatives also reported. Several studies reported that isolated compounds show anti-inflammatory, anti-microbial, anti-feedant, anticancer, antioxidant, immuno-stimulating, hypotensive and other biological activities. These versatile bioactivities can be utilized for the development of novel drugs against multiple diseases. The present chapter summarizes different bioactive compounds and pharmacological properties of *D. scandens.*

13.1 INTRODUCTION

Derris scandens Benth. belongs to family Fabaceae. Synonyms of the plant are *Brachypterum scandens* (Roxb.) Wight, *B. scandens* Benth., *Brachypterum timorense* Benth., *Dalbergia scandens* Roxb., *Dalbergia timoriensis* DC., *Deguelia timoriensis* (DC.) Taub., *Derris timoriensis* (DC.) Pittier, *Galedupa frutescens* Blanco and *Millettia litoralis* Dunn. It is commonly known as Jewel vine, "Gonj" in Hindi and "Thao-wan–priang" in Thailand (Rukachaisirikul et al., 2002; Rao et al., 2007).

It is distributed throughout the Southeast Asia, North Australia, and the Southwest pacific islands (Sekine et al., 1999; Mahabusarakam et al., 2004). In India, it is distributed in the sub-Himalayan tract, and in central and southern India and Andamans. *D. scandens* a spreading, evergreen, woody vine, or climbing shrub. It is also cultivated in gardens. The leaves are imparipinnate compound with length of 3–4 in., channeled rachis, and leaflets are oblong. The inflorescence is axillary raceme with white and pink flowers, calyx is gray, and corolla is silky with vexillary aestivation. Androecium with 10 stamens, united with filaments. Fruit is legume pod with 4–5 reniform, compressed, and dark brown seeds.

The dried stem of *D. scandens* is used in muscular ache as well as in arthritis symptoms (Tiangburanatham, 1996). It is used as folk medicine in Thailand as antidysenteric, antitussive, diuretic, expectorant, and against cachexia and muscular pain; and even Thai people use extract of the roasted stem as a tonic (Sekine et al., 1999; Sriwanthana and Chavalittumrong, 2001). It is also used traditionally in India and Sri Lanka as a fish poison and as an antidote against snake venom, for the treatment of bleeding of wounds, and stomach infections (Chopra et al., 1956; Mohotti et al., 2020). All over the world, it has been reported for controlling of many phytophagous pests (Sreelatha et al., 2010). It is also claimed that the plant acts as a health promotive in cardiovascular diseases, postmenopausal women, and cancer prevention (Kuptniratsaikul et al., 2001).

13.2 BIOACTIVES

The chemical constituents of *D. scandens* are location specific. It varies in different localities of collection sites of plant material (Rao et al., 1994).

13.2.1 Flavanoids

The phytochemical studies of *D. scandens* revealed the presence of different flavanoids, predominantly isoflavones, and coumarins. Few isoflavones and 3-aryl 4-hydroxycoumarins were isolated from the plant extract such as robustic acid, warangalone (scandenone), chandalone, nallanin, lonchocarpic acid (Chandanin), scandenin, lonchocarpenin, scandinone, and osajin (Falshaw et al., 1969 and references therein). Furthermore, the isolation of new isoflavones and 3-aryl coumarins were reported from the chloroform extract of stem-like eturunagarone; 3′-γ,γ-dimethylallyl-wighteone,4,4′-Di-*O*-methyl scandenin, 8-γ,γ-dimethylallyl-wighteone (Rao et al., 1994). The isolation of two new isoflavones named, scandenal and scanderone along with few known compounds like isorobustone, 5-hydroxy-2″,2″-dimethylchromeno-[6,7:5″,6″]-2‴,2‴-dimethylchromeno[3′,4′:5‴,6‴]isoflavone, ulexone A, isochandalone, santal and lupiwighteone, 3′-methylorobol and genistein were reported (Mahabusarakam et al., 2004 and references therein). Recently, the isolation of three unpublished isoflavones—derriscandenon A, B, and C from the leaves extract was reported (Ito et al., 2020).

The isolation of six unreported diprenyl isoflavones, namely derris isoflavones A–F, along with six known isoflavones lupalbigenin, scandinone, erysenegalensein E, lupinisol A, lupinisoflavone G, and 5,7,4′trihydroxy-6, 8-diprenylisoflavone werer eported from the ethanolic extract of stem (Sekine et al., 1999). The 7,8-dihydroxy-4′-methoxyisoflavone has been isolated from the stem extract (Rukachaisirikul et al., 2002). In addition, three new prenylated isoflavones—isoscandinone, scandenin A, and scandenin B were also isolated (Rao et al., 2007).

Similarly, Munikishore et al. (2012) isolated two new flavonoids (2*S*)-3′,4′,5′-trimethoxyflavanone and 2′-hydroxy-2,4-dimethoxy-4′-*O*-[(*E*)-3,7-dimethyl-2,6-octadienyl]chalcone from the seeds. Recently, the two flavonoids dalpanitin and vicenin-3 were isolated from the plant extract (Mohotti et al., 2020) (Figure 13.1).

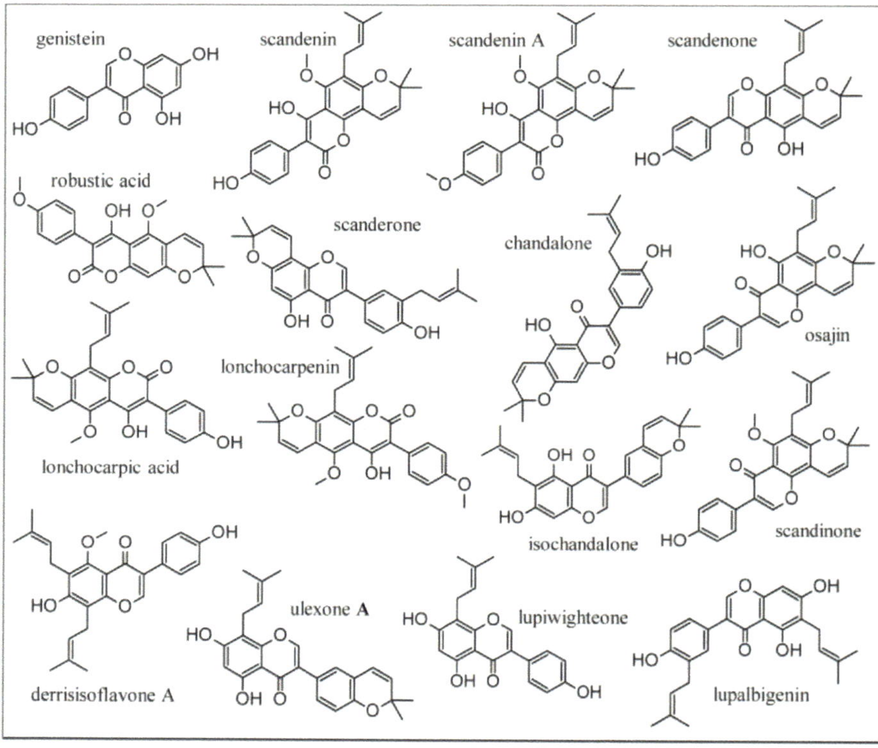

FIGURE 13.1 Isoflavones and coumarins isolated from *D. scandens*

13.2.2 Isoflavone Glycosides

Two isoflavone glycosides—derriscanoside A and derriscanoside B were isolated from the aerial parts of plant (Dianpeng et al., 1999). Few isoflavone glycosides have been isolated from the stem of plant such as derriscandenosides A–E, formononetin 7-*O*-β-glucopyranoside, 8-hydroxy-4′, 7-dimethoxyisoflavone8-*O*-β-glucopyranoside,7-hydroxy-4′,8-dimethoxyisoflavone 7-*O*-β-glucopyranoside, diadzein 7-*O*-[α-rhamnopyranosyl-(1→6)]-β-glucopyranoside, formononetin 7-*O*-[α-rhamnopyranosyl-(1→6)] -β-glucopyranoside and genistein 7-*O*-[α-rhamnopyranosyl-(1→6)]-β-glucopyranoside (Rukachaisirikul et al., 2002 and references cited therein).

13.2.3 *Pterocarpene Derivatives*

Three pterocarpene derivatives flemichapparin B (Munikishore et al., 2012), flemichapparin C, and maackiain have been isolated from the *Derris* (Mahabusarakam et al., 2004 and references therein) (Figure 13.2).

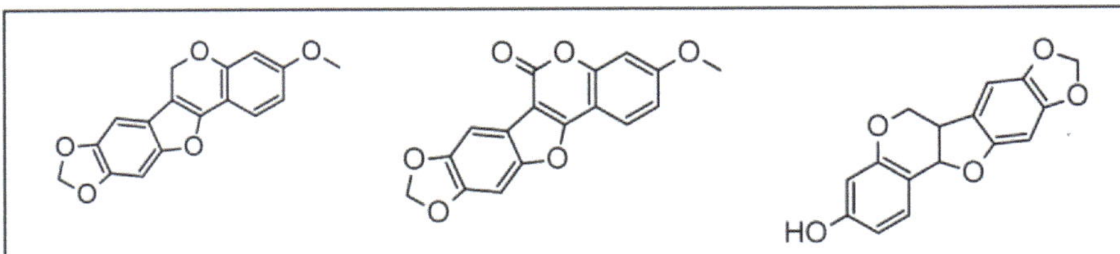

FIGURE 13.2 Pterocarpene derivates isolated from *D. scandens*

13.2.4 *Rotenoid*

One rotenoid compound named dehydrodeguelin was isolated from the seeds (Munikishore et al., 2012).

13.2.5 *Terpenes and Steroids*

The root and stem extracts showed the presence of few terpenes and steroids like betulinic acid, lupeol, β-amyran-3-one, β-amyrin, β-sitosterol, and β-sitosterol glucopyranoside (Hussain et al., 2015).

13.2.6 *Benzoic Acid Derivatives*

Two benzoic acid derivatives—4-hydroxy-3-methoxybenzoic acid and 4-hydroxy-3,5-dimethoxybezoic acid have been isolated from stem extract (Rukachaisirikul et al., 2002 and references there cited).

13.2.7 *Benzil Derivatives*

Two unreported benzil derivatives were isolated from stem extract—scandione (Mahabusarakam et al., 2004) and whole plant extract derrisdione A (Sreelatha et al., 2010).

13.3 PHARMACOLOGY

13.3.1 Anti-Inflammatory Activity

As per the Thailand National List of Essential Medicines, dried stem powder and ethanolic extracts are recommended for the treatment of musculoskeletal pain. The aqueous extract of stem showed the reduction in myeloperoxidase release and eicosanoid production along with potent inhibitory activity against leukotriene B4 generation and strong anti-oxidant activity. Ultimately, these results supported the traditional use of the plant in treatment of arthritis (Laupattarakasem et al., 2003). The root and leaf extracts of the plant as well as isolated flavonoid, ovaliflavone, showed significant anti-inflammatory activity on carrageenan-induced paw edema in rats (Ganapaty et al., 2006). Furthermore, Kuptniratsaikul et al. (2001) tested the efficacy of stem extract on the patients with knee osteoarthritis. They found that the plant extract was efficient and safe in relieving pain as well as improving knee functions. As per the HPLC analysis of the plant extract, genistein-7-*O*-[γ-rhamnopyranosyl-(1→6)-β-glucopyranoside] was found to be the main constituent.

The efficacy of *D. scandens* on musculoskeletal pain was found similar to standard nonsteroidal anti-inflammatory drugs. Indeed, *D. scandens* is safer in comparison to nonsteroidal anti-inflammatory drugs with respect to their side effects (Puttarak et al, 2016). Punjanon (2018) studied the effects of co-administered *D. scandens* extract and diclofenac drug by acetic acid-induced abdominal constriction test in mice; and observed the reduction in abdominal constrictions due to co-administration of extract and diclofenac. He further suggested that, this combination may be used pain treatment by reducing the dose of diclofenac drug and minimizing the side effects of it. In Thai traditional medicine, the mixture of four plants *D. scandens, Zingiber cassumunar, Suregada multiflora,* and *Siphonodon celastrineus* (DZSS formula) is used against muscular pain. The phytochemical studies revealed that 95 % ethanol extract of DZSS has strong anti-inflammatory activity with high concentration of genistein and (E)-4(3′,4′-dimethoxyphenyl)but-3-en-1-ol (Ayameang et al., 2020).

13.3.2 Antimicrobial Activity

The three isoflavone compounds namely, derrisisoflavone C; 5,7,4′-trihydroxy-6,8-diprenylisoflavone; and lupalbigenin isolated from stem showed antifungal activity against *Trichophyton mentagrophytes*, causal organism of skin disease—ringworm (Sekine et al., 1999). The isoflavone—lupalbigenin also showed strongest antimicrobial activity against *Staphylococcus aureus*, both penicillin-sensitive strain ATCC 25923 and methicillin-resistant strain MRSA SK1 (Mahabusarakam et al., 2004). The aqueous extract of plant showed antimicrobial activity against *S. aureus* ATCC 25923, *S. epidermidis* ATCC 12228, and *Escherichia coli* ATCC 25922 (Sittiwet and Puangpronpitag, 2009). The different fractions of root and stem of plant as well as isolated coumarins, scandenin, and scandenin A exhibited high to moderate antibacterial (against *E. coli,* and *Bacillus megaterium*), antialgal (against *Chlorella fusca*), and antifungal (against *Microbotryum violaceum*) activities (Hussain et al., 2015). The antimicrobial activity of isolated flavonoids, vicenin-3 against *S. aureus* and dalpanitin against *E. coli, Pseudomonas aeruginosa,* and *S.aureus* were reported (Mohotti et al. 2020).

13.3.3 Antifeedant Activity and Toxicity Against Pest

Sreelatha et al. (2010) demonstrated the antifeedant activity and toxicity against the castor semilooper pest, *Achaea janata* and they found that few compounds like osajin, scandenin, sphaerobioside, genistein, derrisdione A and one more compound showed good to moderate efficiency. Furthermore, in another study Rani et al. (2013), reported the antifeedant activity and toxicity against larvae of two stored food grain pests, red flour beetle, *Tribolium castaneum* and rice moth, *Corcyra cephalonica.* Among the tested compounds, Osajin, Lupalbigenin, Scandinone, Sphaerobioside, and Genistein revealed 100% contact toxicity and moderate antifeeding activity against both pests as well as lowers the relative growth rate, relative consumption rate, and food utilization of both pests.

13.3.4 Anticancer Activity and Cancer Treatment

Hematulin et al. (2014) reported that the ethanolic extract of plant in combination with gamma radiation synergistically sensitizes the human colon cancer HT-29 cells to radiation-induced cell death. The study suggested the use of the plant extract as a potent radiosensitizer. The anticancer activity of two isoflavones—5,7,4′-trihydroxy-6,8-diprenylisoflavone and lupalbigenin has been reported against the human breast cancer cell lines MCF-7, MDA-MB-231 (Tedasen et al., 2016). Few isoflavones isolated from the plant extracts such as derriscandenon C, derrubone, and glyurallin exhibited the in vitro antiproliferative effect on human cancer cells KB; while derriscandenon B reduced the cell viability of NALM6-MSH+ and KB cancer cells (Ito et al., 2020).

13.3.5 *Immunostimulating Activity*

Sriwanthana and Chavalittumrong (2001) reported the immune stimulating activity of stem extract of the plant. They found that lymphocyte proliferation was increased in the peripheral blood mononuclear cells treated with 10 ng/mL–5 μg/mL concentration of stem extract. The activity of Natural Killer cells was also increased in the peripheral blood mononuclear cell treated cells at the concentration of 10 ng/mL–10 μg/mL; even in HIV-infected donar cells at 10 μg/mL concentration. The induction of IL-2 secretion was also reported in the stem extract-treated PBMS cells (Sriwanthana and Chavalittumrong, 2001).

13.3.6 *Antioxidant Activity*

Three compounds isolated from the plant extract—scandinone, derrisisoflavone A, and santal have been found as a potent antioxidant compounds, even derrisisoflavone A and santal showed greater antioxidant activity than a standard antioxidant—butylated hydroxytoluene (Mahabusarakam et al., 2004). The DPPH free radical scavenging assay showed the moderate antioxidant property of two compounds scandenin B (SC50; 6.18 mg/mL) and scandenin A (SC50; 4.98 mg/mL) (Rao et al., 2007); as well as of ethanolic extract (Nooin et al., 2019).

13.3.7 *HYPOTENSIVE ACTIVITY*

The isoflavone glycoside—derriscanoside B, isolated from crude extract of stem showed the little hypotensive activity with decrease in mean arterial blood pressure (26.7 mm Hg) and heart rate (10 beats/min) in rats at the dose 4 mg/kg of body weight (Rukachaisirikul et al., 2002).

13.3.8 OTHER BIOLOGICAL ACTIVITIES

Rao et al. (2007) tested the enzyme inhibitory activity of chloroform and hexane extract of plant and later isolated three active compounds, namely—4′,5′,7-trihydroxybiprenylisoflavone, scandenone, and scandinone. These compounds revealed strong α-glucosidase inhibitory activity.

KEYWORDS

- ***Derris scandens***
- **flavanoids**
- **anti-inflammatory activity**
- **antimicrobial activity**

REFERENCES

Ayameang, O.; Rattarom, R.; Mekjaruskul, C.; Caichompoo, W. Anti-Inflammatory Activity and Quantitative Analysis of Major Compounds of the Mixtures of *Derris scandens* (DZSS) Formula. *Pharmacogn. J.* **2020,** *12* (4), 828–834.

Chopra, R. N.; Nayar, S. L.; Chopra, I. C. *Glossary of Indian Medicinal Plants*; CSIR, New Delhi; 1956, p 94

Dianpeng, L.; Mingan, O.; Jansakul, C.; Chongren, Y. Two Isoflavonoid Glycosides from *Derris scandens*. *YaoxueXuebao* **1999,** *34* (1), 43–45.

Falshaw, C. P.; Harmer, R. A.; Ollis, W. D.; Wheeler, R. E.; Lalitha, V. R.; Rao, N. S. Natural Occurrence of 3-aryl-4-Hydroxycoumarins. Part II. Phytochemical Examination of *Derris scandens* (Roxb.) Benth. *J. Chem. Soc. C: Org.* **1969,** *3*, 374–382.

Ganapaty, S.; Josaphine, J. S.; Thomas, P. S. Anti-Inflammatory Activity of *Derris scandens*. *J. Nat. Remedies* **2006,** *6* (1), 73–76.

Hematulin, A.; Ingkaninan, K.; Limpeanchob, N.; Sagan, D. Ethanolic Extract from *Derris scandens* Benth Mediates Radiosensitzation via Two Distinct Modes of Cell Death in Human Colon Cancer HT-29 Cells. *Asian Pac. J. Cancer Prev.* **2014,** *15* (4), 1871–1877.

Hussain, H.; Al-Harrasi, A.; Krohn, K.; Kouam, S. F.; Abbas, G.; Shah, A.; et al. Phytochemical Investigation and Antimicrobial Activity of *Derris scandens*. *J. King Saud Univ. Sci.* **2015,** *27*, 375–378.

Ito, C.; Matsui, T.; Miyabe, K.; Hasan, C. M.; Rashid, M. A.; Tokuda, H.; Itoigawa, M. Three Isoflavones from *Derris scandens* (Roxb.) Benth and Their Cancer Chemopreventive Activity and In Vitro Antiproliferative Effects. *Phytochemistry* **2020,** *175*, 112376.

Kuptniratsaikul, V.; Pinthong, T.; Bunjob, M.; Thanakhumtorn, S.; Chinswangwatanakul, P.; Thamlikitkul, V. Efficacy and Safety of *Derris scandens* Benth Extracts in Patients with Knee Osteoarthritis. *J. Altern. Complement. Med.* **2001,** *17* (2), 147–153.

Laupattarakasem, P.; Houghton, P. J.; Hoult, J. R. S.; Itharat, A. An Evaluation of the Activity Related to Inflammation of Four Plants Used in Thailand to Treat Arthritis. *J. Ethnopharmacol.* **2003,** *85*, 207–215.

Mahabusarakam, W.; Deachathai, S.; Phongpaichit, S.; Jansakul, C.; Taylor, W. C. A Benzil and Isoflavone Derivativ es from *Derris scandens* Benth. *Phytochemistry* **2004,** *65* (8), 1185–1191.

Mohotti, S.; Rajendran, S.; Muhammad, T.; Strömstedt, A. A.; Adhikari, A.; Burman, R.; et al. Screening for Bioactive Secondary Metabolites in Sri Lankan Medicinal Plants by Microfractionation and Targeted Isolation of Antimicrobial Flavonoids from *Derris scandens*. *J. Ethnopharmacol.* **2020,** *246*, 112158.

Munikishore, R.; Rammohan, A.; Padmaja, A.; Gunasekar, D.; Deville, A.; Bodo, B. Two New Flavonoids from the Seeds of *Derris scandens*. *Nat. Prod. Commun.* **2012,** *7* (10), 1305–1307.

Nooin, R.; Pitchakarn, P.; Kanchai, C.; Jaikang C. Assessments of Antioxidant, Antilipid Peroxidation, and In Vitro Safety of *Derris scandens* Vine Extracts from Southern Thailand. *Pharmacogn. Res.* **2019,** *11* (1), 60–66.

Punjanon, T. The Additivity Antinociceptive Interactions Between Diclofenac and the *Derris scandens* Extract Drug in Mice. *Asian J. Pharm. Clin. Res.* **2018,** *11* (1), 314–317.

Puttarak, P.; Sawangjit, R.; Chaiyakunapruk, N. Efficacy and Safety of *Derris scandens* (Roxb.) Benth. for Musculoskeletal Pain Treatment: A Systematic Review and Meta-Analysis of Randomized Controlled Trials. *J. Ethnopharmacol.* **2016,** *194*, 316–323.

Rani, P. U.; Hymavathi, A.; Babu, K. S.; Rao, A. S. Bioactivity Evaluation of Prenylated Isoflavones Derived from *Derris scandens* Benth Against Two Stored Pest Larvae. *J. Biopestic.* **2013,** *6* (1), 14–21.

Rao, M. N.; Krupadanam, G. D.; Srimannarayana, G. Four Isolfavones and Two 3-aryl Coumarins from Stems of *Derris scandens*. *Phytochemistry* **1994,** *37* (1), 267–269.

Rao, S. A.; Srinivas, P. V.; Tiwari, A. K.; Vanka, U. M. S.; Rao, R. V. S.; Dasari, K. R.; Rao, M. J. Isolation, Characterization and Chemobiological Quantification of α-Glucosidase Enzyme Inhibitory and Free Radical Scavenging Constituents from *Derris scandens* Benth. *J. Chromatogr. B* **2007,** *855* (2), 166–172.

Rukachaisirikul, V.; Sukpondma, Y.; Jansakul, C.; Taylor, W. C. Isoflavone Glycosides from *Derris scandens*. *Phytochemistry* **2002,** *60* (8), 827–834.

Sekine, T.; Inagaki, M.; Ikegami, F.; Fujii, Y.; Ruangrungsi, N. Six Diprenylisoflavones, *Derris*isoflavones A–F, from *Derris scandens*. *Phytochemistry* **1999,** *52* (1), 87–94.

Sittiwet, C.; Puangpronpitag, D. Antimicrobial Properties of *Derris scandens* Aqueous Extract. *J. Biol. Sci.* **2009,** *9* (6), 607–611.

Sreelatha, T.; Hymavathi, A.; Rao, V. R. S.; Devanand, P.; Rani, P. U.; Rao, J. M.; Babu, K. S. A New Benzil Derivative from *Derris scandens*: Structure-Insecticidal Activity Study. *Bioorganic Med. Chem. Lett.* **2010,** *20* (2), 549–553.

Sriwanthana, B.; Chavalittumrong, P. In Vitro Effect of *Derris scandens* on Normal Lymphocyte Proliferation and Its Activities on Natural Killer Cells in Normals and HIV-1 Infected Patients. *J. Ethnopharmacol.* **2001,** *76* (1), 125–129.

Tedasen, A.; Sukrong, S.; Sritularak, B.; Srisawat, T.; Graidist, P. 5,7,4′-Trihydroxy-6,8-Diprenylisoflavone and Lupalbigenin, Active Components of *Derris scandens*, Induce Cell Death on Breast Cancer Cell Lines. *Biomed. Pharmacother.* **2016,** *81*, 235–241.

Tiangburanatham, W. *Dictionary of Thai Medicinal Plants*; Prachumtong Printing: Bangkok; 1996, pp 349–350.

CHAPTER 14

Remedial Potential of Bioactive Compounds from *Macrotyloma uniflorum* (Lam.) Verdc.

CHETNA FAUJDAR and PRIYADARSHINI*

Department of Biotechnology, Jaypee Institute of Information Technology, Noida, Uttar Pradesh 201309, India

**Corresponding author. E-mail: priyadarshini@jiit.ac.in/ priya.juit@gmail.com*

ABSTRACT

Macrotyloma uniflorum commonly known as horse gram is a nutritional grain legume species. The pulse occupies an important position in human nutrition being a wonderful source of iron, calcium, molybdenum, proteins, natural antioxidants, vitamins and many bioactive compounds with therapeutic potential. Apart from its nutritional value the plant has been explored for its wide range of therapeutic properties against diseases such as hyperglycemia, cardiac diseases, asthma, kidney diseases, bronchitis, leucoderma etc. Bioactive components such as phytic acid, phenolic acid, natural antioxidants, fiber, enzymatic/proteinase inhibitors have been found to be associated with its pharmacological significance. Being an integral part of diet, it reduces the risk of variety of ailments including diabetes, intestinal diseases, coronary heart disease and many others. Potential antioxidant and free radical scavenging activities of its phenolic acids are reported to be responsible for exhibiting cardioprotective effects. Presence of many secondary metabolites in the plant makes it a promising therapeutic agent. Considering the nutritional

Bioactives and Pharmacology of Legumes. T. Pullaiah, PhD (Ed.)

and therapeutic potential of *Macrotyloma uniflorum*, the pulse needs to be exploited to prevent and treat a number of human ailments.

14.1 INTRODUCTION

Macrotyloma uniflorum (Lam.) Verdc. (Syn.: *Dolichos uniflorus* Lam, *Dolichos biflorus sensu auct.*) or horse gram is known as an underutilized food legume with great ethnomedical properties. It is a short-day plant widely cultivated as a food and fodder crop. The significance of this pulse crop was well recognized in folklore or traditional medicine to serve mankind in dealing with health-related complications. In addition to high protein and dietary fiber, it is also a wonderful source of essential amino acids (Bravo et al., 1999), iron, and vitamins including thiamine, riboflavin, niacin, and vitamin C (Sodani et al., 2004). The presence of phytocompounds such as phenolics, flavonoids, alkaloids, carotenoids, phytosterols, tannins, fatty acids, terpenoids, saponins, soluble and insoluble dietary fibers provide the pulse its additional advantage as a therapeutic agent. Traditionally, the seeds of horse gram have been used to treat human ailments including renal calculi, diabetes, obesity, hypertension, leucorrhoea, urinary disorders, and menstrual troubles (Bhuvaneshwari et al., 2014; Ghani, 1998). Dietary antioxidants present in the pulse can be utilized in the treatment of neurodegenerative diseases, such as Alzheimer's and Parkinson's diseases (Nakajima and Ohizumi, 2019).

Historically, bioactive components present in plant extracts led to the development of many plants derived drugs. Considering the high nutritive and therapeutic values of *M. uniflorum*, the pulse can be explored for its clinical significance as a source of potent therapeutic agents. This chapter explains the bioactive compounds present in horse gram and their remedial potential.

14.2 BIOACTIVE PHYTOCONSTITUENTS OF *MACROTYLOMA UNIFLORUM*

Therapeutic potential of horse gram seed extracts is probably due to the presence of bioactive phytoconstituents with significant pharmacological activities. Literature reveals the presence of flavonoids, glycosides, lenoleic acid, polyphenols, β-sitosterol, amino acids, gamma sitosterol, isoflavones, mome inositol, genistein, isoferririn, *p*-coumaric acid, ferulic acid cumesterol, psoralidin, galactosidase, glucosides, and streptogenin as major phytoconsituents (Kawsar et al., 2010; Morris, 2008; Das et al., 2014). Phenolic compounds present in plant extracts are known to reduce oxidative damage and are beneficial in managing diseased conditions arising due to oxidative stress. The main phenolic compound present in *M. uniflorum* seeds are flavonoids quercetin, kaempferol, myricetin, vanillic, ρ-hydroxybenzoic, and ferulic acids (Sreerama et al., 2010). Although a few phytoconstituents such as phenols, tannins, saponins, phytic acid, etc., have been considered as non-nutritional components but now their significance as therapeutic agents is well documented in the literature. Saponins reportedly possess hypocholesterolemic and anticarcinogenic activities (Koratkar and Rao, 1997).

Phenols, tannins (Cardador-Martínez et al., 2002), and phytic acids (Urbano et al., 2000) are also known to possess antioxidant potential. Many medicinal properties have already been attributed to the major component of *M. uniflorum* extract, mome inositol such as antialopecic, anticirrhotic, antineuropathic, cholesterolytic, lipotropic, and sweetener (Kumar et al., 2012; Das et al., 2014). Phenolic acids present in horse gram such as *p*-coumaric acid and ferulic acid have been well investigated for their role as potent antioxidant, cardioprotective, and anticholesterolemic agent (Yogeeta et al., 2006; Yeh et al., 2009; Kawsar et al., 2010). Investigations reveal that linoleic acid is capable of preventing diabetes and diabetic neuropathy along with restoration of normal antioxidant status in tissues (Suresh and Das, 2003; Pitel et al., 2007). Protease inhibitor isolated from *M. uniflorum* has been investigated for antifungal and antimicrobial properties (Mohan et al., 2018). α-Amylase inhibitor from the seeds of *M. uniflorum* has been studied to be used as an effective biological control of *Aedes aegypti*, it was demonstrated that the compound possesses strong larvicidal and a moderate adulticidal activity (Gupta et al., 2011a). Another phytocomponent (3β)-stigmast-5-en-3-ol has been studied to exert insulin-like effect along with cholesterol-lowering efficacy (Sujatha et al., 2010). The main bioactive phytoconstituents present in *M. uniflorum* extract have been explained in Table 14.1 in detail.

14.3 PHARMACOLOGICAL ACTIVITIES OF *MACROTYLOMA UNIFLORUM*

Seed extracts of the pulse have been reported to possess potent antioxidant activity along with trypsin inhibitory and angiotensin-converting enzymatic activities (Siddhuraju and Manian, 2007; Sreerama et al., 2012). *M. uniflorum* has been investigated for a number of pharmacological activities including antioxidant, anti-inflammatory, antilithiatic, antiheliminthic, antihypercholesterolemic, antidiabetic, and many others (Figure 14.1).

TABLE 14.1 Bioactive components of *Macrotyloma uniflorum* and their pharmacological significance.

S.No.	Bioactive component	Pharmacological activity
1	Phenolic acids (*p*-coumaric acid and ferulic acid)	Cardioprotective (Panda et al., 2016), Antioxidant (Siddhuraju and Manian, 2007)
2	Mome inositol	Antialopecic, anticirrhotic, antineuropathic, cholesterolytic, lipotropic, and sweetener (Das et al., 2014; Kumar et al., 2012)
3	Linoleic acid	Antidiabetic, antioxidant (Suresh and Das, 2003; Pitel et al., 2007)
4	(3β)-stigmast-5-en-3-ol	Cholesterol-lowering efficacy (Sujatha et al., 2010)
5	α-Amylase inhibitor	Antihyperglycemic, antidiabetic, (Gupta et al., 2011b), Larvicidal (Gupta et al., 2011a)
6	Dolichin A and Dolichin B	Inhibit HIV replication enzymes reverse transcriptase, protease, Integrase in docking studies (Auxilia et al., 2013)
7	B-Sitosterol	Antidiabetic (Ponnulakshmi et al., 2019), hepatoprotective (Yin et al., 2018)
8	Phytic acid	Anticarcinogenic, hypoglycemic, and prevent lipid peroxidation (Prasad and Singh, 2015)
9	Protease inhibitor	Antimicrobial and antifungal (Mohan et al., 2018), antihyperglycemic (Mohan and Elyas, 2018)
10	DL-Proline, Geranyl geraniol	Antibiotic resistance breaker (Abiraami Gowrie, 2019)
11	Calcium-binding proteins	Antilithiatic (Sharma et al., 2018)
12	Rutaecarpine	Antiurolithiatic (Sharma et al., 2019)

14.3.1 ANTIOXIDANT ACTIVITY

Plants naturally produce antioxidants as a result of experiencing oxidative stress. Seeds of *M. uniflorum* are a rich source of polyphenols and tannins. These phytoconstituents are reported to possess free radicals and oxidase reducing potential leading to the antioxidant activities of the extract (Seyoum et al., 2006). A study conducted by Siddhuraju and Manian explains that not only the phenolic content from raw seeds of horse gram but also from processed seeds of the legume possess potential antioxidant activities (Siddhuraju and Manian, 2007). Reports reveal that extract significantly improved the concentration of enzymes such as superoxide dismutase and catalase along with increased reduced glutathione thus reduced oxidative stress in rabbits (Muthu et al., 2006).

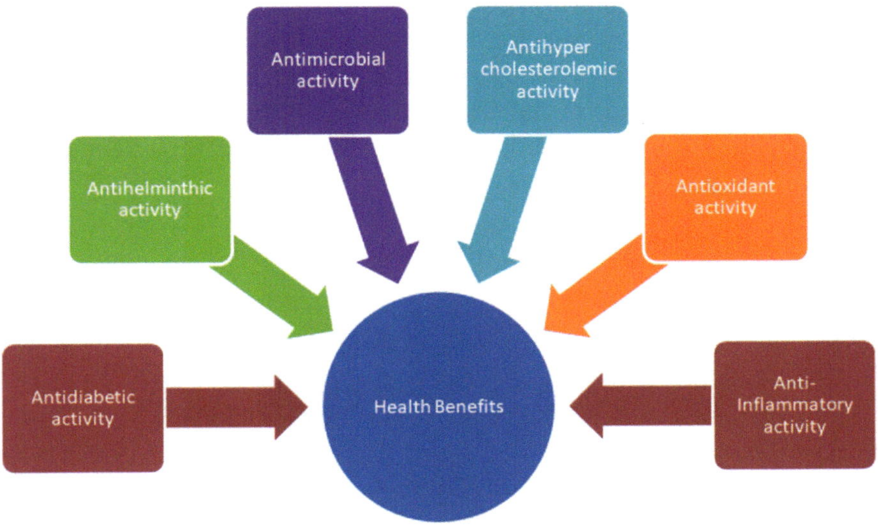

FIGURE 14.1 Health benefits of *Macrotyloma uniflorum.*

14.3.2 ANTIHYPERCHOLESTEROLEMIC EFFECT

In a study, Kumar et al. (2013) demonstrated the antihypercholesterolemic effects of *M. uniflorum* extract. They suggested that phenolic compounds present in the extract were possibly responsible for increased fecal excretion of cholesterol and LDL receptor activity (Chan et al., 1999; Tikkanen et al., 1998). Supersaturation of cholesterol in the bile induces the formation of cholesterol gallstones. Bigoniya et al. (2014) proved that lithogenic diet supplemented with *M. uniflorum* is capable of reducing the occurrence of cholesterol gallstones. Another study claims that horse gram leaf extract effectively increases HDL along with a significant decrease in LDL and triglycerides in an experiment conducted on albino rats (Bharathi and Anand, 2014). Antiobesity potential of horse gram extract has also been reported in the literature (Bhuvaneshwari, 2014).

14.3.3 ANTIDIABETIC ACTIVITY

Diabetes mellitus is a disease characterized by high blood glucose levels. Management of high postprandial glucose levels can be a suitable approach to deal with diabetes mellitus. Gupta et al. (2011b) investigated the anti-hyperglycemic activity of α-amylase inhibitor isolated from *M. uniflorum* seeds. They observed the decreased serum glucose level in the diabetic mice treated with α-amylase inhibitor. Another study reported a protease inhibitor isolated from horse gram seeds as efficient as the drug Metformin against diabetes mellitus (Mohan and Elyas, 2018). The presence of slow digestible starch in seeds results in low postprandial glucose response in diabetic patients (Bravo et al., 1998).

14.3.4 ANTIMICROBIAL ACTIVITY

Currently, the development of natural antimicrobial therapeutic agents to combat antibiotic resistance is a challenge. The development of resistance against antibiotics such as ciprofloxacin, norfloxacin, rifaximin, azithromycin, etc., necessitates the need for the emergence of new antibacterial agents to deal with infections (Abiraami Gowrie, 2019). The presence of naturally occurring antimicrobial phytoconstituents in medicinal plants could be a better approach in the direction of discovering new antimicrobial agents. Abiraami and Gowrie (2019) advocated the use of horse gram sprouts as natural antibiotic resistance breakers on the basis of bioassays and in silico studies. Preparation of silver nanoparticles loaded with horse gram seed extract exhibited good antibacterial potential against both Gram-positive as well as Gram-negative bacteria (Basu et al., 2016). Another study demonstrated that polysaccharides present in the extract possess remarkable antimicrobial activity (Basu et al., 2017). A biosheet incorporated with *M. uniflorum* was prepared by Muthukumar et al. (2013) to provide antimicrobial properties to a wound dressing material.

14.3.5 ANALGESIC AND ANTI-INFLAMMATORY ACTIVITY

Advancement in the field of medicine has resulted in the development of anti-inflammatory drugs to deal with pain and inflammation associated with tissue damage. However, the use of these drugs has been linked with many side effects. Plants-derived analgesics and anti-inflammatory therapeutic agents could be a better strategy to manage pain and inflammation along with reduced side effects. Fatima et al. (2018) reported that *M. uniflorum* fixed oil possesses both analgesic and anti-inflammatory activities. Methanolic extract of *M. uniflorum* seeds exhibited central and peripheral antinociceptive activity along with the potential to inhibit lipoxygenase and or cyclooxygenase (Ashraf et al., 2018).

14.3.6 ANTIHELMINTHIC ACTIVITY

M. uniflorum seed extract possesses worm-eliminating ability (anti helminthic). Alcoholic extract of *M. uniflorum* exhibited good inhibitory potential against *Pheretima posthuma* comparable to the standard albendazole (Sree et al., 2014). In a study, Philip et al. (2009) demonstrated that extract can effectively cause the death of Indian earthworm in a concentration-dependent manner which was comparable to the control, piperazine citrate.

14.3.7 ANTI LITHIATIC ACTIVITY

Horse gram has been investigated for preventing the formation of urinary and renal calculi. The presence of calcium-binding proteins in *M. uniflorum* might play an important role in discovering therapeutic agents against kidney stone formation (Sharma et al., 2018). In silico, studies of bioactive compounds identified from horse gram against Xanthine dehydrogenase revealed the capability of the pulse to prevent urolithiasis. Rutaecarpine was identified as a top bioactive molecule to interact with Xanthine dehydrogenase (Sharma et al., 2019). Anticalcifying activity of *M. uniflorum* has been linked to the presence of heat stable, polar, nontannin inhibitor of crystallization in the seeds (Peshin and Singla, 1994).

14.4 CONCLUSION

Based on the nutritional and therapeutic values of *M. uniflorum*, it can be concluded that the pulse is a rich source of many bioactive phytoconstituents. These phytoconstituents possess high potential for treating many human disorders and diseases. Horse gram has proven cardioprotective, antidiabetic, antibacterial, anti-inflammatory, and antioxidant activities. Thus, the crop can be explored for its therapeutic potential directing the way to the discovery of plant-derived pharmacologically significant compounds.

KEYWORDS

- **proteinase inhibitors**
- **cardioprotective**
- **secondary metabolites**
- **antioxidant**

REFERENCES

Abiraami, V. S.; Gowrie, S. U. Sprouts as Functional Food—An Approach Towards the Identification of Natural Antibiotic Resistance Breakers. *J. Drug Deliv. Ther.* **2019,** *9*, 23–35.

Ashraf, J.; Baig, S. G.; Ahmed, S.; Hasan, M. M. Analgesic, Anti-Inflammatory and Diuretic Activities of *Macrotyloma uniflorum* (Lam.) Verdc. *Pak. J. Pharm. Sci.* **2018,** *31*, 1859–1863.

Auxilia, L. R.; Sundari, T. M.; Daniel, R. R. Molecular Docking Studies of Dolichin A and B, Pterocarpans from Horsegram (*Macrotyloma uniflorum*) Against HIV Replication Enzymes. *Int. J. Comput. Appl.* **2013,** *75*, 19–23.

Basu, S.; Ghosh, M.; Bhunia, R. K.; Ganguly, J.; Banik, B. K. Polysaccharides from *Dolichos biflorus* Linn and *Trachyspermum ammi* Linn Seeds: Isolation, Characterization and Remarkable Antimicrobial Activity. *Chem. Cent. J.* **2017,** *11*, 1–10.

Basu, S.; Maji, P.; Ganguly, J. Biosynthesis, Characterisation and Antimicrobial Activity of Silver and Gold Nanoparticles by *Dolichos biflorus* Linn Seed Extract. *J. Exp. Nanosci.* **2016,** *11*, 660–668.

Bharathi, V.; Anand, V. Antiobesity Effect of *Macrotyloma uniflorum* Leaves Extract on Rats Fed on a Hypercaloric Diet. *World J. Pharm. Pharm. Sci.* **2014,** *3*, 1500–1504.

Bhuvaneshwari, S. Anorectic Activity of *Dolichos biflorus*. *Int. J. Pharm.* **2014,** *5*, 26–28.

Bhuvaneshwari, S.; Gopinathan, G.; Shastri, G.; Kuruvilla, A. Anti-Obesity Activity of *Dolichos biflorus* (Horse Gram) Compared to 5ht in Albino Rats. *Int. J. Pharm. Screen Meth.* **2014,** *4*, 99–101.

Bigoniya, P.; Bais, S.; Sirohi, B. The Effect of *Macrotyloma uniflorum* Seed on Bile Lithogenicity Against Diet Induced Cholelithiasis on Mice. *Anc. Sci. Life.* **2014,** *33*, 242–251.

Bravo, L.; Siddhuraju, P.; Saura-Calixto, F. Effect of Various Processing Methods on the In Vitro Starch Digestibility and Resistant Starch Content of Indian Pulses. *J. Agric. Food Chem.* **1998,** *46*, 4667–4674.

Bravo, L.; Siddhuraju, P.; Saura-Calixto, F. Composition of Underexploited Indian Pulses. Comparison with Common Legumes. *Food Chem.* **1999,** *64*, 185–192.

Cardador-Martínez, A.; Loarca-Piña, G.; Oomah, B. D. Antioxidant Activity in Common Beans (*Phaseolus vulgaris* L.). *J. Agric. Food Chem.* **2002,** *50*, 6975–6980.

Chan, P. T.; Fong, W. P.; Cheung, Y. L.; Huang, Y.; Ho, W. K. K.; Chen, Z. -Y. Jasmine Green Tea Epicatechins Are Hypolipidemic in Hamsters (*Mesocricetus auratus*) Fed a High Fat Diet. *J. Nutr.* **1999,** *129*, 1094–1101.

Das, S.; Vasudeva, N.; Sharma, S. Chemical Composition of Ethanol Extract of *Macrotyloma uniflorum* (Lam.) Verdc. Using GC-MS Spectroscopy. *Org. Med. Chem. Lett.* **2014,** *4*, 1–4.

Fatima, S. A.; Baig, S. G.; Hasan, M. M.; Ahmed, S. Analgesic and Anti-Inflammatory Activities of Fixed Oil of *Macrotyloma uniflorum* (Lam.) Verdc. In Mice and Rats. *Pak. J. Pharm. Sci.* **2018,** *31*, 581–585.

Ghani, A. *Medicinal Plants of Bangladesh: Chemical Constituents and Uses*; Asiatic Society of Bangladesh; 1998, p 467.

Gupta, L.; Deshpande, S.; Tare, V.; Sabharwal, S. Larvicidal Activity of the α-Amylase Inhibitor from the Seeds of *Macrotyloma uniflorum* (Leguminosae) Against *Aedes aegypti* (Diptera: Culicidae). *Int. J. Trop. Insect Sci.* **2011a,** *31*, 69–74.

Gupta, L. H.; Badole, S. L.; Bodhankar, S. L.; Sabharwal, S. G. Antidiabetic Potential of α-Amylase Inhibitor from the Seeds of *Macrotyloma uniflorum* in Streptozotocin-Nicotinamide-Induced Diabetic Mice. *Pharm. Biol.* **2011b,** *49*, 182–189.

Kawsar, S.; Huq, E.; Nahar, N.; Ozeki, Y. Identification and Quantification of Phenolic Acids in *Macrotyloma uniflorum* by Reversed Phase-HPLC. *Am. J. Plant Physiol.* **2010,** *5*, 204–211.

Koratkar, R.; Rao, A. V. Effect of Soya Bean Saponins on Azoxymethane-Induced Preneoplastic Lesions in the Colon of Mice. *Nutr. Cancer* **1997,** *27*, 206–209.

Kumar, D. S.; Prashanthi, G.; Avasarala, H.; Banji, D. Antihypercholesterolemic Effect of *Macrotyloma uniflorum* (Lam.) Verdc (Fabaceae) Extract on High-Fat Diet-Induced hypercholesterolemia in Sprague-Dawley Rats. *J. Diet. Suppl.* **2013,** *10*, 116–128.

Kumar, N. R.; Reddy, J. S.; Gopikrishna, G.; Solomon, K. A. GC-MS Determination of Bioactive Constituents of *Cycas beddomei* cones. *Int. J. Pharm. Bio Sci.* **2012,** *3*, 344–350.

Mohan, M.; Elyas, K. K. Anti-hyperglycemic Properties of a Purified Proteinaceous Protease Inhibitor from *Macrotyloma uniflorum* Seeds. *Curr. Top. Med. Chem.* **2018,** *18*, 2502–2510.

Mohan, M.; Kumari, M.; Elyas, K. K. Antifungal and Antimicrobial Properties of a Purified Protease Inhibitor from *Macrotyloma uniflorum* Seeds. *Curr. Top. Med. Chem.* **2018,** *18*, 2491–2501.

Morris, J. *Macrotyloma axillare* and *M. uniflorum*: Descriptor Analysis, Anthocyanin Indexes, and Potential Uses. *Genet. Resour. Crop. Evol.* **2008,** *55*, 5–8.

Muthu, A. K.; Sethupathy, S.; Manavalan, R.; Karar, P. Antioxidant Potential of Methanolic Extract of *Dolichos biflorus* Linn in High Fat Diet Fed Rabbits. *Indian J. Pharmacol.* **2006,** *38*, 131.

Muthukumar, T.; Senthil, R.; Sastry, T. P. Synthesis and Characterization of Biosheet Impregnated with *Macrotyloma uniflorum* Extract for Burn/Wound Dressings. *Colloid Surf. B* **2013,** *102*, 694–699.

Nakajima, A.; Ohizumi, Y. Potential Benefits of Nobiletin, a Citrus Flavonoid, Against Alzheimer's Disease and Parkinson's Disease. *Int. J. Mol. Sci.* **2019,** *20*, 3380.

Panda, V.; Laddha, A.; Nandave, M.; Srinath, S. Dietary Phenolic Acids of *Macrotyloma uniflorum* (Horse Gram) Protect the Rat Heart Against Isoproterenol-Induced Myocardial Infarction. *Phytother. Res.* **2016,** *30*, 1146–1155.

Peshin, A.; Singla, S. Anticalcifying Properties of *Dolichos biflorus* (Horse Gram) Seeds. *Indian J. Exp. Biol.* **1994,** *32*, 889–891.

Philip, A.; Athul, P.; Charan, A.; Afeefa, T. Anthelmintic Activity of Seeds of *Macrotyloma uniflorum*. *Hygeia* **2009,** *1*, 26–27.

Pitel, S.; Raccah, D.; Gerbi, A.; Pieroni, G.; Vague, P.; Coste, T. C. At Low Doses, a γ-Linolenic Acid-Lipoic Acid Conjugate is More Effective than Docosahexaenoic Acid-Enriched Phospholipids in Preventing Neuropathy in Diabetic Rats. *J. Nutr.* **2007,** *137*, 368–372.

Ponnulakshmi, R.; Shyamaladevi, B.; Vijayalakshmi, P.; Selvaraj, J. In Silico and In Vivo Analysis to Identify the Antidiabetic Activity of Beta Sitosterol in Adipose Tissue of High Fat Diet and Sucrose Induced Type-2 Diabetic Experimental Rats. *Toxicol. Mech. Methods* **2019,** *29*, 276–290.

Prasad, S. K.; Singh, M. K. Horse Gram—An Underutilized Nutraceutical Pulse Crop: A Review. *J. Food Sci. Technol.* **2015,** *52*, 2489–2499.

Seyoum, A.; Asres, K.; El-Fiky, F. K. Structure–Radical Scavenging Activity Relationships of Flavonoids. *Phytochemistry* **2006,** *67*, 2058–2070.

Sharma, N.; Bisht, S. S.; Gupta, S.; Panda, A. K.; Rana, M. Analysis of Proteomic Diversity and Calcium Binding Protein (s) in Seeds of Horse Gram (*Macrotyloma uniflorum*) Cultivars from Uttarakhand. *Int. J. Pharm. Sci. Res.* **2018,** *9*, 3274–3280.

Sharma, N.; Bisht, S. S.; Gupta, S.; Rana, M.; Kumar, A.; Pathak, R. In Silico Study for Antiurolithiatic Activities of Compounds Extracted from Horse Gram (*Macrotyloma uniflorum*) Collected from Uttarakhand (India) As Xanthine Dehydrogenase Inhibitor by Molecular Docking. *Int. J. Recent. Sci.* **2019,** *10*, 33630–33634.

Siddhuraju, P.; Manian, S. The Antioxidant Activity and Free Radical-Scavenging Capacity of Dietary Phenolic Extracts from Horse Gram (*Macrotyloma uniflorum* (Lam.) Verdc.) seeds. *Food Chem.* **2007,** *105*, 950–958.

Sodani, S.; Paliwal, R.; Jain, L. *Phenotypic Stability for Seed Yield in Rainfed Horsegram (Macrotyloma uniflorum [Lam.] Verdc).* Paper Presented in National Symposium on Arid Legumes for Sustainable Agriculture and Trade, November 5–7, 2004.

Sree, V. K.; Soundarya, M.; Ravikumar, M.; Reddy, T. R.; Devi, N. K. D. In Vitro Screening of *Macrotyloma uniflorum* Extracts for Antioxidant and Anthelmintic Activities. *J. Pharmacogn. Phytochem.* **2014,** *3*.

Sreerama, Y. N.; Sashikala, V. B.; Pratape, V. M. Variability in the Distribution of Phenolic Compounds in Milled Fractions of Chickpea and Horse Gram: Evaluation of Their Antioxidant Properties. *J. Agric. Food Chem.* **2010,** *58*, 8322–8330.

Sreerama, Y. N.; Sashikala, V. B.; Pratape, V. M.; Singh, V. Nutrients and Antinutrients in Cowpea and Horse Gram Flours in Comparison to Chickpea Flour: Evaluation of Their Flour Functionality. *Food Chem.* **2012,** *131*, 462–468.

Sujatha, S.; Anand, S.; Sangeetha, K.; Shilpa, K.; Lakshmi, J.; Balakrishnan, A.; Lakshmi, B. Biological Evaluation of (3β)-STIGMAST-5-EN-3-OL as Potent Anti-Diabetic Agent in Regulating Glucose Transport Using In Vitro Model. *Int. J. Diabetes Mellit.* **2010,** *2*, 101–109.

Suresh, Y.; Das, U. Long-Chain Polyunsaturated Fatty Acids and Chemically Induced Diabetes Mellitus: Effect of ω-3 Fatty Acids. *Nutrition* **2003,** *19*, 213–228.

Tikkanen, M. J.; Wähälä, K.; Ojala, S.; Vihma, V.; Adlercreutz, H. Effect of Soybean Phytoestrogen Intake on Low Density Lipoprotein Oxidation Resistance. *Proc. Natl. Acad Sci.* **1998,** *95*, 3106–3110.

Urbano, G.; Lopez-Jurado, M.; Aranda, P.; Vidal-Valverde, C.; Tenorio, E.; Porres, J. The Role of Phytic Acid in Legumes: Antinutrient or Beneficial Function? *J. Physiol. Biochem.* **2000,** *56*, 283–294.

Yeh, Y. -H.; Lee, Y. -T.; Hsieh, H. -S.; Hwang, D. -F. Dietary Caffeic Acid, Ferulic Acid and Coumaric Acid Supplements on Cholesterol Metabolism and Antioxidant Activity in Rats. *J. Food Drug Anal.* **2009,** *17*, 123–132.

Yin, Y.; Liu, X.; Liu, J.; Cai, E.; Zhu, H.; Li, H.; Zhang, L.; Li, P.; Zhao, Y. Beta-Sitosterol and Its Derivatives Repress Lipopolysaccharide/d-Galactosamine-Induced Acute Hepatic Injury by Inhibiting the Oxidation and Inflammation in Mice. *Bioorg. Med. Chem. Lett.* **2018,** *28*, 1525–1533.

Yogeeta, S. K.; Gnanapragasam, A.; Kumar, S. S.; Subhashini, R.; Sathivel, A.; Devaki, T. Synergistic Interactions of Ferulic Acid with Ascorbic Acid: Its Cardioprotective Role During Isoproterenol Induced Myocardial Infarction in Rats. *Mol. Cell Biochem.* **2006,** *283*, 139–146.

CHAPTER 15

Chemical Composition and Biological Properties of *Pterocarpus marsupium* Roxb. (Family: Fabaceae)

SANTHIVARDHAN CHINNI* and RAVILLA JYOTHSNA

Department of Pharmacology, Raghavendra Institute of Pharmaceutical Education and Research, Anantapuramu, Andhra Pradesh, India

Corresponding author. E-mail: chinni0143@gmail.com

ABSTRACT

Pterocarpus marsupium Roxb., also known as Indian Kino tree, has remarkable anti-diabetic, antihyperlipidemic, and anti-inflammatory properties. The pharmacological properties of this tree are majorly attributed to 3 distinguished constituents, pterostilbene, marsupsin, and pterosupin. Several scientific studies have explored the potential mechanisms behind their pharmacology. The current chapter consolidates the findings of peer-reviewed scientific literature available on bioactives and pharmacology of *P. marsupium.*

15.1 INTRODUCTION

Pterocarpus marsupium Roxb. is generally known as Indian Kino tree in English. The other vernacular names of this plant are *Biyo* in Gujarati; *Bijasara, Asana* in Kannada, *Lal Chandeur* in Kashmiri, *Venga* in Malyalam, *Biyala lakda, Bibala* in Marathi, *Chandan lal, Channanlal* in Punjabi, *Pitasala, Asana, Sarfaka, Pijaka, Vengaimaram chakkal, Nengai* in Tamil, *Peddagi Chekka,*

Bioactives and Pharmacology of Legumes. T. Pullaiah, PhD (Ed.)

Yegi, Vegisa in Telugu; *Bijasar* in Urdu. This plant is extensively found in India and Sri Lanka. *P. marsupium* is a medium to large-sized deciduous tree growing up to 30 m in height and 2.5 m in girth. The trees have dark brown to gray bark superficial fissures, the leaves are compound and imparipinnate of 3-7 in. length of 5–8 leaflets, oblong, obtuse, emarginated, glabrous with round, smooth, and waved petioles of 5 or 6 in. Flowers are yellow in terminal panicles and are 1.5 cm in length. Fruits are flat, pods are brown, glabrous, orbicular, winged of 3–6 cm in width; the flowering and fruiting duration is between March and June and the seeds are convex and bony in the shape of kidney, reddish brown, rigid, smooth glossy leathery texture. *P. marsupium* is known for its potential therapeutic activity due to the presence of phytoconstituents like flavonoids, isoflavonoids, pterocarpans, aurones, lignans, stilbenes, sterols, triterpenes, and sesquiterpenes.

15.2 BIOACTIVES

The percentage of constituents in *P. marsupium* are pterostilbene (45%), tannins (5%), alkaloids (0.4%), and other proteins. The main constituents include liquiritigenin, isoliquiritigenin, pterostilbene, pterosupin, epicatechin, catechin, kinotannic acid, kinoin, kino red, β-eudesmol, carsupin, marsupial, marsupinol, pentosan, *p*- hydroxybenzaldehyde. Flavonoids such as liquiritigenin, 6-hydroxy-3,5,7,4′-tetramethoxyflavone-6-O-rhamnopyranoside, 5-deoxykaempferol, garbanzol were isolated from the roots of *P. marsupium*, whereas vijayoside, 8-C-ß-D-glucopyranosyl-3,4′,7-trihydroxyflavone, 8-C-ß-D-glucopyranosyl-3,7,3′,4′-tetrahydroxyflavone, 7,4′-dihydroxyflavone, bianol, pterocarpinol, (2S)-7-hydroxyflavanone were isolated from the wood of *P. marsupium*. 5,7,4′-Trihydroxy-3-(3.-methyl butyl) flavone was isolated from the bark of the tree and from other parts of the plant. Naringenin, 7-hydroxy-6,8-dimethylflavanone-7-O-α-L-arabinopyranoside, and 7,8,4′-trihydroxy-3′,5′-dimethoxyflavanone- 4′-O-β-D-glucopyranoside were isolated (Adinarayana et al. 1982; Bezuidenhoudt et al. 1987; Maurya et al. 2004b; Tripathi and Joshi 1988, 2007).

Isoflavonoids like 5,4′-Dimethoxy-8-methylisoflavone-7-O-α-L-rhamnopyranoside, 5,7-Dihydroxy-6-methoxyisoflavone-7-rhamnoside, Irisolidone-O-α-L-rhamnoside, Retusin-7-O-glucoside, 8-Hydroxy-4′-methoxy isoflavanone-7-O-glucopyranoside, Pteromarsupone, Marsupol (4,4′-dihydroxy-α-methylhydrobenzoin) were isolated from the wood of the *P. marsupium*. Pseudobaptigenin was isolated from the root and remaining Retusin-8-O-α-L-arabinopyranoside, 7-O-α-L-rhamnopyranosyl-oxy-4′-methoxy-5-

hydroxy isoflavone were isolated from other parts of the plant (Anandharajan et al., 2005; Mitra and Joshi 1982, 1983).

(-)-Epicatechin was isolated from the bark of the tree and chalcones like isoliquiritigenin (Seshadri, 1972), pterosupin are isolated from the root (Adinarayana et al., 1982), and coatline A, bijayasaline were isolated from the wood of the tree (Achari et al., 2012; Bezuidenhoudt et al., 1987; Chakraborty et al., 2010; Yahara et al., 2013). All the aurones and Isoaurones of *P. marsupium* are derived from the wood of the tree which included carpusin, marsupsin, 2-α-hydroxy-2-p-hydroxybenzyl-3(2H) benzofuranone-7-C-β-D-glucopyranoside (marsuposide), 2β-hydroxy-2-p-hydroxybenzyl-3(2H)benzofuranone-7-C-β-D-glucopyranoside,6,4′-dihydroxy-7-methylaurone-6-O-α-L-rhamnopyranoside, 4,6,3′,4′-tetrahydroxyaurone-6-O-α-L-rhamnopyranoside, pterocarposide, and pteroisoauroside, respectively (Grover et al., 2004; Handa et al., 2000; Yahara et al. 2013; Mathew and Rao 1984; Maurya et al. 2004a,b; Mohan and Joshi 1989). Lignans like pteroside, piyaline, piyaline methyl ester, and metaline are derived from the wood of *P. marsupium* (Achari et al., 2012; Chakraborty et al., 2010; Yahara et al., 2013; Maurya et al. 2004a, b). Sesquiterpenes like β-eudesmol, selin-4(15)-en-1-ß,11-diol, (+)-pterocarpol were isolated from the root wood of the tree (Adinarayana et al. 1982; Bhargava, 1946). Propterol and propterol B are the phenylpropanoids isolated from wood (Mathew and Rao 1984; Rao et al. 1984) and lupeol, erythrodiol-3-monoacetate triterpenes were remaining phytoconstituents derived from the root wood of the tree (Adinarayana et al., 1982; Maurya et al. 2004a). Yadav and Singh (1988) reported 6-hydroxy-3,5,7,4′-tetramethoxyflavone 6-rhamnoside from roots of *Pterocarpus marsupium*.

Chemical Structures of Important Phytoconstituents

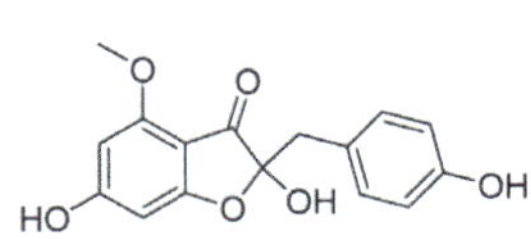

Marsupsin

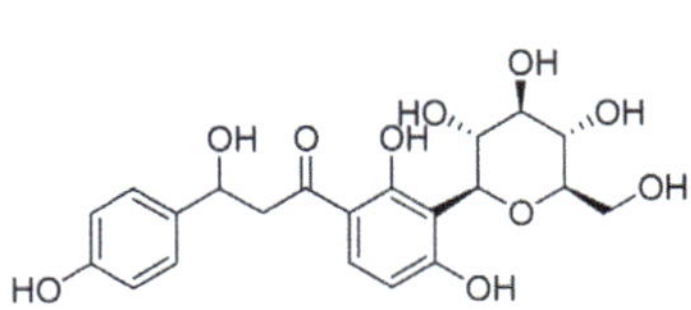

Pterosupin

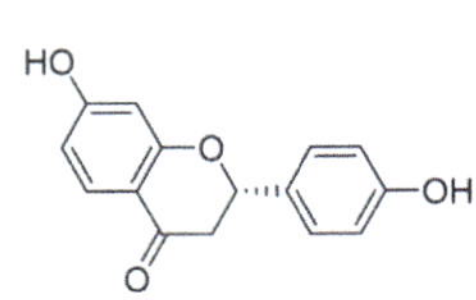

Liquiritigenin

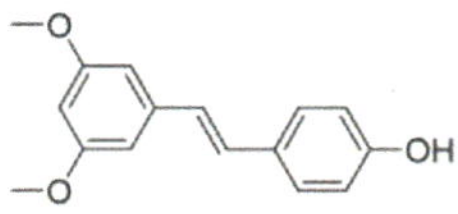

Pterostilbene

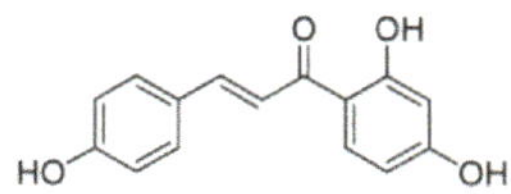

Isoliquiritigenin

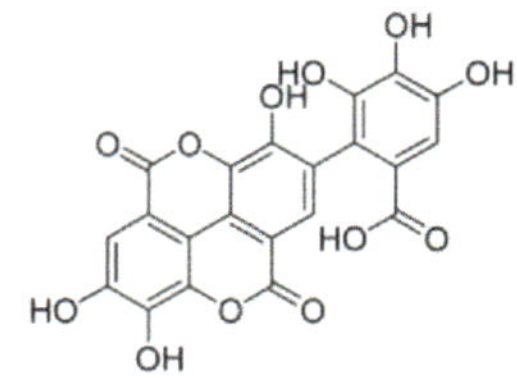

Kinotannic acid

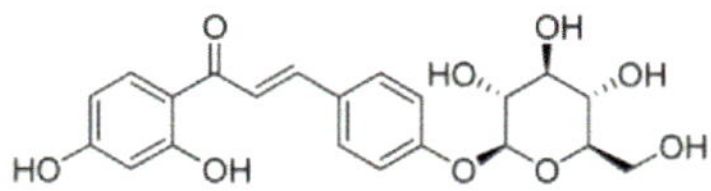

Isoliquiritin

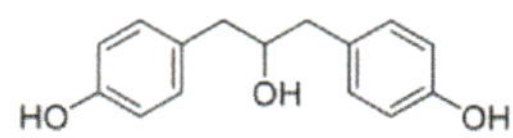

Propterol

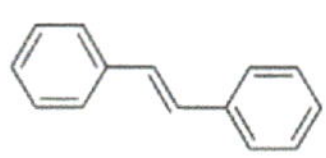

Trans-stilbene

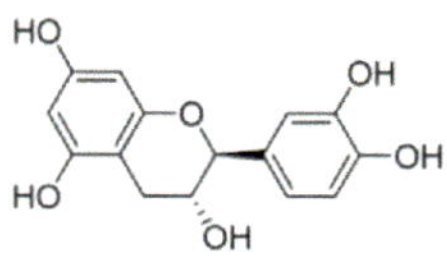

(-)- Catechin

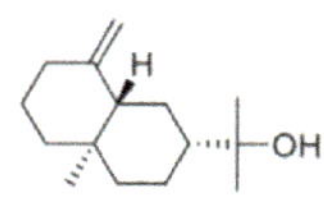

Beta eudesmol

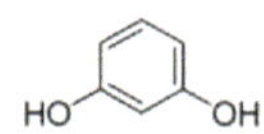

Resorcinol

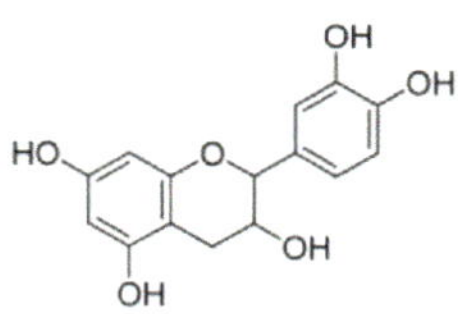

(-)-Epicatechin

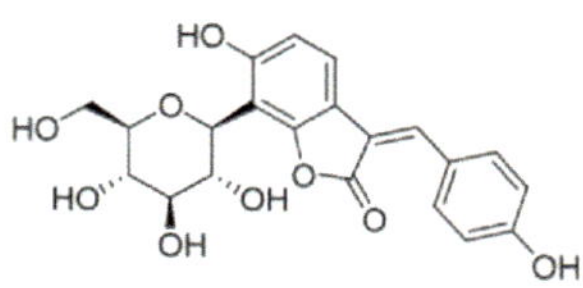

Pterocarposide

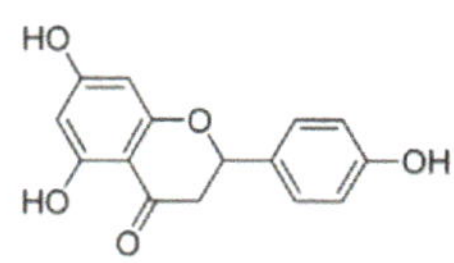

Naringenin

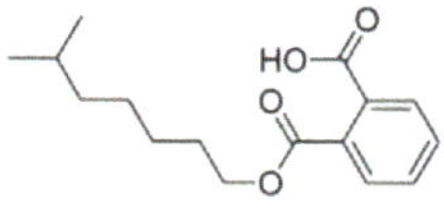

1,2-benzenedicarboxylic acidisooctyl ester

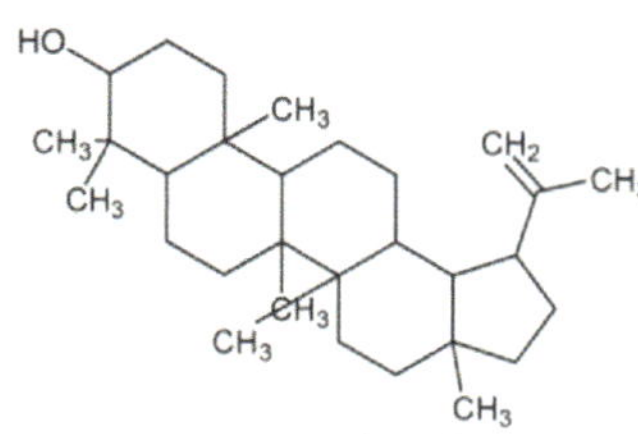

Lupeol

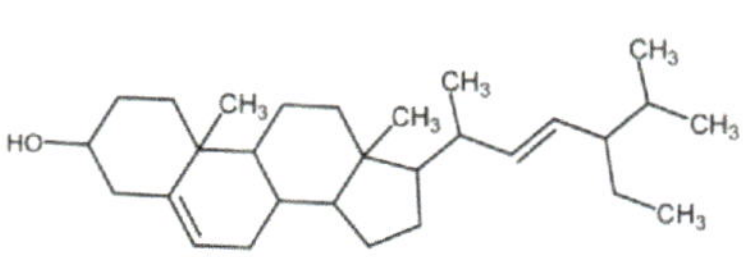

Stigmasterol

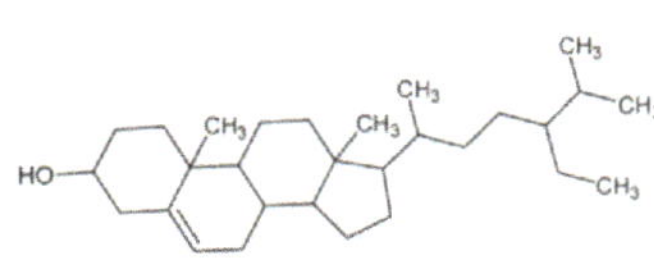

Sitosterol

15.3 PHARMACOLOGY

15.3.1 ANTIDIABETIC ACTIVITY

P. marsupium is an evident antidiabetic agent. *P. marsupium* restores the normal insulin secretion by improving β cells population and refilling the islets (Mishra et al., 2013). A double-blinded multicenter randomized controlled trial conducted at a flexible-dose depicted that *P. marsupium* plant can act as a potent blood sugar reducing agent (Hariharan et al., 2005). When compared with metformin, phenolic compounds like marsupsin and pterostilbene were seemed to be more effective in the treatment of diabetes at a dose of 40 mg/kg b.wt. (Hariharan et al., 2005). A multicenter level study conducted by Indian Council of Medical Research (ICMR) on *P. marsupium* showed that the plant reduced the blood glucose without exhibiting any adverse effects. A similar study on beneficial effects of *P. smarsupium* on diabetes conducted by the ICMR reported a fall in blood glucose level from 151–216 mg/dl to 32–45 mg/dl and HbA_{1c} level from 9.8% to 9.4%, respectively (Philip, 1998). In a streptozotocin-induced model of diabetes, administration of an aqueous extract of *P. marsupium* bark showed a potential effect in diabetes by bringing the glycosylated hemoglobin, LDL cholesterol, triglycerides, total cholesterol to accepted levels and raised enzymes like aspartate transaminase creatine kinase, glutamyl transferase, alkaline phosphatase, alanine transaminase to normal reach (Kannabiran and Gayathri, 2009). (-) epicatechin raised the glutathione content in Type-2 diabetic erythrocytes (Rizvi and Zaid 2001) when studying the impact of insulin and (-) epicatechin on glutathione content in normal and diabetic erythrocytes.

Some of the findings suggest that the antidiabetic potential of *P. marsupium* was possibly due to the capability to interfere with glucose diffusion across the biological membrane. When compared, lowering of blood glucose levels among glimepiride and ethanolic extract of *P. marsupium* at 200 mg/kg b. wt. and 400 mg/kg b. wt. in a period of 180 min, the highest blood-glucose-lowering effect was noted in glimepiride (57.56%) followed by 400 mg/kg b. wt. of extract (55.13%) then 200 mg/kg b. wt. (51.30%) (Pant et al. 2017). Similarly, the antidiabetic property of ethanolic extract of *P. marsupium* heartwood was found in dexamethasone-induced hyperglycemia and hyperinsulinemia in Wistar male albino rats at a dose of 1 gm/kg per oral and 2 gm/kg per oral (Narendar et al., 2016). Aqueous extract of *P. marsupium* at 100 mg/kg and 200 mg/kg b.wt. lowered the cytokine TNF-α levels in streptozocin-induced non-insulin-dependent diabetic Mellitus (Srinivasan et al. 2010). Bioassay showed plant extract had strong antidiabetic properties in vivo and in vitro employing concentration-dependent method (Mohankumar et al., 2012).

15.3.2 ANTIHYPERLIPIDEMIC ACTIVITY

Phytoconstituents of *P. marsupium* like marsupin, pterosupin, and liquiritigenin showed anti-hyperlipidemic activity by reducing serum cholesterol, LDL cholesterol, and atherogenic index and triglyceride level; similarly, ethanolic heartwood extract of the tree was known for its anti-hyperlipidaemic property as it lowered the total cholesterol, serum triglyceride, VLDL- and LDL-cholesterol (Jahromi et al., 2004). Ethanolic extract of bark and wood of *P. marsupium* at a dose of 150 mg/kg b.wt. had lessened the lipid and blood glucose levels in alloxan (150 mg/kg, i.p.) induced diabetic rats for 14 days (Jahromi et al., 2004).

15.3.3 ANTI-INFLAMMATORY ACTIVITY

One of the constituents, pterostilbene, shows anti-inflammatory activity by inhibiting PGE_2 and selective COX-1/2 in LPS-stimulated peripheral blood mononuclear cells (PBMC) (Srinivasan et al. 2010; Salunkhe et al. 2005). In non-insulin-dependent type 2 diabetic rats at specific doses of 100 and 200 mg/kg, an aqueous extract of *P. marsupium* declined the uplifted cytokine levels and tumor necrosis factor TNF-α levels (Srinivasan et al. 2010), and similarly the aqueous and methanolic bark extract at a dose of 50 mg/kg b.wt. 100 mg/kg b.wt. respectively showed anti-inflammatory property in the carrageenan-induced rat paw edema method.

15.3.4 ANTIOXIDANT ACTIVITY

P. marsupium bark ethanolic, aqueous, and ethyl acetate extract showed its free radical scavenging capacity in various 2,2′-azino-bis (3-ethylbenzothiazoline-6-sulfonic acid) (ABTS), NO, OH, SO, 2,2-diphenyl-1-picrylhydrazyl (DPPH) antioxidant models and also suppressed lipid peroxidation in vitro, demonstrating *P. marsupium* antioxidant potential (Mohammadi et al., 2009). In vitro antioxidant activity of *P. marsupium* leaf extract (100 μg/mL) was studied using various assays like 2,2-diphenyl-1-picrylhydrazyl (DPPH), hydroxyl radical scavenging activity, NO radical scavenging activity, ABTS, ferric reducing antioxidant power (FRAP), telomerase repeated amplification protocol (TRAP) assay, hydrogen peroxide (H_2O_2) radical scavenging activity, among this 71% of scavenging activity was found in FRAP.

15.3.5 HEPATOPROTECTIVE ACTIVITY

The methanolic bark extract of *P. marsupium* showed hepatoprotective activity when taken at a dose of 25 mg/kg per day oral for 14 days against carbon tetrachloride (CCl_4) induced hepatoxicity as biochemical parameters like serum protein, total bilirubin, alkaline phosphatase, alanine aminotransaminase, aspartate aminotransaminase were observed normal after the administration of the extract (Krishna et al., 2005). Similarly, in carbon tetrachloride (CCl_4) induced hepatotoxicity study enzymes like aspartate transaminase, alkaline phosphatase, lactate dehydrogenase, bilirubin, and alanine transaminase levels were seemed to be declined in extract-treated groups (Devipriya et al., 2007).

15.3.6 ANTICANCER ACTIVITY

Phytoconstituents like stilbene and pterostilbene are known for their potent anticancer properties as pterostilbene suppressed the cell proliferating factors like Akt, Bcl-2, and stimulated the mitochondrial apoptotic signals such as Bax and caspases. *P. marsupium* extracts are shown to prevent major metastasis inducers like Matrix Metalloproteinase 9 (MMP) and α-Methyl Acyl CoA racemose (AMACR). Pterostilbene induces apoptosis in multifarious target sites for the treatment of prostate and breast cancers (Chakraborty et al., 2010; Pan et al., 2007; Rimando and Suh, 2008). Pterostilbene activity was also assessed in cell lines A-375, HCT-116, Hep-G2, MDAMB- 231, and PC-3 at concentrations 1, 10, 50, and 100 μg/mL via measurement of cell viability. The IC_{50} of each cell line was found to be 3.9 mM (PC-3), 16.0 μM (Hep-G2), 40.6 μM (MDA-MB-231), 45.3 μM (HCT-116), 421 μM (A-375). This study revealed pterostilbene high activity against Hep-G2 (liver) and HCT-116 (colon) cancers and it was not active in PC-3 (Remsberg et al., 2008). Pterostilbene instigated apoptosis in human gastric carcinoma AGS cells by the activation of caspase cascade via the mitochondrial and Fas/FasL pathway, GADD expression, and through modification of cell cycle development and several cycle-regulating proteins (Pan et al., 2007).

15.3.7 ANTIMICROBIAL ACTIVITY

P. marsupium bark methanolic and aqueous extract showed antimicrobial property in vitro, especially methanolic extract suppressed the growth of *A. niger* at 25 μg/mL and *E. faecalis, S. typhi* at 12.5 μg/mL at a zone of inhibition range of 11–22 mm (Ramesh and Aruna, 2017). Through the study of cyclic voltammetry, the ethanolic extract of *P. marsupium* was found to show its antimicrobial activity toward *Bacillus polymyxa, Vibrio cholerae*, and *Candida albicans*. The low anodic current and low anodic peak potential were gained showing the favorable declining ability of the molecules leading to the good antioxidant potential of the extract. The result revealed the marked antimicrobial impact at distinct dosages (Deepa et al., 2014).

15.3.8 ANTIBACTERIAL ACTIVITY

100 mg/mL concentration of methanolic stem extract of *P. marsupium* under paper disc diffusion method suppressed the growth of gram-positive bacteria *Bacillus coagulans* and gram-negative bacteria *Escherichia coli* and the zone of inhibition varied between 11 and 22 mm for different extracts. In vitro studies have shown that the growth of *Pseudomonas aeruginosa, Streptococcus pyrogens*, and *Staphylococcus aureus* has been prevented by *P. marsupium* (Nair et al., 2005). Specifically, antibacterial activity was shown at 50 mg/mL concentration of *P. marsupium* isopropyl alcohol and acetone extract against *Staphylococcus aureus* and *Bacillus cereus* (Gram +ve bacteria) (Pant et al. 2017).

15.3.9 NOOTROPIC ACTIVITY

P. marsupium also showed its efficacy in improving memory. The dose of 25 and 50 mg/kg of methanolic extract administered orally by adult Swiss albino mice enhanced learning and memory, while in scopolamine-induced amnesia it enhanced inflection ratio, lowered transfer latency, and improved the impairment of spatial memory (Katiyar et al., 2016).

15.3.10 CARDIOTONIC ACTIVITY

It was observed that aqueous wood extract of *P. marsupium* showed cardiotonic activity in isolated frog heart perfusion at a low concentration of 0.25

mg/mL. At this concentration, there was a reduced heart rate and increase in the force of contraction, when compared to digoxin when given with the highest concentration of 4 mg/mL cardiac arrest, was observed (Mohire et al. 2007).

15.3.11 ANALGESIC ACTIVITY

Methanolic extract of *P. marsupium* leaves at a dose of 120 mg/kg b.wt. has shown potent analgesic activity in Swiss albino mice through acetic acid-induced writhing assay. In central analgesic activity studied using the hot plate method, compared to drug pentazocine, extract of *P. marsupium* bark had shown a significant effect in reducing the pain limit and increasing reaction time to heat stimulus at a dose of 500 mg/mL (Tippani et al., 2010).

KEYWORDS

- ***Pterocarpus marsupium***
- **Indian Kino**
- **bioactives**
- **pharmacology**
- **pterostilbene**

REFERENCES

Achari, B.; Dutta, P. K.; Roy, S. K.; Chakraborty, P.; Sengupta, J.; Bandyopadhyay, D.; Maity, J. K.l Khan, I. A.l Ding, Y.; Ferreira, D. Fluorescent Pigment and Phenol Glucosides from the Heartwood of *Pterocarpus marsupium*. *J. Nat. Prod.* **2012,** *75* (4), 655–660.

Adinarayana, D.; Syamasundar, K. V.; Seligmann, O.; Wagner, H. Structure Elucidation of Pterosupin from *Pterocarpus marsupium*, the First Naturally Occurring c-Glycosyl-β-Hydroxy-Dihydrochalcone. *Z. Naturforsch. C* **1982,** *37* (3–4), 145–147.

Anandharajan, R.; Pathmanathan, K.; Shankernarayanan, N. P.; Vishwakarma, R. A.; Balakrishnan, A. Upregulation of Glut-4 and PPAR Gamma by an Isoflavone from *Pterocarpus marsupium* on L6 Myotubes: A Possible Mechanism of Action. *J. Ethnopharmacol.* **2005,** *97* (2), 253–260.

Bezuidenhoudt, B. C. B.; Brandt, E. V.; Ferreira, D. Flavonoid Analogues from *Pterocarpus* Species. *Phytochemistry* **1987,** *26* (2), 531–535.

Bhargava, P. N. Chemical Examination of the Fixed Oil Derived from the Wood of *Pterocarpus marsupium* Roxb. *Proc. Indian Acad. Sci. Sect. A* **1946,** *24* (6), Article No. 501.

Chakraborty, A.; Gupta, N.; Ghosh, K.; Roy, P. In Vitro Evaluation of the Cytotoxic, Anti-Proliferative and Anti-Oxidant Properties of Pterostilbene Isolated from *Pterocarpus marsupium*. *Toxicol. in Vitro* **2010,** *24* (4), 1215–1228.

Deepa, R.; Manjunatha, H.; Krishna, V.; Kumara Swamy, B. Evaluation of Antimicrobial Activity and Antioxidant Activity by Electrochemical Method of Ethanolic Extract of *Pterocarpus marsupium* Roxb bark. *J. Biotech. Biomater.* **2014,** *4* (1), 166. doi:10.4172/2155-952X.1000166

Devipriya, D.; Gowri, S.; Nideesh, T. R. Hepatoprotective Effect of *Pterocarpus marsupium* Against Carbon Tetrachloride Induced Damage In Albino Rats. *Anc. Sci. Life,* **2007,** *27* (1), 19–25.

Grover, R. K.; Maurya, R.; Roy, R. Dynamic NMR Investigation of Two New Interconvertible Diasteriomeric Epimers of Natural 2-Benzyl-2-Hydroxybenzofuranone Derivative from *Pterocarpus marsupium*. *Tetrahedron* **2004,** *60* (9), 2005–2010.

Handa, S. S.; Singh, R.; Maurya, R.; Satti, N. K.; Suri, K. A.; Suri, O. P. Pterocarposide, an Isoaurone C-Glucoside from *Pterocarpus marsupium*. *Tetrahedron Lett.* **2000,** *41* (10), 1579–1581.

Hariharan, R.; Venkataraman, S.; Sunitha, P.; Rajalakshmi, S.; Samal, K. Efficacy of Vijayasar (*Pterocarpus marsupium*) in the Treatment of Newly Diagnosed Patients with Type 2 Diabetes Mellitus: A Flexible Dose Double-Blind Multicenter Randomized Controlled Trial. *Diabetol. Croat.* **2005,** *34* (1), 13–20.

Jahromi, M. A. F.; Ray, A. B.; Chansouria, J. P. N. Antihyperlipidemic Effect of Flavonoids from *Pterocarpus marsupium*. *J. Nat. Prod.* **2004,** *56* (7), 989–994.

Kannabiran, K.; Gayathri, M. Antimicrobial Activity of *Hemidesmus indicus, Ficus bengalensis* and *Pterocarpus marsupium* Roxb. *Indian J. Pharmaceut. Sci.* **2009,** *71* (5), 578–581.

Katiyar, D.; Singh, V.; Ali, M. Phytochemical and Pharmacological Profile of *Pterocarpus marsupium:* A Review. *Pharm. Innov. J.* **2016,** *5* (4), 31–39.

Krishna, V.; Manjunatha, B. K.; Vidya, S. M.; Jagadeesh Singh, S. D.; Manohara, Y. N.; Raheman, A.-U.; Avinash, K. R.; Mankani, K. L. Evaluation of Hepatoprotective Activity of Stem Bark of *Pterocarpus marsupium* Roxb. *Indian J. Pharmacol.* **2005,** *37* (3), 165–168

Mathew, J.; Rao, A. V. S. Propterol B, A Further 1,3-Diarylpropan-2-ol from *Pterocarpus marsupium*. *Phytochemistry* **1984,** *23* (8), 1814–1815.

Maurya, R.; Ray, A. B.; Duah, F. K.; Slatkin, D. J.; Schiff, P. L. Constituents of *Pterocarpus marsupium*. *J. Nat. Prod.* **2004a,** *47* (1), 179–181.

Maurya, R.; Singh, R.; Deepak, M.; Handa, S. S.; Yadav, P. P.; Mishra, P. K. Constituents of *Pterocarpus marsupium:* An Ayurvedic Crude Drug. *Phytochemistry* **2004b,** *65* (7), 915–920.

Mishra, A.; Srivastava, R.; Srivastava, S. P.; Gautam, S.; Tamrakar, A. K.; Maurya, R.; Srivastava, A. K. Antidiabetic Activity of Heart wood of *Pterocarpus marsupium* Roxb. and Analysis of Phytoconstituents. *Indian J. Exp. Biol.* **2013,** *51* (5), 363–374.

Mitra, J.; Joshi, T. An Isoflavone Glycoside from the Heartwood of *Pterocarpus marsupium*. *Phytochemistry* **1982,** *21* (9), 2429–2430.

Mitra, J.; Joshi, T. Isoflavonoids from the Heartwood of *Pterocarpus marsupium*. *Phytochemistry* **1983,** *22* (10), 2326–2327.

Mohammadi, M.; Khole, S.; Devasagayam, T. P.; Ghaskadbi, S. S. *Pterocarpus marsupium* Extract Reveals Strong In Vitro Antioxidant Activity. *Drug Discov. Ther.* **2009,** *3* (4), 151–161.

Mohan, P.; Joshi, T. Two Anthochlor Pigments from Heartwood of *Pterocarpus marsupium. Phytochemistry* **1989,** *28* (9), 2529–2530.

Mohankumar, S. K.; O'Shea, T.; McFarlane, J. R. Insulinotrophic and Insulin-Like Effects of a High Molecular Weight Aqueous Extract of *Pterocarpus marsupium* Roxb. hardwood. *J. Ethnopharmacol.* **2012,** *141* (1), 72–79.

Mohire, N. C.; Salunkhe, V. R.; Bhise, S. B.; Yadav, A. V. Cardiotonic Activity of Aqueous Extract of Heartwood of *Pterocarpus marsupium. Indian J. Exp. Biol.* **2007,** *45* (6), 532–537.

Nair, R.; Kalariya, T.; Chanda, S. Antibacterial Activity of Some Selected Indian Medicinal Flora. *Turk. J. Biol.* **2005,** *29* (1), 41–47.

Narendar, K.; Nayak, M.; Jamadar, M.; Patil, A.; Anand, S. Comparison of the Effect of *Pterocarpus marsupium* with Pioglitazone in Dexamethasone-Induced Insulin Resistance. *Asian J. Pharm. Clin. Res.* **2016,** *9* (2), 211–214.

Pan, M.-H.; Chang, Y.-H.; Badmaev, V.; Nagabhushanam, K.; Ho, C.-T. Pterostilbene Induces Apoptosis and Cell Cycle Arrest in Human Gastric Carcinoma Cells. *J. Agric. Food Chem.* **2007,** *55* (19), 7777–7785.

Pant, D.; Pant, N.; Saru, D.; Yadav, U.;Khanal, D. Phytochemical Screening and Study of Anti-Oxidant, Anti-Microbial, Anti-Diabetic, Anti-Inflammatory and Analgesic Activities of Extracts from Stem Wood of *Pterocarpus marsupium* Roxburgh. *J. Intercult. Ethnopharmacol.* **2017,** *6* (2), 170–176.

Philip, S. Flexible Dose Open Trial of Vijayasar in Cases of Newly-Diagnosed Non-Insulin-Dependent Diabetes Mellitus. Indian Council of Medical Research (ICMR), Collaborating Centres, New Delhi. *Indian J. Med. Res.* **19989,** *108*, 24–29.

Ramesh, L. L.; Aruna, L. H. Physicochemical, Phytochemical Profiling and Anti-Microbial Activity of *Pterocarpus marsupium. Int. J. Pharm. Sci. Res.* **2017,** *8* (5), 2177–2183.

Rao, A. V. S.; Mathew, J.; Sankaram, A. V. B. Propterol: A 1,3-Diarylpropan-2-ol from *Pterocarpus marsupium. Phytochemistry* **1984,** *23* (4), 897–898.

Remsberg, C. M.; Yáñez, J. A.; Ohgami, Y.; Vega-Villa, K. R.; Rimando, A. M.; Davies, N. M. Pharmacometrics of Pterostilbene: Preclinical Pharmacokinetics and Metabolism, Anticancer, Antiinflammatory, Antioxidant and Analgesic Activity. *Phytother. Res.* **2008,** *22* (2), 169–179.

Rimando, A.; Suh, N. Biological/Chemopreventive Activity of Stilbenes and Their Effect on Colon Cancer. *Planta Med.* **2008,** *74* (13), 1635–1643.

Rizvi, S. I.; Zaid, M. A. Intracellular reduced Glutathione Content in Normal and Type 2 Diabetic Erythrocytes: Effect of Insulin and (-)Epicatechin. *J. Physiol. Pharmacol.* **2001,** *52* (3), 483–438.

Salunkhe, V.; Yadav, A.; Shete, A.; Kane, S.; Kulkarni, A. Anti-Inflammatory Activity of Hydrogels of Extracts of *Pterocarpus marsupium* and *Coccinia indica. Indian Drugs* **2005,** *42*, 319–321.

Seshadri, T. R. Polyphenols of *Pterocarpus* and *Dalbergia* wood. *Phytochemistry* **1972,** *11*, 881–898.

Srinivasan, B. P.; Halagappa, K.; Girish, H. N. The Study of Aqueous Extract of *Pterocarpus marsupium* Roxb. on Cytokine TNF-α in Type 2 Diabetic Rats. *Indian J. Pharmacol.* **2010,** *42* (6), 392–396.

Tippani, R.; Porika, M.; Allenki, V.; Anreddy, R. N. R.; Yellu, N. R.; Krishna, D. R.; Thammidala, C.; Abbagani, S. Antioxidant and Analgesic Activities of *Pterocarpus marsupium* Roxb. *J. Herbs Spices Med. Plants* **2010,** *16* (1), 63–68.

Tripathi, J.; Joshi, T. Phytochemical Investigation of Roots of *Pterocarpus marsupium.* Isolation and Structural Studies of Two New Flavanone Glycosides. *Z. Naturforsch. C,* **1988,** *43* (3–4), 184–186.

Tripathi, J.; Joshi, T. Flavonoids from *Pterocarpus marsupium. Planta Med.* **2007,** *54* (4), 371–372.

Yadav, R. N.; Singh, R. K. 6-Hydroxy-3,5,7,4′-Tetramethoxyflavone 6-Rhamnoside from Roots of *Pterocarpus marsupium. Phytochemistry* **1988,** *48* (7), 1259–1261.

Yahara, S.; Devkota, H. P.; Joshi, K. R. Chemical Analysis of Heartwood of Bijayasal (*Pterocarpus marsupium* Roxb.). *Nepal J. Sci. Tech.* **2013,** *13* (2), 219–224.

CHAPTER 16

Biomolecules and Therapeutics of *Senna siamea* (Lam.) H. S. Irwin & Barneby (Syn.: *Cassia siamea* Lam.)

VICTORIEN DOUGNON*, BRICE BORIS LEGBA, ESTHER DEGUENON, JERROLD AGBANKPE, HORNEL KOUDOKPON, JEAN ROBERT KLOTOE, HONORÉ BANKOLE and JACQUES DOUGNON

Research Unit in Applied Microbiology and Pharmacology of Natural Substances, Research Laboratory in Applied Biology, Polytechnic School of Abomey-Calavi, University of Abomey-Calavi, Godomey, Benin

Corresponding author. E-mail: victorien.dougnon@gmail.com

ABSTRACT

Medicinal plants still have an important position in the health of populations, for their traditional medicinal use. Research on a large number of these plants has validated their therapeutic properties and highlighted the biomolecules they contain. *Senna siamea* (family Fabaceae) is one of the medicinal plants widely used in many parts of the world. This review aims to provide an overview of the chemical and pharmacological potential of the plant for an optimal valuation. The methodology is based on the analysis of the existing literature on the therapeutic properties and biomolecules isolated or found in the plant. From the results, it appears that *S. siamea* is rich in bioactive compounds such as saponins, reducing sugars, tannin resins, gums, flavonoids, anthraquinones, saponins, and glycosides in its leaves. These

Bioactives and Pharmacology of Legumes. T. Pullaiah, PhD (Ed.)

secondary metabolites account for the interesting therapeutic properties of the plant. Indeed, the antiviral, antibacterial, antioxidant, analgesic and anti-inflammatory, hepatoprotective, anticancer, anti-hyperglycemic and, hyperlipidemia, antiplasmodial and cardioprotective properties of the plant have been demonstrated. These data deserve to be valorised and applied for the development of new treatments.

16.1 INTRODUCTION

Medicinal plants have been used for centuries to ensure the well-being of populations. There are a significant number of medicinal molecules obtained from medicinal plants. *Senna siamea* is one of them. The medicinal plant belongs to the family Fabaceae (Fowler, 2006). It is attributed several common names including Cassodier, Thai copper pod, ironwood, yellow cassia etc. The plant is widely used in all regions of Thailand for ornamentation, cooking, and traditional medicine. In traditional medicine, *S. siamea* is used to treat constipation and insomnia (Monkheang et al., 2011) in Thailand. In Nigeria, it is used to treat Malaria, traditionally (Odugbemi et al., 2007). In the South of Benin (West Africa), *S. siamea* is one of the plants most used to treat salmonellosis (Dougnon et al., 2018). This review focuses on the pharmacological properties and chemical composition of *Senna siamea*. It aims to provide an overview of the chemical and pharmacological potential of the plant for an optimal valuation.

16.2 BIOACTIVES

The identified chemical components in *Senna siamea* and their proportions show slight variability depending not only on the extraction solvents but also on the method used for the chemical characterization. Saponins, reducing sugar, tannins resins, gums, flavonoids, anthraquinones, saponins, and glycosides have been identified in the leaf extracts (Bukar et al., 2009; Mohammed et al., 2013; Momin et al., 2012a,b; Hassan et al., 2015). Barakol is the major constituent of *S. siamea* leaves. The work done by Padumanonda and Gritsanapan (2006) showed, by chromatographic analysis, a composition of 0.4035% w/w barakol in young fresh leaves of *S. siamea*. But, using TLC densitometry, the barakol content was 1.67%, 1.43%, and 0.78% of dry weight, respectively, for young leaves, young flowers, and mature leaves (Padumanonda et al., 2007). The fresh young leaves also contain total anthraquinone glycosides and total anthraquinones, calculated as rhein, in 0.0523% and 0.0910% w/w, respectively (Sakulpanich and Gritsanapan, 2009).

Luteolin was isolated from leaf extracts of *S. siamea* (Ingkaninan et al., 2000).4-(trans)-acetyl-3,6,8-trihydroxy-3-methyldihydronaphthalenone, cassia chromone (5-acetonyl-7-hydroxy-2-methylchromone), 5-acetonyl-7-hydroxy-2-hydroxymethyl-chromone,4-(cis)-acetyl-3,6,8-trihydroxy-3-methyl-dihydronaphthalenone (Ingkaninan et al., 2000), upenone, lupeol, betulinic acid, chrysophanol, physcion and β- sitosterol glucoside (Ogbole et al., 2014), siamaurones A and B (Gao et al., 2013) were also isolated. Wu et al. (2016) isolated three new tricyclic alkaloids, siamalkaloid from the *S. siamea* twigs (Figures 16.1–16.4).

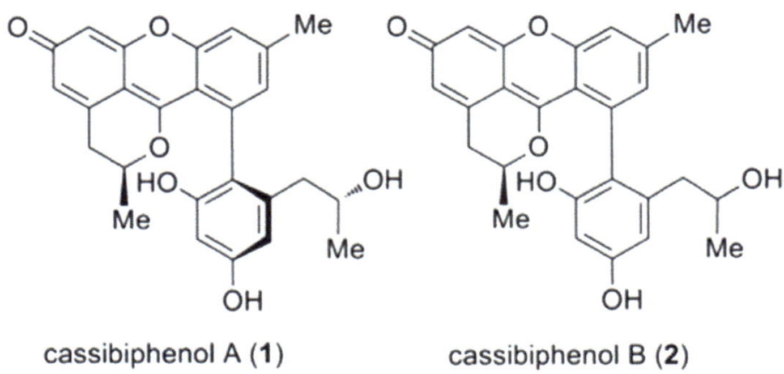

FIGURE 16.1 Cassibiphenol A and cassibiphenol B from *S. siamea* (Deguchi et al., 2014).

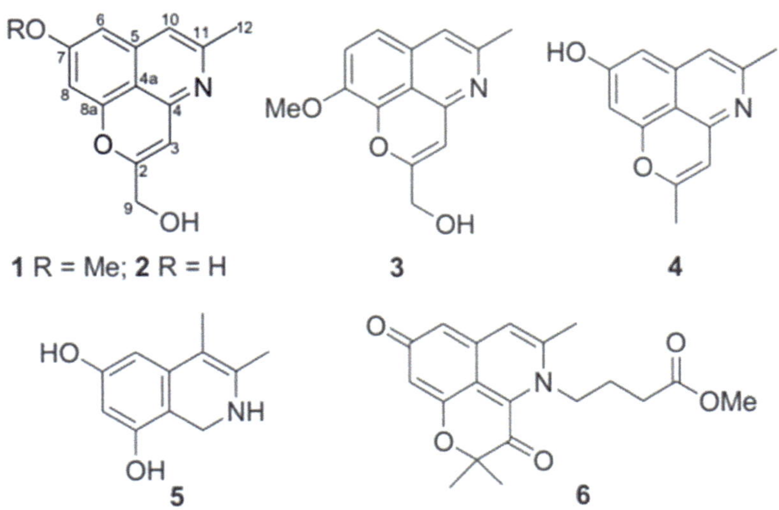

FIGURE 16.2 The structure of alkaloids from the twigs of *S. siamea* (Wu et al., 2016).

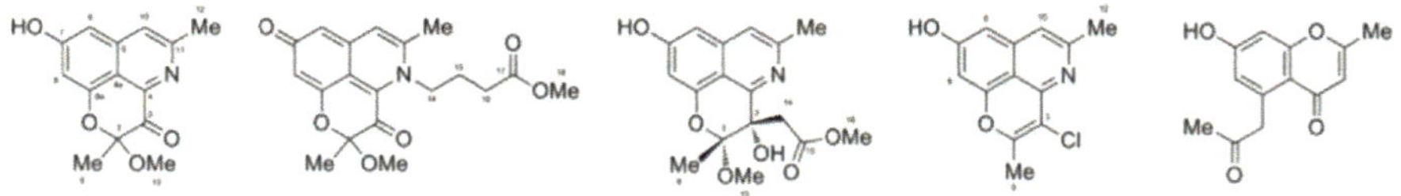

cassiarin G (1) cassiarin H (2) cassiarin J (3) cassiarin K (4) 5-acetonyl-7-hydroxy-2-methylchromone

FIGURE 16.3 Tricyclic Alkaloids, Cassiarins G, H, J, and K from Leaves of *S. siamea* (Deguchi et al., 2012).

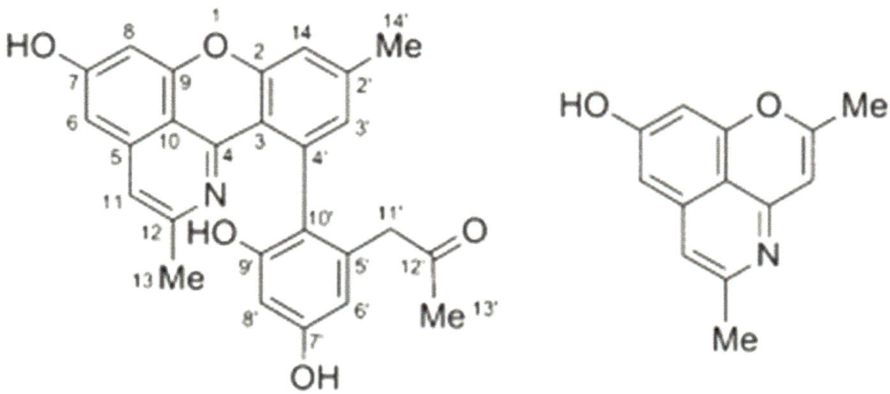

cassiarin F (1) cassiarin A (2)

FIGURE 16.4 Cassiarin F and A from *S. siamea* (Deguchi et al., 2011).

16.3 PHARMACOLOGY

16.3.1 Antiviral Activity

Interesting antiviral activity of triterpenoids and anthraquinones from *S. siamea* on poliovirus has been demonstrated (Ogbole et al., 2014). An IC_{50} of 0.014 μg/mL was obtained with lupeol, making this compound the most active of the triterpenoids tested. Chrysophanol also has an interesting activity (IC_{50} 0.46 μg/mL), associated with a higher selectivity than lupeol (Selective Index (SI) of 32.1) (Ogbole et al., 2014). Furthermore, siamaurones isolated from the plant showed good activity against tobacco mosaic virus (anti-TMV). For B siamaurones, an inhibition rate of 31.8% was observed, which is higher than that of Ningnanmycin, a reference antiviral (26.5%) (Gao et al., 2013).

16.3.2 Antibacterial Activity

Leaf extracts (aqueous, chloroform, and ethanolic extracts) of *S. siamea* were active on *Pseudomonas aeruginosa* at 500 µg/disc and 1000 µg/disc revealing a dose-dependent antibacterial activity (Bukar et al., 2009). Dahiru et al (2013) demonstrated that the different extracts (acetone, ethanol, and water) and ethanolic fractions (chloroform, ethyl acetate, and n-butanol) of *S. siamea* had inhibitory activity on *Pseudomonas aeruginosa, Escherichia coli, Klebsiella pneumoniae*, with inhibition diameter located between 2 and 8 mm. The sensitivity of *Klebsiella pneumoniae, Escherichia coli*, and *Proteus mirabilis* to ethanolic extract of *S. siamea* has been demonstrated in Nigeria. The inhibition diameters ranged from 16 to 20 mm (Abdulrasheed et al., 2015). The crude extracts also inhibited *Bacillus cereus* and *Staphylococcus aureus* (Krasaekoopt and Kongkarnchanatip, 2005).

16.3.3 Antioxidant Efficacy

1,1-diphenyl-2-picrylhydrazyl (DPPH) radical scavenging and antioxidant activities have been observed for *S. siamea* (Barnaby et al., 2016). Its alcoholic flower extract contains a large amount of polyphenols and also exhibited an immense reducing ability. At 500 µg/mL, 42.7, 32.7 and 64.5% of the O_2-, H_2O_2 and NO radicals respectively; and at 250 µg/mL, 96% of the DPPH radicals could be scavenged by the extract (Kaur et al., 2006).

16.3.4 Analgesic and Anti-Inflammatory Activity

Ethanolic and aqueous extracts of the trunk bark have anti-inflammatory and analgesic properties at doses of 100, 200, and 400 mg/kg (Ntandou et al., 2010). A significant inhibition of the writhing reflex of 61.98% was demonstrated by Momin et al. (2012a,b) for the ethanolic extract at the dose of 500 mg/kg.

16.3.5 Hepatoprotective Activity

The subacute toxicity of the aqueous extract of the plant has no significant effect on either the biochemical parameters (serum creatinine, serum urea, sodium ions, potassium, alkaline phosphatase (ALP), alanine aminotransaminase

(ALAT), aspartate aminotransaminase (ASAT)), or the hematological ones (PCV, Hb, WBC, and RBC). There was also no significant effect on tissues (Otimenyin et al., 2010). The aqueous extract of the stem bark of *S. siamea* also gave the same results (Mohammed et al., 2012).

16.3.6 Anticancer Activity

Anthraquinone monomers extracted from *S. siamea* showed higher antitumor activity (Koyama et al., 2001). Emodin and Cassiamine B, obtained from *S. siamea*, showed an antitumor effect on skin tumors in mice (Koyama et al., 2001).

16.3.7 Anti-Hyperglycemic and Anti-Hyperlipidemia Effect

In a study performed by Mohammed and Atiku. (2012), methanol extracts from stem bark and leaves of *S. siamea* possess anti-hyperglycemic and antihyperglycemic properties, after oral administration for 3 weeks. The effective dose was 500 mg/kg b.wt. The increase in triglycerides, LDL-cholesterol, and total cholesterol in diabetic rats was significantly prevented by the two extracts, with concomitant increase in serum HDL-cholesterol.

16.3.8 Antiplasmodial Activity

Cassiarin A, a leaves alkaloid of *S. siamea*, was active against *Plasmodium falciparum* in vitro and against *P. berghei* in vivo (Morita et al., 2009). It had a vasorelaxation effect against the rat aortic ring, which could be due to the production of nitric oxide (NO). In addition, chemicals extracted from the bark stem also have antiplasmodial properties. Emodine and lupeol, for example, have antiplasmodial properties with IC_{50} values of 5 μg/mL, respectively, against the multiresistant phenotype of *Plasmodium falciparum* (K1) (Ajaiyeoba et al., 2008). Cassiarin F (an alkaloid isolated from the flowers) is active against *Plasmodium falciparum* (Deguchi et al., 2011). Cassiarins G, H, J, and K (all alkaloids), isolated from the leaves, are moderately active against *Plasmodium falciparum* 3D7 (Deguchi et al., 2012).

16.3.9 Purgative Activity

The extracts of *S. siamea,* obtained after purification, increased the force of longitudinal smooth muscle contractions (EC_{50} = 0.3 mM) in a dose-dependent way. Saxitoxin (0.3 μM) stopped the stimulating powers of barakol, a result indicating a neural mechanism of action (Deachapunya et al., 2005).

16.3.10 Cardioprotective Activity

Barakol from *S. siamea* leaves reduced the incidence of aconitine-induced ventricular fibrillation and ventricular tachycardia (VT) and (VF) (Chen et al., 1999). The suggested mechanism for the cardioprotective effect of barakol against aconitine-induced toxicity is the prevention of intracellular Na+ accumulation.

16.3.11 Mosquitocidal Activity

Methanol extract of *S. siamea* leaves had 100% mortality against *Anopheles stephensi* (malaria vector) and *Culex quinquefasciatus* (filariasis vector) after 48 h of exposure. The extract can therefore be used in biological control (Kamaraj et al., 2011).

KEYWORDS

- ***Senna siamea***
- **biomolecules**
- **bioactivity**
- **toxicity**

REFERENCES

Abdulrasheed, M.; Isiaka, I. H.; Siddan, I. A. Determining the Phytochemical Constituents and the Antimicrobial Activity of Ethanolic Extract of Acassia Leaf (*Senna siamea*) on Some Enterobacteriaceae. *IOSR J. Pharm.* **2015,** *5* (6), 18–22.

Ajaiyeoba, E. O.; Ashidi, J. S.; Okpako, L. C.; Houghton, P. J.; Wright, C. W. Antiplasmodial Compounds from *Cassia siamea* Stem Bark Extract. *Phytother. Res.* **2008,** *22*, 254–255.

Barnaby AG, Reid R, Warren D. Antioxidant Activity, Total Phenolics and Fatty Acid Profile of *Delonix regia, Cassia fistula, Spathodea campanulata, Senna siamea* and *Tibouchina granulosa. J. Anal. Pharm. Res.* **2016,** *3* (2), 00056. DOI: 10.15406/japlr.2016.03.00056.

Bukar, A.; Mukhtar, M.; Hassan, A. Phytochemical Screening and Antibacterial Activity of Leaf Extracts of *Senna siamea* (Lam) on *Pseudomonas aeruginosa. Bayero J. Pure Appl. Sci.* **2009,** *2*, 139–142.

Chen, F. -S.; Ito, H.; Okabe, F.; Momose, Y.; Yamamura, S.; Ridtitid, W.; et al. Effects of Aarakol on Aconitine-Induced Cardiac Arrhythmias. *Pharm. Biol.* **1999,** *37*, 105–108.

Dahiru, D.; Malgwi, A. R.; Sambo, H. S. Growth Inhibitory Effect of *Senna siamea* Leaf Extracts on Selected Microorganisms. *Am. J. Med. Med. Sci.* **2013,** *3*, 103–107.

Deachapunya, C.; Thongsaard, W.; Poonyachoti, S. Barakol Suppresses Norepinephrine-Induced Inhibition of Spontaneous Longitudinal Smooth Muscle Contractions in Isolated Rat Small Intestine. *J. Ethnopharmacol.* **2005,** *101*, 227–232.

Deguchi, J.; Hirahara, T.; Hirasawa, Y.; Ekasari, W.; Widyawaruyanti, A.; Shirota, O.; et al. New Tricyclic Alkaloids, Cassiarins G, H, J, and K from Leaves of *Cassia siamea. Chem. Pharm. Bull.* **2012,** *60*, 219–222.

Deguchi, J.; Hirahara, T.; Oshimi, S.; Hirasawa, Y.; Ekasari, W.; Shirota, O.; et al. Total Synthesis of a Novel Tetracyclic Alkaloid, Cassiarin F from the Flowers of *Cassia siamea. Org. Lett.* **2011,** *13*, 4344–4347.

Deguchi, J.; Sasaki, T.; Hirasawa Y.; Kaneda T.; Kusumawati I.; Shirota O. et al. Two Novel Tetracycles, Cassibiphenols A and B from the Flowers of *Cassia siamea. Tetrahedron Lett.* **2014,** *55*, 1362–1365.

Dougnon, V.; Legba, B.; Yadouleton, A.; Agbankpe, J. et al. Utilisation des plantes du Sud-Bénin dans le traitement de la fièvre typhoïde :rôle des herboristes. *Ethnopharmacologia* **2018,** *11*.

Fowler, D. G. Traditional Fever Remedies: A List of Zambian Plants. **2006,** 1–61. http://www.giftshealth.org/ritam/news/Traditional.Fever.remedies

Gao, X. M.; Shus, L. -D.; Yang, L. -Y.; Shen, Y. -Q.; Cui M. Z.; Li, X. -M.; Hu, Q. -F. Isoaurones from the Stem of *Cassia siamea* and Their Anti-Tobacco Mosaic Virus (Anti-TMV) Activity. *Heterocycles* **2013,** *87*, 125–131.

Hassan, I. A.; Nasiru, I. A.; Malut, A. M.; Abdulkadir, S. I.; Ali, A. S. Phytochemical Studies and Thin Layer Chromatography of Leaves and Flower Extracts of *Senna siamea* Lam for Possible Biomedical Applications. *J.Pharmacogn. Phytotherapy* **2015,** *7*, 18–26.

Ingkaninan, K.; A. P. IJzerman, A. P.; Verpoorte, R. Luteolin, A Compound with Adenosine A1 Receptor-Binding Activity, and Chromone and Dihydronaphthalenone Constituents from *Senna siamea. J. Nat. Prod.* **2000,** *63* (3), 315–317. DOI: 10.1021/np9904152.

Kamaraj, C.; Rahuman, A. A.; Bagavan, A.; Elango, G.; Zahir, A. A.; Santhoshkumar, T. Larvicidal and Repellent Activity of Medicinal Plant Extracts from Eastern Ghats of South India Against Malaria and Filariasis Vectors. *Asian Pac. J. Trop. Med.* **2011,** *4*, 698–705.

Kaur, G.; Alam, M. S.; Jabbar, Z.; Javed, K.; Athar, M. Evaluation of Antioxidant Activity of *Cassia siamea* Flowers. *J. Ethnopharmacol.* **2006,** *108*, 340–348.

Koyama, J.; Morita, I.; Tagahara, K.; Ogata, M.; Mukainaka, T.; Tokuda, H.; et al. Inhibitory Effects of Anthraquinones and Bianthraquinones on Epstein-Barr Virus Activation. *Cancer Lett.* **2001,** *170*, 15–18.

Koyama, J.; Morita, I.; Tagahara, K.; Nobukuni, Y.; Mukainaka, T.; Kuchide, M.; et al. Chemopreventive Effects of Emodin and Cassiamin B in Mouse Skin Carcinogenesis. *Cancer Lett.* **2002,** *182*, 135–139.

Krasaekoopt, W.; Kongkarnchanatip, A. Anti-Microbial Properties of Thai Traditional Flower Vegetable Extracts. *AU. J. T.* **2005,** *9* (2), 71–74.

Mohammed, A.; Atiku, M. K. Antihyperglycemic and Antihyperglycemic Effect of Leaves and Stem Bark Methanol Extracts of *Senna siamea* in Alloxan Induced Diabetic Rats. *Curr. Res. Cardiovasc. Pharmacol.* **2012,** *1*, 10–17.

Mohammed A, Liman ML, Atiku MK. Chemical Composition of the Methanolic Leaf and Stem Bark Extracts of *Senna siamea* Lam. *J. Pharmacogn. Phytotherapy* **2013,** *5* (5), 98–100.

Mohammed, A., Mada, S. B.; Yakasai, H. M. Sub-Chronic Study of Aqueous Stem Bark Extract of *Senna siamea* in Rats. *Asian J. Biol. Sci.* **2012,** *5*, 314–321.

Momin, M. A. M.; Bellah, S. F.; Afrose, A.; Urmi, K. F.; Hamid, K.; Rana, S. Phytochemical Screening and Cytotoxicity Potential of Ethanolic Extracts of *Senna siamea* Leaves. *J. Pharm. Sci. Res.* **2012a,** *4* (8), 1877–1879.

Momin, M. A. M.; Rana, S.; Khan, M. R.; Bin, T.; Hosen, S. M. Z. Antimicrobial and Peripherally Acting Analgesic Activity of *Senna siamea*. *Mol. Clin. Pharmacol.* **2012b,** *3* (2), 149–157.

Monkheang, P.; Sudmoon, R.; Tanee, T.; Noikotr, K.; Bletter, N.; Chaveerach, A. Species Diversity, Usage, Molecular Markers and Barcode of Medicinal *Senna* Species (Fabaceae, Caesalpnioideae) in Thailand. *J. Med. Plants Res.* **2011,** *5* (260), 6173–6181.

Morita, H.; Tomizawa, Y.; Deguchi, J.; Ishikawa, T.; Arai, H.; Zaima, K.; et al. Synthesis and Structure–Activity Relationships of Cassiarin A as Potential Antimalarials with Vasorelaxant Activity. *Bioorg. Med. Chem.* **2009,** *17*, 8234–8240.

Ntandou, G. F. N.; Banzouzi, J. T.; Mbatchi, B.; Elion-Itou, R. D. G.; Etou-Ossibi, A. W.; Ramos, S, et al. Analgesic and Anti-Inflammatory Effects of *Cassia siamea* Lam. Stem Bark Extracts. *J. Ethnopharmacol.* **2010,** *127*, 108–111.

Odugbemi, T. O.; Akinsulire, O. R.; Aibinu I. E.; Fabeku P. O. Medicinal Plants Useful in Malaria Therapy in Okeigbo Ondo State, Southwest Nigeria. *Afr. J. Trad. Complement. Altern. Med.* **2007,** *4*, 191–198.

Ogbole, O.; Adeniji, J.; Ajaiyeoba, E.; Kamdem, R.; Choudhary, M. Anthraquinones and Triterpenoids from *Senna siamea* (Fabaceae) Lam. Inhibit Poliovirus Activity. *Afr. J. Microbiol. Res.* **2014,** *8*, 2955–2963.

Otimenyin, S. O.; Kolawole, J. A.; Nwosu, M. Pharmacological Basis for the Continual Use of the Root of Senna Siamea in Traditional Medicine. *Int. J. Pharma. Bio. Sci.* **2010,** *1* (3), 1. https://www.cabdirect.org/cabdirect/abstract/20113372346

Padumanonda, T.; Gritsanapan, W. Barakol Contents in Fresh and Cooked *Senna siamea* leaves. *Southeast Asian J. Trop. Med. Public Health* **2006,** *37*, 388–393.

Padumanonda T, Suntornsuk L, Gritsanapan W. Quantitative Analysis of Barakol Content in *Senna siamea* Leaves and Flowers by TLC-Densitometry. *Med. Princ. Pract.* **2007,** *16*, 47–52.

Sakulpanich, A.; Gritsanapan, W. Laxative Anthraquinone Contents in Fresh and Cooked *Senna siamea* Leaves. *Southeast Asian J. Trop. Med. Public Health* **2009,** *40* (4), 835–839.

Wu, H. -Y.; Hu, W. -Y.; Liu, Q.; Yu, Z. -H.; Zhan, J. -B.; Yan, K. -L.; et al. Three New Alkaloids from the Twigs of *Cassia siamea* and Their Bioactivities. *Phytochem. Lett.* **2016,** *15*, 121–124.

CHAPTER 17

Bioactives and Pharmacology of *Senna sophera* (L.) Roxb.

AMBIKA VISWANATHAN PILLAI and V. SURESH*

Post Graduate and Research Department of Botany, Govt. Victoria College, Palakkad, Kerala, India

**Corresponding author. E-mail: sureshmagnolia@gmail.com*

ABSTRACT

Senna sophera (L.) Roxb. (Fabaceae) is an erect shrub which is used in several traditional medicines to cure various diseases. This plant has been known to possess antidiabetic, anti-asthmatic, hepatoprotective, anti-inflammatory, antioxidant, anticancerous, antifungal, anti-ulcer, antibacterial, laxative, diuretic, and analgesic activities. A wide range of chemical compounds including alkaloids, steroids, tannins, saponins, glycosides, phenols, flavonoids, terpenoids, and triterpenes have been isolated from this plant. The present review summarizes the information concerning the phytochemical profile and pharmacological activities of *Senna sophera* (L.) Roxb.

17.1 INTRODUCTION

Plant-based drugs are widely known for their safety, easy availability, and low cost (Iwu et al., 1999). Over the centuries, humans have depended on the nature for our basic needs, like food, clothing shelter, manures, flavorings and fragrances, and medicines (Cragg and Newman, 2005). *Senna sophera* (L.) Roxb. (Syn.: *Cassia sophera* L.) is a medicinal plant in the family Fabaceae and subfamily Caesalpinioideae. It is one of the important medicinal

Bioactives and Pharmacology of Legumes. T. Pullaiah, PhD (Ed.)

plants in the tropical and subtropical regions in Asia, especially in India, Pakistan, Malaysia, Bangladesh, Sri Lanka, and Myanmar. Its local name is "Kasunda" and used in Homeopathy for the treatment of allergic rhinitis, asthma, and osteoarthritis. Ethnobotanical literature also states that it is effective in treating psoriasis, acute bronchitis, cough, diabetes, pityriasis, asthma, and convulsions of children (Roy et al., 2012). This chapter reviews the phytochemistry and pharmacology of *Senna sophera.*

17.2 BIOACTIVES

The leaf extract of *S. sophera* was found to contain large amounts of alkaloids, steroids, tannin, and reducing sugar (Roy et al., 2010). Steroids, glycosides, and tannins were isolated from the bark of *S. sophera* (Nandhini et al., 2016). Qualitative evaluations revealed that this plant has compounds of categories phenols, saponins, flavonoids, terpenoids, and triterpenes and moderate presence of alkaloids, glycosides, and tannins (Akther et al., 2019). Phytochemical analysis of leaves, stems, flowers, and seeds of *S. sophera* showed the presence of flavonoids, alkaloids, and saponins. Fatty acids in the stem, leaves, flowers, and seeds of *S. sophera* analyzed by GC-MS have revealed that leaves contain 22 compounds, 8 from stem, 9 from flowers, and 12 from seeds. Evaluation of petroleum ether extract of the same parts of *S. sophera* showed that it had both saturated and unsaturated fatty acids. The analysis on leaves has shown 73.63% saturated fatty acids and 26.37% unsaturated fatty acids. In stem saturated fatty acids were 38.09% and unsaturated fatty acid content was 50.76%. In flowers the saturated fatty acids were 14.93% and unsaturated fatty acids were 85.07%, whereas in seeds the saturated fatty acids were 32.63% and unsaturated fatty acid content was 67.86% (Aziz and Mitu, 2019) (Figures 17.1 and 17.2). Chemical analysis of seeds revealed the presence of ascorbic acid, dihydroascorbic acid, and beta sistosterol 8 (Rao and Suresh, 2012).

17.3 PHARMACOLOGY

17.3.1 Antidiabetic/Hypoglycemic Activity

Khan and team worked on aqueous extract, methanolic extract, and raw powder of seeds of *S. sophera* and showed its significant hypoglycemic activity in normal as well as in alloxan diabetic rabbits (Khan et al., 2002).

Dolui and team have evaluated the anti-hyperglycemic effect of different extracts of *S. sophera* leaves in rats with alloxan-induced diabetes. Hyperglycemia was induced in the group of diabetic test rats by intraperitoneal injections of alloxan (150 mg/kg body weight). Extracts on chloroform, Petroleum ether, ethyl acetate, and methanol of *S. sophera* leaves were prepared and evaluated for their acute oral toxicity in female rats (Dolui et al., 2012). Nandhini and her team evaluated antidiabetic activity of *S. sophera* bark in Streptozocin-induced diabetic rats and showed that 400 mg/kg dose of ethanolic extract was more effective (Nandhini et al., 2016).

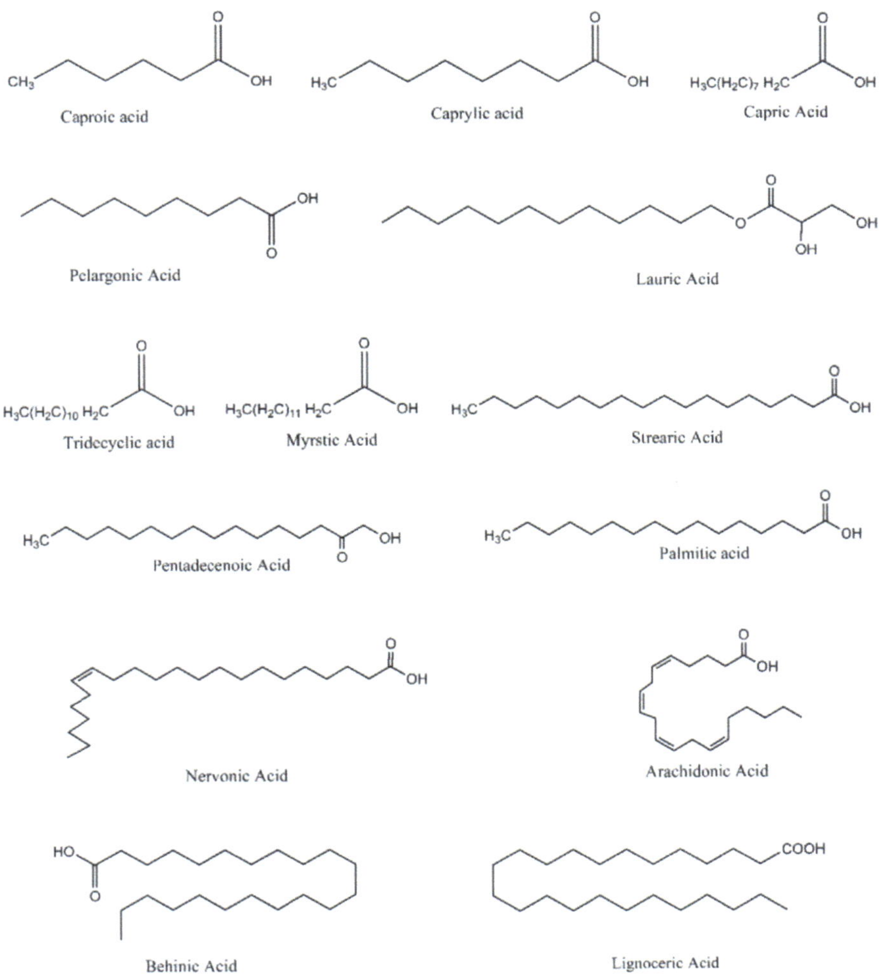

FIGURE 17.1 Chemical structure of the various saturated fatty acids detected by GC-MS on different parts of *Senna sophera* Linn (Redrawn from Aziz and Mitu, 2019).

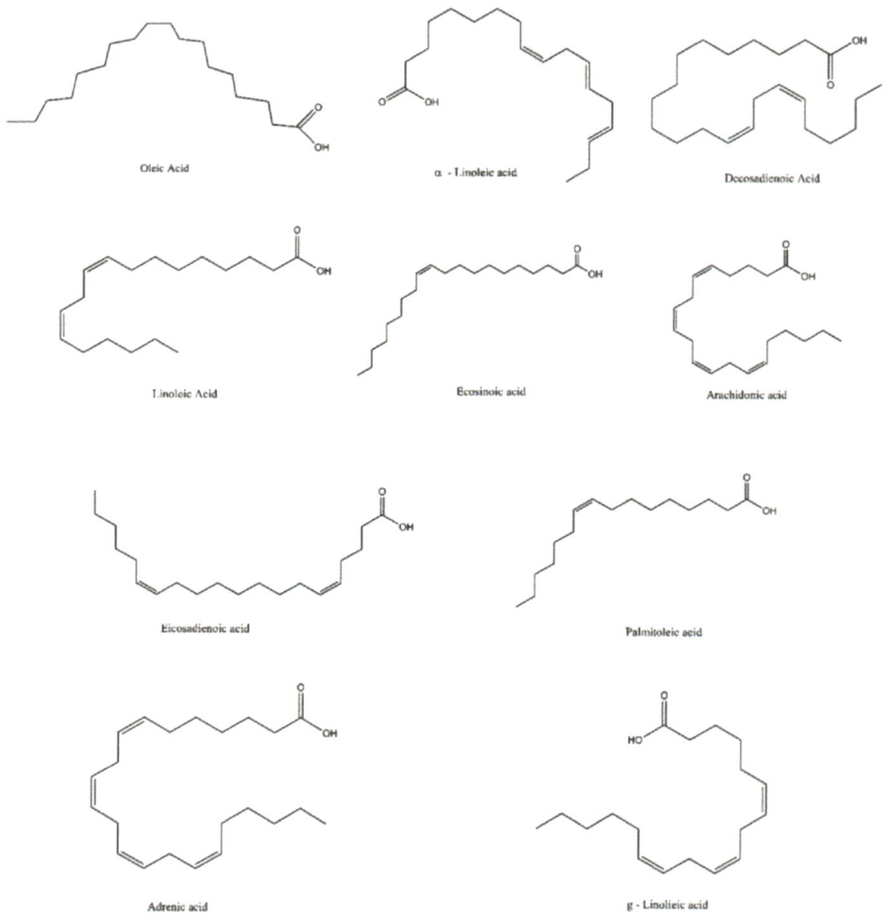

FIGURE 17.2 Chemical structure of unsaturated fatty acid detected from the GC-MS on different parts of *S. sophera* Linn (Redrawn from Aziz and Mitu, 2019).

An experiment conducted on diabetic rats using the methanolic extract of *S. sophera* leaves resulted in a remarkable hypoglycemia, i.e., a dose of 100 mg/kg of MFCS, the sugar level in blood was lowered to 52.33 ± 2.83 mg/dl whereas in vehicle control it was 76.66 ± 3.17 mg/dl; likewise 50 mg/kg of MFCS lowered the sugar level slightly only showing dose dependence, i.e., 72.33 ± 2.42 mg/dl (Kharat et al., 2019). Ascorbic acid, dihydro ascorbic acid, and beta sitosterol present in seed bark infusion of *S. sophera* are effective in diabetic patients (Mazhar et al., 2018).

17.3.2 Anti-asthmatic Effect

Anti-asthmatic effects of ethanol, chloroform, and ethyl acetate extracts of leaves of *S. sophera* were evaluated by Nagore and his team. For the study, swiss albino mouse, Wistar rat, and Guinea pig were subjected to the disease models like histamine-induced bronchoconstriction, carrageenan-induced paw edema, milk-induced leucocytosis, clonidine and haloperidol-induced catalepsy, eosinophilia, and passive paw anaphylaxis (Nagore et al., 2009). Ethanol chloroform and ethyl acetate fractions isolated from the ethanol extract of leaves of *S. sophera* possess anti-asthmatic activity in carrageenan-induced paw edema, milk-induced eosinophilia, leukocytosis, and passive paw anaphylaxis, clonidine and haloperidol-induced catalepsy and histamine-induced bronchoconstriction, animal models at doses 250, 500, and 750 mg/kg and presence of flavonoid may the reason for the same (Taur and Patil, 2011).

17.3.3 Hepatoprotective Activity

Evaluation of the hepatoprotective activity of ethanolic extracts of *S. sophera* leaves on carbon tetrachloride-induced hepatic damaged rats results (200 and 400 mg/kg dose per day) indicates that hepatoprotective activity may be attributed to the presence of flavonoids in the extracts (Mondal et al., 2012). Leaf extract of *S. sophera* showed remarkable hepatoprotective activity against acetaminophen-induced liver damage as judged from the serum markers for liver damage. Rats treated with ethanol extract (500 mg/kg) considerably changed the serum marker enzyme to almost normal level and comparable to the standard drug Liv-52 (5 mL/kg, p.o.) than paracetamol treated rats (Wankhade et al., 2011).

17.3.4 Anti-inflammatory Effect

Mondal and his team studied anti-inflammatory activity of *S. sophera* in rats and found that the compound *O*-methylated flavonol 2-(3, 4-dihydroxy-phenyl)-3, 5-dihydroxy-7-methoxy-chromen-4-one showed the effect (Mondal et al., 2013). Anti-inflammatory activities of alcoholic and aqueous extracts (LD_{50} studies were conducted in different dose levels up to 2 g/kg) of aerial part of *S. sophera* were investigated in rat models with carrageenan and formalin-induced acute paw edema (Ashraf et al., 2013). Carrageenan-induced paw edema and cotton-pellet edema models in Wistar rats and Swiss albino mice were evaluated for anti-inflammatory activities of methanol extract of *S. sophera* root. Doses of 50 mg/kg, 100 mg/kg, and 200 mg/kg resulted in considerable anti-inflammatory activity in carrageenan-induced paw edema model having maximum protection of 65.75% from edema and in cotton-pellet edema model with an inhibition of 57.37% of the granuloma (Hussain et al., 2015).

17.3.5 Antioxidant Activity

The antioxidant activity of *S. sophera* leaf and bark extracts was evaluated using different methods like DPPH assay, Ferric reducing power, and total antioxidant capacity assay. The IC_{50} values by DPPH assay for standard, leaf, and bark extracts were 19.08, 204.44, and 297.37 μg/mL, respectively (Majumder and Afia, 2019). The antioxidant activity of leaf extracts of *S. sophera* in methanol was evaluated using DPPH assay and IC_{50} values for leaves were 487.79 vs 98.30 μg/mL of reference (Akther et al., 2019). Analysis of antioxidant capacity of defatted methanol extracts of the *S. sophera* with Dot-blot assay, DPPH assay, reducing power assay and total antioxidant capacity showed significant free radical scavenging capacity (El-Sayed et al., 2011).

Crude methanolic extract of *S. sophera* was evaluated using DPPH assay and ferric reducing power assays against reference ascorbic acid (Murshid et al., 2014). Leaf extracts of *S. sophera* with petroleum ether, ethyl acetate, and methanol were analyzed for antioxidant activities (Rao et al., 2012).

17.3.6 Anticancer Activity

ID- and 2D-NMR analyses of 1, 6-dihydroxy-3-methyl-9,10-anthraquinone isolated from *S. sophera* showed anticancer activity against several cancers such as HT-29 (colon), U251 (glioma), 786-0 (renal), NCI-ADR/RES (multiple drugs-resistant ovarian), MCF-7 (breast) cells and lung, non-small cells, NCI—H460. The U251 cancer cell line was more susceptible by this compound than others, while effect on non-tumorigenic cells was less than 10% at concentrations equal to or lesser than 2.5 μg/mL (Brahmachari et al., 2017) (Figure 17.3).

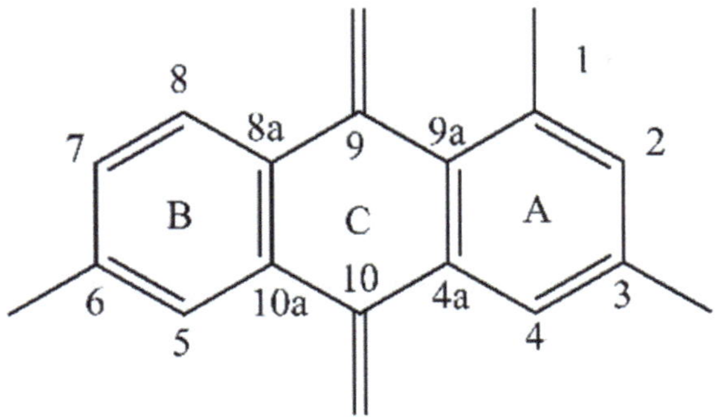

FIGURE 17.3 Structure of 1, 6-dihydroxy-3-methyl-9,10-anthraquinone (Redrawn from Brahmachari et al., 2017).

17.3.7 Antifungal Activity

Analysis of antifungal activity of leaf extracts of *S. sophera* against several fungi including *Candida* and dermatophytes by the poisoned food technique and the agar-well diffusion method showed significant reduction in their growth (Sumangala and Suresh, 2011). Petroleum ether, chloroform, and ethanol extracts of *S. sophera* were evaluated for antifungal activity using *Aspergillus fumigatus* (ATCC No: 102) strain and *Candia albicans* (ATCC No: 10231) strain (Deshpande and Naik, 2016).

17.3.8 Anti-ulcer Activity

Chloroform, ethyl acetate, methanol, and water extracts of *S. sophera* exhibited anti-ulcer potential using ethanol, pyloric ligation, hypothermic restrain stress, and indomethacin-induced ulceration (Nagore et al., 2009).

17.3.9 Antibacterial Activity

The antibacterial activity of essential oil and organic extracts isolated from *S. sophera* was evaluated against *Bacillus* sp. from soil showing the zones of inhibition of 12.6–24.9 mm. Ethanol extract was found to be most active against the bacteria *Bacillus megaterium* (MIC of 62.5 μg/mL) (Rahman et al., 2017). *S. sophera* extracts exhibited anti-mycobacterial activity against multi-drug resistant (MDR) strain of *Mycobacterium tuberculosis* and *M.*

semegmatis. Methanol extract of *S. sophera* produced 20 mm inhibition zone and MIC of 125 μg/mL against *M. semegmatis* and 27 mm inhibition zone and MIC of 250 μg/mL against *M. tuberculosis* (Singh et al., 2013). Petroleum ether, chloroform, and ethanol extracts of *S. sophera* were evaluated for antibacterial effect against the strain *Staphylococcus aureus, Enterococcus faecalis, Klebsiella* sp., and *Escherichia coli*. The leaf extracts of *S. sophera* in ethanol showed increased activity against *E. faecalis* with an MIC of 0.2738 mg/mL. Stem extracts of *S. sophera* in ethanol (MIC 0.3009), petroleum ether (MIC 0.407 mg/mL), and chloroform (MIC 0.4946 mg/mL) showed more activity against *Enterococcus faecalis* (Deshpande and Naik, 2016).

17.3.10 Laxative Activity

Evaluation of laxative activity of ethanolic extract of *S. sophera* showed a superior effect over standard drug at the tested dose level of 200 mg/kg, p.o. (Yele et al., 2010).

17.3.11 Diuretic Activity

Experimental analysis of diuretic activity of ethanolic and aqueous extract of *S. sophera* in rats revealed that ethanolic extract significantly increased the urinary output and electrolyte concentration in urine at a dose of 400 mg/kg, p.o. compared with water extract (Yele et al., 2010).

17.3.12 Analgesic Activity

The phytochemical analysis of seed extract of *S. sophera* exhibited significant analgesic effects (Bilal et al., 2005). The study of peripheral analgesic activity of *S. sophera* leaves in mice model writhing induced by acetic acid and central analgesic activity using heat-induced pain in mice revealed a considerable effect at the doses of 300 and 600 mg/kg body weight (Roy et al., 2012).

17.4 CONCLUSION

Through the current chapter, the phytochemical profile and pharmacological activities of *Senna sophera* were reviewed. *S. sophera* is a rich source of alkaloids, steroids, tannins, saponins, glycosides, phenols, flavonoids, terpenoids, and triterpenes. A detailed literature survey showed that the whole plant extract has antidiabetic, anti-asthmatic, hepatoprotective, anti-inflammatory, antioxidant, anticancerous, antifungal, anti-ulcer, antibacterial, laxative, diuretic, and analgesic activities.

KEYWORDS

- **anti-inflammatory**
- **Fabaceae**
- **flavonoids**
- **Kasunda**
- ***Senna***

REFERENCES

Akther, S.; Rubel, M.; Islam, M. S. In Vitro Antioxidant and Anti-Inflammatory Potential Evaluations of Methanolic Extract of *Cassia sophera* L., A Phytochemical Screening. *J. Clinical Trials Regulations,* **2019,** *2* (1), 1–7.

Ashraf, S.; Munawar, M. S.; Srivastav, M. K.; Sayeed, M.; Ahmed, F.; Lakshman, D. Evaluation of Antiinflammatory Activities with Aerial Part Extracts of *Cassia sophera* (Linn.) in Wistar Rats. *Int. J. Tech. Res. Appl.* **2013,** *1* (3), 103–106.

Aziz, S.; Mitu, T. K. Analysis of Fatty Acid and Determination of Total Protein and Phytochemical Content of *Cassia sophera* Linn Leaf, Stem, Flower and Seed. *Beni Suef Univ. J. Basic Appl. Sci.* **2019,** *8* (1), 3.

Bilal, A.; Khan, N. A.; Ghufran, A.; Inamuddin, H. A. Pharmacological Investigation of *Cassia sophera* Linn. var. *purpurea* Roxb. *Med. J. Islam World Acad. Sci.* **2005,** *15*, 105–109.

Brahmachari, G.; Mondal, A.; Mondal, S.; Modolo, L. V.; Fátima, Â. D.; Ruiz, A. L. T. G.; Carvalho, J. E. D. 1, 6-Dihydroxy-3-Methyl-9, 10-Anthraquinone: An Anti-Cancerous Natural Pigment from *Cassia sophera* Linn. (Caesalpiniaceae). *Indian J. Chem.* **2017,** *56B*, 1251–1255.

Cragg, G. M.; Newman, D. J. Biodiversity: A Continuing Source of Novel Drug Leads. *Pure Appl. Chem.* **2005,** *77* (1), 7–24.

Deshpande, S. R.; Naik, B. S. Evaluation of In Vitro Antimicrobial Activity of Extracts from *Cassia obtusifolia* L. and *Senna sophera* (L.) Roxb. Against Pathogenic Organisms. *J. Appl. Pharm. Sci.* **2016,** *6* (1), 83–85.

Dolui, A.; Das, S.; Kharat, A. Antihyperglycemic Effect of *Cassia sophera* Leaf Extracts in Rats with Alloxan-Induced Diabetes. *Asian J. Trad. Med.* **2012,** *7* (1), 8–13.

El-Sayed, M. M.; Abo-Sedra, S. A.; El-Nahas, H. A.; Abdel-Gawad, M. M.; Abdel-Aziz, M. M.; Abdel-Lateef, E. E.; Abdel-Hameed, S. S. Evaluation of Antioxidant and Antimicrobial Activities of Certain *Cassia* species. *Aust. J. Basic Appl. Sci.* **2011,** *5* (9), 344–352.

Hussain, S. N.; Chaudhry, B. A.; Uzair, M.; Qaisar, M. N. Studies on the Analgesic and Anti-Inflammatory Effects of *Cassia sophera* Roots. *Asian Pac. J. Trop. Dis.* **2015,** *5* (6), 483–488.

Iwu, M. W.; Duncan, A. R.; Okunji, C. O. New Antimicrobials of Plant Origin. In *Perspectives on New Crops and New Uses*; ASHS Press: Alexandria, VA, 1999; pp 457–462.

Khan, A. M.; Khan, A. H.; Shoaib, M.; Ahmed, A. B. Pharmacological Screening of *Cassia sophera* for Hypoglycemic Activity in Normal and Diabetic Rabbits. *Pak. Med. J.* **2002,** *16* (1), 1–4.

Kharat, A. R.; Kharat, K.; Jadhav, M.; Makhija, S. J. Antihyperglycemic, Antihyperlipidemic and Antioxidative Evaluation of Compounds from *Senna sophera* (L.) Roxb. in Streptozotocin-Induced Diabetic Rats. *Nat. Prod. Res.* **2019,** *33* (4), 602–605.

Majumder, S.; Afia, I. J. A Comparative Study on the Antioxidant Activity of *Cassia sophera* (L.) Leaf and Bark Extracts. *J. Drug Deliv. Ther.* **2019,** *9* (4-s), 177–181.

Mazhar, M.; Ahmad, M.; Mumtaz, S. M.; Kumar, Y. Indian Medicinal Herbs—Useful in Diabetes. *Hosp. Pharmacy* **2018,** *13* (1), 33–47.

Mondal, A.; Karan, S. K.; Singha, T.; Rajalingam, D.; Maity, T. K. Evaluation of Hepatoprotective Effect of Leaves of *Cassia sophera* Linn. *Evid. Based Complement. Alternat. Med.* **2012,** *2012*, 1–5.

Mondal, A.; Rajalingam, D.; Maity, T. K. Anti-Inflammatory Effect of O-Methylated Flavonol 2-(3, 4-Dihydroxy-Phenyl)-3, 5-Dihydroxy-7-Methoxy-Chromen-4-One Obtained from *Cassia sophera* Linn. in Rats. *J. Ethnopharmacol.* **2013,** *147* (2), 525–529.

Murshid, G. M. M.; Barman, A. K.; Rahman, M. M. Evaluation of Antioxidant, Analgesic and Cytotoxic Activities of the Aerial Part of *Cassia sophera* L. (Caesalpiniaceae). *Int. J. Phytopharmacol.* **2014,** *5* (5), 383–389.

Nagore, D. H.; Ghosh, V. K.; Patil, M. J. Evaluation of Antiasthmatic Activity of *Cassia sophera* Linn. *Pharmacog. Mag.* **2009,** *5* (19), 109–118.

Nandhini, S.; Geethalakshmi, S.; Selvam, S.; Radha, R.; Muthusamy, P. Preliminary Phytochemical and Anti-Diabetic Activity of *Cassia sophera* Linn. *J. Pharmacogn. Phytochem.* **2016,** *5* (1), 87.

Rahman, M. M.; Sultana, T.; Ali, M. Y.; Rahman, M. M.; Al-Reza, S. M.; Rahman, A. Chemical Composition and Antibacterial Activity of the Essential Oil and Various Extracts from *Cassia sophera* L. Against *Bacillus* sp. from Soil. *Arab. J. Chem.* **2017,** *10*, S2132-S2137.

Rao, S.; Suresh, C. Phytochemical Analysis and In Vitro Efficacy of Two Edible *Cassia species* on Selected Human Pathogens. *Int. J. Pharm. Sci. Res.* **2012,** *3* (12), 4982.

Rao, S.; Vijayadeepthi, T.; Zoheb, M.; Suresh, C. Evaluation of Antioxidant and Antimicrobial Potential of Two Edible *Cassia* Species to Explore Their Nutraceutical Values. *J. Pharm. Res.* **2012,** *5* (3), 1650–1655.

Roy, D.; Dev, S.; Deb, D.; Chandra, M. Phytochemical ad Toxicological Study of Leaf of *Cassia sophera* L. *Pharmacologyonline* **2010,** *1*, 365–374.

Roy, D.; Shill, M. C.; Dev, S.; Deb, D.; Shahriar, M.; Das, A. K.; Choudhuri, M. S. K. Anti-Nociceptive and Antipyretic Activities of Hydroalcoholic Extract of *Cassia sophera* Linn. Leaves. *Bangladesh Pharm. J.* **2012,** *15* (2), 107–111.

Singh, R.; Hussain, S.; Verma, R.; Sharma, P. Anti-Mycobacterial Screening of Five Indian Medicinal Plants and Partial Purification of Active Extracts of *Cassia sophera* and *Urtica dioica*. *Asian Pac. J. Trop. Med.* **2013,** *6* (5), 366–371.

Sumangala, R.; Suresh, C. Comparative In Vitro Efficiency of *Cassia tora* and *Cassia sophera* in Mitigating Fungal Infections. *Adv. Plant Sci.* **2011,** *24* (2), 455–458.

Taur, D. J.; Patil, R. Y. Some Medicinal Plants with Antiasthmatic Potential: A Current Status. *Asian Pac. J. Trop. Biomed.* **2011,** *1* (5), 413–418.

Wankhade, P. W.; Nagore, D. H.; Kotagale, N. R.; Turaskar, A. O.; More, S. M. Hepatoprotective Effect of *Cassia sophera* Leaves Extract Against Paracetamol Induced Hepatic Injury in Rats. *Int. J. Pharm. Bio Sci.* **2011,** *2*, 433–438.

Yele, S. U.; Gokhale, S. B.; Surana, S. J.; Veeranjaneyulu, A. Diuretic and Laxative Activity of *Cassia sophera* Linn, a Prevalent Western Ghat Species. *Pharmacologyonline* **2010,** *1*, 47–52.

CHAPTER 18

Stryphnodendron adstringens (Mart.) Coville: Bioactive Compounds and Pharmacological Actions

MARIA DANIELMA DOS SANTOS REIS*,
JAMYLLE NUNES DE SOUZA FERRO, FELIPE LIMA PORTO,
RAFAEL VRIJDAGS CALADO, TAYHANA PRISCILA MEDEIROS SOUZA,
and EMILIANO BARRETO

Laboratory of Cell Biology, Federal University of Alagoas, Alagoas, Brazil

Corresponding author. E-mail: danielma.reis@icbs.ufal.br

ABSTRACT

Stryphnodendron adstringens (Mart.) Coville is a medicinal plant found in South America whose stem bark is used to treat disorders of the female genitourinary system, gastric ulcers, wounds, and other diseases. Extracts of this plant mainly contain tannins, phenolic acids, and triterpenoids. Important preclinical studies have shown that *S. adstringens* has anti-inflammatory effects in animal models of acute and chronic inflammation, as well as antimicrobial activity against microorganisms of clinical importance. In addition to these effects, this plant showed toxicity in animal models, this should be considered when recommending the popular use of this plant.

18.1 INTRODUCTION

Stryphnodendron adstringens (Mart.) Coville is a plant of the Fabaceae family, included in the Mimosoideae subfamily, typical from South America,

Bioactives and Pharmacology of Legumes. T. Pullaiah, PhD (Ed.)

with distribution from Nicaragua to the south of Brazil (Lima et al., 2017). It can reach 5 m in height, and is composed of bipinnate compound leaves, inflorescences with a variable number of small brown flowers, and a thick and fleshy fruit about 10 cm long (Felfili et al., 1999). The peak of flowers and renovation of the leaves occur between July and October (Felfili et al., 1999). In traditional medicine, the stem bark of *S. adstringens* is used in infusions, macerations, and decoctions preparations to treat infections of the female genitourinary system, gastric ulcers, hemorrhages, diabetes, pain, wound healing, and cancer (Souza-Moreira et al., 2018).

18.2 BIOACTIVES

Plants of *S. adstringens* are rich in tannins (Lopes et al., 2009). Analysis of the stem bark revealed the presence of proanthocyanidins, prodelphinidins, and prorobinetinidins (de Mello et al., 1996a,b; de Mello et al., 1999; Lopes et al., 2008; Nascimento et al., 2013; Baldivia et al., 2018). From the hydroalcoholic extract of the stem bark was isolated several bioactive compounds: gallic acid, caffeic acid, quercetin, catechin, epigallocatechin gallate, rutin, and kaempferol (Pellenz et al., 2018). The caffeic acid was also identified in the total extract of this plant (Meinhart et al., 2017). In the acetone-soluble fraction of *S. adstringens* bark was reported the presence of chalcones, tannins, and triterpenoids (Lima et al., 1998). Phytochemical analyses of the leaves showed the presence of prodelphinidins, gallic acid, flavonoids, saponins, coumarins, and resins (Santos et al., 2002; Oliveira and Figueiredo, 2007). Galactomannans was isolated from the aqueous extract of the seeds (Salvalaggio et al., 2015).

18.3 PHARMACOLOGY

18.3.1 Acaricidal Activity

The ethanolic extract of the bark of *S. adstringens* showed low acaricidal activity against *Sarcoptes scabiei* (Faria et al., 2017).

18.3.2 Angiogenic Properties

A preliminary study demonstrated that the aqueous extract from the stem bark of *S. adstringens* induced blood vessel formation in an in-vivo assay using chorioallantoic membranes of embryonated chicken eggs (Chaves et al., 2016).

18.3.3 Anti-apoptotic Activity

The expression of genes related to apoptosis was decreased in fibroblast and keratinocytes after exposure to the hydroalcoholic extract of *S. adstringens* stem bark (Pellenz et al., 2018). However, the treatment with the aqueous extract of this plant induced apoptosis mediated by reactive oxygen species (ROS) and increased the expression of caspase-3 in melanoma cells (Baldivia et al., 2018).

18.3.4 Antihelmintic Effects

The ethanolic extract from the bark of *S. adstringens* at 0.1%, 0.25%, and 0.5% was able to inhibit in vitro the hatching of *Fasciola hepatica* miracidia (Marques et al., 2020).

18.3.5 Anti-inflammatory Properties

S. adstringens is widely used in folk medicine to treat inflammatory conditions. This effect is corroborated by pre-clinical studies that showed anti-inflammatory effects in animal models of acute and chronic inflammation. Lima and colleagues (1998) showed that the oral administration of the acetone-soluble extract from the bark of this plant reduced rat paw oedema, attenuated the volume and migration of leukocytes in a rat model of pleurisy, and decreased the vascular permeability induced by acid acetic in mice. Importantly, this extract also prevented the hind paw swelling in rats with arthritis induced by the Freund's adjuvant. In another study, the organic and aqueous fractions of the ethanolic extract from the bark administered by oral route was able to reduce leukocyte migration and neutrophil accumulation in the joint of mice with acute arthritis induced by lipopolissacaride (Henriques et al., 2016).

18.3.6 Antimicrobial Activity

Different extracts and isolates from *S. adstringens* presented microbicidal properties in bacteria and fungi involved in human infections. The hydroalcoholic beverage of this plant was able to inhibit *Mycobacterium tuberculosis* growth in vitro (Oliveira et al., 2007). Also, the crude extract from the bark inhibited in vitro the growth and altered the morphology of the *Pythium insidiosum*, but had no effects on the in-vivo infection in a rabbit model of pythiosis (Trolezi et al., 2017).

The *S. adstringens* stem bark dry extract, pure or in a liquid soup formulation, reduced in vitro the growth of *S. aureus, S. epidermidis*, and *E. coli* (Souza et al., 2007a). Stem bark ethanolic extracts showed in-vitro antibacterial activity against *E. coli* and *S. aureus* (Almeida et al., 2017). In a recent work, the aqueous and the ethyl acetate fraction of the stem bark showed in-vitro bactericidal activity against *S. aureus*, either alone or in combination with tannic acid (Trevisan et al., 2020). This effect could be attributed to the bioactive compounds epigallocatechin 3-O-(3-methoxy-4-hydroxy) benzoate and epigallocatechin 3-O-gallate present in these extracts, which the authors found to inhibit enzymes belonging to the metabolic pathway for fatty acid biosynthesis in *S. aureus* (Trevisan et al., 2020).

The hexameric compound extracted from *S. adstringens* stem bark is composed of condensed tannins interfered with cellular morphology, homeostasis, growth, capsule formation, and pigmentation of *Cryptococcus neoformans* (Ishida et al., 2009). An aqueous subfraction of the stem bark, also rich in condensed tannins, exhibited antifungal effects in cultures of *Candida albicans*, with low cytotoxicity in mammals' cells (Ishida et al., 2006). Also, the proanthocyanidin polymer-rich fractions isolated from the stem bark reduced *C. albicans* biofilm and also inhibited *C. tropicalis* growth, biofilm formation, adhesion, and infection in vivo (Luiz et al., 2015; Morey et al., 2016). In another study, a gel formulation containing that fraction reduced vaginal infection by *C. albicans* and *C. glabrata* in mice (de Freitas et al., 2018).

The hydroalcoholic extract from the leaves showed activity against *S. aureus* (Pinho et al., 2012). In other studies, the crude extract and the fractions obtained from leaves, stem, and bark showed antibacterial against *S. aureus, S. epidermidis, E. coli*, and *Pseudomonas aeruginosa* (Audi et al., 2004; Souza et al., 2007b). The hexane extracts from the leaves of this plant could prevent in-vitro growth of the *Trichophyton rubrum* (Silva et al., 2009).

The dry extract from the leaves presented antibacterial and antifungal effects on *S. aureus, E. coli, P. aeruginosa, Klebsiella pneumoniae, C. albicans, C. tropicalis*, and *C. parapsilosis* (Cartaxo-Furtado et al., 2019).

18.3.7 Anti-neoplastic Activity

The treatment in vitro with the proanthocyanidin polymer-rich fraction isolated from the stem bark reduced cervical cancer cells viability, through the augment of oxidative stress and induction of apoptosis and necrosis, along with impaired migration in the wound healing assays. In this same study, the fraction administered in vivo could decrease the volume and weight of Ehrlich solid tumor accompanied by high lipoperoxidation (Kaplum et al., 2018). Moreover, a fraction with high proanthocyanidin content from the leaves reduced breast cancer cells viability and augmented the expression of genes related to apoptosis (Sabino et al., 2018).

18.3.8 Antinociceptive Activity

S. adstringens crude stem bark extract and aqueous and ethyl-acetate fractions showed antinociceptive effects in rats after acid acetic induced-writhing, formalin, and hot plate tests (Melo et al., 2007).

18.3.9 Antioxidant Activity

Cultures of human keratinocytes and human fibroblasts treated with the hydroalcoholic extract of *S. adstringens* stem bark produced less ROS, indicating a potential antioxidant effect of this solution (Pellenz et al., 2018). Also, when this kind of extract was added as a dietary supplement to chicken, it could prevent lipid oxidation and improved meat quality (Lima et al., 2016). On the contrary, the aqueous extract of this plant augmented ROS in melanoma cells, inducing cytotoxicity (Baldivia et al., 2018).

18.3.10 Antivenom Properties

The crude extract, aqueous and subfractions, and diethyl acetate fraction from the bark of *S. adstringens* prevented in vitro and in vivo the toxic effects caused by *Bothrops jararacussu* venom (Pereira Junior et al., 2020).

18.3.11 Antiviral Activities

The aqueous fraction and the ethyl acetate fraction of the *S. adstringens* bark could induce virucidal activity against the bovine herpesvirus type 1 and in the poliovirus type 1 (Sabin strain) (Felipe et al., 2006).

18.3.12 Cytoprotective Effects

The ethyl-acetate fraction of *S. adstringens* stem bark showed protective action on neuroblastoma cells in vitro through the inhibition of mitochondrial depolarization, ROS production, and lipid peroxidation induced by the amyloid $A\beta_{25-35.}$ These effects were accompanied by the regulation of genes involved in Alzheimer disease pathogenesis (Sereia et al., 2019). In gastric lesions induced by acute stress and acidified ethanol in rats, the pretreatment with aceto-aqueous total extract and fractions prevented the gastric tissue damage (Audi et al., 1999). Accordingly, the pretreatment by oral route with the acetone-soluble fraction of the bark decreased gastric lesions induced by ethanol and hypothermic restraint-stress in rats (Martins et al., 2002).

18.3.13 Genotoxicity and Genoprotection

The commercial phytotherapeutic of *S. adstringens* (hydroalcoholic extract, 50% v/v) did not show genotoxicity in *Drosophila melanogaster* somatic mutation and recombination test (SMART), and in the *Drosophila* sex-chromosome loss test (ring-X loss) (de Sousa et al., 2003). No genotoxic effects were observed in bone marrow cells of mice after the oral treatment with proanthocyanidin polymers-rich fraction of the bark (Costa et al., 2010).

The aqueous, water fraction, and ethanolic extracts obtained from the leaves prevented genotoxicity induced by cyclophosphamide in rat bone marrow cells (Santos Filho et al., 2011). The hydroalcoholic extract of the stem bark presented genoprotective effects in non-cellular genomodifier capacity (GEMO) assay, and in keratinocytes and fibroblasts by quantification of 8-deoxyguanosine (DNA-8-OhdG) levels (Pellenz et al., 2018).

18.3.14 Immunomodulatory Activities

The organic and aqueous fractions obtained from bark extracts inhibited the production of tumor necrosis factor-α (TNF-α) by the human monocyte cell line THP-1 under lipopolysaccharide (LPS) stimulus (Henriques et al., 2016). The in-vitro exposure to the hydroethanolic extract from the stem bark impaired murine macrophage polarization, decreasing M1 and augmenting M2 markers (de Carvalho et al., 2020).

18.3.15 Larvicidal Activity

The aqueous extract of *S. adstringens* reduced oviposition and larvae weight in *Plutella xylostella* (Jesus et al., 2011). Similar results were obtained by Fonseca and colleagues (2018), which observed a reduced number of eggs and the number of hatched larvae and food preference index by *P. xylostella* fed with leaves treated with methanolic extracts of the plant.

18.3.16 Molluscicidal Activity

Ethanolic extracts obtained from the bark and leaves induced *Biomphalaria glabrata* mortality (Bezerra et al., 2002).

18.3.17 Protozoacidal Activities

The ethanolic extract of the stem bark from *S. adstringens* reduced *Trypanosoma cruzi* (Y strain) parasitemia in the mice blood (Herzog-Soares et al., 2002). Accordingly, the crude extract and fractions of the bark reduced *Herpetomonas samuelpessoai* (a non-pathogenic trypanosomatid) growth in vitro and induced ultrastructural alterations in these cells (Holetz et al., 2005). A posterior study showed that the crude extract of *S. adstringens* bark inhibited the in-vitro growth of the epimastigote forms of *Trypanosoma cruzi* (Y strain), but it was not effective against promastigote and amastigote forms of *Leishmania amazonensis* (MHOM/BR/75/Josefa strain) (Luize et al., 2005).

18.3.18 Toxicity

Several studies demonstrated the toxicological potential of this plant. In rats, the oral administration of the crude hydroalcoholic from *S. adstringens* (LD_{50} = 4992.8 mg/kg) seeds caused malnutrition, dehydration, oedema, leucopenia, lymphopenia, hypercalcemia, higher levels of hepatic enzymes, bilirubin, phosphorus, urea, and dextrose (Bürger et al., 1999). The authors also demonstrated that the oral use of this extract reduced the number of live fetuses and the weight of the uterus in female rats.

The total extract of the stem bark showed low toxicity when administrated orally for 7 days in male mice, with LD50 of 2699 mg/kg, but it could induce an augment in the glucose and aspartate aminotransferase levels in the plasma of rat after subchronic treatment for 30 days (Rebecca et al., 2002). In this same study, the extract reduced the body and thymus weight in rats treated for 30 days with 800 and 1600 mg/kg, respectively. The hepatotoxicity showed in this study may be related to the effects of the *S. adstringens* extract on hepatocytes energy metabolism, as demonstrated by Rebecca and colleagues (2003).

Corroborating with these results, the orally administrated stem bark extract did not show acute toxicity in rats; however, animals that received 800 mg/kg once a day, for 30 days, presented hematological, biochemical, and histopathological alterations, indicating anemia, renal and liver deregulation, and minimal hydropic hepatocellular degeneration (Almeida et al., 2017). The proanthocyanidin polymer-rich fraction (heptamer compound) of the stem bark showed low toxicity in acute (LD_{50} = 3015 mg/kg) and repeated-dose oral studies in rats (Costa et al., 2010, 2013). On the contrary, the crude extract from the bark did not show acute toxicity in rats and rabbits when administrated orally at a concentration of 90 mg/day (Trolezi et al., 2017).

In vitro, the hexameric compound extracted from the stem bark treatment did not affect melanin synthesis in murine melanoma B16-F10 cells (Ishida et al., 2009), while the ethanol extract presented inhibition of the tyrosinase, an important enzyme in melanogenesis (Souza et al., 2012). Also, the hydroalcoholic extract of the stem bark treatment did not alter the viability of human fibroblasts and keratinocytes (Pellenz et al., 2018), while the aqueous fraction and the ethyl acetate fraction of the stem bark did not affect the viability of Vero and HaCaT cells (Trevisan et al., 2020).

The flowers, methanolic and dichloromethane extracts of flowers, and peduncles showed toxic effects on two bee species, *Apis mellifera* and *Scaptotrigona postica*, that can act as pollinators (Cintra et al., 2003; de Souza et al., 2006)

18.3.19 Wound-healing Properties

The ethyl-acetate fraction extracted from *S. adstringens* stem bark (1%) was tested in a rat model of excisional wounds as a component of an ointment for topical application. In the wounds treated with the cream, there was an augment in the keratinocyte proliferation; however, no effect was observed in the length of the epithelium or the contraction of the wounds (Hernandes et al., 2010).

18.3.20 Clinical Trials

The cream with 6% of *S. adstringens* bark extract applied in a double-blind, randomized, placebo-controlled trial showed 60.98% of efficiency in the reduction and reversion of terminal hair growth in female patients with excess of terminal hair (Vicente et al., 2009).

KEYWORDS

- ***Stryphnodendron adstringens***
- **Fabaceae**
- **pharmacology**
- **inflammation**
- **traditional medicine**
- **tannins**

REFERENCES

Almeida, A. C.; Andrade, V. A.; Fonseca, F. S.; Macêdo, A. A.; Santos, R. L.; Colen, K. G.; et al. Acute and Chronic Toxicity and Antimicrobial Activity of the Extract of *Stryphnodendron adstringens* (Mart.) Coville. *Pesqui. Vet. Bras.* **2017,** *37* (8), 840–846.

Audi, E. A.; Toledo, C. M.; Santos, F. S.; Bellanda, P. R.; Alves-do-Prado, W.; Ueda-Nakamura, T.; et al. Biological Activity and Quality Control of Extract and Stem Bark from *Stryphnodendron adstringens*. *Acta farmacéutica bonaerense* **2004,** *23*, 328–333.

Audi, E. A.; Toledo, D. P.; Peres, P. G.; Kimura, E.; Pereira, W. K.; de Mello, J. P.; et al. Gastric Antiulcerogenic Effects of *Stryphnodendron adstringens* in Rats. *Phytother. Res.* **1999,** *13* (3), 264–266.

Baldivia, D. D. S.; Leite, D. F.; Castro, D. T. H. D.; Campos, J. F.; Santos, U. P. D.; Paredes-Gamero, E. J.; et al. Evaluation of In Vitro Antioxidant and Anticancer Properties of the Aqueous Extract from the Stem Bark of *Stryphnodendron adstringens*. *Int. J. Mol. Sci.* **2018,** *19* (8), 2432.

Bezerra, J. C. B.; Silva, I. A.; Ferreira, H. D.; Ferri, P. H.; Santos, S. C. Molluscicidal Activity Against *Biomphalaria glabrata* of Brazilian Cerrado Medicinal Plants. *Fitoterapia* **2002,** *73* (5), 428–430.

Bürger, M. E.; Ahlert, N.; Baldisserotto, B.; Langeloh, A.; Schirmer, B.; Foletto, R. Analysis of the Abortive and/or Infertilizing Activity of *Stryphnodendron adstringens* (Mart. Coville). *Braz. J. Vet. Res. Anim. Sci.* **1999,** *36* (6), https://doi.org/10.1590/S1413-95961999000600003

Cartaxo-Furtado, N. A. O.; Brandao, D. O.; Júnior, F. J. D. L. R.; Silva, K. M. A.; Macêdo, R. O. Investigation of Thermal and Kinetic Behavior of the *Stryphnodendron adstringens* Dry Extract with Antimicrobial Activity. *J. Therm. Anal. Calorim.* **2019,** *138* (5), 3781–3788.

Chaves, D. A.; Lemes, S. R.; Araujo, L. A.; Sousa, M. A. M.; Freitas, G. B.; Lino-Junior, R. S.; Mrue, F.; Melo-Reis, P. R. Avaliação da atividade angiogênica da solução aquosa do barbatimão (*Stryphnodendron adstringens*). *Revista Brasileira de Plantas Medicinais* **2016,** *18* (2), 524–530.

Cintra, P.; Malaspina, O.; Bueno, O. C. Toxicity of Barbatimão to *Apis mellifera* and *Scaptotrigona postica,* Under Laboratory Conditions. *J. Apicult. Res.* **2003,** *42* (1–2), 9–12.

Costa, M. A.; Ishida, K.; Kaplum, V.; Koslyk, E. D.; de Mello, J. C.; Ueda-Nakamura, T.; Dias Filho, B. P.; Nakamura, C. V. Safety Evaluation of Proanthocyanidin Polymer-Rich Fraction Obtained from Stem Bark of *Stryphnodendron adstringens* (BARBATIMAO) for Use as a Pharmacological Agent. *Regul. Toxicol. Pharmacol.* **2010,** 58 (2), 330–335.

Costa, M. A.; Palazzo de Mello, J. C.; Kaneshima, E. N.; Ueda-Nakamura, T.; Dias Filho, B. P.; Audi, E. A.; Nakamura, C. V. Acute and Chronic Toxicity of An Aqueous Fraction of the Stem Bark of *Stryphnodendron adstringens* (Barbatimão) in Rodents. *Evid. Based Complement. Alternat. Med.* **2013,** *2013*.

de Carvalho, J. T. G.; Agudelo, J. S. H.; Baldivia, D. D. S.; Carollo, C. A.; Silva, D. B.; de Picoli Souza, K.; et al. Hydroethanolic Stem Bark Extracts of *Stryphnodendron adstringens* Impair M1 Macrophages and Promote M2 Polarization. *J. Ethnopharmacol.* **2020,** *254*, 112684.

de Freitas, A. L. D.; Kaplum, V.; Rossi, D. C. P.; da Silva, L. B. R.; Melhem, M. D. S. C.; Taborda, C. P.; et al. Proanthocyanidin Polymeric Tannins from *Stryphnodendron adstringens* Are Effective Against *Candida* spp. Isolates and for Vaginal Candidiasis Treatment. *J. Ethnopharmacol.* **2018,** *216*, 184–190.

de Mello, J. P.; Petereit, F.; Nahrstedt, A. Flavan-3-ols and Prodelphinidins from *Stryphnodendron adstringens*. *Phytochemistry* **1996a,** *41* (3), 807–813.

de Mello, J. P.; Petereit, F.; Nahrstedt, A. Prorobinetinidins from *Stryphnodendron adstringens*. *Phytochemistry* **1996b,** *42* (3), 857–862.

de Mello, J. C. P.; Petereit, F.; Nahrstedt, A. A Dimeric Proanthocyanidin from *Stryphnodendron adstringens*. *Phytochemistry* **1999,** *51* (8), 1105–1107.

de Sousa, N. C.; de Carvalho, S.; Spanó, M. A.; Graf, U. Absence of Genotoxicity of a Phytotherapeutic Extract from *Stryphnodendron adstringens* (Mart.) Coville in Somatic and Germ Cells of *Drosophila melanogaster*. *Environ. Mol. Mutagen.* **2003,** *41* (4), 293–299.

de Souza, T. F.; Cintra, P.; Malaspina, O.; Bueno, O. C.; Fernandes, J. B.; da Silva Almeida, S. S. M. Toxic Effects of Methanolic and Dichloromethane Extracts of Flowers and Peduncles of *Stryphnodendron adstringens* (Leguminosae: Mimosoideae) on *Apis mellifera* and *Scaptotrigona postica* Workers. *J. Apic. Res.* **2006,** *45* (3), 112–116.

Faria, A. M.; Santos, A. D. S.; Ferreira, L. L.; Bastos, T. S. A.; Louly, C. C. B.; Matos, M. P. C.; et al. Acaricidal Activity of Ethanolic Extracts of *Stryphnodendron adstringens* and *Lafoensia pacari* and Oil Resins from *Copaifera* sp. and *Pterodon emarginatus* Against *Sarcoptes scabiei* Var. *suis*. *Pesquisa Veterinária Brasileira* **2017,** *37* (12), 1411–1415.

Felfili, J.; Silva Júnior, M. C. D.; Dias, B. J.; Rezende, A. V. Estudo fenológico de *Stryphnodendron adstringens* (Mart.) Coville no cerrado sensu stricto da fazenda Água Limpa no distrito federal, Brasil. *Braz. J. Bot.* **1999,** *22* (1), 83–90.

Felipe, A. M.; Rincão, V. P.; Benati, F. J.; Linhares, R. E.; Galina, K. J.; de Toledo, C. E.; Lopes, G. C.; de Mello, J. C.; Nozawa, C. Antiviral Effect of *Guazuma ulmifolia* and *Stryphnodendron adstringens* on Poliovirus and Bovine Herpesvirus. *Biol. Pharm. Bull.* **2006,** *29* (6), 1092–1095.

Fonseca, J.; Couto, I. F. S.; da Silva, R. M.; Fioratti, C. A. G.; Pereira, F. F.; Mauad, M.; et al. Efeito de extratos metanólicos de *Stryphnodendron adstringens* (MART) coville na alimentação e reprodução de *Plutella xylostella* L. (Lepidoptera: Plutellidae). *Interciencia* **2018,** *43* (3), 182–187.

Henriques, B. O.; Corrêa, O.; Azevedo, E. P. C.; Pádua, R. M.; Oliveira, V. L. S. D.; Oliveira, T. H. C.; et al. In Vitro TNF-Inhibitory Activity of Brazilian Plants and Anti-Inflammatory Effect of *Stryphnodendron adstringens* in an Acute Arthritis Model. *Evid. Based Complement. Alternat. Med.* **2016,** 2016.

Hernandes, L.; Pereira, L. M. D. S.; Palazzo, F.; Mello, J. C. P. D. Wound-Healing Evaluation of Ointment from *Stryphnodendron adstringens* (barbatimão) in Rat Skin. *Braz. J. Pharm. Sci.* **2010,** *46* (3), 431–436.

Herzog-Soares, J. D.; Alves, R. K.; Isac, E.; Bezerra, J. C. B.; Gomes, M. H.; Santos, S. C.; Ferri, P. H. Atividade tripanocida in vivo de *Stryphnodendron adstringens* (barbatimão verdadeiro) e *Caryocar brasiliensis* (pequi). *Revista Brasileira de Farmacognosia* **2002,** *12,* 1–2.

Holetz, F. B.; Ueda-Nakamura, T.; Dias Filho, B. P.; Mello, J. C. P. D.; Morgado-Díaz, J. A.; Toledo, C. E. M. D.; Nakamura, C. V. Biological Effects of Extracts Obtained from *Stryphnodendron adstringens* on *Herpetomonas samuelpessoai*. *Memórias do Instituto Oswaldo Cruz* **2005,** *100* (4), 397–401.

Ishida, K.; de Mello, J. C. P.; Cortez, D. A. G.; Filho, B. P. D.; Ueda-Nakamura, T.; Nakamura, C. V. Influence of Tannins from *Stryphnodendron adstringens* on Growth and Virulence Factors of *Candida albicans*. *J. Antimicrob. Chemother.* **2006,** *58* (5), 942–949.

Ishida, K.; Rozental, S.; de Mello, J. C. P.; Nakamura, C. V. Activity of Tannins from *Stryphnodendron adstringens* on *Cryptococcus neoformans*: Effects on Growth, Capsule Size and Pigmentation. *Ann. Clin. Microbiol. Antimicrob.* **2009,** *8* (1), 1–10.

Jesus, F. G.; de Paiva, L. A.; Gonçalves, V. C.; Marques, M. A.; Boiça Junior, A. L. Efeito de plantas inseticidas no comportamento e biologia de *Plutella xylostella* (Lepidoptera: Plutellidae). *Arquivos do Instituto Biológico* **2011,** *78* (2), 279–285.

Kaplum, V.; Ramos, A. C.; Consolaro, M. E.; Fernandez, M. A.; Ueda-Nakamura, T.; Dias-Filho, B. P.; et al. Proanthocyanidin Polymer-Rich Fraction of *Stryphnodendron adstringens* Promotes In Vitro and In Vivo Cancer Cell Death via Oxidative Stress. *Front. Pharmacol.* **2018,** *9*, 694.

Lima, C. B.; Migotto, D. L.; Oliveira, G. R.; Souza, T. C.; Santana, R. O.; Castejon, F. V.; et al. Dietary Supplementation of Barbatimão (*Stryphnodendron adstringens*) and Pacari (*Lafoensia pacari*) Extracts on the Oxidative Stability and Quality of Chicken Meat. *Braz. J. Poult. Sci.* **2016,** *18* (4), 669–676.

Lima, J. C. D. S.; Martins, D. T. O.; de Souza Jr, P. T. Experimental Evaluation of Stem Bark of *Stryphnodendron adstringens* (Mart.) Coville for Antiinflammatory Activity. *Phytother. Res.* **1998,** *12* (3), 218–220.

Lima, T. C. D. D.; Cardoso, M. V.; Modesto, T.; Oliveira, A. L. D. B.; Silva, M. N. D.; Monteiro, M. C. Breve revisão etnobotânica, fitoquímica e farmacologia de *Stryphnodendron adstringens* utilizada na Amazônia. *Revista Fitos.* **2017,** *10* (3), 329–338.

Lopes, G. C.; Machado, F. A. V.; de Toledo, C. E. M.; Sakuragui, C. M.; de Mello, J. C. P. Chemotaxonomic Significance of 5-Deoxyproanthocyanidins in *Stryphnodendron* Species. *Biochem. Syst. Ecol.* **2008,** *12* (36), 925–931.

Lopes, G. C.; Sanches, A. C. C.; Toledo, C. E. M. D.; Isler, A. C.; Mello, J. C. P. D. Determinação quantitativa de taninos em três espécies de *Stryphnodendron* por cromatografia líquida de alta eficiência. *Braz. J. Pharm. Sci.* **2009,** *45* (1), 135–143.

Luiz, R. L. F.; Vila, T. V. M.; de Mello, J. C. P.; Nakamura, C. V.; Rozental, S.; Ishida, K. Proanthocyanidins Polymeric Tannin from *Stryphnodendron adstringens* Are Active Against *Candida albicans* Biofilms. *BMC Complement. Altern. Med.* **2015,** *15* (1), 1–11.

Luize, P. S.; Tiuman, T. S.; Morello, L. G.; Maza, P. K.; Ueda-Nakamura, T.; Dias Filho, B. P.; et al. Effects of Medicinal Plant Extracts on Growth of *Leishmania* (*L.*) *amazonensis* and *Trypanosoma cruzi*. *Revista Brasileira de Ciências Farmacêuticas* **2005,** *41* (1), 85–94.

Marques, L. T.; Guedes, R. A.; Rodrigues, W. D.; Archanjo, A. B.; Severi, J. A.; Martins, I. V. F. Chemical Composition of Various Plant Extracts and Their In Vitro Efficacy in Control of *Fasciola hepatica* eggs. *Ciência Rural* **2020,** *50* (5), e20190363.

Martins, D. T.; Lima, J. C.; Rao, V. S. The Acetone Soluble Fraction from Bark Extract of *Stryphnodendron adstringens* (Mart.) Coville Inhibits Gastric Acid Secretion and Experimental Gastric Ulceration in Rats. *Phytother. Res.* **2002,** *6* (5), 427–431.

Meinhart, A. D.; Damin, F. M.; Caldeirão, L.; da Silveira, T.; Filho, J. T.; Godoy, H. T. Chlorogenic Acid Isomer Contents in 100 Plants Commercialized in Brazil. *Food Res. Int. (Ottawa, Ont.)* **2017,** *99* (Pt 1), 522–530.

Melo, J. O. D.; Endo, T. H.; Bersani-Amado, L. E.; Svidzinski, A. E.; Baroni, S.; Mello, J. C. P. D.; Bersani-Amado, C. A. Effect of *Stryphnodendron adstringens* (barbatimão) Bark on Animal Models of Nociception. *Revista Brasileira de Ciências Farmacêuticas* **2007,** *43* (3), 465–469.

Morey, A. T.; de Souza, F. C.; Santos, J. P.; Pereira, C. A.; Cardoso, J. D.; de Almeida, R. S.; Costa, M. A.; de Mello, J. C.; Nakamura, C. V.; Pinge-Filho, P.; Yamauchi, L. M.; Yamada-Ogatta, S. F. Antifungal Activity of Condensed Tannins from *Stryphnodendron adstringens*: Effect on *Candida tropicalis* Growth and Adhesion Properties. *Current Pharm. Biotechnol.* **2016,** *17* (4), 365–375.

Nascimento, A. M. D.; Guedes, P. T.; Castilho, R. O.; Vianna-Soares, C. D. *Stryphnodendron adstringens* (Mart.) Coville (Fabaceae) Proanthocyanidins Quantitation by RP-HPLC. *Braz. J. Pharm. Sci.* **2013,** *49* (3), 549–558.

Oliveira, A. L. S.; Figueiredo, A. D. L. Prospecção Fitoquímica das Folhas de *Stryphnodendron adstringens* (Mart.) Coville (Leguminosae-Mimosoidae). *Revista Brasileira de Biociências* **2007,** *5* (S2), 384–386.

Oliveira, D. G.; Prince, K. A.; Higuchi, C. T.; Santos, A. C. B.; Lopes, L. M. X.; Simões, M. J. S.; Leite, C. Q. F. Antimycobacterial Activity of Some Brazilian Indigenous Medicinal Drinks. *Revista de Ciências Farmacêuticas Básica e Aplicada* **2007,** *28* (2),165–169.

Pellenz, N. L.; Barbisan, F.; Azzolin, V. F.; Duarte, T.; Bolignon, A.; Mastella, M. H.; Teixeira, C. F.; Ribeiro, E. E.; da Cruz, I.; Duarte, M. Analysis of In Vitro Cyto- and Genotoxicity of Barbatimão Extract on Human Keratinocytes and Fibroblasts. *BioMed Res. Int.* **2018,** *2018*, 1942451.

Pereira Junior, L. C. S.; Coriolano de Oliveira, E.; Valle Rorig, T. D.; Pinto de Araújo, P. I.; Sanchez, E. F.; Garrett, R.; Palazzo de Mello, J. C.; Fuly, A. L. The Plant *Stryphnodendron adstringens* (Mart.) Coville as a Neutralizing Source Against Some Toxic Activities of Bothrops Jararacussu Snake Venom. *Toxicon* **2020,** *186*, 182–190.

Pinho, L. D.; Souza, P. N. S.; Macedo Sobrinho, E.; Almeida, A. C. D.; Martins, E. R. Atividade antimicrobiana de extratos hidroalcoolicos das folhas de alecrim-pimenta, aroeira, barbatimão, erva baleeira e do farelo da casca de pequi. *Ciência Rural* **2012,** *42* (2), 326–331.

Rebecca, M. A.; Ishii-Iwamoto, E. L.; Grespan, R.; Cuman, R. K. N.; Caparroz-Assef, S. M.; de Mello, J. C. P.; Bersani-Amado, C. A. Toxicological Studies on *Stryphnodendron adstringens*. *J. Ethnopharmacol.* **2002,** *83* (1–2), 101–104.

Rebecca, M. A.; Ishii-Iwamoto, E. L.; Kelmer-Bracht, A. M.; Caparroz-Assef, S. M.; Cuman, R. K. N.; Pagadigorria, C. L. S.; et al. Effect of *Stryphnodendron adstringens* (barbatimão) on Energy Metabolism in the Rat Liver. *Toxicol. Lett.* **2003,** *143* (1), 55–63.

Sabino, A.; Eustáquio, L.; Miranda, A.; Biojone, C.; Mariosa, T. N.; Gouvêa, C. *Stryphnodendron adstringens* ("Barbatimão") Leaf Fraction: Chemical Characterization, Antioxidant Activity, and Cytotoxicity Towards Human Breast Cancer Cell Lines. *Appl. Biochem. Biotechnol.* **2018,** *184* (4), 1375–1389.

Salvalaggio, M. de O.; de Freitas, R. A.; Franquetto, E. M.; Koop, H. S.; Silveira, J. L. M. Influence of the Extraction Time on Macromolecular Parameters of Galactomannans. *Carbohydr. Polym.* **2015,** *116*, 200–206.

Santos Filho; Plínio R. dos; Ferreira, Lidiane A.; Gouvêa Cibele M. C. Paiva. Protective Action Against Chemical-Induced Genotoxicity and Free Radical Scavenging Activities of *Stryphnodendron adstringens* ("barbatimão") Leaf Extracts. *Revista Brasileira de Farmacognosia* **2011,** *21* (6), 1000–1005.

Santos, S. C.; Costa, W. F.; Ribeiro, J. P.; Guimarães, D. O.; Ferri, P. H.; Ferreira, H. D.; Seraphin, J. C. Tannin Composition of Barbatimão Species. *Fitoterapia* **2002,** *73* (4), 292–299.

Sereia, A. L.; de Oliveira, M. T.; Baranoski, A.; Marques, L. L. M.; Ribeiro, F. M.; Isolani, R. G.; et al. In Vitro Evaluation of the Protective Effects of Plant Extracts Against Amyloid-Beta Peptide-Induced Toxicity in Human Neuroblastoma SH-SY5Y Cells. *PloS one* **2019,** *14* (2), e0212089.

Silva, F. M.; De Paula, J. E.; Espindola, L. S. Evaluation of the Antifungal Potential of Brazilian Cerrado Medicinal Plants. *Mycoses* **2009,** *52* (6), 511–517.

Souza, P. M.; Elias, S. T.; Simeoni, L. A.; de Paula, J. E.; Gomes, S. M.; Guerra, E. N. S.; et al. Plants from Brazilian Cerrado with Potent Tyrosinase Inhibitory Activity. *PLoS One* **2012,** *7* (11), e48589.

Souza, T. M.; Moreira, R. R.; Pietro, R. C.; Isaac, V. L. Avaliação da atividade anti-séptica de extrato seco de *Stryphnodendron adstringens* (Mart.) Coville e de preparação cosmética contendo este extrato. *Revista Brasileira de Farmacognosia* **2007a,** *17* (1), 71–75.

Souza, T. M.; Severi, J. A.; Silva, V. Y. A.; Santos, E.; Pietro, R. C. L. R. Bioprospecção de atividade antioxidante e antimicrobiana da casca de *Stryphnodendron adstringens* (Mart.) Coville (Leguminosae-Mimosoidae). *Revista de Ciências Farmacêuticas Básica e Aplicada* **2007b,** *28* (2), 221–226.

Souza-Moreira, T. M.; Queiroz-Fernandes, G. M.; Pietro, R. *Stryphnodendron* Species Known as "Barbatimão": A Comprehensive Report. *Molecules* **2018,** *23* (4), 910, https://doi.org/10.3390/molecules23040910.

Trevisan, D.; da Silva, P. V.; Farias, A.; Campanerut-Sá, P.; Ribeiro, T.; Faria, D. R.; de Mendonça, P.; de Mello, J.; Seixas, F.; Mikcha, J. Antibacterial Activity of Barbatimão (*Stryphnodendron adstringens*) Against *Staphylococcus aureus*: In Vitro and In Silico Studies. *Lett. Appl. Microbiol.* **2020,** *71* (3), 259–271.

Trolezi, R.; Azanha, J. M.; Paschoal, N. R.; Chechi, J. L.; Dias Silva, M. J.; Fabris, V. E.; Vilegas, W.; Kaneno, R.; Fernandes Junior, A.; Bosco, S. M. *Stryphnodendron adstringens* and Purified Tannin on *Pythium insidiosum*: In Vitro and In Vivo Studies. *Ann. Clin. Microbiol. Antimicrob.* **2017,** *16* (1), 7.

Vicente, R. A.; Baby, A. R.; Velasco, M. V.; Bedin, V. Double-Blind, Randomized, Placebo-Controlled Trial of a Cream Containing the *Stryphnodendron adstringens* (Martius) Coville Bark Extract for Suppressing Terminal Hair Growth. *J. Eur. Acad. Dermatol. Venereol.* **2009,** *23* (4), 410–414.

CHAPTER 19

Tamarindus indica and Its Bio-activities: An Important Fruit Tree

POORNANANDA MADHAVA NAIK[1] and VINAYAK UPADHYA[2*]

[1]*Department of Botany, Karnatak University, Dharwad, Karnataka 580003, India*

[2]*Department of Forest Products and Utilization, College of Forestry (University of Agricultural Sciences, Dharwad), Banavasi Road, Sirsi, Karnataka 581401, India*

[*]*Corresponding author. E-mail: sirsivinayak@gmail.com*

ABSTRACT

Tamarindus indica (family–Fabaceae) is a cultivated fruiting tree in tropical regions and the fruits are used as food source, nutraceuticals and medicine. Medicinally Tamarind fruit pulp is used both in codified and non codified systems of traditional medicine to treat number of human disorders. Different parts of the tamarind plant such as leaves, bark, twigs, seeds and roots are also used to treat throat infection, rheumatism, cough, fever, intestinal worms, conjunctivitis, asthma, amenorrhea, febrifuge, sores, rashes, ulcers, boils, scurvy, dysentery and diarrhea. Tamarind is a rich source of diverse bioactive compounds such as proteins, starch, sulfur amino acids, phenolic antioxidants, fatty acids, pectin, alkaloids, flavonoids, saponins, tannins, vitamins such as A, B1, B2, B3, B5, B6, C, K, organic acids, minerals and many more. Hence a wide range of bioactivities namely antimicrobial, antioxidant, hepatoprotective, antidiabetic, antivenom, anti-inflammatory, analgesic activity and anticancer properties were proved scientifically. Due

Bioactives and Pharmacology of Legumes. T. Pullaiah, PhD (Ed.)

to the wide range of compounds present and bioactivities shown, makes Tamarind as an important fruit tree of food and medicine.

19.1 INTRODUCTION

Tamarindus indica is an evergreen tropical tree that belongs to the family Fabaceae. The tree is native to Africa and Southern Asia. Tamarindus is an important food source in tropical region and also serves as nutraceuticals and traditional medicine (De Caluwé et al., 2010). Tamarind fruit pulp is similar to that of dried fruits of date, so the word tamarind may be originated from the Arabic word "Tamar-u'l-Hind," which means date of India (Azad, 2018). The cultivation of the tree species was documented in Egypt as early as 400 B.C. and now it is cultivated in India, Sri Lanka, China, Nepal, Taiwan, Bangladesh, and Pakistan. Tamarind is an evergreen moderate-to-large in size tree, shows wide spread branches with an almost-round canopy, and reaches a height of 18–25 m. The tree parts such as seeds, fruit, leaves, roots, and bark have rich nutritional value along with broad medicinal, chemical, industrial, and timber usage (Bhadoriya et al., 2011; Kuru, 2014). *T. indica* has a rich source of proteins, carbohydrates, amino acids, and phytochemicals. This chapter aims to bestow a detailed account of the phytochemicals compositions and pharmacological activity of *T. indica.*

Conventionally, tamarind fruits are commonly used in the preparation of curries, jam, pickles, sauce, syrup, fruit drink, etc., and as flavoring stabilizer and binder in food preparations. This plant is mentioned in codified traditional medicinal systems such as Ayurveda, Unani, and Siddha and also used by non-codified traditional practitioners around the world as a medicine. Detailed description of tamarind was mentioned in the Ayurvedic pharmacopeia of India (Anonymous, 2004). In Ayurveda, tamarind fruit pulp is used to treat anorexia, stomach problems, digestion problems, ear pain, pilonidal sinus, fatigue, etc. It is used to prepare Sankha Dravaka and Sankhavati. Uses of tamarind fruits are mentioned in the Unani system of medicine to cure bilious vomiting, paralytic stomach, toxicity of alcohol, gastric paralysis, chronic diarrhea, piles, etc. (Ahmad et al., 2018). In the Siddha system, it is mentioned in the treatment of pallor on the skin (Paheerathan et al., 2018). Tamarindus is also used in a polyherbal preparation of Siddha formulation to treat diabetes (Sofia et al., 2014). Different parts of the tamarind plant such as leaves, bark, twigs, and roots are used by local healers; leaf decoction is used to treat throat infection, rheumatism, cough, fever, intestinal worms, and conjunctivitis; bark decoction is used for asthma and amenorrhea, as a febrifuge and a digestive tonic, to treat sores, rashes, ulcers, and boils; pulp is a mild laxative and is used to treat scurvy; and seeds are used to cure dysentery and diarrhea (Santosh et al., 2011). In Madagascar, the fruits of tamarind are used as medicine to reduce fever (Norscia and Borgognini-Tarli, 2006), in Ghana, tamarind leaves are used in the treatment of malaria (Asase et al., 2005).

19.2 BIOACTIVE COMPOUNDS

Plants possess diverse bioactive compounds and have antibacterial and antioxidant properties that make them fight against diseases (Abdelrahman and Mariod, 2019). Tamarind is one such tree that has a rich source of bioactive compounds. A wide range of research has been conducted with the different parts of the plants to evaluate their efficacy against pharmacological effects (Menezes et al., 2016). The chemical constituent of root bark of *T. indica* includes n-hexacosane, eicosanoic acid, octacosanylferulate, β-sinosterol, 21-oxobehenic acid, and (+)-pinitol (Bhadoriya et al., 2011). The stem bark contains saponins, tannins, flavonoids, cardiac glycosides, alkaloids, peroxidase, and lipids. The stem bark is used to make tea to treat sore throat. Bioactive compounds derived from the stem bark has analgesic, antimicrobial, hypoglycemic, and spasmogenic activities (Abdelrahman and Mariod, 2019). Tamarind leaves are the major source of lipid, fiber, protein, thiamine, riboflavin, niacin, β-carotene, ascorbic acid, triterpenes, lupanone, and lupeol (Bhadoriya et al., 2011). Leaf oil of tamarind contains 13 essential compounds of which limonene and benzyl benzoate are highly concentrated (Pino et al., 2002).

The fruit is considered as the most essential part of the plant and is the harbor for a variety of bioactive compounds. The seed as part of fruit is a major source of protein, starch, sulfur amino acids, phenolic antioxidants, campesterol, β-amyrin, arabinose, xylose, galactose pectin, glucose and uronic acid, cellulose, albuminoid. amyloids, phytohemagglutinins, chitinase, and also fatty acids such as palmitic acid, oleic, acid, linoleic acid, and eicosanoic acid (Menezes et al., 2016; Zohrameena et al., 2017). Tamarind pericarp is comprised of rich source of polyphenolics such as apigenin, catechin, epicatechin, taxifolin, eriodictyol, naringenin, procyanidin dimer, and procyanidintrimer (Bhadoriya et al., 2011; Figure 19.1).

The fruit pulp of the tamarind contains carbohydrates; proteins; pectin; alkaloids; flavonoids; protein; fat; saponins and tannins; vitamins such as A, B1, B2, B3, B5, B6, C, and K; organic acids (tartaric acid, malic acid, acetic acid, citric acid, formic acid, succinic acid, and amino acids); minerals (calcium, phosphorus, magnesium, potassium, sodium, and selenium), some pyrazines; and some thiazoles as fragrant (Bhadoriya et al., 2011; Menezes et al., 2016; Ferrara, 2019). Due to wide range of bioactive compounds present in fruit pulp, it makes them an important pharmacological source and has various bioactivities: antifluorose, analgesic, hepatoregenerativa, hypolipidemic activity, antioxidant, and antispasmodic (Menezes et al., 2016).

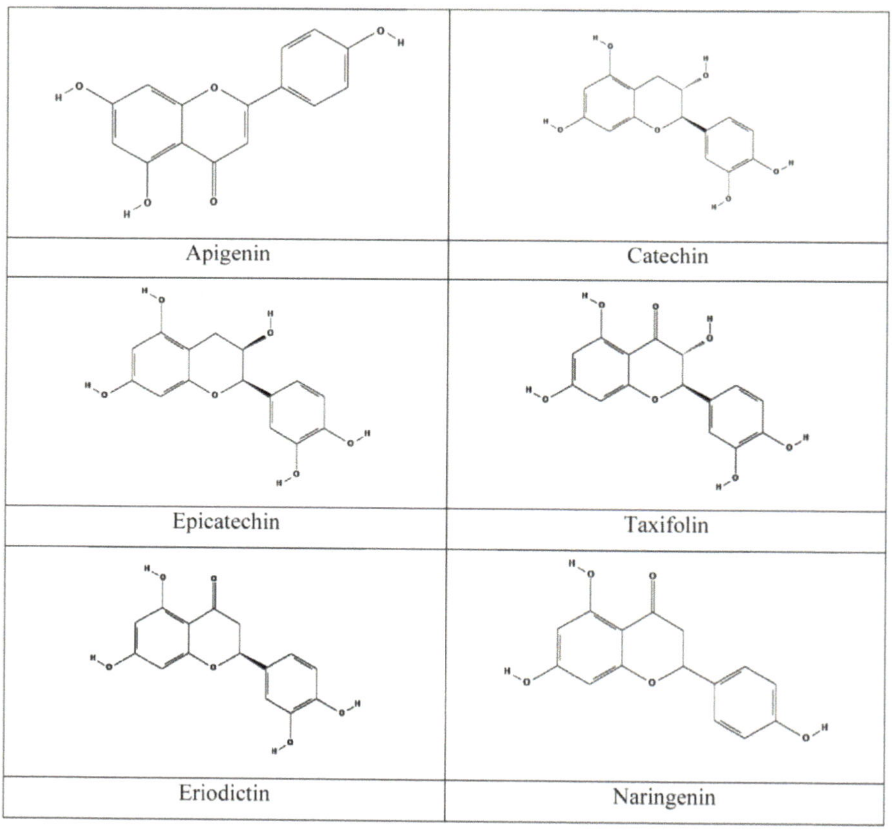

FIGURE 19.1 Polyphenolic compounds from *T. indica* pericarp.

19.3 PHARMACOLOGICAL ACTIVITY

Since times immemorial, tamarind is used by local practitioners as part of traditional medicine to cure a number of diseases; now, modern science research is corroborating with the application of the plant with respect to their therapeutic properties (Ferrara, 2019). Recent research findings showed a number of pharmacological activities and are been discussed in the following sections.

19.3.1 Antimicrobial Activity

Tamarind has a wide range of antimicrobial activities and this property of the plant is due to the presence of important bioactive compounds. Antibacterial activity of methanolic leaf extract of *T. indica* is tested against *Burkholderia pseudomallei* and its in-vitro inhibitory activity suggests that there is need of further animal research to know the important role of *T. indica* in treating melioidosis (Muthu et al., 2005). Paper disk diffusion method was used to assess the antimicrobial activity of aqueous, ethanolic, and acetone-concentrated tamarind extracts against bacteria and fungi, and the test results showed potential effects against *Bacillus subtilis*, *Salmonella typhi*, *Salmonella paratyphi*, and *Staphylococcus aureus* (Doughari, 2006). Vaghasiya and Chanda (2009) found that methanol and acetone extract of *T. indica* has a strong antimicrobial activity, which was done by the agar disk diffusion method against *Klebsiella pneumoniae*. The activity was compared with antimicrobial standards amikacin and piperacillin. Antibacterial activity of *T. indica* studied by Biswas and Shinha (2015) showed that there is promising effect on bacteria, isolated from urinary tracts of infected women. *Escherichia coli* and *Shigella* spp., isolated from the stools of pregnant women were tested against the leaves and fruit extract of *T. indica*; among the different solvent extracts, ethanol extract showed maximum activity (Abdallah and Muhammad, 2018).

19.3.2 Antioxidant Activity

Tamarind is rich in phenolic contents, this leads them to possess potential antioxidant properties and studies showed all the extract of tamarind particularly seed and pericarp holds good antioxidant properties (Sudjaroen et al., 2005). Martinello et al. (2006) reported that the crude ethanolic extract of tamarind fruit pulp has significant antioxidant and hypolipidemic activities in hypercholestrolemic hamsters. Tamarind seed coat ethanolic extract was tested for antioxidant activity by DPPH (2,2-diphenyl-1 picrylhydrazyl) free radical scavenging method using standard ascorbic acid (Vyas et al., 2009). The study revealed that compound present in the extract has capacity to enhance the antioxidant defense system and so can prevent the cell damage. Application of tamarind seed powder into cookies and mango juice showed positive association, in increasing their bioactive phytochemical content with potential antioxidant activity and also effective against the oxidative damage (Natukunda et al., 2016).

19.3.3 Hepatoprotective Activity

Aqueous extracts of various parts such as fruits, leaves and unroasted seeds of tamarind were tested in rats by paracetamol induced hepatotoxicity and found it is an effective hepato-regenerative (Pimple et al., 2007). Tayade et al. (2009) investigations showed methanolic extract of leaves of tamarind is having anti-asthmatic and hepatoprotective effect and also had significant stabilizing activity of antihistaminic, adaptogenic, and mast cell in laboratory animals. Ethanolic extract of stem bark was investigated for hepatoprotective activity against the drug induced hepatic damage during chemotherapy in Sprague Dawley rats, and showed effective against hepatic damage, also confirmed as hepatoprotective against development of drug induced damage (Meena et al., 2018).

19.3.4 Antidiabetic Activity

Several researchers showed tamarind seed has antidiabetic property. Aqueous extract of tamarind seeds exhibit antidiabetogenic activity, where diabetic and severe diabetic male rat was subjected to aqueous extract of tamarind seed, and results found there was a significantly reduced hyperglycemial state which was counter checked by measuring various fasting blood glucose levels (Maiti et al., 2004). The seed coat aqueous extract application on streptozotocin-induced diabetic male rats showed reduced hyperlipidemia (Maiti et al., 2005). Bhadoriya et al. (2017) investigated the effect of hydro-ethanolic extract of seed coat in alloxan-induced diabetic rats which is rich in polyphenolic content and showed the potential antidiabetic property.

19.3.5 Antivenom Property

Tamarind has a widely grouped bioactive compounds that may neutralize snake venom; hence, Indian traditional healers used its seed extract as a remedy for snake bite (Soni and Singh, 2019). Tamarind extract can neutralize indirect hemolysis caused by snake venom and degradation of human fibrinogen β-chain. Administration of different doses of the extract showed prolonged duration of clotting, significant neutralization of myotoxic effects, and can be used as substitute for the serum therapy (Ushanandini et al., 2006).

19.3.6 Anti-inflammatory and Analgesic Activity

Tamarind is traditionally used to treat pain. The presence of tannins, flavonoids, and saponins in seeds shows anti-inflammatory activity (Suralkar et al., 2012). Bhadoriya et al. (2012) conducted an experiment on carrageenan-induced hind paw edema in male Wistar albino rats by applying the hydro-ethanolic extract of tamarind leaves to investigate the in-vitro anti-inflammatory effect. The study results revealed that the oral administration of extracts is highly dose dependent in the treatment of anti-inflammatory disorder. Dighe et al. (2009) extracted the tamarind bark using various solvents (petroleum, ether, and ethanol) to study the analgesic activity of phytochemicals (sterol and triterpenes) using the hot plate method on mice. The study results showed that there was significant effect of petroleum ether bark extract. Authors report that the analgesic effect may be due to the involvement of opioid receptors that exhibit analgesic activity.

19.3.7 Anticancer Activity

Scientific research has proved that tamarind plant has various therapeutically important phytochemicals, which has the capacity to fight against cancer. Polysaccharide PST001 isolated from tamarind seed kernel showed the antitumor and immunopotentiating activity in mice (Aravind et al., 2012). Methanolic extract of tamarind seed was investigated for the cytotoxicity potential of two cancer cell lines, human lymphoma cell line (SR) and rhabdomyosarcoma cancer (RD). The results revealed that the seed extract of tamarind holds potentially strong cytotoxicity effects against cancer cell lines (Hussein et al., 2017). Srinivas et al. (2018) performed an experiment to study the ethanolic extracts of tamarind bark to assess the anticancer potential of tamarinds by using MTTs assay on human colorectal adenocarcinoma cell line (HT29) and the bark extract exhibited strong cytotoxic effect over human cancer cell line.

KEYWORDS

- ***Tamarindus indica***
- **Fabaceae**
- **fruit tree**

REFERENCES

Abdallah, M. S.; Muhammad, A. Antibacterial Activity of Leaves and Fruit Extract of *Tamarindus indica* Against Clinical Isolates of *Escherichia coli* and *Shigella* at Potiskum Yobe State. *Nigerian J. Anal. Pharm. Res.* **2018,** *7* (5), 606–609.

Abdelrahman, G. H.; Mariod, A. A. *Tamarindus indica*: Phytochemical Constituents, Bioactive Compounds and Traditional and Medicinal Uses. In *Wild Fruits: Composition, Nutritional Value and Products*; Mariod, A. A., Ed.; Springer Nature: Switzerland; 2019, pp 229–238.

Ahmad, A.; Ahmad, W.; Zeenat, F.; Sajid, M. Therapeutic, Phytochemistry and Pharmacology of *Tamarindus indica*: A Review. *Int. J. Unani Integrat. Med.* **2018,** *2* (2), 14–19.

Anonymous. *The Ayurvdeic Pharmacopoeia of India*, Part 1, Vol. 4; Government of India, Ministry of Health and Family Welfare, Department of ISM&H (AYUSH): New Delhi; 2004, pp 18–19.

Aravind, S. R.; Joseph, M. M.; Varghese, S.; Balaram, P.; Sreelekha. T. T. Antitumor and Immuno Potentiating Activity of Polysaccharide PST001 Isolated from the Seed Kernel of *Tamarindus indica*, an In Vivo Study in Mice. *Sci. World J.* **2012,** *2012*, 361382.

Asase, A.; Oteng-Yeboah, A. A.; Odamtten, G. T.; Simmonds, M. S. J. Ethnobotanical Study of Some Ghanaian Anti-Malarial Plants. *J. Ethnopharmacol.* **2005,** *99*, 273–279.

Azad, M. S. Tamarindo—*Tamarindus indica*. In *Exotic Fruits Reference Guide;* Rodrigues, S.; Silva, D. E. O.; De Brito E. S., Eds.; Academic Press: London, United Kingdom; 2018, pp 403–412. doi:10.1016/B978-0-12-803138-4.00055-1.

Bhadoriya, S.; Ganeshpurkar, A.; Bhadoriya, R.; Sahu, S. K.; Patel, J. R. Antidiabetic Potential of Polyphenolic Rich Fraction of *Tamarindus indica* Seed Coat in Alloxan-Induced Diabetic Rats. *J. Basic Clinic. Physiol. Pharmacol.* **2017,** *29* (1), 37–45.

Bhadoriya, S. S.; Ganeshpurkar, A.; Narwaria, J.; Rai, G.; Jain, A. P. *Tamarindus indica*: Extent of Explored Potential. *Pharmacogn. Rev.* **2011,** *5*, 73–81.

Bhadoriya, S. S.; Mishra, V.; Raut, S.; Jain, S. Antiinflammatory and Antinociceptive Activities of Hydroethanolic Extract of *Tamarindus indica* Leaves. *Sci. Pharm.* **2012,** *80* (3), 685–700.

Biswas, K.; Shinha, S. N. Antimicrobial Activity of *Tamarindus indica* Linn. Against Bacteria Causing Urinary Tract Infection. *J. Bio. Sci.* **2015,** *23*, 47–55.

De Caluwé, E.; Halamová, K.; Van Damme, P. *Tamarindus indica* L.: A Review of Traditional Uses, Phytochemistry and Pharmacology. *Afrika Focus.* **2010,** *23* (1), 53–83.

Dighe, N. S.; Pattanal, S. R.; Nirmal, S. A.; Kalkotwar, R. S.; Gware, V. M.; Hole, M. B. Analgesic Activity of *Tamarindus indica*. *Res. J. Pharmacogn. Phytochem.* **2009,** *1*, 69–71.

Doughari, J. H. Antimicrobial Activity of *Tamarindus indica* Linn. *Trop. J. Pharm. Res.* **2006,** *5*, 597–603.

Ferrara, L. Nutritional and Pharmacological Properties of *Tamarindus indica* L. *J. Nutr. Food Sci. Forecast.* **2019,** *2* (2), 1012.

Hussein, S. I.; Yaseen, N. Y.; Jawad, S. Q.; Abd, T. S. Seeds of *Tamarindus indica* as Anti-Cancer in Some Cell Line. *Int. J. Adv. Biol. Res.* **2017,** *7* (2), 360–362.

Kuru, P. *Tamarindus indica* and Its Health Related Effects. *Asian Pac. J. Trop. Biomed.* **2014,** *4* (9), 676–681.

Maiti, R.; Das, U. K.; Ghosh, D. Attenuation of Hyperglycemia and Hyperlipidemia in Streptozotocin-Induced Diabetic Rats by Aqueous Extract of Seeds of *Tamarindus indica. Biol. Pharm. Bull.* **2005,** *28*, 1172–1176.

Maiti, R.; Jana, D.; Das, U. K.; Ghosh, D. Antidiabetic Effect of Aqueous Extract of Seed of *Tamarindus indica* in Streptozotocin-Induced Diabetic Rats. *J. Ethnopharmacol.* **2004,** *92*, 85–91.

Martinello, F.; Soaresh, S. M.; Franco, J. J.; Santos, A. C.; Sugohara, A.; Garcia, S. B.; Curti, C.; Uyemura, S. A. Hypolipemic and Antioxidant Activities from *Tamarindus indica* Pulp Fruit Extract in Hypercholesterolemic Hamsters. *Food Chem. Toxicol.* **2006,** *44*, 810–818.

Meena, S. J.; Rahman, Md.A.; Bagga, P.; Mujahid, Md. Hepatoprotective Activity of *Tamarindus indica* Linn. Stem Bark Ethanolic Extract Against Hepatic Damage Induced by Co-Administration of Antitubercular Drugs Isoniazid and Rifampicin in Sparague Dawley Rats. *J. Basic Clin. Physiol. Pharmacol.* **2018,** *30* (1), 131–137.

Menezes, A. P.; Cerini, S. C.; Barbalho, S. M.; Guiguer, E. L. *Tamarindus indica* L. A Plant with Multiple Medicinal Purposes. *J. Pharmacogn. Phytochem.* **2016,** *5* (3), 50–54.

Muthu, S. E.; Nandakumar, S.; Roa, U. A. The Effect of Methanolic Extract of *Tamarindus indica* on the Growth of Clinical Isolates of *Burkholderia pseudomallei*. *Indian J. Med. Res.* **2005,** *122*, 525–528.

Natukunda, S.; Mayunga, J. H.; Mukisa, I. M. Effect of Tamarind (*Tamarindus indica* L.) Seed on Antioxidant Activity, Phytocompounds, Physiochemical Characteristics, and Sensory Acceptability of Enriched Cookies and Mango Juice. *Food Sci. Nutr.* **2016,** *4* (4), 494–507.

Norscia, I.; Borgognini-Tarli, S. M. Ethnobotanical Reputation of Plant Species from Two Forests of Madagascar: A Preliminary Investigation. *S. Afr. J. Bot.* **2006,** *72*, 656–660.

Paheerathan, V.; Piratheepkumar, R.; Kokilan, K. Efficacy of *Tamarindus indica* in the Management of Pandu Noei. *Siddha Papers* **2018,** *13* (1).

Pimple, B. P.; Kadam, P. V.; Badgujar, N. S.; Bafna, A. R.; Patil, M. J. Protective Effect of *Tamarindus indica* Linn. Against Paracetamol Induced Hepatotoxicity in Rats. *Indian J. Pharm. Sci.* **2007,** *69*, 827–831.

Pino, J. A.; Escalona, J. C.; Licea, I.; Perez, R.; Aguero, J. Leaf Oil of *Tamarindus indica* L. *J. Essent. Oil Res.* **2002,** *14*, 187–188.

Santosh, S. B.; Aditya, G.; Jitendra, N.; Gopal, R.; Alok, P. J. *Tamarindus indica*: Extent of Explored Potential. *Pharmacogn. Rev.* **2011,** *5* (9), 73–81.

Sofia, H. N.; Merlin Kumari, V. H.; Walter, T. M.; Senthil Kumar S. G. Anti-Diabetic Polyherbal Siddha Formulation Atthippattaiyathi Kasayam: A Review. *Int. J. Pharm. Sci. Rev. Res.* **2014,** *28* (2), 169–174.

Soni, N.; Singh V. K. Traditional, Nutraceutical and Pharmacological Approaches of *Tamarindus indica* (Imli). *Eur. J. Biol. Res.* **2019,** *9* (3), 141–154.

Srinivas, G.; Naru, R. R.; Malarselvi, S.; Rajakumar, R. In Vitro Anticancer Activity of the Ethanol Bark Extractsof *Tamarindus indica* Linn. Against HT29 Cancer Cell Line. *Int. J. Curr. Adv. Res.* **2018,** *7* (10B), 15820–15823.

Sudjaroen, Y.; Haubner, R.; Wurtelw, G.; Hull, W. E.; Erben, G.; Spiegelhalder, B. Isolation and Structure Elucidation of Phenolic Antioxidant from Tamarind (*Tamarindus indica* L.) Seed and Pericarp. *Food Chem. Toxicol.* **2005,** *43*, 1673–1682.

Suralkar, A. A.; Rodge, K. N.; Kamble, R. D.; Maske, K. S. Evaluation of Anti-Inflammatory and Analgesic Activities of *Tamarindus indica* Seed. *Int. J. Pharm. Sci. Drug Res.* **2012,** *4*, 213–217.

Tayade, P. M.; Ghaisas, M. M.; Jagtap, S. A.; Dongre, S. H. Anti-Asthmatic Activity of Methanolic Extract of Leaves of *Tamarindus indica* Linn. *J. Pharm. Res.* **2009,** *2*, 944–947.

Ushanandini, S.; Nagaruju, S.; Harish, K. K. The Antisnake Venom Properties of *Tamarindus indica* (Leguminosae) Seed Extract. *Phytother. Res.* **2006,** *20*, 851–858.

Vaghasiya, Y.; Chanda, S. Screening of Some Traditionally Used Indian Plants for Antibacterial Activity Against *Klebsiella pneumonia. J. Herbal Med. Toxicol.* **2009,** *3*, 161–614.

Vyas, N.; Gavatia, N. P.; Gupta, B.; Tailing, M. Antioxidant Potential of *Tamarindus indica* Seed Coat. *J. Pharm. Res.* **2009,** *2* (11), 1705–1706.

Zohrameena, S.; Mujahid, M.; Bagga, P.; Khalid, M.; Noorul, H.; Nesar, A.; Saba, P. Medicinal Uses & Pharmacological Activity of *Tamarindus indica. World J. Pharm. Sci.* **2017,** *5* (2), 121–133.

CHAPTER 20

Pharmacological Significance of *Uraria picta* (Jacq.) DC. in the Prevention and Treatment of Diseases

CHETNA FAUJDAR[1], ABHISHEK NEGI[1], and PRIYADARSHINI[1*]

[1]*Department of Biotechnology, Jaypee Institute of Information Technology, Noida, Uttar Pradesh 201309, India*

[*]*Corresponding author. E-mail: priyadarshini@jiit.ac.in/ priya.juit@gmail.com*

ABSTRACT

Uraria picta which is commonly known as Prishnaparni, is a perennial herb that belongs to the plant family Leguminosae. The herb has been extensively used in Ayurvedic preparations to treat various ailments. Almost every part of the plant has been explored for ethnomedical properties. Roots of the plant have already been used for formulation of well known drug Dashmularishta Bioactive phytoconstituents present in *Uraria picta* are of great significance owing to its therapeutic potential. *Uraria picta* has been reported to possess therapeutic abilities such as antimicrobial, fracture healing properties, antioxidant, antibacterial, antiinflammatory etc. The herb also possesses analgesic activity and is helpful in painful urination, vaginal pain and gonorrhea. Ability to reduce oxidative damage and lipid peroxidation makes it a wonderful agent to prevent or cure diseases that arise due to oxidative stress. Many research reports reveal the health promoting benefits of *Uraria picta*. These abilities of *Uraria picta* seem to be of considerable medical significance and can be exploited to design drug molecules against different diseases.

Bioactives and Pharmacology of Legumes. T. Pullaiah, PhD (Ed.)

20.1 INTRODUCTION

According to traditional medicinal system, most of the human diseases are the result of dys-functioning of mind and inner physiological processes. Ayurveda has a well-established accumulated knowledge of managing or preventing diseased conditions with the help of naturally occurring medicinal plants. *Uraria picta* (Jacq.) DC., commonly known as Prishnaparni is an erect, perennial herb belonging to the plant family Fabaceae. The herb has been extensively used in ayurvedic preparations to treat various ailments. It is one of the main ingredients of established ayurvedic formulation "Dashmula" (Yadav et al., 2009). Therapeutic values of medicinal plants lie in bioactive phytocompounds present in different parts of the plants. Secondary metabolites such as alkaloids, essential oils, flavonoids, tannins, terpenoids, saponins, phenolic compounds, cardiac glycosides, etc., are mainly responsible for ethnomedical properties of the *Uraria picta* (Saxena et al., 2014). *U. picta* has been reported to possess therapeutic abilities such as antimicrobial, fracture healing properties, antioxidant, antibacterial, antipyretic, anti-inflammatory, etc. (Chhipa et al., 2021). The presence of macro/trace elements in the herbal extracts offers an additional advantage in the form of serving as key components for vital biochemical reactions. In a study researcher from Tropical Forest Research Institute, India performed an in-depth phytochemical and elemental screening in different parts of *U. picta* plant. They screened for various secondary metabolites that are backbone of modern medicine. *U. picta* is known to possess considerable amount of macro elements such as potassium, calcium, phosphorus, magnesium, and sodium (Saxena et al., 2014). Inflammation and oxidative stress play a crucial role in the generation of several health related complications. Literature suggests the significance of *U. picta* acting as a wonderful anti-inflammatory and antioxidant agent (Mohan et al., 2019; Hem et al., 2017).

20.2 SOURCE OF IMPORTANT CLASSES OF PHYTOCHEMICALS

Literature reveals the presence of different classes of phytochemicals in the extracts from different parts of *U. picta*. These classes include alkaloids, flavonoids, cardiac glycosides, steroids, terpenoids, phenols, and saponins. Various pharmacological activities such as antioxidant, analgesic, lipid lowering, antimicrobial, antitumor, anti-inflammatory, antihypertensive, antipyretics, have been attributed to these phytochemicals (Saxena et al., 2014).

Cardiac glycosides belong to a group of naturally derived compounds that are capable of inhibiting Na^+/K^+-ATPase. Apart from their applications in treating cardiac problems these compounds have also been suggested as a potential drug component against many diseases including neurological dysfunctions. According to recent research, tumor cells show susceptibility for cardiac glycosides. Such studies provides the supporting data for the use of these compounds as probable antitumor agents (Prassas and Diamandis, 2008; Ayogu and Odoh, 2020). The presence of therapeutically important compounds in *U. picta* provides the plant an additional advantage to be used as a broad spectrum drug candidate (Figure 20.1).

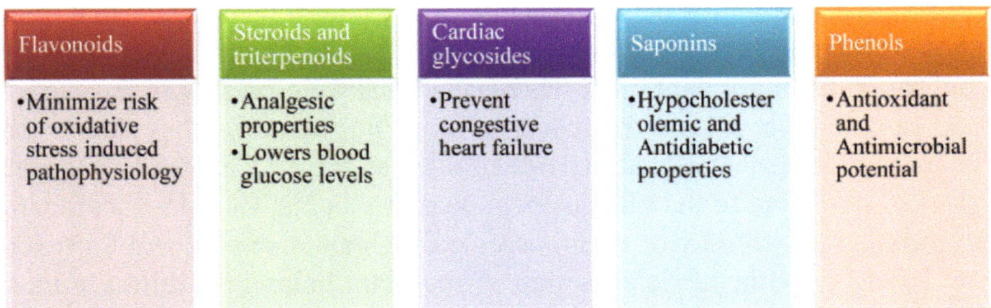

FIGURE 20.1 Different classes of phytochemicals present in *U. picta* and their bioactivity.

20.3 BIOACTIVE PHYTOCONSTITUENTS PRESENT IN *U. PICTA* AND THEIR PHARMACOLOGICAL SIGNIFICANCE

Phytomedicinal property of any part of plant depends mainly on bioactive phytocompounds that leads to specific chain of action in human system aiding in restoring vital biochemical functions. In a study conducted by Rahman et al. (2007). Two isoflavanones, 5,7-dihydroxy-2'-methoxy-3',4'-methylenedioxyisoflavanone (2) and 4',5-dihydroxy-2',3'-dimethoxy-7-(5-hydroxyoxychromen-7yl)-isoflavanone (4) along with six known compounds including isoflavanones, triterpenes and steroids were isolated from *U. picta*. They investigated the antimicrobial properties of these compounds and found them to be effective against *Staphylococcus aureus* (Rahman et al., 2007). Flavonoids have been linked with the number of pharmacologically significant activities of medicinal plants. Rhoifolinis, a flavonoid, which is also known as Apigenin-7-*o*-neohesperidoside, is a bioactive compound isolated and quantified from aerial parts of *U. picta* (Yadav et al., 2009). Study suggests that Rhoifolin is capable of regulating inflammatory markers thus reducing inflammation (Eldahshan and Azab, 2012). Other studies also confirm the antiproliferative (Eldahshan, 2013), antihypertensive, antidiabetic, and hemodynamic (Refaat et al., 2015) potential of Rhoifolin. The compound is also reported to exhibit vasorelaxing effect (Yadav et al., 2009).

In a study conducted by Mohan et al. (2007), they determined phyto-constituents from ethanolic and aqua-ethanolic root extract using GC-MS. They found N-capric acid isopropyl ester and 9-octadecenoic acid-, hexyl ester in ethanolic extract and propanoic acid, 2-hydroxy-, pentyl ester, α-d-mannofuranoside, 1-nonyl and 2-bromopropionic acid, pentadecyl ester (Mohan et al., 2020). In a study researcher from Tropical Forest Research Institute, India performed an in-depth phytochemical and elemental screening in different parts of *U. picta* plant. They screened for various secondary metabolites that are backbone of modern medicine. Their study showed the presence of alkaloids, flavonoids, steroids, terpenoids, phenols, and saponins in all plant parts whereas tannins were present in only leaves and cardiac glycosides were absent in roots. Furthermore, they determined quantitatively essential macroelements such as Na, K, Ca, Mg, P in various parts of plant and found the leaves to be richer in Na, Ca, Mg as compared to root and ste. Moreover stem possessed high amount of K and P (Saxena et al., 2014). Mineral rich fraction of the plant helps in fulfilling mineral requirements of human body thus used for herbal health drinks. Due to its enriched phytochemical profile, roots of *U. picta* are used for treating bleeding pile, fracture in bone, ulcer, asthma, inflammation in chest and diarrhea. Prasad et al. (1965) demonstrated its bone fracture healing property in rat model. In their study, they used 130 rats with fractured humerus and administered half of the rat population with alcohol extract of *U. picta* and rest with distilled water. Rats were injected with ^{45}Ca or ^{32}P 24 h before death and they found higher radioactivity of bone in the rats treated with Uraria extract compared to not treated one (Prasad et al., 1965).

Plant also possess antithrombotic property and used as antidote against bite of certain vipers and scorpions. Furthermore, its leaves are used for antiseptic properties, genitourinary infections, urinary disorders, and gonorrhea (Khare, 2007). In an in vivo study conducted by Hem et al. (2017), they used methanol extract of *U. picta* root to investigate its anti-inflammatory and hepatoprotective activity. In their study, they induced injury in rat liver thereby causing increased level of alanine transaminase, alkaline phosphatase, and aspartate aminotransferase enzymes in the blood. Rats were administered with *U. picta* extract and they observed significant reduction in all three enzymes level in blood that was comparable to standard drug silymarin 100 mg/kg (Hem et al., 2017). Apart from conventional medicinal use of *U. picta* for humans, it has high ethnomedical use and in some countries like Nigeria it is used to control ectoparasites in men and domestic animals. Igboechi et al. in their study demonstrated acaricidal activity of methanolic extract of the plant on *Ixodes ricinus* (Igboechi et al., 1989).

Acaricidal activity of the plant is due to the biochemical composition of plant that is rich in various flavonoid, phenolics, glycosides and tannins derivatives. Due to these phytochemicals the methanolic extract shows antioxidant and anti-inflammatory activity. Chhipa et al. (2021) in their study on mice found IC_{50} value of methanolic extract of *U. picta* and cisplatin to be 436.92 µg/mL and 8.75 µg/mL, respectively. Furthermore, they reported induction of apoptosis evident through DAPI staining and reduction in intracellular ROS through DCFH-da assay. Literature reveals that inhibition of cholinesterase enzymes might be an effective approach to manage Alzheimer's disease. Aqueous extract of *U. picta* has been found to inhibit acetylcholinesterase and butyrylcholinesterase enzymes in a concentration dependent manner. The study supports the use of *U. picta* as a preventive therapy for Alzheimer's disease (Odubanjo et al., 2013) (Table 20.1).

TABLE 20.1 Bioactive Phytoconstituents Present in *Uraria picta* and Their Pharmacological Activity.

Sr. No.	Bioactive component	Pharmacological activity
1	Isoflavanones	Antimicrobial activity (Rahman et al., 2007)
2	Rhoifolin	Antiproliferative effect, Anti-inflammatory, Antihypertensive, Antidiabetic and hemodynamic (Eldahshan, 2013; Saxena et al., 2014)
3	N-Capric acid isopropyl ester	Flavor and fragrance agents (Mohan, 2020)
4	2-Bromopropionic acid, Pentadecyl ester	Antibacterial activity (Kumar et al., 2011)
5	Propanoic acid, 2-hydroxy-, pentyl ester	Flavour and fragnance (Mohan et al., 2020)

20.4 CONCLUSION

Due to increasing demand of safe and natural alternatives to combat everyday health complications, medical potential of phytoconstituents present in medicinal plants is being investigated thoroughly. Since *U. picta* has already been used in ayurvedic formulations such as Dashmularishta, its efficacy as a remedy to various human ailments is approved in traditional medicinal system. The presence of number of bioactive phytoconstituents in different parts of the plant offers an opportunity to discover potential therapeutic agents in *U. picta*. Many preclinical investigations reveal the potential of *U.*

picta to be used as an antibacterial, antioxidant, anti-inflammatory and anti cancer agent.

KEYWORDS

- **Phytoconstituents**
- **Therapeutic**
- **Analgesic**
- **Lipid peroxidation**

REFERENCES

Ayogu, J. I.; Odoh, A. S. Prospects and Therapeutic Applications of Cardiac Glycosides in Cancer Remediation. *ACS Comb. Sci.* **2020,** *22*, 543–553.

Chhipa, A. S.; Baksi, R.; Nivsarkar, M. Anticancer Studies on Methanolic Extract of Aerial Parts of *Uraria picta* (Jacq.) DC. *Futur. J. Pharm. Sci.* **2021,** *7*, 1–12.

Eldahshan, O. A. Rhoifolin; A Potent Antiproliferative Effect on Cancer Cell Lines. *J. Pharm. Res. Int.* **2013,** *3* (1), 46–53.

Eldahshan, O. A.; Azab, S. S. Anti-Inflammatory Effect of Apigenin-7-Neohesperidoside (Rhoifolin) in Carrageenin-Induced Rat Oedema Model. *J. Appl. Pharm. Sci.* **2012,** *2*, 74.

Igboechi, A.; Osazuwa, E.; Igwe, U. Laboratory Evaluation of the Acaricidal Properties of Extracts from Uraria Picta (Leguminosae). *J. Ethnopharmacol.* **1989,** *26*, 293–298.

Khare, C. *Indian Medicinal Plants: An Illustrated Dictionary*; Springer-Verlag: Berlin; 2007, pp 699–700.

Kumar, V.; Bhatnagar, A.; Srivastava, J. Antibacterial Activity of Crude Extracts of *Spirulina platensis* and Its Structural Elucidation of Bioactive Compound. *J. Med. Plant Res.* **2011,** *5*, 7043–7048.

Mohan, B. Identification of Phytoconstituents In Ethanolic and Aqua-Ethanolic Extracts of *Solanum indicum* L. Through GC-MS. *Chem. Sci. Rev. Lett.* **2020,** *9*, 693–699.

Mohan, B.; Saxena, H.; Kakkar, A.; Mishra, M. Determination of Antioxidant Activity, Total Phenolic and Flavonoid Contents in Leaves, Stem and Roots of *Uraria picta* Desv. *Environ. Conserv. J.* **2019,** *20*, 1–8.

Mohan, B.; Saxena, H. O.; Parihar, S.; Kakkar, A. Gas Chromatography-Mass Spectrometry (GC-MS) Determination of Phytoconstituents from Ethanolic and Aqua-Ethanolic Root Extracts of *Uraria picta* Desv.(Fabaceae). *Pharm. Innov.* **2020,** *9*, 463–467.

Odubanjo, V. O.; Oboh, G.; Ibukun, E. O. Antioxidant and Anticholinesterase Activities of Aqueous Extract of *Uraria picta* (Jacq.) DC. *Afr. J. Pharm. Pharmacol.* **2013,** *7*, 2768–2773.

Prasad, G.; Sankaran, P.; Desh Pande, P. Studies on Fracture Healing by Using Radioactive P-32 and Ca-45 Under the Influence of *Uraria picta. Indian J. Med. Res.* **1965,** *53*, 645–650.

Prassas, I.; Diamandis, E. P. Novel Therapeutic Applications of Cardiac Glycosides. *Nat. Rev. Drug Discov.* **2008,** *7*, 926–935.

Rahman, M. M.; Gibbons, S.; Gray, A. I. Isoflavanones from *Uraria picta* and Their Antimicrobial Activity. *Phytochemistry* **2007,** *68*, 1692–1697.

Refaat, J.; Desoukey, S. Y.; Ramadan, M. A.; Kamel, M. S. Rhoifolin: A Review of Sources and Biological Activities. *Int. J. Pharmacogn.* **2015,** *2*, 102–109.

Saxena, H. O.; Soni, A.; Mohammad, N.; Choubey, S. K. Phytochemical Screening and Elemental Analysis in Different Plant Parts of *Uraria picta* Desv.: A Dashmul Species. *J. Chem. Pharm. Res.* **2014,** *6*, 756–760.

Hem, K.; Singh, N. K.; Singh, M. K. Anti-Inflammatory and Hepatoprotective Activities of the Roots of *Uraria picta. Int. J. Green Pharm.* **2017,** *11*, S166.

Yadav, A. K.; Yadav, D.; Shanker, K.; Verma, R. K.; Saxena, A. K.; Gupta, M. M. Flavone Glycoside Based Validated RP-LC Method for Quality Evaluation of Prishniparni (*Uraria picta*). *Chromatographia* **2009,** *69*, 653–658.

CHAPTER 21

Acacia ferruginea DC. (Safed Khair): Bioactive Compounds and Pharmacological Activity

NEETESH K. JAIN[1], YOGESH CHAND YADAV[2*], SUMEET DWIVEDI[3], and PANKAJ YADAV[3]

[1]*OCPR, Faculty of Pharmacy, Oriental University, Indore, Madhya Pradesh, India*

[2]*Faculty of Pharmacy, Uttar Pradesh University of Medical Sciences, Saifai, Etawah, Uttar Pradesh 206130, India*

[3]*UIP, Faculty of Pharmacy, Oriental University, Indore, Madhya Pradesh, India*

**Corresponding author. E-mail: drycy31@gmail.com*

ABSTRACT

Acacia ferruginea is a species of legume in the family Fabaceae. It consists of bioactive compound like alkaloids, flavonoids, saponins, anthraquinones, tannins, glycosides, triterpenoids and phenolic compounds. There are 12 components including catechin, procyanidin B1, quercetin, ellagic acid, rosmanol from GC/MS and LC/MS analysis of acetone extracts of *A. ferruginea* bark. *A. ferruginea* has good antioxidant potential. It has various pharmacological activities such as antibacterial, antifungal, anti-cancer, antihaemorrhoidal, analgesic, antiinflammatory etc.

Bioactives and Pharmacology of Legumes. T. Pullaiah, PhD (Ed.)

21.1 INTRODUCTION

Acacia ferruginea DC. [Sin. *Senegalia ferruginea* (DC.) Pendley and *Mimosa ferruginea* Roxb.], a member of the legume family, typically originate in India and Sri Lanka. *Acacia ferruginea* is a 20 m high tree and 10–12 mm thick, with 0.56 mm long leaves and light yellow hermaphrodite flowers.

21.2 BIOACTIVES

A. ferruginea contains carbohydrates, proteins, alkaloids, flavonoids, saponins, tannins, triterpenoids, phenols, gums, etc. (Deb et al., 2015). The aerial portions of plants did not give positive tests above mentioned chemical constituents except carbohydrate and proteins. Devi and Prasad (1991) reported that 50% dimethyl sulfoxide leaf extract contains phenol and tannins 14.8% and 13.9%, respectively, which are higher than 70% acetone bark extract contains 5.6%, 3.7% (Sakthivel and Guruvayoorappan, 2013; Devi and Prasad, 1991). Phenol content (mg/GAE/100 L) of *A. ferruginea* was found 445.4 ± 8.6 in methanol extract, 165.2 ± 6.0 in aqueous extract while 14.4 in pods (Thippeswamy et al., 2015; Kota, 2017). The seed extract of pressure-cooked *A. ferruginea* had a lower total phenol and tannin content and increased in iron chelating effect than raw seed extract (Loganayaki et al., 2011). Sowndhararajan et al. (2016) reported that *A. ferruginea* has found 12 components including catechin, procyanidin B1, quercetin, ellagic acid, and rosmanol and major components have been identified as derivatives of quinone (37.3%), quinoline (22.9%), imidazolidine (6.4%), pyrrolidine (4.5%), and cyclopentenone (3.5%) Hexadecenoic acid, pyridine, pyrazole, and pyrimidine derivatives have also been isolated in the methanol extract of the aerial fraction of *A. ferruginea*. LC/MS examination showed carboxamidine, imidazole, thiazole, catechin, and coumarin derivatives.

21.3 PHARMACOLOGY

21.3.1 Antioxidant Activity

In the human body, there are diseases caused due to excess generation of oxidants and these are trapped resulting in oxidative stress. Antioxidants that are developed from plants can ameliorate disorders produced by oxidative stress such as cardiovascular disease, cancer, aging, diabetes,

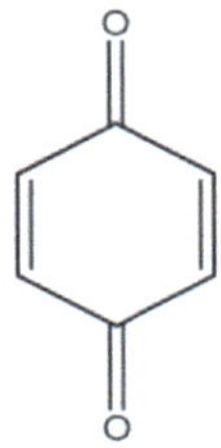

1,4-benzoquinone

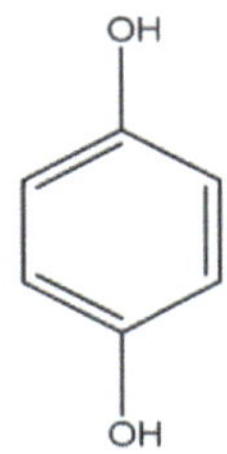

Hydroquinone

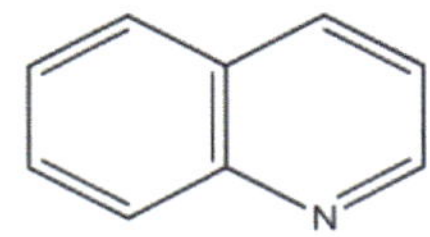

Quinoline

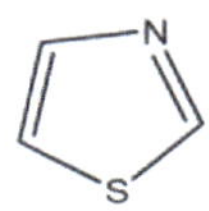

Thiazole

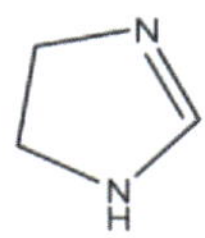

Imidazoline

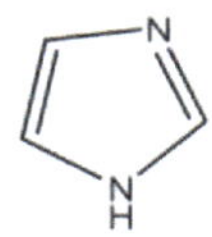

Imidazole

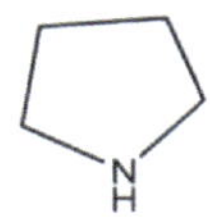

Pyrrolidine

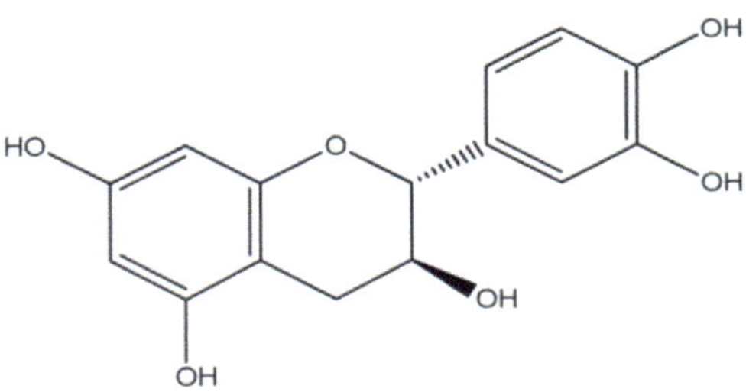

Catechin

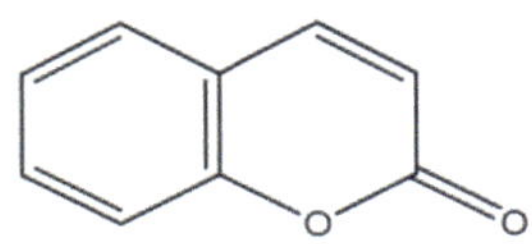

Coumarin

neurodegenerative diseases, and overweight. Ingestion of fruits, herbs, and vegetables diminish the risk of several disorders due to their high levels of antioxidant phytochemicals (Zhang et al., 2015). Sowndhararajan et al. (2015) proposed *A. ferruginea* bark contained in dietary supplements and functional foods due to its high total phenol content (11.70 g) RE/100 g and high content of antioxidants and nutrients. The flavonoids present in the acetone extract of *A. ferruginea* bark showed more significant effects compared with controls such as butylhydroxyanisole and α-tocopherol. The extract antioxidant doses decrease and reserve of peroxidation (DPPH •, ABTS • and OH •, FRAP, metal chelation, phosphomolybdenum). Inhibitory activity of polyherb extract containing *A. ferruginea* was verified at a concentration of 20.06 mcg/mL (Loganayaki et al., 2011). Thippeswamy et al. (2015) determined the antioxidant activity of *A. ferruginea* can be used for the treatment of various oxidative stress disorders. *A. ferruginea* seeds are rich antioxidant activity against positive standard such as ascorbic acid (30.12 mg), pyrogallol (3.86 mg), catechol (9.50 mg), etc. *Acacia* species acetone extract can excellently reduce

H_2O_2-induced oxidative stress leading to disorders (Sowndhararajan et al., 2015). Methanolic extract of *A. ferruginea* has ulcerative colitis protective effect due to the presence of catechin and coumarin derivatives in the extract (Kota, 2017; Sowndhararajan and Kang, 2013).

21.3.2 Antimicrobial and Antifungal Activity

Aqueous and methanol extracts of *A. ferruginea* leaves showed significant antibacterial activity against *Escherichia coli, Klebsiella pneumoniae, Proteus vulgaris, Pseudomonas aeruginosa,* and *Staphylococcus aureus* (Thippeswamy et al., 2013). Alcoholic extract of leaves inhibits many bacteria such as *S. aureus, E. coli, Aspergillus niger, Aspergillus flavus,* inhibition zones are 18.0 ± 0.20, 13.0 ± 0.44, 13.0 ± 0.72, 23.0 ± 0.52, respectively (Modi et al., 2016). Aqueous extract of *A. ferruginea* leaves showed moderate antifungal activity while the methanol extract exhibited good activity that inhibited 52.8 ± 0.29% (Kota et al., 2017).

21.3.3 Antitumor Activity

In vivo study of *A. ferruginea* extract showed significant antitumor activity (Sakthivel and Guruvayoorappan, 2018). It has significantly reduced many increased level parameters such as hydroxyproline, hexosamines, uronic acid, sialic acid, gamma glutamil transpeptidase levels, and suppressed iNOS and COX2 levels and also reduces significant nuclear factor kappa B (NFκB) and pro-inflammatory cytokine levels. Angiogenesis studies have observed a decrease in the number of tumors directed to capillaries that help fight melanoma with effective treatment (Sakthivel and Guruvayoorappan, 2018).

21.3.4 Antidiabetic Activity

The methanol extract of *A. ferruginea* stem bark possesses good antidiabetic activity. It showed significantly reduced glucose in blood as compared with metformin (250 mg/kg) (Sakthivel and Guruvayoorappan, 2018).

21.3.5 Anti-Inflammatory and Analgesic Effect

Inflammation is a host's neighborhood protection reaction to cellular impairment and long-time period infection can also additionally result in many issues and pathogenesis. Recognizing medicinal plant bioactive that could inhibit inflammatory mediators. A total of 400 mg/kg HCl extract of *A. ferruginea* leaves had significantly (74.68%) inhibited the inflammation as compared to indomethacin (82.8%). HCl extracts of aerial part of *A. ferruginea* significantly decreases the paw-edema-induced inflammation (Faujdar et al., 2016). Sakthivel and Guruvayoorappan (2014) reported that *A. ferruginea* extract showed protective effect against ulcerative colitis (UC). Methanol extract of *A. ferruginea* extract inhibited nitric oxide (NO) synthesis and also free radical scavenge in DPPH and NO radical producing assays (Sakthivel and Guruvayoorappan, 2016). The IC_{50} values of acetone extracts of *A. ferruginea* aerial components were found to be 51.87 mg/mL (Faujdar et al., 2019).

21.3.6 Anti-Hemorrhoidal Activity

Faujdar et al. (2019) reported that the HCl extract of *A. ferruginea* bark significantly decreased the inflammatory cytokines and showed as an anti-hemorrhoidal agent with the aid of using molecular docking and pharmacophore mapping. This capability interest can be accredited to the remarkable antioxidants potential due to flavonoids contents. It showed significantly inhibited inflammatory markers like interleukins, leukotrienes and prostaglandins.

21.3.7 Other Activities

Anticancer drug cyclophosphamide (CTX) induces toxic effect such as immunotoxicity and urotoxicity. *A. ferruginea* extract was able to inhibit CTX-induced toxicity in mice. 10 mg/kg ip daily methanol extract of *A. ferruginea* reduced CTX (25mg/kg ip)-induced toxicity while administering for 10 successive days and the extract additionally averted decreases in organs (liver, kidney, spleen, and thymus). This extract significantly inhibited urotoxicity brought about with the aid of using CTX because extract has good antioxidant potential. Furthermore, the usage of the extract considerably accelerated WBCs counts and protected the hematopoietic system (Sakthivel and Guruvayoorappan, 2015). Vahitha et al. (2002) mentioned that *A. ferruginea* extract confirmed larvicidal effect on *Culex quinquefaciatus* via LC_{50} values as 362.6 ppm.

KEYWORDS

- ***Acacia ferruginea***
- **bioactive**
- **pharmacological activity**
- **antibacterial**
- **antifungal**
- **anticancer**

REFERENCES

Deb, J.; Singh, A.; Rathore, D. S.; Dash, G. K.; Deb, N. K. Studies on Antidiabetic Activity of *Acacia ferruginea* DC. Stem Bark. *Indian J. Pharm. Biol. Res.* **2015,** *3*, 9267.

Devi, S. R.; Prasad, M. N. V. Tannins and Related Polyphenols from Ten Common *Acacia* Species of India. *Bioresour. Technol.* **1991,** *36*, 189–192.

Faujdar, S.; Sati, B.; Sharma, S.; Pathak, A. K.; Paliwal, S. K. Phytochemical Evaluation and anti-Hemorrhoidal Activity of Bark of *Acacia ferruginea* DC. *J. Tradit. Complement. Med.* **2019,** *9*, 85–89.

Faujdar, S.; Sharma, S.; Sati, B.; Pathak, A. K.; Paliwal, S. K. Comparative Analysis of Analgesic and Anti-Inflammatory Activity of Bark and Leaves of *Acacia ferruginea* DC. *Beni-Suef Univ. J. Basic Appl. Sci.* **2016,** *5*, 70–78.

Kota, C. S. Evaluation of Antioxidant Activity in Polyherbal Extract. *World J. Pharm. Res.* **2017,** *6*, 1422–1427.

Kota, C. S.; Kumar, V. H.; Sunitha, M. Activity of Polyherbal Extract Against Bacteria Causing Skin Infections in Diabetic Patients. *Asian J. Pharm. Clin. Res.* **2017,** *10*, 147–149.

Loganayaki, N.; Siddhuraju, P.; Manian, S. A. Comparative Study on In Vitro Antioxidant Activity of the Legumes *Acacia auriculiformis* and *Acacia ferruginea* with a Conventional Legume *Cajanus cajan*. *CyTA J. Food* **2011,** *9*, 8–16.

Modi, R. K.; Seetharam, M.; Pratima, M. Y. N. In Vitro Antimicrobial Screening of a Few Ethno-Medicinal Plants of Mimosoideae of Gulbarga-Karnataka India. *Int. J. Pharm. Life Sci.* **2016,** *7*, 4860–4863.

Sakthivel, K. M.; Guruvayoorappan, C. *Acacia ferruginea* Inhibits Tumor Progression by Regulating Inflammatory Mediators-(TNF-a, iNOS, COX-2, IL-1ß, IL-6, IFN-γ, IL-2, GM-CSF) and Pro-Angiogenic Growth Factor-VEGF. *Asian Pac. J. Cancer Prev.* **2013,** *14*, 3909–3919.

Sakthivel, K. M.; Guruvayoorappan, C. Protective Effect of *Acacia ferruginea* Against Ulcerative Colitis via Modulating Inflammatory Mediators, Cytokine Profile and NF-κB Signal Transduction Pathways. *J. Environ. Pathol. Toxicol. Oncol.* **2014,** *33*, 83–98.

Sakthivel, K. M.; Guruvayoorappan, C. *Acacia ferruginea* Inhibits Cyclophosphamide-Induced Immunosuppression and Urotoxicity by Modulating Cytokines in Mice. *J. Immunotoxicol.* **2015,** *12*, 154–163.

Sakthivel, K. M.; Guruvayoorappan, C. *Acacia ferruginea* Inhibits Inflammation by Regulating Inflammatory iNOS and COX-2. *J. Immunotoxicol.* **2016,** *13*, 127–135.

Sakthivel, K. M.; Guruvayoorappan, C. Targeted Inhibition of Tumor Survival, Metastasis and Angiogenesis by *Acacia ferruginea* Mediated Regulation of VEGF, Inflammatory Mediators, Cytokine Profile and Inhibition of Transcription Factor Activation. *Regul. Toxicol. Pharmacol.* **2018,** *95*, 400–411.

Sowndhararajan, K.; Hong, S.; Jhoo, J. W.; Kim, S.; Chin, N. L. Effect of Acetone Extract from Stem Bark of Acacia Species (*A. dealbata, A. ferruginea* and *A. leucophloea*) on Antioxidant Enzymes Status in Hydrogen Peroxide-Induced HepG2 Cells. *Saudi J. Biol. Sci.* **2015,** *22*, 685–691.

Sowndhararajan, K.; Kang, S. C. Protective Effect of Ethyl Acetate Fraction of *Acacia ferruginea* DC. Against Ethanol-Induced Gastric Ulcer in Rats. *J. Ethnopharmacol.* **2013,** *148*, 175–181.

Sowndhararajan, K.; Santhanam, R.; Hong, S.; Jhoo, J. W.; Kim, S. Suppressive Effects of Acetone Extract from the Stem Bark of Three *Acacia* Species on Nitric Oxide Production in Lipopolysaccharide-Stimulated RAW 264.7 Macrophage Cells. *Asian Pac. J. Trop. Biomed.* **2016,** *6*, 658–664.

Thippeswamy, S.; Abhishek, R. U.; Manjunath, K.; Mohana, D. C. Evaluation of Antibacterial and Antioxidant Properties of Some Traditional Medicinal Plants from India. *Int. J. Green Pharm.* **2015,** *9*, 50–57.

Thippeswamy, S.; Mohana, D. C.; Abhishek, R. U.; Manjunath, K. Effect of Plant Extracts on Inhibition of *Fusarium verticillioides* Growth and Its Toxin Fumonisin B 1 Production. *J. Agric. Technol.* **2013,** *9*, 889–900.

Vahitha, R.; Venkatachalam, M. R.; Murugan, K.; Jebanesan, A. Larvicidal Efficacy of *Pavonia zeylanica* L. and *Acacia ferruginea* DC. Against *Culex quinquefasciatus* Say. *Bioresour. Technol.* **2002,** *82*, 203–204.

Zhang, Y. J.; Gan, R. Y.; Li, S.; Zhou, Y.; Li, A. N.; et al. Antioxidant Phytochemicals for the Prevention and Treatment of Chronic Diseases. *Molecules* **2015,** *20*, 21138–21156.

CHAPTER 22

Acacia modesta Wall. (Phulai): Bioactive Compounds and Pharmacological Activity

YOGESH CHAND YADAV[1*], NEETESH K. JAIN[2], SUMEET DWIVEDI[2], and PANKAJ YADAV[1]

[1]*Faculty of Pharmacy, Uttar Pradesh University of Medical Sciences, Saifai, Etawah, Uttar Pradesh 206130, India*

[2]*OCPR, Faculty of Pharmacy, Oriental University, Indore, Madhya Pradesh, India*

[3]*UIP, Faculty of Pharmacy, Oriental University, Indore, Madhya Pradesh, India*

[*]*Corresponding author. E-mail: drycy31@gmail.com*

ABSTRACT

Acacia modesta Wall., belonging to the family Fabaceae, is commonly known as phulai and is distributed in Afghanistan, India and Pakistan. Flavonoids, alkaloids, terpenoids and tannins and many bioactive compounds such as octacosanol, nonaeicosanol, hentriacontanol, hentriacontane, octacosane 4-hydroxy benzoic acid, palmitone quercetin, and Kaempferol have been reported from this plant. Many pharmacological activities of *A. modesta* have been reported such as antiobesity, anti-diabetic, antifungal, analgesic, antiinflammatory, anti-platelet haemagglutination, and spasmolytic.

Bioactives and Pharmacology of Legumes. T. Pullaiah, PhD (Ed.)

22.1 INTRODUCTION

Acacia modesta Wall. (Syn. *Sengalia modesta* Wall., *Mimosa obovata* Roxb., *Mimosa dumosa* Roxb.) belongs to the family Fabaceae. It is commonly known as phulai and is distributed in Afghanistan, India, and Pakistan (Nadkarni, 1976). *A. modesta* is a deciduous tree. Pods are flat, straight, mucronate, three- to five-seeded

22.2 BIOACTIVES

Aerial parts of *A. modesta* showed the presence of flavonoids, terpenoids, alkaloids, and tannins (Khan, 2011; Bukhari et al., 2010). There are many bioactive compounds such as octacosanol, nonaeicosanol and hentriacontanol, hentriacontane, octacosane 4-hydroxy benzoic acid, and palmitone were also reported (Khan, 2011; Joshi et al., 1975), whereas quercetin and kaempferol were extracted from *A. modesta* by Khan (2004).

Terpenoids: Lupeol was extracted from *A. modesta* (Khan, 2011). Betulin and α-amyrin were present in benzene extract of the stem bark. All of these three compounds are pentacyclic triterpenes. A phytosterol, β-sitosterol (Mahmood et al., 2004), was also obtained from *A. modesta* (Joshi et al., 1975). When aerial parts of the plant were explored for essential oils, 38 components were obtained (Ahmad et al., 2012).

Nonprotein amino acids These are frequently present in seeds and leaves of Mimosoideae plants (Krauss and Reinbothe, 1973). Neurolathyrogen/α-amino-β-oxalylaminopropionic acid, was obtained from *A. modesta* (Quereshi et al., 1977). Triglycerides mainly contain linoleic and oleic acids. Both are unsaturated fatty acids. Seed of *A. modesta* contained 26% and 27% linoleic acid and oleic acid, respectively (Seigler, 2003).

Cyclitols Pinitol is an important cyclitol having antidiabetic effects (Bates et al., 2000). It was found in ethanolic extract of *A. modesta* heartwood (Joshi et al., 1975).

22.3 PHARMACOLOGY

22.3.1 Antibacterial Activity

The antibacterial effects of *A. modesta* extract were reported by Ahmad et al. (2009) and Rios et al. (1988). Methanol extract and its various

Octacosane

Octacosanol

1-Eicosanol

Hentriacontane

Hentriacontanol

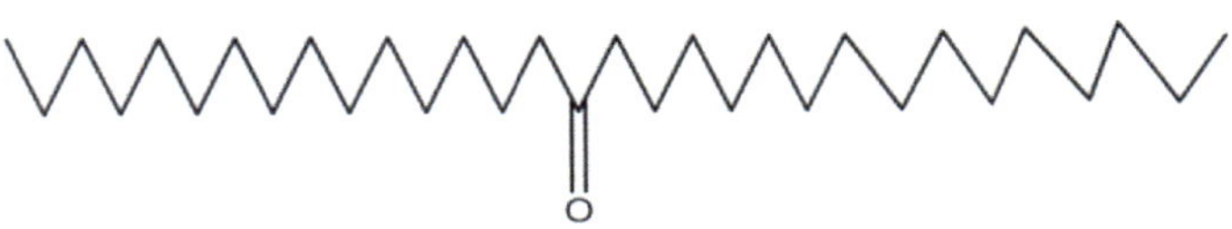

Palmitone

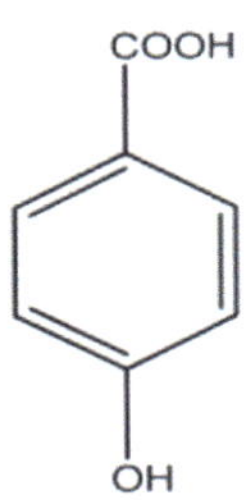

4-hydroxybenzoic acid

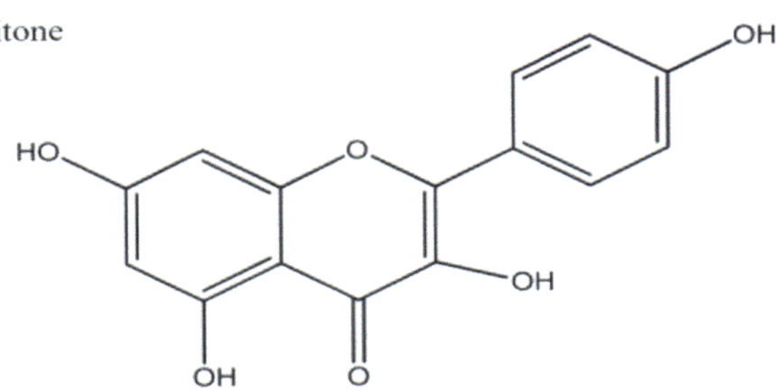

Kaempferol

fractions showed less activity against Gram positive (*Streptococcus pneumoniae*, *Staphylococcus epidermidis*, and *Staphylococcus aureus*) and Gram-negative *Enterobacter aerogenes*. While moderate activity was seen against Gram negative (*P. aeruginosa*, *S. typhi*, *E. coli*) and the

Gram positive *Bacillus pumilus*. The *n*-hexane and ethyl acetate fraction produced substantial activity against *Klebsiella pneumoniae*. Nevertheless, MIC values was between 2400 and 3900 μg/mL (Ahmad et al., 2011a,b). Bacterial growth inhibition was dependent on the concentration of leaves methanol extract (Napar et al., 2012).

Ethanol leaves extract exhibited a significant activity against *P. aeruginosa, E. coli, S. typhi, Proteus mirabilis, S. aureus, S. pneumoniae, B. cereus* and *B. subtilis*. Likewise, ethanol:H_2O (1:1) extract also showed good activity against the abovementioned organisms. When MICs were calculated, it was 4.570–26.500 μg/mL for ethanol extracts and 8.90–32.50 μg/mL for ethanol:water extracts (Jawla et al., 2012). Hot- and cold-water extracts showed substantial activity against *S. aureus, Enterococcus faecalis, B. subtilis,* and *P. aeruginosa* (Khalid et al., 2013). Ethanol extract of root showed an increase of activity from Gram positive, *S. aureus* and *β-Streptococcus* to Gram negative, *E. coli,* and *K. pneumoniae* (Rashid and Hashmi, 1999). No activity was produced by essential oils of *A. modesta* (Khan, 2011).

22.3.2 Antifungal Activity

Antifungal activity was reported in *A. modesta*. Methanol extract of leaves showed a growth inhibition of 11.53% and 0.8% against *A. niger* and *A. fumigatus*, respectively (Napar et al., 2012). Ethyl alcohol and ethyl alcohol:water (1:1) extracts of *C. albicans* and *Cryptococcus albidus* showed positive response. Ethyl alcohol extract exhibited MIC_{50} of 0.0055 mg/mL against *C. albicans* and against *A. flavus*, a very low effect was produced by methyl alcohol aerial parts extract and its fractions. Antifungal activity was concentration dependent. Root extract ethanol had more potential against *S. cerevisiae* than hot method extract. The same ethanol extracts also showed inhibitory effects against *Fusarium* sp. and *Rhizoctonia solani* (Khan, 2011). Essential oils had revealed variable inhibitory actions on fungal species. *A. modesta* oils also showed moderate effects (40.0%) against *Microsporum canis*. While a less activity against *A. flavus* and *F. saloni* and no activity against *C. albicans* and *C. glabrata* were observed (Ahmad et al., 2012).

22.3.3 Anti-Hyperglycemic Activity

When ethanol or ethyl alcohol:water (1:1) extract of the *A. modesta* leaves were given to rats, a significant lowering of blood glucose level was observed. The activity was similar to glibenclamide. LD_{50} of these extracts were greater than 5000 mg/kg. Ethanol extract, at dose of 100 mg/kg, was 12.34% extra potent than glibenclamide (0.20 mg/kg). While no sign of toxicity was detected in tested rats (Jawla et al., 2011).

22.3.4 Analgesic Activity

Different pain models were used to investigate the *A. modesta* methanol extract. Writhing, induced by acetic acid, was considerably diminished by injection (i.p) of the extract to the mice. The result was similar to standard diclofenac sodium. In formalin model, injection of the extract reversed both phases (neurogenic and inflammatory) of licking response in mice. These results were related to centrally acting morphine. During hot plate test, analgesia was produced by methanol extract which was again similar to morphine. Thiopental-induced hypnosis test was employed to note the sedative effects of the *A. modesta* extract. In this test, pretreatment of mice with extract significantly increased the sleep time. This outcome was comparable to standard diazepam (Bukhari et al., 2010).

22.3.5 Anti-Inflammatory Activity

Inflammation induced by carrageenan in rat paw, methanol extract of the *A. modesta* exhibited striking antiphlogistic response. This effect was analogous to diclofenac sodium (Rios et al., 1988). Literature reports indicated that components involved in reversal of carrageenan tempted inflammation was also active in cyclooxygenase inhibition (Selvam et al., 2004).

22.3.6 Antiplatelet Activity

Usually COX inhibitors and compounds having anti-inflammatory effects are also active against platelet aggregation. In case of *A. modesta* MeOH extract, a dose-dependent inhibition was seen against arachidonic-acid-induced platelet aggregation. IC_{50} was 0.80 mg/mL at a dose of 2.50 mg/mL (Saeed et al., 1995; Siess et al., 1983).

22.3.7 Antioxidant Activity

A. modesta MeOH extract and its ethyl acetate, *n*-hexane, aqueous, and chloroform fractions showed the modest NO radical hunting effect at 1.5 mg/mL concentration. Another model, DPPH assay, also exhibited that plant has antioxidant potential (Ahmad et al., 2011a, b; Hou et al., 1999; Taylor et al., 1997).

22.3.8 Brine Shrimp Cytotoxicity

The methanolic extract of *A. modesta*, weak toxic activity of 16.6% was seen at a concentration of 1 mg/mL with LD_{50} of 4251653.00. With the aqueous and chloroform fractions, lethal effects of 40% were observed at 1000 μg/mL conc. While ethyl acetate and *n*-hexane fractions showed mortality of 26.6% and 20.0%, respectively, at 1 mg/mL concentration. High toxicity was shown by essential oils of the plant at a concentration of 0.10 mg/mL. When test was performed on 30 shrimps, the number of survivors was 0. This shows that oils can be utilized as cytotoxic product.

22.3.9 Hemagglutination Activity

Lectins have been used to demonstrate the functional and structural roles of sugar components present on cell surfaces (Sharon and Lis, 1972). They have also been employed for characterization and isolation of glycoconjugates (Kuroki et al., 1991). Hemagglutination activity of *A. modesta* plant extract was determined by using human's RBCs of different blood groups. Weak or no activity was observed in all blood groups except O (+)ive where ethyl acetate fraction exhibited moderate activity at 1:2 dilution.

22.3.10 Spasmolytic Activity

When experiment was performed on isolated tissue of rabbit's jejunum by using methanolic extract of *A. modesta*, a dose-dependent relaxing effect was observed. With the increase in dose, a decrease was seen in persistent KCl-induced contractions and spontaneous movements. In case of KCl-produced contractions, activity was detected at a dose of 0.1 mg/mL with

EC_{50} of 6.8 ± 0.53. This proposed the muscle relaxant effect of the extract (Khan, 2011).

KEYWORDS

- ***Acacia modesta***
- **antimicrobial**
- **antiobesity**
- **anti-diabetic**
- **antifungal**
- **analgesic**
- **antiinflammatory and anti-platelet**

REFERENCES

Ahmad, B.; Bashir, S.; Azam, S.; Ali, N. Screening of *Acacia modesta* for Antifungal, Anti-Termite, Nitric Oxide Free Radical Scavenging Assay and Brine Shrimp Cytotoxic Activities. *J. Med. Plants Res.* **2011a,** *5*(15), 3380–3386.

Ahmad, B.; Khan, I.; Azam, S.; Bashir, S.; Ahmad, J.; Hussain, F. Screening of *Acacia modesta* for Haemagglutination, Antibacterial, Phytotoxic and Insecticidal Activities. *J. Med. Plants Res.* **2011b,** *5*(14), 3090–3096.

Ahmad, B.; Khan, I.; Bashir, S.; Azam, S. Chemical Composition and Antifungal, Phytotoxic, Brine Shrimp Cytotoxicity, Insecticidal and Antibacterial Activities of the Essential Oils of *Acacia modesta*. *J. Med. Plants Res.* **2012,** *6*, 4653–4659.

Bates, S. H.; Jones, R. B.; Bailey, C. J. Insulin-Like Effect of Pinitol. *Br. J. Pharmacol.* **2000,** *130* (8), 1944–1948.

Bukhari, I. A.; Khan, R. A.; Gilani, A. H.; Ahmed, S.; Saeed, S. A. Analgesic, Anti-Inflammatory and Anti-Platelet Activities of the Methanolic Extract of *Acacia modesta* Leaves. *Inflammopharmacol* **2010,** *18* (4), 187–196.

Hou, Y. C.; Janczuk, A.; Wang, P. G. Current Trends in the Development of Nitric Oxide Donors. *Curr. Pharm. Des*. **1999,** *5* (6), 417–442.

Jawla, S.; Kumar, Y.; Khan, M. S. Y. Antimicrobial and Antihyperglycemic Activities of *Acacia modesta* Leaves. *Pharmacologyonline* **2011,** *2*, 331–347.

Joshi, K. C.; Tholia, M. K.; Sharma, T. Chemical Examination of *Acacia modesta. Planta Med.* **1975,** *27*, 281–283.

Khalid, A.; Rehman, U.; Sethi, A.; Khilji, S.; Fatima, U.; Khan, M. I.; et al. Antimicrobial Activity Analysis of Extracts of *Acacia modesta, Artemisia absinthium, Nigella sativa* and *Saussurea lappa* Against Gram Positive and Gram Negative Microorganisms. *Afr. J. Biotech.* **2013,** *10* (22), 4574–4580.

Khan, A. Phytochemical and Pharmacological Studies on *Acacia modesta*. M.Phil. Thesis, University of Peshawar, Peshawar, Pakistan, 2004.

Khan, I. Phytochemical Evalution, Bioassay Screening and Standardization of *Zizyphus jujuba* and *Acacia modesta*. PhD. Thesis, University of Peshawar, Khyber Pakhtunkhwa, Pakistan, 2011.

Krauss, G. J.; Reinbothe, H. Die freien aminosäuren in samen von Mimosaceae. *Phytochemistry* **1973,** *12* (1), 125–142.

Kuroki, T.; Kubota, A.; Miki, Y.; Yamamura, T.; Utsunomiya, J. Lectin Staining of Neoplastic and Normal Background Colorectal Mucosa in Nonpolyposis and Polyposis Patients. *Dis. Colon Rectum* **1991,** *34* (8), 679–684.

Mahmood, T.; Khan, M. A.; Ahmad, J., Ahmad, M. Ethnomedicinal Studies of Kala Chitta Hills of District Attock, Pakistan. *Asian J. Plant Sci.* **2004,** *3* (3), 335–339.

Nadkarni, K. M. *Indian Materia Medica with Ayurvedic, Unani-Tibbi, Siddha, Allopathic, Homeopathic, Naturopathic and Home Remedies, Appendices and Indexes;* Popular Prakashan: Bombay, 1976; pp 278–279.

Napar, A. A.; Bux, H.; Zia, M. A.; Ahmad, M. Z.; Iqbal, A.; Roomi, S.; Muhammad, I.; Shah, S. H. Antimicrobial and Antioxidant Activities of Mimosaceae Plants; *Acacia modesta* Wall (Phulai), *Prosopis cineraria* (Linn.) and *Prosopis juliflora* (Swartz). *J. Med. Plants Res.* **2012,** *6* (15), 2962–2970.

Quereshi, M. Y.; Pilbeam, D. J.; Evans, C. S.; Bell, E. A. The Neurolathryogen, α-Amino-β-Oxalylaminopropionic Acid in Legume Seeds. *Phytochemistry* **1977,** *16* (4), 477–479.

Rashid, A.; Hashmi, H. In Vitro Susceptibility of Some Gram Positive and Gram Negative Strains of Bacteria and Fungi to Root Extracts of *Acacia modesta. Pak. J. Pharmacol. Sci.* **1999,** *2* (3), 746–749.

Rios, J. L.; Recio, M. C.; Villar, A. Screening Methods for Natural Products with Antimicrobial Activity: A Review of the Literature. *J. Ethnopharmacol.* **1988,** *23* (2), 127–149.

Saeed, S. A.; Simjee, R. U.; Shamim, G.; Gilani, A. H. Eugenol: A Dual Inhibitor of Platelet-Activating Factor and Arachidonic Acid Metabolism. *Phytomedicine* **1995,** *2* (1), 23–28.

Seigler, D. S. Phytochemistry of *Acacia*-Sensu Lato. *Biochem. Syst. Ecol.* **2003,** *31* (8), 845–873.

Sharon, N.; Lis, H. Lectins: Cell-Agglutinating and Sugar-Specific Proteins. *Science* **1972,** *177* (4053), 949–959.

Siess, W.; Cuatracasas, P.; Lepentina, E. G. A Role for Cyclooxygenase Products in the Formation of Phosphatidic Acid I Stimulated Platelets. *J. Pharmacol. Chem.* **1983,** *258* (8), 4683–4686.

Taylor, B. S.; Kim, Y. M.; Wang, Q. I.; Shapiro, R. A.; Billiar, T. R.; Geller, D. A. Nitric Oxide Down-Regulates Hepatocyte–Inducible Nitric Oxide Synthase Gene Expression. *Arch. Surg.* **1997,** *132* (11), 1177–1183.

CHAPTER 23

Bioactive Compounds and Pharmacological Activity of *Acacia macrostachya* Rchb. ex DC. and *Acacia leucophloea* (Roxb.)Willd.

NEETESH K. JAIN[1], SUMEET DWIVEDI[2], and
YOGESH CHAND YADAV[3*]

[1]*OCPR, Faculty of Pharmacy, Oriental University, Indore, Madhya Pradesh, India*

[2]*UIP, Faculty of Pharmacy, Oriental University, Indore, Madhya Pradesh, India*

[3]*Faculty of Pharmacy, Uttar Pradesh University of Medical Sciences, Saifai, Etawah, Uttar Pradesh 206130, India*

[*]*Corresponding author. E-mail: drycy31@gmaail.com*

ABSTRACT

Acacia macrostachya is a species of legume in the family Fabaceae. 4 Oleanane-type saponins, macrostachyaosides A, B, C, and D (1–4) were isolated from the roots of *A. macrostachya*. It has good antioxidant potential. The pharmacological activities of *A. macrostachya* reported include anti-obesity, anti-diabetic, lipase inhibition and glucosidase, antiplasmodial, and antiradical. The bioactive compound of *Acacia leucophloea* are anthraquinones, tannins, phenolic compounds catechin, quercetin. It has also good antioxidant potential and its pharmacological activities include gastrointestinal, respiratory and anti-inflammatory.

Bioactives and Pharmacology of Legumes. T. Pullaiah, PhD (Ed.)

23.1 INTRODUCTION

Acacia leucophloea (Roxb.) Willd. [Syn. *Mimosa leucopholea* Roxb., *Vachellia leucophloea* (Roxb.) Maslin et al.] belongs to the family Fabaceae that grows in India, Nepal, Pakistan, Sri Lanka, Myanmar, Thailand, Vietnam, and Indonesia. Trees of *A. leucophloea* (Roxb.) are 20 m high and bark yellow to yellowish-brown, leaves are 1 cm long, and flowers 1 mm, yellow or yellowish-white in color.

Acacia macrostachya Rchb. ex DC. is a species of legume family Fabaceae. A prickly shrub of 4–5 m high or a small tree which is found in West Africa from Senegal to Nigeria, and in Sudan and Angola. It is recognizable by its rose-like thorns (plants.jstor.org).

23.2 BIOACTIVES

The two flavonoids, rutin and quercetin in the extracts of leaves and bark of *A. leucophloea* exhibit notable pharmacological activities. Through the HPTLC method important constituent present in plant, that is, rutin and quercetin is estimated. Lipids, protein, and carbohydrates are found in the leaves and seeds. Tannin and oxalic acid are found in the seeds. The oil content of the kernel is approximately 17–20%. Mimosine, a poisonous nonprotein chemical found in the leaves and seeds, is also present (Deshmukh and Bhajipale, 2018). From the plant, two pimarane derivatives have been identified (Bansal et al., 1980; Perales et al., 1980; Rojas et al., 2001). The seeds' nutritional properties have also been reported, that is, seed oil fatty acids (Banerji et al., 1988; Devra et al., 2000), proteins and amino acids in seeds (Vijayakumari et al., 1994; Siddhuraju et al., 1997). The roots contain anthraquinones (Saxena and Srivastava, 1986).

Four novel oleanane-type saponins, macrostachyaosides A, B, C, and D (1–4), were isolated from the roots of *A. macrostachya*. Extensive 1D- and 2D-NMR data, as well as HR-ESI-MS analysis, were used to deduce their structures. None of the chemicals examined caused a substantial growth decrease in HL60 cells at doses of 100 M (Tchoukoua et al., 2017). Four new oleanane-type saponins were also isolated from *A. macrostachya* (Tchoukoua et al., 2017).

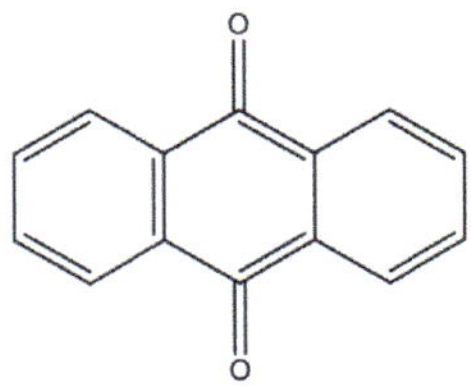

Anthraquinone

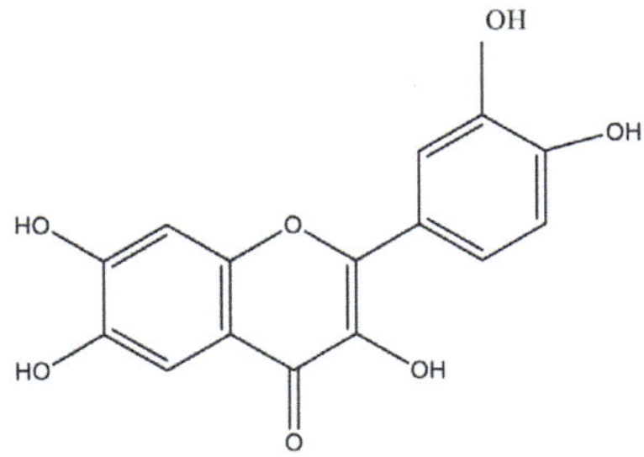

Quercetin

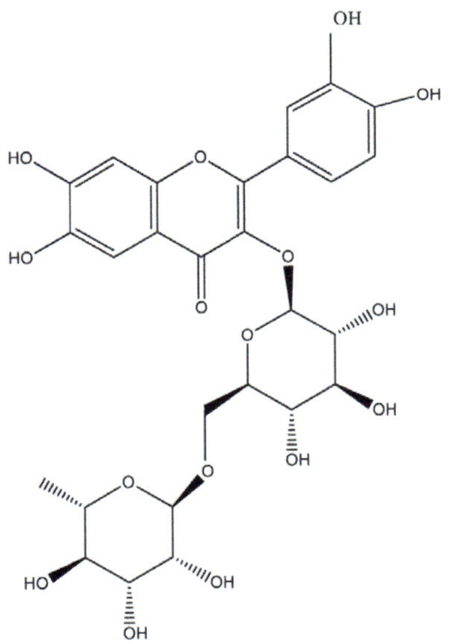

Rutin

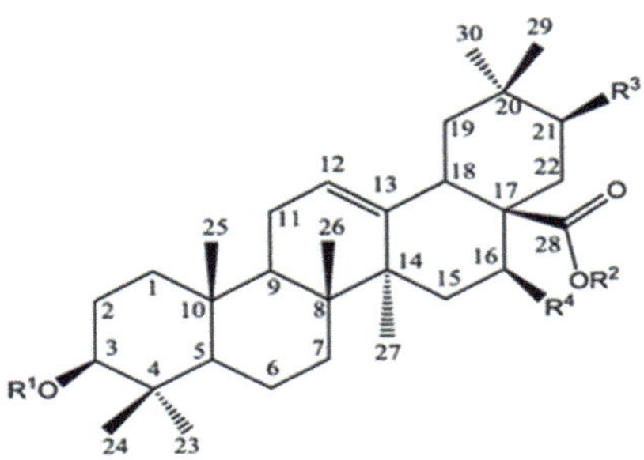

1 R^1 = Xyl(1→2)Ara(1→4)Xyl(1→2)Gal, R^2=Rha, R^3=OH, R^4=H
2 R^1=Xyl(1→4)Ara(1→4)[*Ara* (1→4)Xyl(1→2)]Gal, R^2=R^3=H, R^4=OH
3 R^1=Glc(1→2)Ara(1→6)[*Ara*(1→2)]Gal, R^2= R^3= R^4=H
4 R^1=Glc(1→3)[Xyl(1→4)[*Xyl*(1→2)]Ara(1→2)]*Ara*, R^2= R^3= R^4=H

23.2.1 Pharmacology of A. leucophloea

23.2.1.1 Gastrointestinal and Respiratory Activities

In isolated rabbit jejunum preparations, a methanol crude extract of the plant's barks elicited a concentration-dependent relaxation (0.1–3 mg/mL) in a manner similar to that of nifedipine and dicyclomine, implying a Ca^{2+} channel-blocking mechanism in addition to an anticholinergic activity. The extract generated a parallel shift in the Ach curves in the guinea-pig ileum without suppressing maximum contractile response, followed by a nonparallel shift with suppression of maximum contractile response at higher concentrations, comparable to dicyclomine. Furthermore, at a dose in between 0.1578 and 0.734 mg/mL and 0.46–0.94 mg/mL, it produced the relaxation of carbachol (1 M) and high K^+-induced contractions in rabbit trachea. These findings suggest that the extract has spasmolytic and bronchodilator properties, which may be mediated by Ca^{2+} channel blocking, suggesting its usage as a treatment for diarrhoea and asthma. The methanol extract of *A. leucophloea* had a dose-dependent (100–500 mg/mL) preventive effect against castor-oil-induced diarrhoea (Imran et al., 2011).

23.2.1.2 Antioxidant and Free Radical Scavenging Activities

In an acidic media, the FRAP assay assesses antioxidants' ability to decrease the TPTZ–Fe(III) complex to the highly blue colored TPTZ–Fe(II) complex. The antioxidant capability of acetone and methanol extracts of *Acacia* species' barks is expressed as the concentration of compounds with ferric-TPTZ reducing capacity equal to 1 mmol Fe concentration (II). According to the data on the ferric reduction capacity of *Acacia* species extracts, *A. leucophloea* acetone extract had the highest FRAP activity (2953 mmol Fe(II)/g extract). Small phenolics, such as flavonoids and phenolic acids, have been studied extensively for their antiradical and antioxidant properties. •+ a protonated radical, ABTS having a distinctive absorbance peak at 734 nm, which diminishes when proton radicals are scavenged. Incubation of ABTS with potassium persulfate yielded ABTS•+. ABTS radical scavenging activity was high in all of the samples, but the acetone extract of *A. leucophloea* bark had the highest activity (63,359 mol/g) (Kandhasamy et al., 2013).

23.2.1.3 Anti-Inflammatory Activity

The anti-inflammatory efficacy of stem bark methanol extract was tested against carrageenin-induced paw oedema and cotton-pellet-induced granuloma at doses of 100 and 200 mg/kg b.w. The methanol extract had strong anti-inflammatory action, which was comparable to that of phenylbutazone, a common anti-inflammatory medicine. At doses of 200 mg/kg, the stem bark extract inhibited carrageenin-induced paw oedema by 52.54%, which was comparable to the standard treatment (62.71%). At doses of 200 mg/kg, the stem bark extract inhibited cotton-pellet-induced granuloma by 55.42%, which was comparable to standard medication treatment (57.53%). In extract and phenyl, there was a dose-dependent reduction in granular tissue development (www.pharmatutor.org).

23.2.2 Pharmacology of *A. macrostachya*

23.2.2.1 Antioxidant Activity

The DPPH test was used to determine antioxidant activity. The antioxidant properties of *A. macrostachya* extracts were dosage dependent. The antioxidant effects of extracts (Cl_2CH_2, Cl_2CH_2-MeOH, MeOH, MeOH-W, and W) were evaluated and compared to quercetin, which had an IC_{50} of 2.63 ± 1.26 g/mL. The extract is more active if the 50% inhibitory concentration (IC_{50}) is lower (Issa et al., 2016).

23.2.2.2 Antiplasmodial Activity

Plasmodium falciparum strains exposed to *A. macrostachya* extracts for 72 h demonstrated negative effects on their overall growth. There was a dose–response relationship between the effects. The lower the IC_{50}, the less active the extract is. Overall, leaf extracts had the best antiplasmodial potential (Issa et al., 2016).

23.2.2.3 Antidiabetic and Antiradical Activity

The antidiabetic effectiveness of three *A. macrostachya* extracts was examined using an in vitro model to track their inhibitory effect. These extracts' antiradical activity was also assessed. The enzyme activity of α-glucosidase was significantly inhibited by methanol extracts of root and stem barks, with IC_{50} values of 2.487 +/− 0.441 g/mL and 1.650 +/− 0.229 g/mL, respectively. The same extracts had the strongest antiradical activity, scavenging the radical DPPH with IC_{50} values of 9.307 +/− 0.262 g/mL and 5.242 +/− 0.068 g/mL, respectively. The cationic radical ABTS had an IC_{50} of 45.049 +/− 0.730 g/mL in methanolic root bark extract and 14.136 +/− 0.161 g/mL in methanolic stem bark extract (Ganame et al., 2020).

KEYWORDS

- ***Acacia macrostachya***
- ***Acacia leucophloea***
- **bioactive**
- **gastrointestinal, respiratory and antiinflammatory activities**

REFERENCES

Banerji, R.; Chowdhury, A. R.; Misra, G.; Nigam, S. K. Chemical Composition of *Acacia* Seeds. *J. Am. Oil Chem. Soc.* **1988,** *65*, 1959–1960.

Bansal, R. K.; Garcia-Alvarez, M. C.; Joshi, K. C.; Rodriguez, B.; Patni, R. Diterpenoids from *Acacia leucophloea. Phytochemistry* **1980,** *19*, 1979–1983.

Deshmukh, S. P.; Bhajipale, N. S. A Review on *Acacia* Species of Therapeutics Importance. *Int. J. Pharm. Biol. Sci. Arch.* **2018,** *6* (4), 24–34.

Devra, A.; Mathur, A.; Sindal, R. S.; Sherwani, M. R. K. Chemical Examination of Wild Plant Seed Oils from Arid Land of Rajasthan. *Orient. J. Chem.* **2000,** *21*, 295–298.

Ganame, H. T.; Karanga, Y.; Ilboudo, O.; Nikiema, W. K. H. C.; Sawadogo, RW.; Tapsoba, I. α-Glucosidase Inhibitory and Antiradical Properties of *Acacia macrostachya. Eur. J. Med. Health Sci.* **2020,** *2* (5), 30–35.

Imran, I.; Hussain, L.; Zia-Ul-Haq, M.; Janbaz, K. H.; Gilani, A. H.; Feo, V. D. Gastrointestial and Respiratory Activities of *Acacia leucophloea. J. Ethnopharmacol.* **2011,** *138* (3), 676–682.

Issa, T.; Souleymane, F.; Charlemagne, G.; Eloi, P.; Roger, H. C. N.; et al. Antiplasmodial and DPPH Radical Scavenging Effects in Extracts from *Acacia macrostachya* (Mimosaceae) DC. *World J. Pharm. Res.* **2016,** *5* (8), 219–233.

Kandhasamy, S.; Joseph, J. M.; Sellamuthu, M. Antioxidant and Free Radical Scavenging Activities of Indian *Acacias*: *Acacia leucophloea* (Roxb.) Willd., *Acacia ferruginea* DC., *Acacia dealbata* Link. and *Acacia pennata* (L.) Willd. *Int. J. Food Prop.* **2013,** *16* (8), 1717–1729.

Perales, A.; Martinez-Ripoll, M.; Fayos, J.; Bansal, R. K.; Joshi, K. C.; Patni, R.; Rodriguez, B. Leucoxol, a New Diterpenoid from *Acacia leucophloea*. X-Ray Structure Determination. *Tetrahedron Lett.* **1980,** *21*, 2843–2844.

Rojas, M. C. A.; Cano, F. H.; Rodriguez, B. Diterpenoids from *Acacia leucophloea*: Revision of the Structures of Leucophleol and Leucophleoxol. *J. Nat. Prod.* **2001,** *64*, 899–902.

Saxena, M.; Srivastava, S. K. Anthraquinones from the Roots of *Acacia leucophloea. J. Nat. Prod.* **1986,** *49*, 205–209.

Siddhuraju, P.; Vijayakumari, K.; Janardhanan, K. Tryptophan Contents of Total (True) Proteins and Protein Fractions, Albumins and Globulins of Tribal/Little Known/Under-Exploited Indian Legumes. *J. Food Sci. Tech.* **1997,** *34*, 140–142.

Tchoukoua, A.; Tabopda, T. K.; Usukhbayar, N.; Kimura, K.; Kwon, E.; Momma, H.; Koseki, T.; Shiono, Y.; Ngadjui, B. T. New Triterpene Saponins from the Roots of *Acacia macrostachya* (Mimosaceae). *Biosci. Biotech. Biochem.* **2017,** *81* (12), 2261–2267.

Vijayakumari, K.; Siddhuraju, P.; Janardhanan, K. Nutritional Assessment and Chemical Composition of the Lesser Known Tree Legume, *Acacia leucophloea* (Roxb.) Willd. *Food Chem.* **1994,** *50*, 285–288.

CHAPTER 24

Bioactives and Pharmacology of *Flemingia strobilifera* (L.) W.T. Aiton

SANDIP KISAN GAVADE[1*] and MANOJ MADHWANAND LEKHAK[2]

[1]*Department of Botany, Dattajirao Kadam Arts, Science and Commerce College, Ichalkaranji, Maharashtra, India*

[2]*Angiosperm Taxonomy Laboratory, Department of Botany, Shivaji University, Kolhapur, Maharashtra, India*

[*]*Corresponding author. E-mail: skgavadenogra@gmail.com*

ABSTRACT

Flemingia strobilifera is one of the important medicinal plants and shows presence of carbohydrates, flavonoid, glycosides, saponins, and steroids. This species is commonly found in Asia, Australia and America. It is used in various diseases like diarrhea, dysentery, epilepsy, insomnia, tuberculosis, hysteria, rheumatism and to reduce body pain. In this review we have provided an overview of pharmacological activities of *F. strobilifera* and its bioactives. We have also provided some important structures of bioactives of *F. strobilifera*.

24.1 INTRODUCTION

Flemingia strobilifera (L.) W.T. Aiton (family Fabaceae), an undershrub, is an important medicinal plant. *Hedysarum strobiliferum* L., *Zornia strobilifera* Pers., *Maughania strobilifera* (L.) J.St.-Hil. ex Kuntze are the synonyms of

Bioactives and Pharmacology of Legumes. T. Pullaiah, PhD (Ed.)

F. strobilifera. It is called as Luck Plant or Wild Hops (English); Makhiyati (Assamese); Poptyo (Gujarati); Kanphuta (Hindi); Kumalu, Nari Baalada Honne (Kannada); Kanalam, Kumaanchedi, Kumalu, Ponthorila, Theri (Malayalam); Bandar, Kanphuti, Klipti (Marathi); Bhatmase, Bhatwasi (Nepali); Nalla Baddu (Telugu).

The plant is erect shrub, branched, up to 1.2–3.2 m tall. Leaves unifoliolate, broad ovate, rounded or cordate at base, apex acuminate, glabrous on both surfaces, gland-dottted. Inflorescence an axillary and terminal raceme. Flowers 8–10 mm long; enclosed in membranous bracts; bracts reniform, papery. Fruits a pod, beaked, turgid, sparsely gland-dotted; Seeds 2, rounded, mottled, shiny. This species is distributed in Asia, Australia (Queensland) and Central America (Gavade et al., 2020).

The Santhal tribes use the roots of *F. strobilifera* in insomnia, hysteria, epilepsy, and to relieve pain and the leaves are used as vermifuge for children in Java (Anonymous, 1993). The roots of *F. strobilifera* are used to treat epilepsy in Myanmar. A small piece of the root of *F. strobilifera* is used by Assamese to induce sleep and it is also used even in great pain for heavy sleep (Kirtikar and Basu, 1993). The tribal people from Jorhat, Assam, use plant parts of *F. strobilifera* to cure diarrhea, dysentery, epilepsy, insomnia, tuberculosis, hysteria, rheumatism, and to relieve body pain etc. (Das, 2018).

24.2 BIOACTIVES

Ethanolic leaf extract of *F. strobilifera* shows the presence of two crystalline glycosides naringin (naringenin-5-rhamno-glucoside) and phloridzin (2-phloretin-β-glucoside) (Saxena et al., 1976). Madan et al. (2008) reported flemingiaflavanone (8,3′-diprenyl-5,7,4′-trihydroxy flavanone), Genistin (5,4′-dihydroxyisoflavone 7-*O*-glucoside) and β-sitosterol-D glucoside from the root extracts of *F. strobilifera*. A new isoflavone 5,7,4′-trihydroxy 8,2′,5′-tri(3-methylbut-2-enyl) was isolated and its structure elucidated by Madan et al. (2009). Gahlot et al., (2011) reported quercitin, rutin, quercimetrin, leptosidin, leptosin, phloridzin and naringin, chalcones, *n*-triacontane, sitosterol, 3,6-dihydroxy 2,4,5,4-tetramethoxychalcone. Ethanolic extract of its root contains flavonoids and carbohydrates whereas the chloroform extract has flavonoids, steroids, proteins and fats; petroleum ether extract of root contains fixed oils and fats and aqueous extract of root contains tannins and carbohydrates (Kumar et al., 2011b). The roots of *F.*

strobilifera showed the presence of terpenoids, tannin, saponins, quinine, phenol, flavanoids, coumarins, and carbohydrate (Bhatt, 1975; Mahajon et al., 2014). GC-MS analysis showed the presence of eight compounds, namely, eicosane, heptacosane, hexacosane, hexanedioic acid, bis(2-ethylhexyl) ester, thiocyanic acid ethyl ester, 3,5-bis(1,1-dimethylethyl) phenol, nonadecane, and limonene (Pizon et al., 2016). Yang et al. (2016) isolated and elucidated the structure of 12 compounds, namely, betulinic acid, daucosterol, daidzein, emodin, flemichpparin C, genistein, genistin, naringin, *p*-methoxyphenylpropionic acid, quercetin, salicylic acid, and stigmasterol. Tikadar et al. (2017) reported alkaloid, flavonoid, glycoside, phenol, saponin, tannin, and terpenoid from *F. strobilifera*. Preliminary phytochemical screening of leaf and roots of *F. strobilifera* showed the presence of alkaloids, amino acids, flavonoids, phenols, saponins, tannin, and terpenoids (Das, 2018). The stems and leaves showed the presence of 17β-estradiol, 2-hydroxy genistein, 2′,3′,4′,6′-tetramethoxychalcone, cajanin, genistein, and pisatin (Jeong et al., 2018). Physicochemical screening of methanol and water-soluble extract of *F. strobilifera* showed the presence of carbohydrates, flavonoid, glycosides, saponins, and steroids (Shreedevi et al., 2018). GC-MS analysis of the hexane extract roots of *F. strobilifera* showed the presence (Z, Z)-9, 12-octadecadienoic acid; 1,4a-dimethyl-7-(propan-2-ylidene) decahydronaphthalen-1-ol (Juniper camphor); 2-((2-ethylhexyl)oxy) carbonyl) benzoic acid; 2-phenyldecane; 2-phenyldodecane; 3-phenyldecane; 3-phenyldodecane; 4-(4-methoxyphenyl)-2-butanone; 4-phenyldecane; 4-phenyldodecane; 5-phenyldecane; 5-phenyldodecane; 6-phenyldodecane; butyl octyl phthalate; *n*-hexadecanoic acid; octadecanoic acid and oleic acid. Aqueous methanolic extract showed the presence of five compounds, namely, phthalicanhydride; *n*-hexadecanoic acid; 3,5-dihydroxy-6-methyl-2, 3-dihydro-4H-pyran-4-one; 1-heptadecene and octadecanoic acid (Nemkul et al., 2019) (Figure 24.1).

24.3 PHARMACOLOGY

24.3.1 Analgesic Activity

Methanolic extract of roots *F. strobilifera* showed significant analgesic activity (Kumar et al., 2011a). For the study of analgesic activity authors divided Albino mice in five groups. A total of 300, 500, and 1000 mg/kg doses of extract were given to group I, II, and III, respectively. A standard

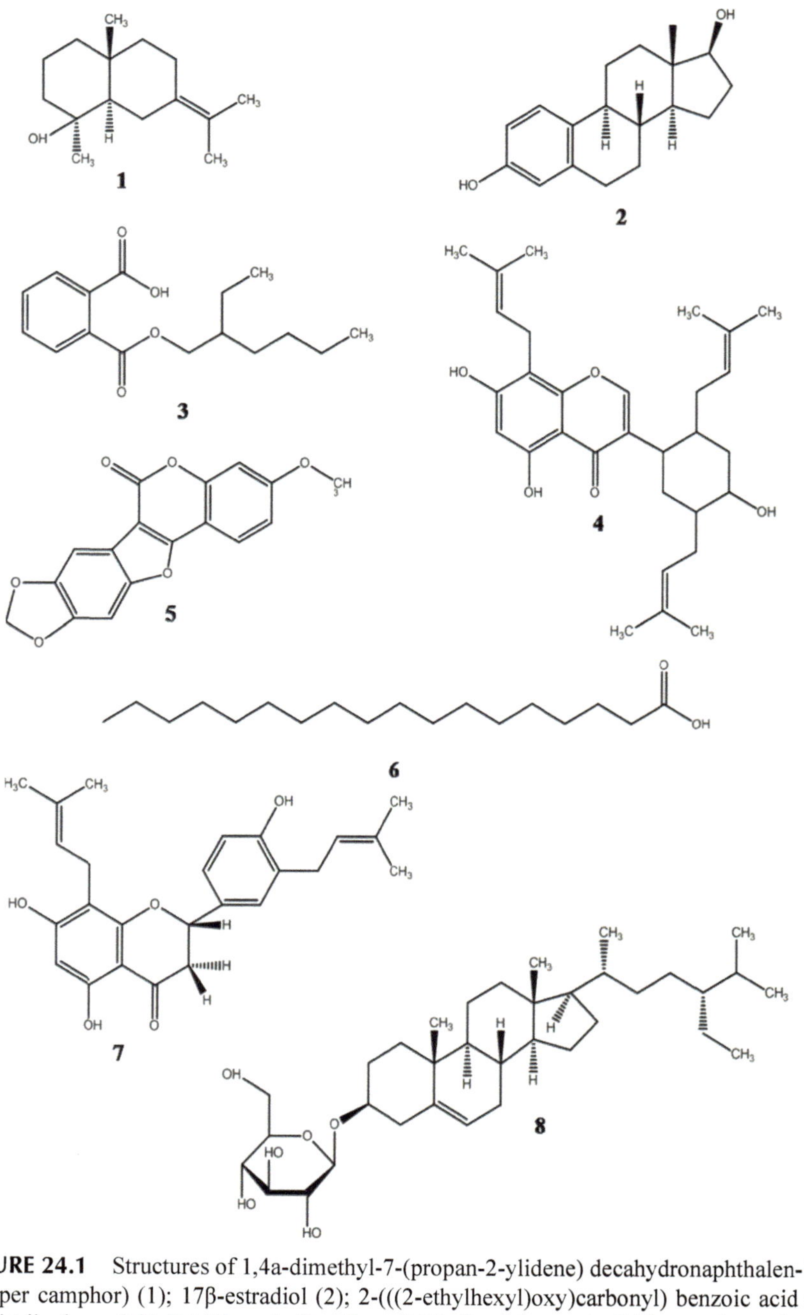

FIGURE 24.1 Structures of 1,4a-dimethyl-7-(propan-2-ylidene) decahydronaphthalen-1-ol (Juniper camphor) (1); 17β-estradiol (2); 2-(((2-ethylhexyl)oxy)carbonyl) benzoic acid (3); 5,7,4′-trihydroxy 8,2′, 5′-tri(3-methylbut-2-enyl) (4); flemichapparin C (5); octadecanoic acid (6); 8,3′-diprenyl-5,7,4′-trihydroxy flavanone (7); β-sitosterol-D glucoside (8)

drug, acetylsalicylic acid in the dose of 300 mg/kg was given to group IV. For the control, group V received 0.9% NaCl (saline) solution. The findings revealed a significant analgesic activity of root extract of *F. strobilifera* of 300 mg/kg, a low dose at the first 60 min of the test and it was comparable to the activity produced by acetylsalicylic acid. The dose at 1000 mg/kg of plant extract exhibited the highest analgesic activity at +180 min and its time period and intensity of analgesia were also greater as compared to acetylsalicylic acid (Kumar et al., 2011a).

Kumar et al. (2011b) evaluated the analgesic activity of ethanolic extract of *F. strobilifera* roots by Tail flick method. The albino rats were divided into six groups. The control group I received an oral suspension of 0.2 mL of 2% w/v carboxy methyl cellulose for seven days. Group II and III received 400 and 600 mg/kg body weight of ethanol extract of *F. strobilifera* orally for 7 days, respectively. Group IV and V received 400 and 600 mg/kg body weight of aqueous extract of *F. strobilifera* orally for 7 days, respectively. For the standard drug, Diclofenac sodium in dose of 10 mg/kg of body weight was given intraperitoneally to group VI for 7 days. The ethanol extract of roots *F. strobilifera* indicated significant analgesic activity as compared to the aqueous extract (Kumar et al., 2011b).

24.3.2 Anthelmintic Activity

The alcoholic, aqueous, chloroform, and petroleum ether leaf extracts of *F. strobilifera* showed a significant anthelmintic activity in alcoholic as well as in chloroform extracts as compared to other extracts (Kumar et al., 2011c). The drug, piperazine citrate, was used as a standard in this study.

24.3.3 Antianxiety Activity

Aqueous extract of roots of *F. strobilifera* showed antianxiety activity and it was assessed by zero maze test and mirror chamber test by using Swiss albino mice (Mahajon et al., 2017). The results showed that the latency period to enter in open arm was reduced significantly in elevated zero maze test. There was significant duration required in open arm and open head dip as compared to the control group. A significant increase in the number of partial entries and a marked increase in duration required in the mirror-chamber and the number of the full entries in mirror chamber test was obtained as compared to the normal control (Mahajon et al., 2017).

24.3.4 Anticholesterol Activity

Quevedo et al. (2015) studied the effect of whole extract *F. strobilifera* (paying payang) on blood cholesterol levels of albino mice. Results showed the ethanolic crude extract of leaf, root, and stem of *F. strobilifera* lowered blood cholesterol level of albino mice but the leaf extract showed the highest anticholesterol activity (Quevedo et al., 2015).

24.3.5 Anticonvulsant Activity

Ethyl acetate fraction and the crude ethanolic extract of roots of *F. strobilifera* showed a central nervous system depressant activity and works as a potential anticonvulsant (Gahlot et al., 2013). Authors used Swiss albino mice of weight ranging in between 12 and 35 g for studying the anticonvulsant activity of ethanolic extracts and its solvent partitioned fractions. The 50 mA for 0.2 s electroconvulsive shock by using electroconvulsometer was given to animals for inducing seizure. The results showed 200 and 300 mg/kg, p.o. of doses of ethyl acetate fraction and 400 and 600 mg/kg, p.o. of 95% ethanol crude extract of roots of *F. strobilifera* reduced the duration of seizure which was induced using maximal electroshock. Similar dose also protected from pentylenetetrazole induced tonic seizures. There is also a significantly delayed the onset of tonic seizures. While aqueous fractions, ethyl acetate, petroleum ether, and chloroform at any of the doses was used (100, 200, 300 mg/kg, p.o.) does not show any significant effect on pentylenetetrazole and maximal electroshock-induced convulsions. The ethyl acetate fraction and crude ethanol extract treatment showed signs of central nervous system depressant activity in locomotor activity test. It was confirmed by the potentiating of sodium pentobarbital sleeping period (Gahlot et al., 2013).

24.3.6 Antidiabetic Activity

Hsieh et al. (2010) evaluated antidiabetic effects of the aqueous root extracts of *F. strobilifera.* In vitro α-glucosidase and aldose reductase inhibitory methods were used to study the antidiabetic activities. The IC_{50} values of the aqueous extracts of *F. strobilifera* for α-glucosidase inhibition and aldose reductase inhibition was 1468.60 ± 2.10 μg/mL and 112.12 ± 2.32 μg/mL, respectively (Hsieh et al., 2010).

Thalugula and Yellu (2019) studied the antidiabetic and pancreatic protective effects of methanol extract of whole plant of *F. strobilifera* in a high-fat diet and streptozotocin-induced type 2 diabetes model. The treatment given to male Sprague Dawley (SD) rats with extract resulted in a significant decrease in diabetes incidence with a marked reduction in the blood glucose levels. After treatment with plant extracts with a high dose of 100 mg/kg, the glucose tolerance was greatly improved and the beneficial effects were similar to the standard drug mediated. Increase in beta-cell number and size was observed in histopathological studies indicating pancreatic beta-cell protective effects of *F. strobilifera* (Thalugula and Yellu, 2019).

24.3.7 Anti-Inflammatory Activity

Mohammad and Itankar (2009) studied the anti-inflammatory activity of 200 and 400 mg/kg body weight of hydro-alcoholic and methanolic aerial part extracts of *F. strobilifera* in Swiss albino mice. The effect of hydro-alcoholic and methanolic extracts of *F. strobilifera* on the subacute and acute phases of inflammation studied by two methods, that is, cotton-pellet-induced granuloma method and carrageenan-induced paw edema method. The hydro-alcoholic extracts showed the highest inhibition of 41.17 and 37.06% at the dose of 400 mg/kg in subacute phase and acute phase of inflammation, respectively. The results revealed that hydro-alcoholic extracts showed potential of anti-inflammatory activity (Mohammad and Itankar, 2009).

The anti-inflammatory activity of the ethanolic extract of roots of *F. strobilifera* which was done by carrageenan-induced paws edema method by Kumar et al. (2011b). Albino rats used for the study and grouped into six groups of six rats each. For the control, group I received oral suspension of 0.2 mL of 2% w/v carboxy methyl cellulose for 7 days. Group II and III received 400 and 600 mg/kg body weight of ethanolic extract of *F. strobilifera*, respectively, whereas group IV and V received 400 and 600 mg/kg body weight of aqueous root extract of *F. strobilifera* orally for 7 days, respectively. For the standard drug, group VI received 10 mg/kg of body weight of Indomethacin intraperitoneally for 7 days. All groups were injected 0.1 mL of 1% w/v carrageenan in the subplantar part in the right-hind paw of rat to induction of acute inflammation. The results showed ethanolic root extract had significant anti-inflammatory activity, that is, 50.94–64.20% inhibition as compared to the aqueous root extracts that exhibited 4.52–7.21% inhibition (Kumar et al., 2011b).

24.3.8 Antimicrobial Activity

Antibacterial activity of the flavanone isolated from roots extracts was studied against Gram positive bacteria, namely, *Staphylococcus aureus*, *S. epidermidis*, methicillin resistant *S. aureus*, Gram negative bacteria, namely, *Escherichia coli* and *Pseudomonas aeruginosa* and the fungus (*Candida albicans*). The results showed minimum inhibitory concentration (MIC) of 17 μg/mL for *S. aureus, S. epidermidis*, methicillin resistant *S. aureus, E. coli, P. aeruginosa*, and *C. albicans* (Madan et al., 2008).

Madan et al. (2009) studied antimicrobial activity of roots of *F. strobilifera* against some bacteria and fungi. Authors revealed that the 5,7,4′-trihydroxy 8,2′,5′-tri(3-methylbut-2-enyl), 5,7,2′,4′ tetrahydroxy isoflavone, 5,7,4′-trihydroxy isoflavone showed moderate activity against Gram positive bacteria, namely, *S. aureus*, *S. epidermidis* and MIC of 28 and 34 μg/mL, respectively. The chloroform extract showed potent activity with MIC of 2.1 μg/mL against *S. aureus*, *S. epidermidis*, methicillin resistant *S. aureus* (Gram positive bacteria) and MIC of 34 and 17 μg/mL against *P. aeruginosa*, *E. coli* (Gram negative bacteria), respectively, and 17 μg/mL of MIC against the fungus *C. albicans* (Madan et al., 2009).

Anil Kumar et al. (2011) found that chloroform root extract showed activity against Gram positive bacteria *S. aureus* as well as Gram negative bacteria *E. coli* at a concentration of 10 mg/mL.

Nemkul et al. (2019) evaluated antimicrobial activity (in vitro) and phytochemicals of hexane and aqueous methanolic root extracts by chemical tests and GC-MS analysis. The GC-MS analysis of the hexane extract showed the presence of 27 compounds accounting 99.37% of the total constituents. Of the eight bacteria studied, aqueous-methanolic extract of roots revealed moderate antimicrobial activity against *Shigella dysenteriae* (zone of inhibition (ZOI) = 10.33 ± 0.33) and significant against *E. coli* (ZOI = 16.5 ± 0.67) (Nemkul et al., 2019).

24.3.9 Antioxidant Activity

Madan et al. (2009) tested the compounds from methanol root extract for antioxidant activity (in vitro) by 2,2-diphenyl-1-picrylhydrazyl (i.e., DPPH) radical scavenging assay. The compound 5,7,2′,4′-tetrahydroxyisoflavone from the extract of *F. strobilifera* indicated the IC_{50} value of 32.7 μg/mL.

The antioxidant activity of two root extracts (methanolic and butanolic) was assessed by DPPH radical scavenging method, nitric oxide radical inhibition assay, and a method of scavenging of hydroxyl radical by *p*-NDA. Methanolic root extract showed an effective DPPH radical scavenging activity with low IC_{50} value of 11.4 μg/mL and leaf extract showed IC_{50} value of 38.0 μg/mL. Butanolic root extract showed significant nitric oxide radical inhibition activity with IC_{50} of 150.0 μg/mL in a dose-dependent manner. In hydroxyl radical scavenging by *p*-NDA method methanolic root extract showed IC_{50} value of 378.33 μg/mL. The quantification of methanolic root extract showed the presence of 2.14% and 3.26% w/w of flavonol and 13.75% and 8.84% w/w phenolics (calculated as gallic acid equivalent) (Madan et al., 2010).

Hsieh et al. (2010) examined antioxidant activity, α-glucosidase inhibition and aldose reductase inhibition of the aqueous extracts of four *Flemingia* species found in Taiwan including *F. strobilifera*. The results revealed that the extraction yields in the aqueous extracts of *F. strobilifera* was 11.67%. Total antioxidant activity assessed as [TEAC (trolox equivalent antioxidant capacity)] was 0.32 ± 0.01 mM/mg extract and the ferric reducing antioxidant power 0.42 ± 0.01 μmol Fe^{2+}/mg extract. IC_{50} value of the aqueous extracts of *F. strobilifera* in DPPH radical scavenging activity was 197.97 ± 1.41 μg/mL.

Anil Kumar et al. (2011) used DPPH extensively as a free radical to evaluate free radical scavenging activity of chloroform root extracts of *F. strobilifera*. The activity of root extract showed the IC_{50} value 193.38 μgm/mL. As a reference standard, gallic acid was used and its IC_{50} value was 2.7 μgm/mL.

Antioxidant activity and total phenolics content were evaluated from the aqueous and hydro methanol (80%) extracts of leaves *F. strobilifera* by Pizon et al. (2016). The GC-MS analysis showed the presence of eight compounds, namely, eicosane, heptacosane, hexacosane, hexanedioic acid, bis(2-ethylhexyl) ester, limonene, nonadecane, phenol, 3,5-bis(1,1-dimethylethyl) and thiocyanic acid ethyl ester. DPPH radical scavenging activity of the aqueous and 80% methanol leaf extracts of *F. strobilifera* indicated a low IC_{50} values of <0.25 and 0.299 mg/mL, respectively. The total phenolics content of the aqueous extract was higher as compared to the 80% methanolic extract with gallic acid equivalence values, that is, 102.98 ± 0.003 and 12.49 ± 0.020, respectively (Pizon et al., 2016).

24.3.10 Antiulcer Activity

Flemingia strobilifera is used in folk medicine in the treatment of peptic ulcer (Anil Kumar et al., 2010a). Authors used indomethacin to induce ulcer in rats and studied the gastro-protective effect of chloroform root extract. Isolated guinea pig ileum preparation was used to study the effect of root extracts on gastrointestinal motility as well as the activity on contraction evoked by standard agonists. Chloroform root extract did not produce spasmogenic effect on isolated guinea pig ileum preparation. It produced the dose-related inhibition of contractile responses to acetylcholine and histamine with IC_{50} 34.95 and 12.5 μg/mL, respectively, while increasing concentrations. In indomethacin-induced ulcer model activity of the extract at a dose 15 mg and 30 mg/kg body weight reduced the ulcer indices significantly as compared to the control (Anil Kumar et al., 2010a).

The root extract can be used in peptic ulcers and it showed antiulcerogenic activity (Anil Kumar et al., 2011). Chloroform root extract at a dose of 300 mg/kg body weight was safe orally in female Wistar rats. A water immersion stress induced peculiar lesions in the glandular region of stomach of rat. The results showed that the characteristic lesions reduced in a dose-dependent manner in case of pretreatment with chloroform root extract as compared to the control. The total protein content and gastric mucosal glutathione level increased significantly in case of pretreated rats at a dose of 15 and 30 mg/kg body weight as compared to the control group. There is a significant reduction in gastric mucosal malonaldehyde levels as compared to the control (Anil Kumar et al., 2011).

24.3.11 Estrogenic Activity

Jeong et al. (2018) examined the estrogenic activity of methanol extract of leaves and stems of *F. strobilifera* (in vitro and in vivo) for the first time. They provided insights into the estrogenic effects of phytoestrogens obtained from it via estrogen receptors. For this study, human breast cancer cell line (MCF-7) and SD rats were used. The application of extracts led to the proliferation of MCF-7 cells. The uterine weights in immature female rats increased indicating that the extracts could be useful in postmenopausal therapy (Jeong et al., 2018). The results demonstrated that methanol extract of *F. strobilifera* and its bioactive compounds (Cajanin, Genistein, Pisatin, 20, 30, 40, 60-tetramethoxychalcone, 20-hydroxygenistein) showed estrogenic properties (Jeong et al., 2018).

24.3.12 Hepatoprotective Activity

Anil Kumar et al. (2010a,b) evaluated the antioxidant activity and hepatoprotective of the chloroform extract of *F. strobilifera* leaves is in albino

Wistar rats having hepatotoxicity induced by paracetamol. It was concluded that the chloroform leaf extract of *F. strobilifera* does not show effective hepatoprotective and antioxidant action against paracetamol-induced hepatic injury in rats (Anil Kumar et al., 2010b).

A similar study was done to evaluate the antioxidant and hepatoprotective effect of the chloroform extract of *F. strobilifera* on Wistar albino rats with hepatotoxicity induced by ethanol-carbon tetrachloride. Aspartate aminotransferase, alkaline phosphatase, alanine aminotransferase and total bilirubin level were increased and the total protein level decreased in rats treated with ethanol-CCl_4. The chloroform extract of leaf of *F. strobilifera* and both the doses did not restore a normalcy of total protein and did not decrease the elevated levels of all these biochemical parameters effectively. The study showed that the chloroform extract of leaves of *F. strobilifera* did not show effective hepatoprotective and antioxidant action (Anil Kumar and Gowda, 2011).

ACKNOWLEDGEMENT

We are grateful to the Head, Department of Botany, Shivaji University, Kolhapur and the Principal, Swami Vivekanand Shikshan Santha's, Dattajirao Kadam Arts, Science and Commerce College, Ichalkaranji for providing research facilities. SKG thanks the Science and Engineering Research Board, New Delhi, for financial assistance vide sanction letter SERB/F/201/2014–15 dated 15.05.2014.

KEYWORDS

- **Fabaceae**
- **Flemingia**
- **India**
- **isoflavone**
- **medicine**

REFERENCES

Anil Kumar, K. V.; Dewan, B.; Rama, T. Evaluation of Anti-Ulcerogenic Properties from the Root of *Flemingia strobilifera*. *J. Basic Clin. Pharm.* **2011,** *2* (1), 33–39.

Anil Kumar, K. V.; Dewan, B.; Rama, T. Evaluation of Gastro-Protective Effects of *Flemingia strobilifera* R.Br. (Fabaceae) Root Extract. *Arch. Pharm. Sci. Res.* **2010a,** *2* (2), 347–353.

Anil Kumar, K. V.; Gowda, V. K. Evaluation of Hepatoprotective and Antioxidant Activity of *Flemingia strobilifera* R.Br. Against Experimentally Induced Liver Injury in Rats. *Int. J. Pharm. Pharm. Sci.* **2011,** *3* (3), 115–119.

Anil Kumar, K. V.; Satish, R.;Rama, T.; Kumar, A.; Babul, D.; Samhitha, J. Hepatoprotective Effect of *Flemingia strobilifera* R.Br. on Paracetamol Induced Hepatotoxicity in Rats. *Int. J. Pharm. Tech. Res.* **2010b,** *2* (3), 1924–1931.

Anonymous. *The Wealth of India, Raw Materials*, Vol. 4; Publication and Information Directorate, CSSIR: New Delhi, **1993,** pp 45–48.

Bhatt, S. Chalkones and Some Other Constituents of *Flemingia strobilifera*. *Indian J. Chem.* **1975,** *13*, 1105–1108.

Das, K. Ethnobotanical Studies and Phytochemical Analysis of *Flemingia strobilifera* (L.) W. T. Aiton of Jorhat, Assam, India. *Int. J. Sci. Res. Rev.* **2018,** *7* (7), 556–565.

Gahlot, K.; Lal, V. K.; Jha, S. Anticonvulsant Potential of Ethanol Extracts and Their Solvent Partitioned Fractions from *Flemingia strobilifera* Root. *Pharmacogn. Res.* **2013,** *5* (4), 265–270.

Gahlot, K.; Lal, V. K.; Jha, S. Phytochemical and Pharmacological Potential of *Flemingia* Roxb. ex W. T. Aiton (Fabaceae). *Int. J. Phytomedicine* **2011,** *3*, 294–307.

Gavade, S.K., van der Maesen, L.J.G., Lekhak, M.M. Taxonomic revision of the genus Flemingea (Leguminosae: Papilionoideae) in India. Webbia 2020; *75*(2), 141–218.

Hsieh, P. C.; Huang, G. J.; Ho, Y. L.; Lin, Y. H.; Huang, S. S.; Chiang, Y. C.; Tseng, M. C.; Chang, Y. S. Activities of Antioxidants, α-Glucosidase Inhibitors and Aldose Reductase Inhibitors of the Aqueous Extracts of Four *Flemingia* Species in Taiwan. *Bot. Stud.* **2010,** *51*, 293–302.

Jeong, S. Y.; Chang, M.; Choi, S.; Oh, S. R.; Wu, H. H.; Zhu, Y.; Gao, X.; Wang, X.; Zhang, B.; Lim, D. S.; Lee, J. Y.; Kim, S. D.; Song, Y. S. Estrogenic Effects of Phytoestrogens Derived from *Flemingia strobilifera* in MCF-7 Cells and Immature Rats. *Arch. Pharm. Res.* **2018,** *41* (5), 519–529.

Kirtikar, K. R.; Basu, B. D. *Indian Medicinal Plants*, Vol. 1; **1993,** pp 814–815.

Kumar, A.; Dora, J.; Gahlot, K.; Tripathi, R. Anthelmintic Activity of *Flemingia strobilifera* (R. Br). *Int. J. Res. Pharm. Biomed. Sci.* **2011c,** *2* (3), 1077–1078.

Kumar, A.; Gahlot, K.; Dora, J.; Singh, P. Analgesic Activity of Methanolic Extract of *Flemingia strobilifera* (R. Br). *Int. J. Res. Pharm. Chem.* **2011a,** *1* (4), 825–827.

Kumar, A.; Gahlot, K.; Dora, J.; Tripathi, R. Anti-Inflammatory and Analgesic Activities of *Flemingia strobilifera* (Linn). *Res. J. Pharm. Biol. Chem. Sci.* **2011b,** *2* (3), 541–546.

Madan, S.; Singh, G. N.; Kohli, K.; Ali, M.; Kumar, Y.; Singh, R. M.; Prakash, O. Isoflavonoids from *Flemingia strobilifera* (L) R. Br. Roots. *Acta Pol. Pharm. Drug Res.* **2009,** *66* (3), 297–303.

Madan, S.; Singh, G. N.; Kumar, Y.; Kohli, K. Phytochemical Analysis and Free-Radical Scavenging Activity of *Flemingia strobilifera* (Linn) R. Br. *Res. J. Pharm. Biol. Chem. Sci.* **2010,** *1* (4), 183–190.

Madan, S.; Singh, G. N.; Kumar, Y.; Kohli, K.; Singh, R. M.; Mir, SR.; Ahmad, S. A New Flavanone from *Flemingia strobilifera* (Linn) R.Br. and Its Antimicrobial Activity. *Trop. J. Pharm. Res.* **2008,** *7* (1), 921–927.

Mahajon, B.; Nath, R.; Thakur, S.; Remadevi, R.; Ravi Shankar, B. Anxiolytic Effect of Root Extract of Amazing Folklore Herb *Flemingia strobilifera* (L.) W. T. Aiton In Albino Mice. *Int. J. Curr. Res.* **2017,** *9* (11), 60216–60219.

Mahajon, B.; Remadevi, R; Sunil Kumar, K. N.; Ravishankar, B. Preliminary Analysis of Botanical and Phytochemical Features of Kamalu—Root of *Flemingia strobilifera* (L.) W. T. Aiton. *J. Homeopath. Ayurvedic Med.* **2014,** *3* (4), 1–6.

Mohammad, T. A.; Itankar, P. R. Evaluation of Anti-Inflammatory Activity of *Flemingia strobilifera* Linn. Fabaceae. *Res. J. Pharm. Tech.* **2009,** *2* (4), 865–867.

Nemkul, C. M.; Bajracharya, G. B.; Shrestha, I. Phytochemical Evaluation and In Vitro Antimicrobial Activity of the Roots of *Flemingia strobilifera* (L.) R. Br. *J. Plant Resour.* **2019,** *17* (1), 98–103.

Pizon, J. R. L.; Nuñeza, O. M.; Uy, M. M.; Senarath, W. T. P. S. K. GC-MS Analysis and Evaluation of *In Vitro* Antioxidant Potential and Total Phenolics Content of Wild Hops (*Flemingia strobilifera* (L.) W. T. Aiton). *Int. J. Biosci.* **2016,** *8* (1), 25–32.

Quevedo, K. L. R. G.; Mia, H. R. S.; San Sebastian, J. S. Effect of *Flemingia strobilifera* Linn. (paying payang) Extract on Blood Cholesterol Levels of Albino Mice. *Int. J. Curr. Sci.* **2015,** *14,* 37–43.

Saxena, V. K.; Nigam, S. S.; Singh, R. B. Glycosidic Principles from the Leaves of *Flemengia strobilifera. Planta Med.* **1976,** *29* (1), 94–97.

Shreedevi; Patel, B. R.; Shukla, V. J.; Harisha C. R. Pharmacognostical and Phytochemical Evaluation of leaf of *Flemingia strobilifera* W. Ation. *Pharma Sci. Monit.* **2018,** *9* (1), 326–333.

Thalugula, S. and Yellu, N. R. Pancreatic Beta Cell Protective Effects of *Flemengia strobilifera* Extracts in Combination Model of High Fat Diet and Streptozotocin Induced Type 2 Diabetes. *Int. J. Pharm. Sci. Res.* **2019,** *10* (12), 5723–5732.

Tikadar, P.; Palita, S. K.; Panda, D. Phytochemical Analysis of Medicinal Plants Used for Treatment of Dysentery and Diarrhoea by the Paraja Tribe of Koraput, Odisha, India. *Int. J. Herb. Med.* **2017,** *5* (2), 01–04.

Yang, R. Y.; Wang, X. Z.; Wei, B.; Zhou, D. X.; Xu, W. F.; Liang, H.; Chen, Z. F.; Li, J. Chemical Constituents of the Stems of *Flemingia strobilifera. Chem. J. Nat. Compd.* **2016,** *52* (1), 139–141.

CHAPTER 25

Biomolecules and Pharmacology of *Flemingia vestita* Benth. ex Baker

SANDIP KISAN GAVADE[1*] and MANOJ MADHWANAND LEKHAK[2]

[1]*Department of Botany, Dattajirao Kadam Arts, Science and Commerce College, Ichalkaranji, Maharashtra, India*

[2]*Angiosperm Taxonomy Laboratory, Department of Botany, Shivaji University, Kolhapur, Maharashtra, India*

[*]*Corresponding author. E-mail: skgavadenogra@gmail.com*

ABSTRACT

Flemingia vestita is very important tuber crop growing in India, Laos, Nepal, Philippines, southern China and Vietnam. The tubers of *F. vestita* are nutritious and also effective to cure worm infections. This species is an important source of Genistein, a phytoestrogen. We have provided a detailed review on bioactives and pharmacological activities of *F. vestita*. We have aslo provided some important structures of bioactives of *F. vestita*.

25.1 INTRODUCTION

Flemingia vestita Benth. ex Baker is an important herbaceous, tuberous, medicinal plant belonging to the family Fabaceae. The tubers are consumed by the local people from the northeast India to cure intestinal worm infections caused by *Artyfechinostomum sufrartyfex, Ascaris suum, Fasciolopsis buski*, and *Raillietina echinobothrida.* The locals from the Khasi hills (Meghalaya)

Bioactives and Pharmacology of Legumes. T. Pullaiah, PhD (Ed.)

cultivate this species as a tuber crop. Tubers are sold in the local market and eaten raw as a source of starch. This species is distributed in Assam, Himachal Pradesh, Meghalaya, Uttarakhand and some parts of southeast Asia. It grows on slopes of mountains at altitudes of ca. 1800–2100 m above sea level (Gavade et al., 2019).

25.2 BIOACTIVES

The tuber peels of *F. vestita* are the source of isoflavanoids such as daidzein, formononetin, genisteinand, and pseudobaptigenin (Rao and Reddy, 1991). Genistein, a phytoestrogen is a main active isoflavone compound occurring in *F. vestita* in a range of 0.56–1.346 mg/g (Shailajan et al., 2014). Standal et al. (1985) compared the nutrient content of tribal foods *F. vestita* and *Perilla frutescens* from India and while studying they reported various amino acids like valine, tyrosine, tryptophan, cystine, alanine, arginine, aspartic acid, glutamic acid, gistidine, leucine, glycine, phenylalanine, isoleucine, lysine, proline, methionine, serine, threonine, and ammonia. Recently, Marboh and Mahanta (2020) analyzed the nutritional and antioxidant properties of *F. vestita* and reported ascorbic acid, caffeic acid, catechin, chlorogenic acid, coumaric acid, ferulic acid, gallic acid, quercetin, rutinhydrate, sinapic acid, and isoflavones: biochanin-A, daidzein, daidzin, formononetin, genistein, and genistin.

Structures

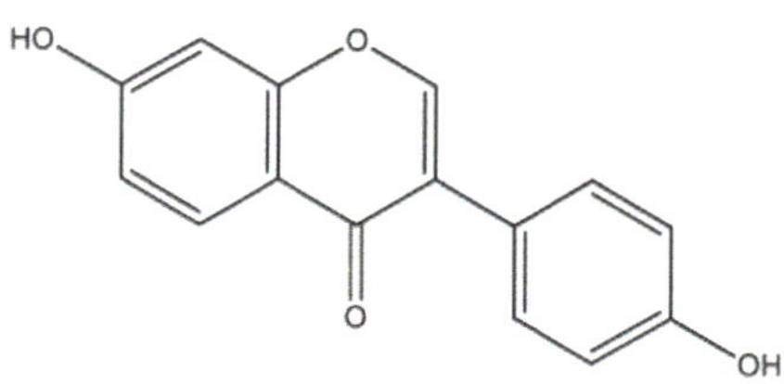

Daidzein

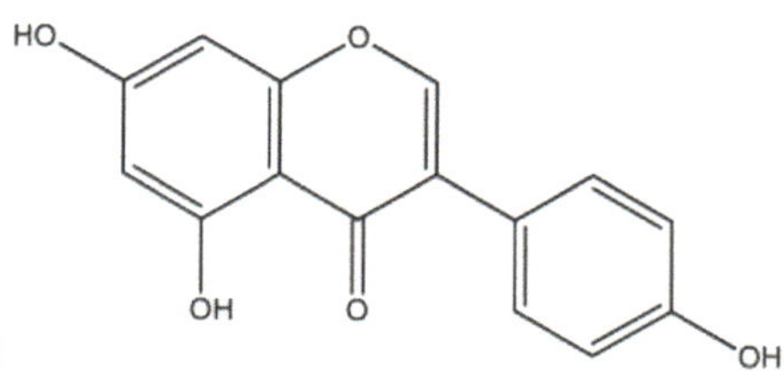

Genistein

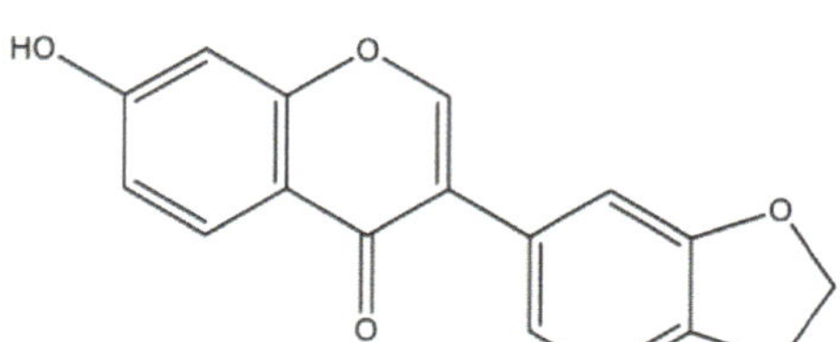

Pseudobaptigenin

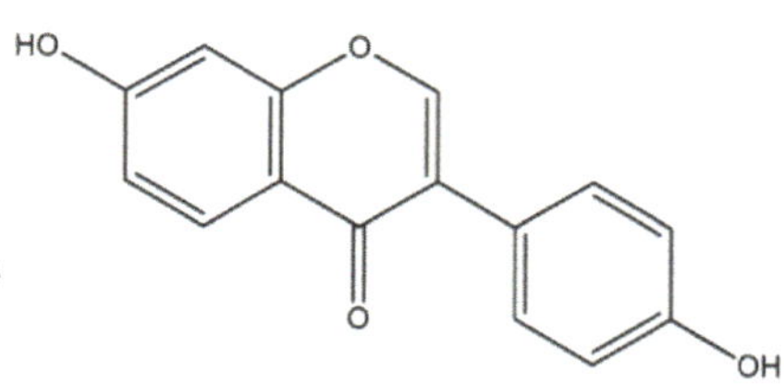

Formononetin

25.3 PHARMACOLOGY

25.3.1 Anthelmintic Activity

Consumption of tubers of *F. vestita* is very common and effective against many worms. A lot of work has been done on anthelmintic activity of *F. vestita* (Yadav et al., 1992; Roy and Tandon, 1996; Tandon et al., 1997, 2003; Pal and Tandon, 1998a,b; Kar et al., 2002, 2004; Das et al., 2004a,b,c, 2006, 2007, 2009; Tandon and Das, 2007). Yadav et al. (1992) studied the anthelmintic activity (in vitro) of extract of fresh tuber of *F. vestita* on the motility and histomorphology of *Ascaris suum* under light and electron microscopy. This worm (*A. suum*) was treated with 0.36% of tuber extract. It was found that within 5–8 h of treatment the worm got paralyzed. In comparison to the antiparasitic medicine Mabex, the tuber extract exhibited faster action.

The root-tuber extract of *F. vestita* has been treated against intestinal digenetic worms *A. sufrartyfex* and *Fasciolopsis buski* found in pigs and these worms were tested under the scanning electron microscope by Roy and Tandon (1996). While studying the efficacy they treated *F. buski* and *A. sufrartyfex* in vitro with ethanolic extract of *F. vestita* at a concentration of 5, 10, and 20 mg/mL of phosphate-buffered saline. They noted that *A. sufrartyfex* got paralyzed within 1.1–1.4, 0.8–1.0, and 0.3–0.5 h, and *F. buski* within 3.0–3.6, 1.5–2.0, and 0.6–0.8 h (Roy and Tandon, 1996).

To study the in vitro anthelmintic activity of root-tuber extract of *F. vestita*, Tandon et al. (1997) used various live parasites. They collected *Ascaridia galli* and *Heterakis gallinarum* (from domestic fowl), *R. echinobothrida* (from domestic fowl), *A. suum* (from pigs), *A. lumbricoides* (from humans), and *Paramphistomum* species (from cattle). All the parasites were collected in phosphate-buffered saline (0.9%) and these were maintained at 37 ± 1 °C. These parasites treated in vitro condition with the plants extract (50 mg/mL) in phosphate-buffered saline showed complete immobilization of the *Paramphistomum* species and *R. echinobothrida* in ca. 43 and 20 min, respectively (Tandon et al., 1997).

Pal and Tandon (1998a) studied the anthelmintic efficiency of crude root-tuber peel extract and genistein in relation to the esterase activity in *R. echinobothrida* a fowl tapeworm. They exposed *R. echinobothrida* to plant extract (50 mg/mL) of the incubation medium and with genistein (0.5 mg/mL). They found that there is a great reduction in action of cholinesterase (ChE) and nonspecific esterases in sites of nervous system and tegument of

R. echinobothrida. It was found that the total acetylcholinesterase (AChE) activity was reduced to 49.07% after treating the tapeworm with plant extract and 56.77% treating the with genistein (Pal and Tandon 1998a). Pal and Tandon (1998b) also assessed the anthelmintic efficiency of crude root-tuber peel extract and genistein with reference to the tegumental ultra-structure of *R. echinobothrida.* This work showed alternation in tegument structure of *R. echinobothrida* leading to deformed microtriches (Pal and Tondon, 1998b)

Kar et al. (2002) examined the effect of crude peel extract of *F. vestita* and genistein on the enzyme nitric oxide synthase (NOS) and the neuronal messenger nitric oxide (NO) of *F. buski. F. buski* the large intestinal fluke found in swine and human host. The results showed the increase in NOS action in treated parasites and the NADPH-diaphorasehistochemical staining (marker for NOS in neuronal tissues) was demonstrable in neuronal cell bodies (Kar et al., 2002).

Further, the effect of the crude peel extract and pure genistein on the glycogen metabolism occurring in the fowl tapeworm, *R. echinobothrida* was studied. After treating with 5 mg/mL of crude peel extract of root of *F. vestita* and 0.2 mg/mL of genistein the glycogen concentration was found to decrease by 15–44%. There is also an increase of action of active form of glycogen phosphorylase by 29–39% and decrease of action of the active form of glycogen synthase by 36–59% in treated parasites than untreated controls (Tandon et al., 2003).

Seven glycolytic enzymes, namely phosphofructokinase, hexokinase (HK), pyruvatekinase, phosphoenolpyruvatecarboxykinase (PEPCK), malic enzyme (ME), malatedehydrogenase (MDH), and lactate dehydrogenase (LDH) of the *R. echinobothrida* were used to study the alternation in enzyme activity after treating the worm with the genistein, praziquantel and crude root-peel extract of *F. vestita* (Das et al., 2004a). This study showed the activities of phosphofructokinase, LDH, PEPCK, and HK were increased by 41–125%, 55–67%, 44–49%, and 33–39%, respectively, and that of pyruvatekinase was decreased by 14–26% in the paralyzed parasite. The ME and MDH activities of the tissue homogenate were also found to be high by 28–59% and 22–43%, respectively, in the treatments and the activity of mitochondrial MDH and cytosolic increased by 43–73% and 33–58%, respectively. The cytosolic ME activity was found to be increased by 33–39%. A significant enhancement in the mitochondrial ME activity was not found. Histochemically, the enhancement in the activities of MDH, LDH, and HK was clearly discernible. A glycolytic activity seems to be function of anthelmintic stress. This stress is caused due to the phytochemicals (Das et al., 2004a).

Das et al. (2004b) treated cestode, *R. echinobothrida* with extract of peel of tuber of *F. vestita* and its active component, genistein, to undertstand the effect on glucose metabolism. *R. echinobothrida* was incubated at 39 ± 1 °C in concentrations of the 5 mg/mL plant extract, 0.2 mg/mL genistein, and 1 Ag/mL praziquantel in phosphate-buffered saline along with 1% of dimethyl sulphoxide with controls. The study revealed that there was a decline in the glucose content and glycogen and a noteworthy rise in the malate content. The energy demand in the parasite possibly gets enhanced under stress and it is toward aerobic degradation of glucose (Das et al., 2004b).

The effects of phytochemicals of *F. vestita* on enzymes of gluconeogenesis and glucose 6-phosphate dehydrogenase (G6PDH) in *R. echinobothrida* was studied by Das et al. (2004c). While studying they tested the crude root-peel extract, pure genistein, and praziquantel on enzymes of gluconeogenesis-phosphoenolpyruvate carboxykinase (PEPCK), pyruvate carboxylase, and fructose 1,6-bisphosphatase (FBPase) and glucose 6-phosphate dehydrogenase (G6PDH) and in *R. echinobothrida.* Based on their study they postulated that the phytochemicals of *F. vestita*, especially genistein, influence both gluconeogenesis and hexose monophosphate pathway in the *R. echinobothrida*, which seems to be a response to high-energy demanded by parasite when it was under anthelmintic stress (Das et al., 2004c).

Kar et al. (2004) treated a giant intestinal trematode *F. buski* with crude root-peel extract of *F. vestita* and flukicide oxyclozanide to find out the anthelmintic efficacy of *F. vestita*. The main aim of their study was to know the alterations in ammonia levels and free amino acid pool in *F. buski*. While studying they came to know that there is a decrease in the levels of alanine, arginine, glycine, isoleucine, leucine, ornithine, proline, serine, taurine, threonine, tyrosine, valine treated with of crude tuber peel extract of *F. vestita* (20 mg/mL), genistein (0.5 mg/mL), and oxyclozanide drug (20 mg/mL) and increase in the levels of citrulline, glutamine, glutamic acid, γ-amino butyric acid, and phosphoserine (Kar et al., 2004).

Genistein from *F. vestita* showed effect on the Ca^{2+} homeostasis and some metal ions occurring in *R. echinobothrida.* This effect was studied by treating the *R. echinobothrida* to alcoholic crude root-peel extract of *F. vestita* and pure genistein (Das et al., 2006). They found the alterations in the structural architecture of the tegumental interface and metabolic activity in *R. echinobothrida*. The final results showed a significant amount of Ca^{2+} (~400 μg/g dry tissue weight), while cadmium, chromium, iron, lead, magnesium, nickel, whilst manganese, and zinc were below the level of detection. As compared with the controls, the Ca^{2+}concentration of decreased by 39–49% in the parasite tissue, and there was increase in Ca^{2+} efflux by 91–160% into the culture medium. All these changes (Ca^{2+} homeostasis) may be related to the quick muscular contraction and resultant paralysis in the parasite tissue because of the anthelmintic stress caused by the phytochemicals of *F. vestita* (Das et al., 2006). Das et al. (2007) also studied the activity of NOS, nitric oxide efflux, and the Cyclic GMP concentration in *R. echinobothrida* after treating it with crude peel extract of *F. vestita* and pure genistein. The parasite got paralyzed and the activity of NOS showed a notable increase (35–46%). A 2-fold increase of nitric oxide efflux into incubation medium in treated worms was noted when compared to the respective controls. There was 46–84% increase in Cyclic GMP concentration in the parasite tissue. This increase in Cyclic GMP concentration was responsible for downstream signal pathways (Das et al., 2007).

Tandon and Das (2007) tested the anthelmintic efficacy root peel extract of *F. vestita* in respect with carbohydrate metabolism in *R. echinobothrida*. The results showed increase in the activity of phosphoenolpyruvate carboxykinase whereas there was a decrease in the activity of pyruvate kinase. Finally, it was proved that during the paralytic time glucose breakdown follow the phosphoenolpyruvate carboxykinase malate pathway in parasite for getting a high energy demand because of the anthelmintic stress (Tandon and Das, 2007).

In order to find out the anthelmintic properties of a crude tuber peel extract and purified fraction of *F. vestita* and a crude extract of rhizome of *Stephania glabra* (Menispermaceae), the effect of these extracts on the activity of NOS, nitric oxide efflux and Cyclic GMP concentration in *R. echinobothrida* was studied. This study showed that the phytochemicals genistein from *F. vestita* and tetrahydropalmatine from *S. glabra* disturbs the downstream signaling pathway of nitric oxide as shown by the change in Cyclic GMP concentration in parasite tissue (Das et al., 2009).

25.3.2 Anticancer Activity

F. vestita is a one of the plant sources of natural flavonoids (genistein). Genistein shows chemo-therapeutic and chemopreventive exercises in numerous tumors, particularly in hormone-dependent cancers. The in vitro and in vivo studies revealed its antineoplastic effects in prostate, breast, urinary, colon, and skin cancer (Hazafa et al., 2019).

25.3.3 Antioxidant Properties

The antioxidant and nutritional properties of *F. vestita* were studied by Marboh and Mahanta (2020). They used the flour of cultivated species (CS) and market species (MS) of *F. vestita* for total flavonoid content (TFC), antioxidant activities (AA), and total phenolic content (TPC). Their results revealed that the flour of CS is having better nutritional properties TFC, AA, and TPC than the flour of MS. The TPC of CS flour samples extracted by ultrasound-assisted extraction, microwave-assisted extraction, and conventional-assisted extraction ranged from 917.65 to 1189.41 mg GAE/100 g dry sample, while in MS flour it ranged from 776.67 to 940.52 mg GAE/100 g dry sample. In case of TFC, the tuber flours ranged from 654.42 to 769.22 mg QE/100 g in CS flour and 537.04 to 645.56 mg QE/100 g in MS flour on dry basis. In all the extraction process, CS flour sample showed notably higher amount of antioxidant properties than MS flour sample except for 2,2-diphenyl-1-picrylhydrazyl (DPPH), 2,2′-azino-bis, 3-ethylbenzothiazoline-6-sulphonic acid (ABTS) and metal chelating capacity (MCC) of conventional-assisted-extraction extracted sample where no significant differences were observed (Marboh and Mahanta, 2020).

25.3.4 Nematicidal Properties

F. vestita, *S. glabra*, and *Trichosanthes multiloba* (Cucurbitaceae) exhibited nematicidal properties with respect to hookworm, *Ancylostoma ceylanicum* collected from the intestine of *Mesocricetus auratus* (golden hamsters) (Lyndem et al., 2008). In their study Lyndem et al. (2008) exposed the parasite to root peel extract (100 mg/mL) and mebendazole, a nematicidal drug. They noticed that the parasite got paralyzed 6.18 ± 0.04 h after exposing it to root peel extract (100 mg/mL) of *F. vestita*, whereas major active principle genistein (10 mg/mL) from *F. vestita* caused paralysis in 0.93 ± 0.03 h, which was much faster than the effect of mebendazole (2.5 ± 0.12 h) at similar concentration. Also some changes were noted in the surface cuticlular features of the parasite observed under scanning electron microscope (Lyndem et al., 2008).

Preliminary study had shown that the powder extract of tubers of *F. vestita* (mistakenly referred as *F. procumbens* by the authors) exhibit nematicidal properties against a common root-knot nematode, *Meloidogyne incognita* (Gawade et al., 2019). It was tested in vitro against juveniles second-stage (J2s) of *M. incognita* and three different type of powder extracts found effective and it causes varied mortality in J2s of *M. incognita* with range of 88–95% at 100 mg/mL (after 72 h exposure) (Gawade et al., 2019).

25.3.5 Estrogenic Action

The 500 mg/kg body weight of ethanol extract of tubers of *F. vestita* may exhibit an effective estrogenic action in ovariectomized rat model (Shailajan et al., 2016). The study showed rats treated with the ethanol extract of *F. vestita* tubers led to increase in glycogen levels, uterine weight, estrogen, progesterone levels, and G6PDH levels as compared with the ovariectomized control. Out of the three dose levels was evaluated in ovariectomized rat model, the high dose of tuber extract indicated a notable increase in uterine glycogen content, uterine weight, 17-βestradiol and progesterone levels, G6PDH and LDH levels as well as major decrease in HDL and triglycerides levels compared to ovariectomized control. *F. vestita* can also be used as an alternative therapy for menopausal syndrome (Shailajan et al., 2016).

ACKNOWLEDGMENT

We are grateful to the Head, Department of Botany, Shivaji University, Kolhapur and the Principal, Swami Vivekanand Shikshan Santha's, Dattajirao Kadam Arts, Science and Commerce College, Ichalkaranji for providing research facilities. SKG thanks the Science and Engineering Research Board, New Delhi, for financial assistance vide sanction letter SERB/F/201/2014–15 dated 15.05.2014.

KEYWORDS

- ***Flemingia vestita***
- **anticancer**
- **biomolecules**
- **pharmacology**
- **antioxidant**

REFERENCES

Das, B.; Tandon, V.; Saha, N. Anthelmintic Efficacy of *Flemingia vestita* (Fabaceae): Alteration in the Activities of Some Glycolytic Enzymes in the Cestode, *Raillietina echinobothrida. Parasitol. Res.* **2004a,** *93* (4), 253–261.

Das, B.; Tandon, V.; Saha, N. Anthelmintic Efficacy of *Flemingia vestita* (Fabaceae): Alterations in Glucose Metabolism of the Cestode, *Raillietina echinobothrida. Parasitol. Int.* **2004b,** *53* (4), 345–350.

Das, B.; Tandon, V.; Saha, N. Effects of Phytochemicals of *Flemingia vestita* (Fabaceae) on Glucose 6-Phosphate Dehydrogenase and Enzymes of Gluconeogenesis in a Cestode (*Raillietina echinobothrida*). *Comp. Biochem. Physiol. Part C: Toxicol. Pharmacol.* **2004c,** *139*, 141–146.

Das, B.; Tandon, V.; Saha, N. Effect of Isoflavone from *Flemingia vestita* (Fabaceae) on the Ca^{2+} Homeostasis in *Raillietina echinobothrida*, the Cestode of Domestic Fowl. *Parasitol. Intern.* **2006,** *55* (1), 17–21.

Das, B.; Tandon, V.; Saha, N. Genistein from *Flemingia vestita* (Fabaceae) Enhances NO and Its Mediator (cGMP) Production in a Cestode Parasite, *Raillietina echinobothrida. Parasitology* **2007,** *134* (10), 1457–1463.

Das, B.; Tandon, V.; Lyndem, L. M.; Gray, A. I.; Ferro, V. A. Phytochemicals from *Flemingia vestita* (Fabaceae) and *Stephania glabra* (Menispermeaceae) Alter cGMP Concentration in the Cestode *Raillietina echinobothrida. Compar. Biochem. Physiol. Part C: Toxicol. Pharmacol.* **2009,** *149* (3), 397–403

Gavade, S. K.; Surveswaran, S.; Van der Maesen, L. J. G.; Lekhak, M. M. Taxonomic Revision and Molecular Phylogeny of *Flemingia* Subgenus *Rhynchosioides* (Leguminosae). *Blumea* **2019,** *64*, 253–271.

Gawade, B. H.; Pandey, A.; Khan, Z.; Nivedhitha, S.; Dubey, S. C. First Report on Nematocidal Properties of *Flemingia procumbens* Against *Meloidogyne incognita. Indian Phytopathol.* **2019,** *72* (3), 551–553.

Hazafa, A.; Rehman, K.; Jahan, N.; Jabeen Z. The Role of Polyphenol (flavonoids) Compounds in the Treatment of Cancer Cells, Nutrition and Cancer. *Nutr. Cancer* **2019,** *72* (3), 386–397.

Kar, P. K.; Tandon, V.; Saha, N. Anthelmintic Efficacy of *Flemingia vestita*: Genistein-Induced Effect on the Activity of Nitric Oxide Synthase and Nitric Oxide in the Trematode Parasite, *Fasciolopsisbuski. Parasitol. Int.* **2002,** *51* (3), 249–257.

Kar, P. K.; Tandon, V.; Saha, N. Anthelmintic Efficacy of Genistein, the Active Principle of *Flemingia vestita* (Fabaceae): Alterations in the Free Amino Acid Pool and Ammonia Levels in the Fluke, *Fasciolopsis buski. Parasitol. Int.* **2004,** *53* (4), 287–291.

Lyndem, L. M.; Tandon, V.; Das, B. Anthelmintic Efficacy of Medicinal Plants from Northeast India Against Hookworms: An In Vitro Study on *Ancylostoma ceylanicum. Pharmacology online* **2008,** *3*, 697–707.

Marboh, V.; Mahanta, C. Characterisation and Antioxidant Activity of Sohphlang (*Flemingia vestita*), a Tuberous Crop. *J. Food Sci. Technol.* **2020,** *57* (10), 3533–3544.

Pal, P.; Tandon, V. Anthelmintic Efficacy of *Flemingia vestita* (Fabaceae): Genistein-Induced Alterations in the Esterase Activity in the Cestode, *Raillietina echinobothrida. J. Biosci.* **1998a,** *23* (1), 25–31.

Pal, P.; Tandon, V. Anthelmintic Efficacy of *Flemingia vestita* (Fabaceae): Genistein-Induced Alternations in the Ultrastructure of the Tegument in the Cestode, *Raillietina echinobothrida. J. Parasit. Dis.* **1998b,** *22*, 104–109.
Rao, H. S. P.; Reddy, K. S. Isoflavones from *Flemingia vestita. Fitoterapia* **1991,** *62* (5), 458.
Roy, B.; Tandon, V. Effect of Root-Tuber Extract of *Flemingia vestita*, a Leguminous Plant, on *Artyfechinostomum sufrartyfex* and *Fasciolopsis buski*: A Scanning Electron Microscopy Study. *Parasitol. Res.* **1996,** *82* (3), 248–252.
Shailajan, S.; Kumaria, S.; Pednekar, S.; Menon, S.; Joshi, H.; Matani, A. Chromatographic Evaluation of a Phytoestrogen Genistein from *Flemingia vestita* Benth: An Endemic Plant of Northeast India. *Pharmacogn. Commun.***2014,** *4* (4), 2–9.
Shailajan, S.; Kumaria, S.; Pednekar, S.;Menon, S.; Choudhury, H.; Matani, A. Estrogenic Potential of *Flemingia vestita* Benth. Tubers in Ovariectomized Rat Model. *Pharmacogn. J.* **2016,** *8* (1), 44–49.
Standal, B. R.; Ako, H.; Standal, G. S. S. Nutrient Content of Tribal Foods from India: *Flemingia vestita* and *Perilla frutescens. J. Plant Foods* **1985,** *6* (3), 147–153.
Tandon, V.; Das, B. In vitro testing of anthelmintic efficacy of *Flemingia vestita* (Fabaceae) on carbohydrate metabolism in *Rallietina echinobothrida. Methods.* **2007,** *42*(4), 330–338.
Tandon, V.; Das, B.; Saha, N. Anthelmintic Efficacy of *Flemingia vestita* (Fabaceae): Effect of Genistein on Glycogen Metabolism in the Cestode, *Raillietina echinobothrida. Parasitol. Int.* **2003,** *52* (2), 179–183.
Tandon, V.; Pal, P.; Roy, B.; Rao, H. S. P.; Reddy, K. S. In Vitro Anthelmintic Activity of Root-Tuber Extract of *Flemingia vestita*, an Indigenous Plant in Shillong, India. *Parasitol. Res.* **1997,** *83* (5), 492–498.
Yadav, A. K.; Tandon, V.; Rao, H. S. P. In Vitro Anthelmintic Activity of Fresh Tuber Extract of *Flemingia vestita* Against *Ascaris suum. Fitoterapia* **1992,** *63* (5), 395–398.

CHAPTER 26

Psoralea corylifolia: A Herbal Drug with Promising Metabolites and Pharmacology Activity

SUPRIYA THALE[1*] and NUTAN MALPATHAK[2]

[1]*Department of Botany, V. G. Vaze College of Arts, Science and Commerce (Autonomous), Maharashtra, Mumbai 400081, India*

[2]*Department of Botany, Savitribai Phule Pune University, Pune 411007, India*

[*]*Corresponding author. E-mail: supriyathale@vazecollege.net; s.sups1983@gmail.com*

ABSTRACT

Psoralea corylifolia L. (Leguminosae) is one of the important medicinal plants in Chinese medicines and Ayurvedic formulations. It is a rich source of various metabolites like flavonoids, isoflavonoids, coumarins and terpenes. The analysis of various metabolites is carried out by using different types of secondary metabolite analysis technique like LC-MS, HPLC, DART-MS etc. This plant possesses various activities like antibacterial, antioxidant, antifungal, anti-depressant, and antitumor. This chapter gives a comprehensive overview of the various metabolites and pharmacological activities of *P. corylifolia*, which will be helpful for further research.

26.1 INTRODUCTION

Psoralea corylifolia L. (Indian bread root), commonly known as Babachi, is a herbaceous medicinal plant distributed in India. *P. corylifolia* is also known

Bioactives and Pharmacology of Legumes. T. Pullaiah, PhD (Ed.)

as Karkokil, Karkokilari, Kaurkoalari (Malayalam), Latakasturi, Kakuch, Barachi (Bengal), Babchi, Bavacha, Babichi (Gujrati), Bavanchigida, Somaraji, Karbekhiga (Karnataka), Karpokarishi, Karpuva-arishi, Karpuvanshi, Karpogalarisi (Tamil), and Babachi, Bavachi Bavachya, Bavanchi (Marathi) (Khusboo et al., 2010).

The plant is erect, perennial growing from 30 to 160 cm in height with firm branches bearing simple, broadly elliptical leaves, and rounded with mucronate apex. The leaves show presence of epidermal outgrowth in the form of white hairs. The leaf base is cuneate having 0.6–2.5 cm long petiole. It bears flowers in axillary racemes consisting of 10–30 purple blue flowers. It produces single-seeded, kidney-shaped fruits with slightly compressed black smooth seeds adhering to the pericarp.

P. corylifolia is one of the most valued medicinal plant with several pharmaceutical and nutraceutical applications. The plant is a rich source of various secondary metabolites principally flavonoids, isoflavonoids, and coumarins. The work on phytochemical constituents of *P. corylifolia* has led in the isolation and characterization of different active components and the evaluation of their biological properties.

26.2 BIOACTIVE CONSTITUENTS FROM *P. CORYLIFOLIA*

Liu et al. (2004) separated psoralen and isopsoralen by high-speed countercurrent chromatography and the identification of these compounds was carried out by NMR. The analysis of 70% ethanol extracts of seeds of *P. corylifolia* led to the identification of twelve components psoralen, isopsoralen, neobavaisoflavone, bovachin, bavaisoflavone, bavachromene, psoralidin, corylifolinin, bavachinin, bavachalcone, and an unknown compound (Zhao et al., 2005). The isolation of benzofuran glycosides psoralenoside and isopsoralenoside was done from the fruits of *P. corylifolia* and their structural elucidation was analyzed by 2D-NMR (Qiao et al., 2006). The dried fruits of *P. corylifolia* when extracted with 95% ethanol and through column chromatographic studies showed the presence of a new isoflavone corylinin along with the previous isolated compounds (Ruan et al., 2007).

Direct Analysis in Real Time Mass Spectroscopy was performed to analyze secondary metabolites from populations of *P. corylifolia* (seed, stem, root, and leaves) from Maharashtra region. Psoralen/isopsoralen, bakuchiol, daidzein, and genistein were the major metabolites present in the samples of *P. corylifolia*. The Orthogonal partial least square discriminatory analysis

analysis showed clear distinction in all four plant parts (Baikar, 2013). This is represented in Figures 1 and 2.

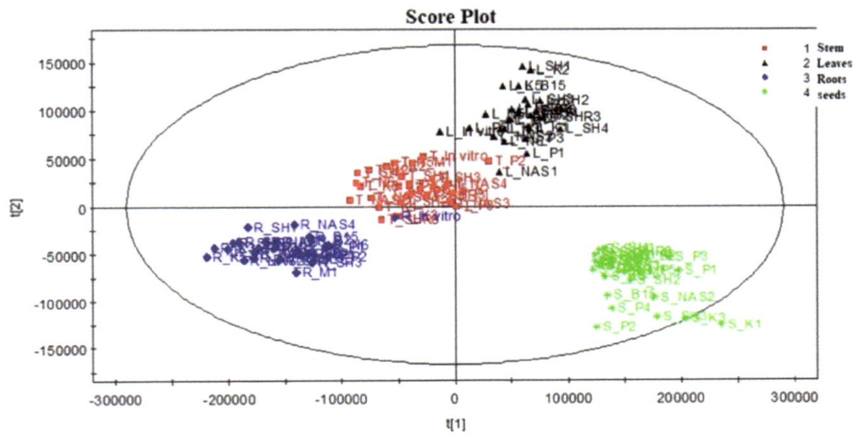

FIGURE 26.1 Classification study among four plant parts of *P. coryliforia* by OPLS/O2PLS-DA.

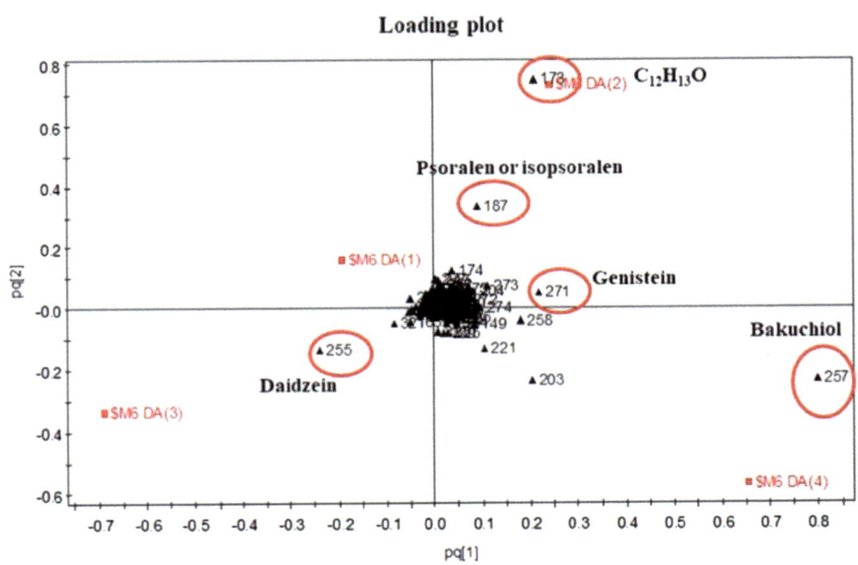

FIGURE 26.2 Loading plot of distinguishing metabolites from different plant parts of *P. coryliforia* nu OPLS/O2PLS-DA

Figures 1 and 2 are drawn by author.

The mass spectrometry analysis showed the presence of number of metabolites which were identified by matching the peaks with the spectral libraries. The mass spectral data is represented in Table 26.1.

TABLE 26.1 List of different metabolites from *P. corylifolia* reported through DART-MS analysis.

Compounds	Plant parts			
	Seed	Leaves	Stem	Roots
Daidzein	+	+	+	+
Genistein	+	+	+	+
Corylin	+	+	+	+
Corylifol A	+	+	+	+
Bavachromene	+	–	+	+
Bavachinin	+	–	+	–
7-methoxybava-chinin	–	–	–	+
Psoralidin	–	–	–	+
Bavachin	+	–	+	–
Psoralester	+	–	–	–
Psorachromene	+	–	–	–

26.3 PHARMACOLOGY

26.3.1 Photobiological Activity

Psoralen isolated from *P. corylifolia* is interesting from pharmaceutical point of view because of its photosensitizing, photobiological, and phototherapeutic property (Frank et al., 1998). It has been used cosmetically in tanning products and dermatologically for photochemotherapy for vitiligo and skin disorders like psoriasis, mycosis fungoides, and atopic eczema (Pathak and Fiztpatrick, 1992; Yones et al., 2005).

26.3.2 Antimicrobial Activity

Bakuchiol was reported to possess significant antibacterial activity against oral microorganisms (Kastura et al., 2001). Yin et al. (2004) reported that the prenylcorylifols A-C from *P. corylifolia* as antibacterial against pathogenic bacteria *Staphylococcus aureus* and *Staphylococcus epidermis*.

The antimicrobial activity of the active constituent's psoralen, psoralidin, backuchicin, and angelicin extracted from the seeds of *P. corylifolia* showed significant results on various Gram-positive and Gram-negative bacteria (Khatune et al., 2004).

Chanda et al. (2011) reported that the antibacterial activity of seed extract (dioxan extract) of *P. corylifolia*. Chopra et al. (2013) observed that the presence of bakuchiol in the extract of *P. corylifolia* seeds were effective against both Gram +ve bacteria and Gram−ve bacteria.

The significant effect of various extracts, purified fractions, and isolated compounds of *P. corylifolia* as potent antimicrobial against pathogenic strains of bacteria was reported by Baikar (2013). The experiments showed that the MIC of psoralen against *P. mirabilis* (pus cells) was 0.482 and 12.5 μg/mL against *Pseudomonas*, *Klebsiella* and *P. mirabilis* (urine cells). The MIC of stem fraction 1 against *E. coli* (pus and urine cells) was 12.5 and 50 μg/mL for *Salmonella typhi*. The antimicrobial assay showed that the compounds corylifol A, bakuchiol, and corylin purified from seed extracts of *P. corylifolia* showed minimum inhibitory concentration 25 μg/mL against all the tested bacteria.

Sharath et al. (2016) reported that the hexane and methanol extracts of *P. corylifolia* seed were effective against *S. aureus, Candida albicans*, and *Aspergillus niger*. The results showed unique characters of this plant in inhibiting microbial growth. He et al. (2018) reported the potential of corylifolinin isolated from *P. corylifolia* against *S. aureus*, Methcillin-resistant *S. aureus*, and extended spectrum β-lactamases *S. aureus*.

The investigation carried out by Li et al. (2019) showed the potentials of ethanol extracts of *P. corylifolia* seeds as effective antimicrobials against food-borne pathogens. The minimum inhibitory concentration obtained was 50 and 100 μg/mL, respectively, against both analyzed pathogens. The probable antimicrobial action of ethanol extracts of *P. corylifolia* seeds could be due to the depletion of energy by hampering the activity of adenosine triphosphate or by shuffling the proton gradient which could result in damaging of the membranes.

26.3.3 Antioxidant Activity

Bakuchiol isolated from seeds of *P. corylifolia* acted as a potent antioxidant in protecting the red blood cells from oxidative haemolysis. Thus these phenolic compounds from *P. corylifolia* acted as protectants against oxidative damage to biological membranes (Haraguchi et al., 2002). The antioxidant potential of compounds bakuchiol, psoralidin, and corylin isolated from seed extracts *P. corylifolia* were investigated in lard at 100°C by using Oxidative Stability Instrument demonstrated the strong antioxidant properties (Jiangning et al. 2005). Xaio et al. (2010) investigated the potent antioxidant compounds isolated from *P. corylifolia* by using a combination of different chromatographic techniques.

The results of plasmid nicking assay showed that crude extracts of seed stem and roots as well as isolated compounds from *P. corylifolia* are DNA protectors which act as antioxidants to prevent DNA damage by scavenging the reactive oxygen species. The results of DNA prevention assay indicate that the extracts, compounds, and fractions disrupt formation of hydroxyl radicals and acts as DNA protectors (Baikar, 2013).

Nabi et al. (2017) reported radical scavenging activity and antioxidant properties of *P. corylifolia* seed extract, which contains large amounts of phenol and flavonoid compounds. The in vitro antioxidant assay showed that *P. corylifolia* seed extract act as a promising source of natural antioxidant, which may help in preventing the harmful effects of oxidative stresses.

26.3.4 Hepatoprotective Activity

Dai et al. (2009) reported the effect of *P. corylifolia* extracts on the bile secretion and contraction of gall bladder muscles. Cho et al. (2001) reported the EC_{50} values of bakuchiol, bakuchincin, and psoralen 1.0, 47.0, and 50.0 μg/mL, respectively, for HepG2 human liver cells in the cytotoxicity-induced assay by tacrine. Silymarin that was used as the positive control showed EC_{50} value 5.0 μg/mL.

26.3.5 Antidepressant Activity

Furanocoumarins, where reported to possess antidepressant effect by controlling monoamine oxidase activity, hypothalamic-pituitary-adrenal axis function, and oxidative stress (Chen et al., 2005, 2007, 2008). Xu et al. (2008) investigated that Psoralen could substantially reduce the immobility and increase swimming without altering climbing in the mouse forced swimming test (FST). They also showed that psoralen altered the effect of serotonin (5-HT) and 5-hydroxyindoleacetic acid levels in frontal cortex and hippocampus in mice in the FST. Psoralen has also effectively reduced the FST-induced rise in serum) and corticosterone concentrations to regularise the HPA axis activity.

26.3.6 Antitumor Activity

The work on the anticancer property of *P. corylifolia* extracts has been evaluated by many researchers (Yang et al., 1996; Guo et al., 2003; Bapat et al., 2005; Wang et al., 2011). Lee et al. (2009) reported that psoralidin, possessed the activity of Quinone Reductase in Hepa- 1c1c7 murine hepatoma cell line. The IC_{50} values of psoralidin were reported to be 0.3, 0.4, 53, and 203 μg/mL against HT-29 (colon) human cancer cell line, MCF-7 (breast) human cancer cell line, SNU- 1 carcinoma cell line, and SNU-16 carcinoma cell line (Yang et al. 1996; Latha et al., 1999, 2000). Another study showed that isobavachalcone/corylifolin may be used as an effective and harmless drug as it induced the apoptotic cell death by modifying the mitochondrial pathway in neuroblastoma as it had no cytotoxicity against normal cells (Nishimura et al., 2007). Wu et al. (2008) reported the effect *O*-Methyl- bakuchiols and *O*-ethyl-bakuchiols on inhibition of HIF-1 (IC_{50} values: 8.7 and 26.3 μM, resp.) and NF-κB (IC_{50} values: 5.7 and 12.2 μM, respectively) without significantly hampering the viability of the human gastric cancer cell and human cervical adenocarcinoma. The ethanolic extract consisting bakuchiol as the activity constituent of *P. corylifolia* was found to be cytotoxic against L929-cells in cell culture (Iwamura et al., 1989; Kubo et al., 1989).

26.3.7 Topoisomerase Inhibitory Activity

Sun et al. (1998) reported daidzein and bakuchicin as weak inhibitors of topoisomerase II, which is the only report on action of metabolites isolated from *P. corylifolia* on inhibitiory activity of topoisomerase II activity.

Topoisomerase I inhibitors from *P. corylifolia* are reported (Baikar, 2013). The experiments revealed that compounds psoralen, bakuchiol, corylifol A, and corylin are the potent inhibitors of topoisomerase I (*Escherichia coli*). Bakuchiol, corylifol A, and corylin are the potent inhibitors of human topoisomerase I but did not inhibit human topoisomerase II activity. Stem fractions and root fractions proved effective in inhibiting the activity of human topoisomerase I but did not inhibit topoisomerase I (*E. coli*). Thus it can be concluded that compounds, semipurified fractions, and ethanol stem and root extracts are potent inhibitors of topoisomerase I (human and *E. coli*) and weak inhibitors of topoisomerase II.

26.3.8 DNA Polymerase Inhibitory Activity

Sun et al. (1998) reported that bakuchiol present in ethanol seed extracts of *P. corylifolia* may be one of the major constituent in inhibiting polymerase activity of Simian virus 40 along with corylifolin, daidzein, and neobavaisoflavone, which showed minor contribution in inhibiting polymerase activity.

The experiments carried out on inhibition of DNA polymerase (Klenow fragment) activity by using a fluorescence dye revealed that bakuchiol and psoralen were the major constituents while daidzein and genistein showed minor contribution in inhibition of DNA polymerase activity. The experiments also indicated that crude extracts of *P. corylifolia* were significant inhibitors of polymerase as compared to semipurified fractions and isolated compounds. The order of potency of DNA polymerase inhibition for analyzed samples was crude extracts > semipurified fraction > isolated compounds > CPT > reference compound (daidzein and genistein) (Baikar, 2013).

KEYWORDS

- **DART-MS**
- **topoisomerase**
- **DNA polymerase**
- **psoralen**
- **daidzein**
- **genistein**

REFERENCES

Baikar, S. K. Diversity Analysis of *Psoralea corylifolia* Linn. Syn *Cullen corylifolia* and Screening for DNA Polymerase I Inhibitor Activity. Ph.D Thesis, Savitribai Phule Pune University, 2013. http://hdl.handle.net/10603/104212.

Bapat, K.; Chintalwar, G. J.; Pandey, U.; Thakur, V. S.; Sarma, H. D.; Samuel, G.; Pillai, M. R. A.; Chattopadhya, S.; Venkatesh, M. Preparation and in vitro evaluation of Radio Iodinated Bakuchiol as an Anti Tumor Agent. *Appl. Radiat. Isotopes* **2005,** *62* (3), 389–393.

Chanda, S.; Kaneria, M.; Nair, S. Antibacterial Activity of *Psoralea corylifolia* Seed and Aerial Parts with Various Extraction Methods. *Res. J. Microbiol.* **2011,** *6* (2), 124–131.

Chen, Y.; Kong, L. D.; Xia, X.; Kung, H. F.; Zhang, L. Behavioral and Biochemical Studies of Total Furocoumarins from Seeds of *Psoralea corylifolia* in the Forced Swimming Test in Mice. *J. Ethnopharmacol.* **2005,** *96* (3), 451–459.

Chen, Y.; Wang, H. D.; Xia, X.; Kung, H. F., Pan, Y.; Kong, L. D. Behavioral and Biochemical Studies of Total Furocoumarins from Seeds of *Psoralea corylifolia* in the Chronic Mild Stress Model of Depression in Mice. *Phytomedicine.* **2007,** *14* (7–8), 523–529.

Chen, Y. C.; Cheung, Y. T.; Kong, L. D.; Ng, T. B.; Qiao C. F.; Mo, S. F., Xu, H, X. X.; Kung, H. F. Transcriptional Regulation of Corticotrophin Releasing Factor Gene by Furocoumarins Isolated from Seeds of *Psoralea corylifolia. Life Sci.* **2008,** *82* (21–22), 1117–1121.

Cho, H.; Jun, J. Y.; Song, E. K.; Kang, K. H.; Ko, Y. S.; Kim, Y. C. Bakuchiol: A Hepatoprotective Compound of *Psoralea corylifolia* on Tacrine-Induced Cytotoxicity in Hep G2 Cells. *Planta Medica.* **2001,** *67* (8), 750–751.

Chopra, B.; Dhingra, A. K.; Dhar, K. L. Antimicrobial Activity of *Psoralea corylifolia* Linn. (Baguchi) Seeds Extracts by Organic Solvents and Supercritical Fluids. *Intern. J. Pharma. Clin. Res.* **2013,** *5* (1), 13–16.

Dai, Z.; Zhao, C.; Zhou, K.; Qu, C. Q.; Hu, L. M. Effect of fructus *Psoraleae* on Rats' Bile Secretion and Contractile Activity of Gallbladder Smooth Muscle of Guinea-Pigs. *Strait Pharma. J.* **2009,** *22* (4), 37–38.

Guo, J. N.; Wu, H.; Weng, X. C.; Yan, J. H.; Bi, K. S. Studies on Extraction and Isolation of Active Constituents from *Psoralen corylifolia* L. and the Antitumor Effect of the Constituents In Vitro. *J. Chinese Medicinal Mater.* **2003,** *26* (3), 185–187.

Haraguchi, H.; Inoue, J.; Tamura, Y.; Mizutani, K. Antioxidative Components of *Psoralea corylifolia* (Leguminosae). *Phytother. Res.* **2002,** *6*, 539–544.

He, N.; Zhou, J.; Hu, M.; Ma, C.; Kang, W. The Mechanism of Antibacterial Activity of Corylifolinin Against Three Clinical Bacteria from *Psoralea corylifolia* L. *Open Chem.* **2018,** *16*, 882–889.

Iwamura, J.; Dohi, T.; Tanaka, H.; Odani, T.; Kubo, M. Cytotoxicity of Corylifoliae Fructus. II. Cytotoxicity of Bakuchiol and the Analogues. *J. Pharma. Soc. Jpn.* **1989,** *109* (12), 962–965.

Jiangning, G.; Xinchu, W.; Hou, W.; Qinghua, L.; Kaishun, B. Antioxidants from a Chinese Medicinal Herb—*Psoralea corylifolia* L. *Food Chem.* **2005,** *91*, 287–292.

Katsura, H.; Tsukiyama, R.I; Suzuki, A.; Kobayashi, M. In Vitro Antimicrobial Activities of Bakuchiol Against Oral Microorganisms. *Antimicrob. Agents Chemother.* **2001,** *45* (11), 3009–3013.

Khatune, N. A.; Islam, M. E.; Haque, M. E.; Khondkar, P.; Rahman, M. M. Antibacterial Compounds from the Seeds of *Psoralea corylifolia. Fitoterapia.* **2004,** *75*, 228–230.

Khushboo, P. S.; Jadhav, V. M.; Kadam, V. J.; Sathe, N. S. *Psoralea corylifolia* Linn.–"Kushtanashini". *Pharmacogn Rev.* **2010,** *4* (7), 69–76.

Kubo, T.; Dohi, T.; Odani, H.; Tanaka, H; Iwamura, J. Cytotoxicity of Corylifoliae Fructus. I. Isolation of the Effective Compound and the Cytotoxicity. *J. Pharma. Soc. Jpn.* **1989,** *109* (12), 926–931.

Latha, P. G.; Panikkar, K. R. Inhibition of Chemical Carcinogenesis by *Psoralea corylifolia* Seeds. *J. Ethnopharmacol.* **1999,** *68* (1–3), 295–298.

Latha, P. G.; Evans, D. A.; Panikkar, K. R.; Jayavardhanan, K. K. Immunomodulatory and Antitumor Properties of *Psoralea corylifolia* Seeds. *Fitoterapia* **2000,** *71*, 223–231.

Lee, S. J.; Nam, K. W.; Mar, W. Induction of Quinone Reductase Activity by Psoralidin Isolated from *Psoralea corylifolia* in Mouse Hepa 1c1c7 Cells. *Arch. Pharmacal Res.* **2009,** *32* (7), 1061–1065.

Li, H. N.; Wang, C. Y.; Wang, C. L.; Chou, C. H.; Leu, Y. L.; Chen, B. Y. Antimicrobial Effects and Mechanisms of Ethanol Extracts of *Psoralea corylifolia* Seeds Against *Listeria monocytogenes* and Methicillin-Resistant *Staphylococcus aureus*. *Food borne Pathogens Dis.* **2019,** *16* (8), 573–580.

Liu, R.; Li, A.; Sun, A.; Kong, L. Preparative Isolation and Purification of Psoralen and Isopsoralen from *Psoralea corylifolia* by High-Speed Counter-Current Chromatography. *J. Chromatogr. A.* **2004,** *1057*, 225–228.

Nabi, N.G; Shrivastava, M. Phytochemical Screening and Antioxidant Activity of Ethanol Extract of *Psoralea corylifolia* Seeds. *UK J. Pharma. Biosci.* **2017,** *5* (2), 1–7.

Nishimura, R.; Tabata, K.; Arakawa, M.; Ito, Y.; Kimura, Y.; Akihisa, T.; Nagai, H.; Sakuma, A.; Kohno, H.; Suzuki, T. Isobavachalcone, a Chalcone Constituent of *Angelica keiskei*, Induces Apoptosis in Neuroblastoma. *Biol. Pharma. Bull.* **2007,** *30* (10), 1878–1883.

Pathak, M. A.; Fitzpatrick, T. B. The Evolution of Photochemotherapy with Psoralens and UVA (PUBA): 2000 BC to 1992 AD. *J. Photochem. Photobiol. B.* **1992,** *14*, 3–22.

Qiao, C. F.; Han, Q. B.; Mo, S. F.; Song, J. Z.; Xu, L. J.; Chen, S. L.; Yang, D. J.; Kong, L. D.; Kung; Xu, H. X. Psoralenoside and Isopsoralenoside, Two Benzofuran Glycosides from *Psoralea corylifolia*. *Chem. Pharm. Bull.* **2006,** 54(5), 714–716.

Ruan, B.; Kong, L. Y.; Takaya. Y.; Niwa, M. Studies on the Chemical Constituents of *Psoralea corylifolia* L. *J. Asian Nat. Prod. Res.* **2007,** *9* (1), 41–44.

Sharath, K.; Krishna, M. G.; Sandhya, R. M.; Kowmudi, V.; Suresh, N.; Chintanippula, S. R. R. Evaluation of Antibacterial and Anti-Fungal Activity of Hexane and Methanol Extracts of *Psoralea corylifolia* Seed. *Intern. J. Appl. Pharma. Sci. Res.* **2016,** *1* (1), 25–30.

Sun, N. J.; Woo, S. H.; Cassady, J. M.; Snapka, R. M. DNA Polymerase and Topoisomerase II Inhibitors from *Psoralea corylifolia*. *J. Nat. Prod.* **1998,** *61*, 362–366.

Wang, Y.; Hong, C.; Zhou, C.; Xu, D.; Qu, H. B. Screening Antitumor Compounds Psoralen and Iso-Psoralen from *Psoralea corylifolia* L. Seeds. *Evidence-Based Complem. Altern. Med.* **2011,** Article ID 363052.

Wu, C. Z.; Hong, S. S.; Cai, X. F.; Dat, N. T.; Nan, X. J.; Hwang, B. Y.; Lee, JJ.; Lee, D. Hypoxia-Inducible Factor- 1 and Nuclear Factor-κB Inhibitory Meroterpene Analogues of Bakuchiol, a Constituent of the Seeds of *Psoralea corylifolia*. *Bioorg. Medicinal Chem. Letters.* **2008,** *18* (8), 2619–2623.

Xiao, G.; Li, G.; Chen, L.; Zhang, Z.; Yin, J.J; Wu, T.; Cheng Z.; Wei, X.; Wang, Z. Isolation of Antioxidants from *Psoralea corylifolia* Fruits Using High-Speed Counter Current Chromatography Guided by Thin Layer Chromatography Antioxidant Autographic Assay. *J. Chromatogr. A,* **2010,** *1217* (34), 5470–5476.

Xu, Q.; Pan, Y.; Yi, L. T.; Li, C. Y.; Mo, S. F.; Jiang, F. X.; Qiao, C. F.; Xu, H. X.; Lu, X. B.; Kong, L. D.; Kung, H. F. Antidepressant-Like Effects of Psoralen Isolated from the Seeds of *Psoralea corylifolia* in the Mouse Forced Swimming Test. *Biol. Pharma. Bull.* **2008,** *31* (6), 1109–1114.

Yang, Y. M.; Hyun, J. W.; Sung, M. S.; Chung, SH.; Kim, B. K.; Paik, W. H.; Kang W. H.; Park B. K. The Cytotoxicity of Psoralidin from *Psoralea corylifolia*. *Planta Medica.* **1996,** *62* (4), 353–354.

Yin, S.; Fan, C. Q.; Wang, Y.; Dong, L.; Yue J. M. Antibacterial Prenylflavone Derivatives from *Psoralea corylifolia,* and Their Structure–Activity Relationship Study. *Bioorg. Medicinal Chem.* **2004,** *12*, 4387–4392.

Yones, S. S.; Palmer, R.; Kuno, K.; Hawk, J. L. M. Audit of the Use of Psoralen Photochemotherapy (PUVA) and Narrowband UVB Phototherapy in the Treatment of Psoriasis. *Dermatol. Treat.* **2005,** *16*, 108–112.

Zhao, L.; Huang, C.; Shan, Z.; Xiang, B.; Mei, L. Fingerprint Analysis of *Psoralea corylifolia* L. by HPLC and LC–MS. *J. Chromatography B*. **2005,** *821*, 67–74.

CHAPTER 27

Albizia lebbeck (L.) Benth.: *Shirish* Tree with Its Phytochemistry, Ethnobotany and Pharmacology

DIGAMBAR N. MOKAT* and TAI D. KHARAT

Department of Botany, Savitribai Phule Pune University, Pune 411007, Maharashtra, India

**Corresponding author. E-mail: parimalsai@gmail.com*

ABSTRACT

The bark, leaves, seeds and flowers of *A. lebbeck* are used in traditional systems of medicine for treating various diseases and disorders. Ethnobotanically leaf paste, bark and latex are used for treating eczema, acne and conjunctivitis respectively. The different parts the tree shows numerous pharmaceutical activities such as antimicrobial, anti-inflammatory, immune-modulatory, antiarthritic, antiallergic, antioxidant, anticancer, antivenom, antihistamine, antitumor and etc. The phytochemistry, ethnobotany and pharmacology of the tree *A. lebbeck* are given in this chapter.

27.1 INTRODUCTION

Albizia lebbeck (L.) Benth. is a large deciduous legume tree, commonly known as *Shirish*, belonging family Fabaceae. It has synonyms such as *Acacia lebbeck* (L.) Willd., *Acacia macrophylla* Bunge, *Albizia latifolia* Boivin, *Acacia speciosa* (Jacq.) Willd., *Albizia speciosa* (Jacq.) Benth. etc. It is a very significant plant for production of herbal medicines and having

Bioactives and Pharmacology of Legumes. T. Pullaiah, PhD (Ed.)

several industrial uses. It is found in tropical and subtropical parts of India, Andaman Island, Myanmar, Tropical Africa, Asia, and Northern Australia. The bark is gray-brown in color and usually cracked. The leaves are bipinnate, leaflets subopposite with two to four pairs and ultimate leaflets opposite with six to eight pairs as well as oblong. The showy flowers are rounded in clusters nearby stem tips, color ranges from cream or yellowish-white. Fruit a flat as well as linear pod with many seeds; dried pods persistent after leaf-fall on tree.

27.2 ETHNOBOTANY

Leaf paste of *A. lebbeck* is applied to cure eczema (Thirumalai et al., 2011). The bark is ground with roots of *Curcuma longa* (Haldi) with water and used to cure acne (Gul et al., 2012; Ahmed, 2007). Latex of *A. lebbeck* is mixed with milk of sheep and used as eye drops to cure conjunctivitis (Galav et al., 2013).

27.3 PHYTOCHEMISTRY

The phytochemical investigation of *A. lebbeck* revealed the presence of alkaloids, terpenoids, saponins, flavonoids, steroids as well as glycosides in the methanolic, petroleum ether, and ethyl acetate extracts of *A. lebbeck* leaves. There are total 10 different alkaloids present in petroleum ether extracts of leaves detected by HPTLC analysis (Bobby et al., 2012). Two new tri-O-glycoside flavonols such as kaempferol and quercetin 3-O-alpha-rhamnopyranosyl (1->6)-beta-glucopyranosyl (1->6)-beta- galactopyranosides were isolated and identified of leaves of *A. lebbeck* (El–Mousallamy, 1998). Triterpenoid saponins were found in different *Albizia* species. Lebbekanin-E was also found in *A. lebbeck* (Varshney et al., 1976). A hexaglycosylated saponin albizziahexoside was isolated from *A. lebbeck* leaves (Veda et al., 2003).

Major antinutrient saponin is present in fruits and seeds. Potassium was found in the maximum amount and copper in the meager amount. Amino acid, lysine and arginine were found in extreme amounts in seeds, though aspartic acid and glutamic acid are present in the uppermost concentrations in the pods. The linoleic acid was noticed as the main fatty acid in fruit and seed oils. α-tocopherol was determined as the main tocopherol constituent in oil, which showed potent antioxidant potential (Zia-Ul-Haq et al., 2013).

The oil from leaves contains 24 compounds, which constituted about 91.2% of the oil and was dominated by aromatic hydrocarbons (35.9%), oxygenated monoterpenes (32.3%), n-hydrocarbons (13.7%), alkanals (8.1%), and sesquiterpenes (0.4%). The oil of flowers showed the presence of 14 compounds, which constituted about 96.2% of the oil and was dominated by n-hydrocarbons (54.4%), aromatic hydrocarbons (14.1%), alkanals (13.2%), and oxygenated monoterpenes (8.2%) (Aiyelaagbe et al., 2010).

Fatty acid profiles of seed oils in *A. lebbeck* and *Albizia saman* reported in various literatures. The oils were analyzed by GC-MS, GC, and NMR. The maximum fatty acid in both oils is linoleic acid (30–40%), followed by palmitic acid and oleic acid of *A. lebbeck* and oleic acid and behenic acid of *A. saman,* respectively. Both oils contain slightly more than 30% saturated fatty acids with eicosanoic, tetracosanoic, and stearic acids present as well as odd-numbered saturated fatty acids reported in less amounts (Knothe et al., 2015). Phytochemical analysis of *A. lebbeck* stem bark in different solvents showed the presence of tannins, alkaloids, terpenoids, glycosides, phenols, steroids, flavonoids, saponins, proteins, and carbohydrates (Aisha et al., 2020).

27.4 PHARMACOLOGY

The plant possesses a wide range of pharmaceutical activities such as antimicrobial, anti- inflammatory, immunomodulatory, antiarthritic, antiallergic, antioxidant, anticancer, antivenom, steroid-genic, antihistaminic, antitumor, etc. *A. lebbeck* extract inhibited the passive cutaneous anaphylaxis, mast cell degranulation in rat dose powerlessly and could protect the informed guinea pig from antigen-induced anoxic convulsion (Baruach et al., 1997). *A. lebbeck* is very significant medicinal plant by tradition. Various Ayurvedic preparation contains *A. lebbeck* parts such as "*Sirisa twak kvatha,*" "*Vasadikwath,*" antiasthma "*kadha,*" etc. *A. lebbeck* contains alkaloids, tannins, flavonoids, and saponins that have beneficial value. Hence it is used pharmacologically on numerous isolated unadulterated compounds (Mohammad et al., 2012).

27.4.1 Antiallergic Activity

Previous studies have detected the antiallergic activity of *A. lebbeck* extract by using catechin as a standard compound at IC_{50} value at conc. of 85 μg/mL. This inhibitory potential of catechin from *A. lebbeck* may be due to modulation of two very important effector's functions, histamine discharge and cytokine appearance of antigen -IgE stimulated mast cells (Venkatesh et al., 2010).

Histamine plays the most important roles in sensitized (allergic) disease and its stroke is mediated mostly through histamine H1 receptor. Authors verified that histamine signaling-related histamine H1 receptor and histidine decarboxylase genes are allergic syndromes sensitive genes and their appearance level disturbs severity of the allergic indications. Therefore, compounds that overpower histamine signaling must be situated capable applicants as antiallergic drugs. Islam et al. (2011) studied the influence of *A. lebbeck* bark extract, one of the constituents of Ayruvedic medicines, on histamine H1 receptor, and histidine decarboxylase gene expression through toluene-2,4-diisocyanate sensitized allergy model rats and HeLa cells expressing endogenous histamine H1 receptor. Administration of the *A. lebbeck* extract significantly reduced the records of sneezing and adenoidal release. Pretreatment with the *A. lebbeck* extract suppressed toluene-2,4-diisocyanate-induced histamine H1 receptor and histidine decarboxylase mRNA raises and [3H] mepyramine binding, histidine decarboxylase activity, and histamine content in the nasal mucosa. *A. lebbeck* extract also repressed toluene-2,4-diisocyanate-induced up-regulation of IL-4, IL-5, and IL-13 mRNA. In HeLa cells, *A. lebbeck* extract suppressed phorbol-12-myristate-13-acetate- or histamine-induced upregulation of histamine H1 receptor mRNA. So that these data suggest that *A. lebbeck* alleviated nasal symptoms by inhibiting histamine signaling in toluene-2,4-diisocyanate-sensitized rats through suppression of histamine H1 receptor and histidine decarboxylase gene transcriptions. Suppression of Th2-cytokine signaling by *A. lebbeck* also proposes that it could affect the histamine–cytokine network.

Studies showed that the bark of *A. lebbeck* has been in use by Ayurvedic physicians for respiratory asthma and eczema. The outcome of *A. lebbeck* was investigated on the degranulation rate of sensitized peritoneal mast cells of albino rats as soon as defied with antigen (horse serum). Disodium cromoglycate and prednisolone were used for assessment. Medicines were given during the 1st or 2nd week of sensitization and the mast cells investigated at the end of the 2nd or 3rd week. Serum from these rats was used to passively sensitize recipient rats whose peritoneal mast cells were then studied. The *in-vitro* effects of *A. lebbeck* and Disodium cromoglycate on the degranulation rate of the sensitized mast cells were studied. The outcomes display that *A. lebbeck* has an important cromoglycate-like action on the mast cells. In addition, it inhibits the sensitization and synthesis of reaginic-type antibodies.

If *A. lebbeck* is given during the 1st week of sensitization it decidedly inhibits the initial sensitizing procedures, although if given through the second week it suppresses antibody production (Tripathi et al., 1979).

27.4.2 In Neurodegenerative Disorder: Parkinson's Disease

A. lebbeck improved the motor functions and endurance in Parkinson's disease *in-vivo* (Saleem et al., 2019).

27.4.3 Antivenom Property

The seed extracts were used to screen antivenom properties and the results revealed that *A. lebbeck* methanolic extract has principal role in antivenom activity (Amog et al., 2016).

27.4.4 Antimicrobial Activity

Phosphate-buffered saline (PBS) protein extract of *A. lebbeck* showed the highest antibacterial activity against *Bacillus subtilis* and *Bacillus cereus*. Antibacterial activity of these protein extracts suggests that these extracts can be used against the bacterial infection to improve human health (Habiba et al., 2021). Methanolic extracts of roots, flowers, pods, and seeds of *A. lebbeck* and *A. leucophloea* were tested *in-vitro* for their antibacterial and antifungal activities. The antibacterial study was performed against six bacterial species of both positive and negative gram types such as *Escherichia coli, Staphylococcus aureus, P. aeruginosa, Staphylococcus typhi, Proteus mirabilis,* and *B. subtilis*, when associated with gentamicin and gatifloxacin, presented that considered plants have strong activity against all of the tested microorganisms. The antifungal activity of the extracts was tested against six fungal strains such as *Aspergillus parasiticus, Aspergillus niger, Candida albicans, Aspergillus effusus, Fusarium solani,* and *Saccharomyces cerevisiae* and paralleled with Itraconazole and Amphoteracin-B. The extracts displayed important activity against all fungal strains. The order of antibacterial and antifungal activity, communicated as Minimum inhibitory concentration (MIC) detected was seed>fruit>flower > roots for all bacterial and fungal strains were tested (Shahid and Firdous, 2012).

27.4.5 Anti-Inflammatory Activity

The flowers of the plant showed estrogenic and anti-inflammatory activities rising in Saudi Arabia (Farag et al., 2013). The extracts were found by using different solvent such as petroleum ether, chloroform as well as ethanol at different concentrations, that is, 100, 200, and 400 mg/kg of body weight. The petroleum ether and ethanol extracts at 400mg/kg, displayed extreme inhibition of inflammation encouraged by carrageenan (petroleum ether-48.6%; ethanol-59.57%), dextran (petroleum ether-45.99%; ethanol-52.93%), cotton pellet (petroleum ether-34.46%; ethanol-53.57%), and Freund's adjuvant (petroleum ether-64.97%; ethanol-68.57%). The noticeable inhibitory result on paw edema displays that *A. lebbeck* possesses notable anti-inflammatory activity (Babu et al, 2009).

A. lebbeck bark extract with a mixture of equal amounts of ethyl acetate, methanol, and petroleum ether were selected for pharmacological showing. In rat paw edema model induced by carrageenan, the extract at 400 mg/kg dose level indicated 36.68% inhibition of edema volume at the end of 4 h. In the acetic acid-induced writhing test, the extract at 200 and 400 mg/kg dose level displayed 39.9% and 52.4% inhibition of writhing, individually. In radiant heat tail-flick method, the crude extract formed 40.74% and 61.48% elongation of tail flicking time 30 min after oral administration at 200 and 400 mg/kg dose level (Saha and Ahmed, 2009).

27.4.6 Ovicidal Activity

The methanol extract exhibited ovicidal and adulticidal activity. The maximum adulticidal activity was detected in the leaf methanol extract against *Anopheles stephensi* where the LC_{50} and LC_{90} values were 65.12 and 117.70 ppm (Govindarajan et al., 2015).

27.4.7 Antioxidant Activity

The ethanolic root extract exhibited antioxidant activity possessing IC_{50} values (H_2O_2 scavenging activity), 345.94 (nitric oxide scavenging activity), 945.76 (DPPH free radical scavenging activity). In ferric reducing assay, the absorbance was found to increase with increase in concentration of the extract, revealing its reducing power (Priyanka et al., 2013). Experimental study discovered that *A. lebbeck* caused antidiabetic consequence at doses

25–125 mg/kg body weight and 75 mg was found to be the effective dose (Reshmi et al., 2006).

27.4.8 Anticancer Activity

Ethanolic extract has a powerful antitumor activity, because it is effective for inhibition of tumor growth both *in-vivo* and *in-vitro* cancer cell lines (Kavitha et al., 2021). Phyto-constituents arbitrated synthesis of gold nanoparticles (AuNPs) of *A. lebbeck* aqueous leaf extract by the green method revealed the potential anticancer activity against colon cancer (HCT-116) cells (Malaikolundhan et al., 2020).

27.4.9 Anti-Passive cutaneous anaphylaxis (PCA) Activity

Hot aqueous extracts and butanolic fractions of *A. lebbeck* were investigated for the anti-PCA activity in mice and rats used as guinea pig and rat antisera. 50 mg/kg p.o., dose rate, there was 74% and 66% activity, respectively. The effect of saponin containing with n-butanolic fraction extracted from dried bark of *A. lebbeck* was deliberate on cognitive behavior and anxiety in albino mice. A raised plus maze was used for the valuation of both nootropic and anxiolytic activity (Barucah et al., 2000).

27.4.10 Immunomodulatory Activity

The immunomodulatory influence of the *A. lebbeck* bark was assessed by learning humoral and cell-mediated immune replies. The hot aqueous extract as well as butanolic fraction were administered once on day-to-day for one week in mice immunized formerly using red blood cells of sheep. At the dose levels tested (6.25, 12.5, and 25 mg/kg, p.o.), *A. lebbeck* treated mice developed maximum serum antibody titres related to the vehicle-treated group and the influence was similar to the standard drug muramyl dipeptide. Late nature hypersensitivity response was suppressed in red blood cells of sheep immunized mice treated with *A. lebbeck* extract. These outcomes are deliberated in the light of possible immune-potentiating effects of *A. lebbeck* (Barua et al., 2000).

Immunomodulator activity of aqueous and ethanolic extracts of bark and leaves of *A. lebbeck* were studied in Swiss albino mice used by swim endurance test and acetic acid-induced writhing test model. The aqueous and ethanolic extracts of bark and leaves of *A. lebbeck* were managed to the investigational animals amongst which the ethanolic extract of *A. lebbeck* leaves revealed to show strong immunomodulator effect by increased the swimming or survival time and reduced the writhing formed by glacial acetic acid. The extreme increase in swimming or survival time was renowned in mice getting test and standard drugs, which were expressively more than control. Test and standard drugs offered extreme defense against acetic acid-induced writhes by dropping frequency of writhes per minute (Chaudhary et al., 2012).

27.4.11 Spermicidal Activity

The 2% concentration of ethanolic extract of pods and roots, which contains saponins, lebbekanin-E exhibited spermicidal activity in human and rat's semen (Ganguli and Bhatt, 1993). Electrophoretic changes were experiential on the protein profiles of somniferous tubules fluid and epididymis fluid from caput and caudo areas afterward the administration of alcoholic extract of *A. lebbeck* dry seed (Singh et al., 1991).

Oral administration of saponins isolated by *A. lebbeck* bark of 50 mg/kg/b.w. dose level per day for 60 days to male rats brought about a significant decrease in the weights of testes, epididymides, seminal vesicle, and ventral prostate. The production of round spermatid was condensed by 73.04% in *A. lebbeck* treated rats. The population of spermatogonia and preleptotene spermatocytes were condensed by 47.48% and 65.07% and secondary spermatocytes by 73.41%, respectively. Cross sectional surface area of Sertoli cells and the cell totals were resulted to be depleted importantly. Leydig cell nuclear zone and total figure of matured Leydig cells were reduced by 57.47% and 54.42%, respectively. Sperm motility and sperm thickness were reduced significantly. *A. lebbeck* abridged the fertility of male rats by 100%. There were no significant modifications in Red Blood Cell and White Blood Cell total, haemoglobin, haematocrit, and glucose in the blood and protein, cholesterol, phospholipid, and triglyceride in the serum. The cholesterol, protein, and glycogen contents of the testes, protein in epididymides, and fructose in the seminal vesicle were significantly reduced. Histoarchitecture of the testes showed vacuolization at primary spermatocytes stage. Extremely condensed seminiferous tubular diameter and increased intertubular space were also detected when as compared with controls (Gupta et al., 2005).

A. lebbeck methanolic extract of bark when directed orally at the dose 100 mg/day /rat to male rats of verified fertility for 60 days did not cause any significant harm in their weights of body nonetheless the weights of reproductive organs, that is, epididymides, testis ventral prostate, and seminal vesicle were reduced in a significant manner as compared to controls. Motility of sperm as well as density of sperm were decreased significantly, which resulted in decrease of male fertility by 100%. Noticeable weakening in the germ cell population was observed. Population of pachytene, preleptotene, secondary spermatocytes, and step-19 spermatid were weakened by 65.81%, 60.86%, 71.56%, and 66.55%, respectively. Cross-sectional surface zones of sertoli cells and the cells totals number were found to be tired significantly. Leydig cells nuclear zone and number of matured Leydig cells were condensed by 60.03% and 51.56%, separately. Serum testosterone ranks exhibited significant decrease after *A. lebbeck* extract serving. Oral administration of the extract didn't affect red and white blood cell count, haemoglobin, haematocrit, and glucose in the blood and protein, cholesterol, phospholipid, and triglyceride in the serum. In assumption, *A. lebbeck* bark extracts administration captures spermatogenesis in male rats lacking conspicuous any side-effects (Gupta et al., 2006).

27.4.12 Nootropic Activity

The effect of saponin having n-butanolic fraction extracted from *A. lebbeck* dried leaves on learning and memory was investigated in albino mice by used passive shockwave prevention paradigm and the raised up plus maze. Important development was observed in the retaining capability of the normal and amnesic mice as related to their particular panels and also studied the effects of n-butanolic fraction on the behavior influenced by serotonin (5-HT), dopamine, and noradrenaline. The brain levels of gamma-aminobutyric acid, serotonin and dopamine were also assessed to compare the behavior with neurotransmitter stages. The brain meditations of gamma-amino-butyric acid and dopamine were reduced, while the 5-HT level was improved. This result indicates the envelopment of monoamine neurotransmitters in the nootropic action of n-butanolic fraction of *A. lebbeck* (Chintawar et al., 2002).

27.4.13 Antiasthmatic Activity

Kumar et al. (2010) study showed antiasthmatic activity of *A. lebbeck* on 81 patients. Patients were given *A. lebbeck* stem bark decoction in a dose of 50 mL 3 times a day for 6 weeks with light diet. The effects were evaluated in terms of clinical recovery, indicative relief, and pulmonary function progress. The result of the treatment was evaluated based on subjective as well as objective parameters. A significant increase in Peak Expiratory Flow Rate (PEFR) and considerable reduction in total leukocyte count, eosinophil count, and E.S.R. replace with 'Erythrocyte Sedimentation Rate (ESR) were observed. Overall, 56% of cases have shown good response, 38% cases showed fair response, 6% cases showed poor response, and 31% cases were dropouts. The consequences demonstrate that the preparation can be used as an actual drug in bronchial asthma however comprehensive observational studies are necessary to determine the effect of drug at molecular level. The effects of the decoction of the bark and flower of *A. lebbeck* were studied for its antiasthmatic as well as antianaphylactic activity. The decoctions sheltered the guinea pig in contradiction of histamine and acetylcholine encouraged broncho-spasm. Chronic usage with the bark decoction has also sheltered the prepared guinea pigs contrary to antigen challenge. Though, the drug has not important effect on the rat mesenteric mast cell count and has not sheltered the mast cell from the disturbance induced by compound 48/80. The drug reserved the rate of disturbance of mast cells induced by antigen in alerted albino rats. It has no outcome on the adrenal, thymus, and spleen weight and adrenal ascorbic-acid, nevertheless the cholesterol content was significantly condensed (Tripathi and Das, 1977). *A. lebbeck* at a dose of 1000 mg 4 times a day with water for 30 days provided marked improvement in 40% patients, mild enhancement in 20% patients. Decoction of the inflorescence meaningfully protected the guinea pig from bronchospasm induced by histamine. The activity could be due to smooth muscle reduction (Shaw and Bera, 1986).

27.4.14 Antianaphylactic Activity

A. lebbeck bark decoction described to possess cromoglycate similar action on mast cells of albino rats. Studies showed the antianaphylactic activity is due to inhibition of the synthesis antibodies and suppression of T-lymphocytes (Mukhopadhyay et al., 1992).

27.4.15 Free Radical Scavenging and Antiarthiritic Activity

Rheumatoid arthritis is a predominant and incapacitating disease that affects the joints. Penetration of blood-derived cells in the pretentious joints upon initiation generates reactive oxygen or nitrogen species, resultant in an oxidative pressure. One method to respond this oxidative pressure is the used by antioxidants as therapeutic agents. The methanolic extract of *A. lebbeck* that showed important anti-inflammatory activity, was assessed for the conceivable mode of action by reviewing its antioxidant potential in adjuvant-induced arthritic rats. The biological defense system establishing the superoxide dismutase, catalase level displayed important increase though the lipid peroxide content was initiate to reduction to a large range on *A. lebbeck* treatment there by representative the extracts showed free radical scavenging property. During 21st day of treatment, paw edema in disease control inflamed paw is increase in time dependent manner and all administration groups significantly inhibited the development of joint swelling induced by complete Freund's adjuvant. It was concluded that *A. lebbeck* methanolic extract owns strong antiarthritic and antioxidant property (Pathak et al., 2010).

KEYWORDS

- ***Albizia lebbeck***
- **phytochemistry**
- **ethnobotany**
- **pharmacology**

REFERENCES

Ahmed, S. S. B. Medicinal wild plants from Lahore-Islamabad Motorway. *Pak. J. Bot.* **2007,** *39* (2), 355–375.

Aisha, M.; Kalid, A.; John, A. O. Preliminary Phytochemical Screening of *Albizia lebbeck* stem bark. *Int. J. Integr. Educ.* **2020,** *3* (12), 112–116.

Aiyelaagbe, O. O.; Oyewole, A. E.; Oladosu, A. I. The Chemistry and Cytotoxicity Effects of Essential Oils of Leaves and Flowers of *Albizia lebbeck* (L.) Benth (Fabaceae) Growing in Nigeria. *J. Essent Oil Bear. Plants* **2010,** *13* (5), 644–649.

Amog, P. U.; Manjuprasanna, V. N.; Yariswamy, M.; Nanjaraj Urs, A. N.; Joshi, V.; Suvilesh, K. N.; Gowda, T. V. *Albizia lebbeck* Seed Methanolic Extract as a Complementary Therapy to Manage Local Toxicity of *Echis carinatus* Venom in a Murine Model. *Pharm. Biol.* **2016,** *54* (11), 2568–2574.

Babu, N. P.; Pandikumar, P.; Ignacimuthu, S. Anti-Inflammatory Activity of *Albizia lebbeck* Benth., an Ethnomedicinal Plant, in Acute and Chronic Animal Models of Inflammation. *J. Ethnopharmacol.* **2009,** *125*, 356–360.

Barua, C. C.; Gupta, P. P.; Patnaik, G. K.; Mishra, B. S.; Goel, R. K.; Kulshestra, D. K.; Dubey, M. P.; Dhawan, B. N. Immunomodulatory Effect of *Albizzia lebbeck*. *Pharm Biol.* **2000,** *38*, 161–166.

Baruach, C. C.; Gupta, P. P.; Patnaik, G. K.; Amarnath; Kulshreshtha, D. K.; Dhawan, B. N. Anti Anaphylactic and Mast Cell Stabilizing Activity of *Albizzia lebbeck. Indian Vet. Med. J.* **1997,** *21* (2), 127–132.

Baruah, C. C.; Gupta, P. P.; Patnaik, G. K.; Dubey, M. P.; Goel, R. K.; Dhawan, B. N. Comparative Study of the Anti-PCA and Mast-Cell Stabilizing Activity Fractions of *Albizzia lebbeck*: A Traditional Medicinal Plant. *J. Med. Arom. Plants Sci.* **2000,** *22*, 42.

Bobby, M. N.; Wesely, E. G.; Johnson, M. High Performance Thin Layer Chromatography Profile Studies on the Alkaloids of *Albizia lebbeck. Asian Pac. J. Trop. Biomed.* **2012,** *2* (1), S1-S6.

Chaudhary, M.; Sharma, A. K.; Kumar, R.; Chauhan, B.; Kaushik, K.; Agarwal, V. Comparative Immunomodulator Activity of Leaves and Bark of *Albizia lebbeck* (Linn.) Benth. *Int. J. Res. Dev. Pharm. L. Sci.* **2012,** *1* (1), 21–23.

Chintawar, S. D.; Somani, R. S.; Kasture, V. S.; Kasture, S. B. Nootropic Activity of *Albizzia lebbeck* in Mice. *J. Ethnopharmacol.* **2002,** *81* (3), 299–305.

El–Mousallamy, A. M. D. Leaf flavanoids of *Albizzia lebbeck. Phytochemistry* **1998,** *48* (4), 759–61.

Farag, M.; El Gamal, A.; Kalil, A.; Al-Rehaily, A.; El Mirghany, O.; El Tahir, K. Evaluation of Some Biological Activities of *Albizia lebbeck* Flowers. *J. Pharm. Pharmacol.* **2013,** *4* (6), 473–477.

Galav, P.; Jain, A.; Katewa, S. S. Ethnoveterinary Medicines Used by Tribals of Tadgarh-Raoli Wildlife Sanctuary, Rajasthan, India. *Indian J. Tradit. Knowl.* **2013,** *12* (1), 56–61.

Ganguli, N. B.; Bhatt, R. M. Mode of Action of Active Principles from Stem Bark of *Albizzia lebbeck. Indian J. Exp. Biol.* **1993,** *31* (2), 125–129.

Govindarajan, M.; Rajeswary, M. Ovicidal and Adulticidal Potential of Leaf and Seed Extract of *Albizia lebbeck* (L.) Benth. (Family: Fabaceae) Against *Culex quinquefasciatus*, *Aedes aegypti*, and *Anopheles stephensi* (Diptera: Culicidae). *Parasitol. Res.* **2015,** *114* (5), 1949–1961.

Gul, F.; Shinwari, Z. K.; Afzal, I. Screening of Indigenous Knowledge of Herbal Remedies for Skin Diseases Among Local Communities of North West Punjab, Pakistan. *Pak. J. Bot.* **2012,** *44* (5), 1609–1616.

Gupta, R. S.; Chaudhary, R.; Yadav, R. K.; Verma, S. K.; Dobhal, M. P. Effect of Saponins of *Albizzia lebbeck* (L.) Benth. Bark on the Reproductive System of Male Albino Rats. *J. Ethnopharmacol.* **2005,** *96* (1–2), 31–36.

Gupta, R. S.; Kachhawa, J. B.; Chaudhary, R. Antispermatogenic, Antiandrogenic Activities of *Albizia lebbeck* (L.) Benth. Bark Extract in Male Albino Rats. *Phytomed.* **2006,** *13*, 277–283.

Habiba, U.; Nisar, J.; Choohan, M. A.; Shah, S. M. A.; Nisar, Z.; Mustafa, I. Antibacterial Activity of Tris NaCl and PBS Buffer Protein Extract of *Cassia fistula*, *Saccharum officinarum*, *Albizia lebbeck* and *Cymbopogon citratus* Against Bacterial Strains. *Dose-Response* **2021,** *19* (1), 1559325821992239.

Islam, M. N.; Hiroyuki, M.; Masum, S. *Albizia lebbeck* Suppresses Histamine Signaling by the Inhibition of Histamine H1 Receptor and Histidine Decarboxylase Gene Transcriptions. *Int. Immunopharmacol.* **2011,** *11* (11), 1–7.

Kavitha, C. N.; Raja, K. D.; Rao, S. K. Antitumor Activity of *Albizia lebbeck* L. Against Ehrlich Ascites Carcinoma *In-vivo* and HeLa and A549 Cell Lines *In-vitro*. *J. Cancer Res. Ther.* **2021,** *17* (1). DOI: 10.4103/jcrt.JCRT_454_19

Knothe, G.; Phoo, Z. W. M. M.; de Castro, M. E. G.; Razon, L. F. Fatty Acid Profile of *Albizia lebbeck* and *Albizia saman* Seed Oils: Presence of Coronaric Acid. *Eur. J. Lipid Sci. Technol.* **2015,** *117* (4), 567–574.

Kumar, S.; Bansal, P.; Gupta, V.; Sannd, R.; Rao, M. M. The clinical effect of *Albizia lebbeck* Stem Bark Decoction on Bronchial Asthma. *Int. J. Pharma Sci. Drug Res.* **2010,** *2* (1), 48–50.

Malaikolundhan, H.; Mookkan, G.; Krishnamoorthi, G.; Matheswaran, N.; Alsawalha, M.; Veeraraghavan, V. P.; Di, A. Anticarcinogenic Effect of Gold Nanoparticles Synthesized from *Albizia lebbeck* on HCT-116 Colon Cancer Cell Lines. *Artif. Cells Nanomed. Biotechnol.* **2020,** *48* (1), 1206–1213.

Mohammad F.; Singh, P. P.; Irchhaiya, R. Review on *Albizia lebbeck* a Potent Herbal Drug. *Int. Res. J. Pharm.* **2012,** *3* (5), 63–68.

Mukhopadhyay, B.; Nagaraja, K.; Sharma, K. R. *Albizia lebbeck*—A Remedy for Allergic Conjunctivitis. *J. Res. Edu. Indian Med.* **1992,** *11* (4), 17–23.

Pathak, N. L.; Patel, N.; Kasture, S., Jivani, N.; Bhalodia, Y.; Malavia, S. Free Radical Scavenging Activity of *Albizia lebbeck* Methanolic Extract in Arthritic Rats. *Intern. J. Pharma. Res. Dev.* **2010,** *1* (1), 1–8.

Priyanka, B.; Anitha, K.; Shirisha, K.; Sk, J.; Dipankar, B.; Rajesh, K. Evaluation of Antioxidant Activity of Ethanolic Root Extract of *Albizia lebbeck* (L.) Benth. *Int. J. Pharm. Biol. Sci.* **2013,** *3* (2), 93–101.

Reshmi, C. R.; Venukumar, M. R.; Latha, M. S. Antioxident Activity of *Albizia lebbeck* (Linn.) Benth. in Alloxan Diabetic Rats. *Indian J. Physiol. Pharmacol.* **2006,** *50* (3), 297–302.

Saha, A.; Ahmed, M. The Analgesic and Antiinflammatory Activities of the Extract of *Albizia lebbeck* in Animal Model. *Pak. J. Pharm. Sci.* **2009,** *22* (1), 74–77.

Saleem, U.; Raza, Z.; Anwar, F.; Chaudary, Z.; Ahmad, B. Systems Pharmacology Based Approach to Investigate the *In-vivo* Therapeutic Efficacy of *Albizia lebbeck* (L.) in Experimental Model of Parkinson's Disease. *BMC Complem. Altern. Med.* **2019,** *19* (1), 1–16.

Shahid, S. A.; Firdous, N. Antimicrobial Screening of *Albizia lebbeck* (L.) Benth. and *Acacia leucophloea* (Roxb.). *African J. Pharm. Pharmacol.* **2012,** *6* (46), 3180–3183.

Shaw, B. P.; Bera, B. Treatment of Tropical Pulmonary Eosinophilia with Sirish Flower (*Albizzia lebbeck* Benth.) Churna. *Nagarjuna* **1986,** *29* (6), 1–3.

Singh, Y. N.; Bisht, H.; Panday, D. Effect of Dry Seed Extract of a Medicinal *Plant Albizzia lebbeck* on Testicular and Epididymal Protein Profiles of Rat. *Himalayan J. Env. Zool.* **1991,** *5* (2), 94–98.

Thirumalai, T.; Kelumalai, E.; Senthilkumar, B.; David, E. Ethnobotanical Study of Medicinal Plants Used by the Local People in Vellore District, Tamilnadu, India. *Ethnobot. Leaflets* **2009,** *13*, 1302–11.

Tripathi, R. M.; Sen, P. C.; Das, P. K. Studies on the Mechanism of Action of *Albizzia lebbeck*, an Indian Indigenous Drug Used in the Treatment of Atopic Allergy. *J. Ethnopharmacol.* **1979,** *1* (4), 385–96.

Tripathi, R. M.; Das, P. K. Studies on Anti-Asthmatic and Antianaphylactic Activity of *Albizzia lebbeck*. *Indian J. Pharmacol.* **1977,** *9*, 189–194.

Varshney, I.; Pal, R.; Vyas, P. Studies on Lebbekanin E, a New Saponin from *Albizia lebbeck* Benth. *J. Indian Chem. Soc.* **1976,** *53*, 859–860.

Veda, M.; Tokunaga, T.; Okazaki, M.; Sata, N. U. Albizziahexoside, a Hexaglycosylated Saponin Isolated from Leaves of *Albizzia lebbeck. Nat. Prod. Res.* **2003,** *17* (5), 329–335.

Venkatesh, P.; Mukherjee, P. K.; Kumar, N. S.; Bandyopadhyay, A.; Fukui, H.; Mizuguchi, H.; Islam, N. Anti-Allergic Activity of Standardized Extract of *Albizia lebbeck* with Reference to Catechin as a Phytomarker. *Immunopharmacol. Immunotoxicol.* **2010,** *32* (2), 272–276.

Zia-Ul-Haq, M.; Ahmad, S.; Qayum, M.; ERCİŞLİ, S. Compositional Studies and Antioxidant Potential of *Albizia lebbeck* (L.) Benth. Pods and Seeds. *Turk. J. Biol.* **2013,** *37* (1), 25–32.

CHAPTER 28

Phytochemical and Pharmacological Activities of *Senna hirsuta* (L.) H. S. Irwin & Barneby

K. S. SHANTHI SREE[1*], B. KAVITHA[2], A. SUVARNA LATHA[1], and P. LAKSHMI PADMAVATHI[1]

[1]*Department of Biosciences and Sericulture, Sri Padmavathi Mahila Visvavidyalayam, Tirupati, Andhra Pradesh, India*

[2]*Department of Botany, Rayalaseema University, Kurnool, Andhra Pradesh 518007, India*

[*]*Corresponding author. E-mail: shanthi4botany@gmail.com*

ABSTRACT

Senna hirsuta is an important medicinal plant which has been extensively studied for its therapeutic activities. Different parts of *Senna* have numerous phytochemical constituents. In vitro, in vivo and clinical studies revealed that *S. hirsuta* have strong antioxidant and antimicrobial activities. The phytochemical compounds contribute to the biological activity of *S. hirsuta*. The review emphasizes bioactive compounds, antibacterial, antioxidant and hepatoprotective activities of *S. hirsuta*.

28.1 INTRODUCTION

Senna hirsuta (L.) H.S. Irwin & Barneby belongs to the Family Fabaceae (Leguminosae) and Sub-family Caesalpinioideae. Synonyms of this species include *Cassia hirsuta* L., *C. caracasana* Jacq., *C. venenifera* Rodschied ex G. Meyer, *C. hirsuta* L., *C. tomentosa* Arn., *C. venenifera* G. Mey., *Senna*

Bioactives and Pharmacology of Legumes. T. Pullaiah, PhD (Ed.)

hirsuta var. hirsuta., Ditremexa hirsuta (L.) Britton et Wilson (http://www.theplantlist.org/tpl1.1/record/ild-1037). It is a native of tropical America and now distributed in Thailand, California, Malaysia, Brazil, Indo-China, New Mexico, Laos, Java, India, Asian, and African tropics (Holm et al., 1979). In India it is widely available in the Deccan in the Bababudan Hills of Mysore, Ramanadrug, Bellary and also in the Carnatic near Madras. It is commonly called slim pod glaberrima senna, shower tree senna, sickle pod, hairy senna, woolly wild, stinking cassia, woolly senna, sensitive plant (https://keys.lucidcentral.org/keys/v3/eafrinet/weeds/key/weeds/Media/Html/Senna_hirsuta (Hairy_Senna).htm).

It is a large upright perennial, terrestrial, shrub up to 150 cm in tall, stem solid, glabrous, round, plant is softly tomentose; branches groved; leaves 15–20 cm long with a gland at the base of petiole, stipules leaner, leaves four to six pairs, acuminate, ovate to elliptic, cuneate at the base. Its flowers are irregular, yellow orange, borne in small clusters axillary or terminal recemes, bract lanceolate, acuminate and flowering period is from September to December. Stamens–10, unequal in length, 7 of them fertile, Ovary—subsessile, hirsute, style—glabrous, short, and stigma small. Pod is a long legume, flat, slender, 10–20 × ca. 0.5 cm and fruiting can be seen in November to January. Seeds—numerous, obovoid, flat, 3–4 mm (Pacific Island Ecosystems at Risk, 2011; Vellingiri et al., 2011).

S. hirsuta has been used as a remedy for lowering diarrhoea, cholesterol levels, high blood pressure, malaria, typhoid fever, and skin rashes (Henderson, 2001). It is also used against herpes (Neuwinger, 2000). Decoction of the leaves of *S. hirsuta* is used for skin irritation in Thailand. The seeds are used as a substitute for coffee in Loas. They are also beneficial for the eyes (Oliver, 2005; Sofowora, 2008). *S. hirsuta* contains a water-soluble gum that contains medicinally important bioanthraquinone (Sangat-Roemantoyo, 1997).

28.2 BIOACTIVES

S. hirsuta revealed the presence of different active phytochemical constituents. The phytochemical screening of chloroform and ethanol extracts of seeds of *S. hirsuta* showed presence of amino acids, proteins, carbohydrates, steroids, glycosides, flavonoids, alkaloids, phenolic compounds, and tannins. Apart from steroids all other constituents are also present in chloroform extract. Soxhlet extraction of *S. hirsuta* was carried out using three different solvents and tested for the presence of bioactive chemical compounds. Among the extracts methanolic extract proved to be more in quantity and presence of more number of phytochemical constituents when compared with diethyl ether and hexane. Flavonoids, Alkaloids, steroids, terpenoids, carbohydrates, and phenols are the phytochemical constituents that are present in *S. hirsuta*. The quantitative analysis of methanolic extract shows 4.20% of alkaloid content, 0.84% of flavonoid, and 28.76% of phenol content. It proves that methanolic extract of *S. hirsuta* has free radical scavenging potential that may be helpful in treating human diseases related to reaction of free radicals (Madankumar and Pari, 2019).

S. hirsuta seeds contain phytochemical compounds like amino acids, carbohydrates, proteins, saponins, glycosides, alkaloids, phenolic compounds, phyto steroids, flavonoids, and tannins. The estimation of total phenolic content was done using Folin Ciocalteu's method and total flavonoid content was estimated using the aluminum chloride colorimetric assay. Results show that more number of phytochemical constituents were present in the ethanol and chloroform extracts of the seeds.

The total phenolic content of 13.7 ± 0.4187 and 7.367 ± 0.2987 mg of gallic acid equivalent weight/g of extract was found in the ethanol and chloroform extracts. The ethanol and chloroform extracts of seeds were found to have total flavonoid content of 114.6 ± 13.33 and 99.56 ± 11.6 mg of quercetin equivalent weight/g of the extract (Shalavadi et al., 2019).

The main components isolated from *S. hirsuta* fruit oil were benzyl benzoate (24.7%), τ-cadinol (18.9%), 2,5-dimethoxy-*p*-cymene (14.6%), and β-caryophyllene (5.1%) (Essien et al., 2018).

Chemical structures of some isolated compounds from *Senna hirsuta*

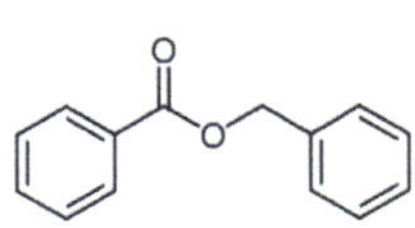

Benzyl benzoate

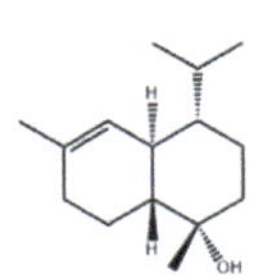

τ-cadinol

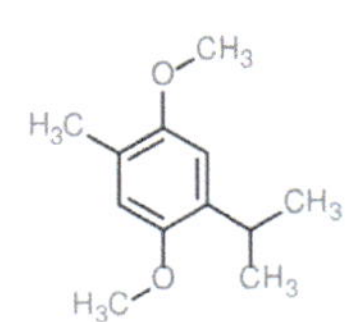

2, 5-dimethoxy-p-cymene

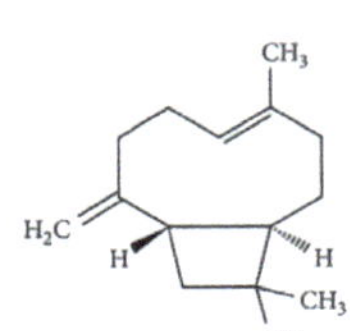

β-caryophyllene

28.3 PHARMACOLOGICAL ACTIVITIES

28.3.1 Antibacterial Activity

The aqueous and ethanol extracts of the leaves, bark, and roots of *S. hirsuta* were examined for a possible source of antimicrobial activity. The preliminary screening of both the extracts exhibited appreciable inhibitory activities on the selected pathogenic bacterial isolates at a concentration of 20 mg m/L. *S. hirsuta* showed more therapeutic potency where the leaves aqueous extract displayed inhibitory activity (30–37 mm) in diameter on the tested bacterial isolates. Leaves ethanol extract showed inhibitory zone between 30 and 38 mm in diameter (Akharaiyi and Bolatito, 2009).

The leaf ethanolic extract of *S. hirsuta* showed significant zone of inhibition against Gram-positive *Bacillus cereus* (13 ± 1.05 mm) and *B. megaterium* (9 ± 0.85 mm), and Gram-negative *Vibrio cholerae* (20 ± 0.68 mm), *Escherichia coli* (8 ± 0.92 mm), *Pseudomonas aeruginosa* (15 ± 1.03 mm), *Salmonella paratyphi* (9 ± 0.79 mm), and *Shigella dysenteriae* (8 ± 0.88 mm) (Rahman et al., 2013).

28.3.2 Antioxidant Activity

S. hirsuta leaf ethanol extract was tested by the DPPH (1, 1-diphenyl-2-picrylhydrazyl) free radical scavenging method. The extract showed a significant effect compared with standard ascorbic acid. The IC_{50} value of leaf ethanol extract and ascorbic acid was found to be 200.96 and 1.25 µg/mL, respectively. The extract value indicated a significant antioxidant activity of the plant (Dey et al., 2012).

In another study the antioxidant activity measured by the DPPH (2,2-diphenyl-1-picrylhydrazyl) free radical scavenging method in the ethanolic extracts of *S. hirsuta* leaf showed a significant radical scavenging effect (IC_{50}, 200.96 ± 0.85 µg/mL) in comparison with ascorbic acid ($IC_{50,}$ 1.24 ± 0.08 µg/mL) (Rahman et al., 2013).

28.3.3 Cytotoxicity Assay

To test the cytotoxicity brine shrimp lethality bioassay was done by using the *S. hirsuta* leaf ethanolic extract. The extract showed the LC_{50} value 315.5 µg/mL which was statistically significant compared to reference cytotoxic agent vincristine sulfate (LC_{50} −38.99 µg/mL) (Rahman et al., 2013).

28.3.4 *Hepatoprotective Activity*

The results of the hepatoprotective activity of leaf ethanolic extract of *S. hirsuta* showed that the enzyme levels of Alanine aminotransferase (ALT), Aspertate aminotransferase (AST), and Alkaline phosphatase in the serum of the rats treated with CCl_4 (positive control) increased significantly. Dose-related differences in all parameters assayed were observed when the rats were administered varying doses (500–2000 mg/kg body weight) of the extract. There was a marked significant difference in the levels of the enzyme markers (AST, ALT, and Alkaline phosphatase). The GST activity decreased significantly in the serum of the test animals exposed to CCl_4 (positive control) when compared with the negative control. The rats exposed to both the extracts at 500 mg/kg body weight showed a significant difference and the toxicant as compared with the positive control. Histopathological studies revealed that exposure (i.p) to CCl_4 has led to the induction of hepatocellular necrosis. The liver section of untreated (negative control) rat showed central vein of hepatocytes with radiating cords. The liver section of rat treated with low dose (500 mg/kg body weight) of extract after exposure (i.p) to CCl_4 showed degeneration (vacuolation) of hepatocytes around hepatic vein of the centrilobular area (Joshua and Nwodo, 2010).

28.4 CONCLUSION

The present review concludes that, by using the information presented on the phytochemical compounds and various biological properties of the extracts, further studies on *S. hirsuta* can be carried out to isolate bioactive compounds that can be useful in preparation drugs.

KEYWORDS

- ***Senna hirsuta***
- **phytochemical constituents**
- **therapeutic activities**
- **medicinal plant**
- **antibacterial**

REFERENCES

Akharaiyi, F. C.; Bolatito, B. Antibacterial and Phytochemical Evaluation of Three Medicinal Plants. *J. Nat. Prod.* **2009,** *3* (2010), 27–34.

Dey, T. K.; Emran, T. B.; Saha, D.; Rahman, M. A.; Hosen, S. M. Z.; Chowdhury, N. Antioxidant Activity of Ethanol Extract of *Cassia hirsuta* (L.) Leaves. *Bull. Pharma. Res.* **2012,** *2* (2), 78–82.

Essien, E. E.; Thomas, P. S.; Ascrizzi, R.; Setzer, W. N.; Flamini, G. *Senna occidentalis* (L.) Link and *Senna hirsuta* (L.) H. S. Irwin & Barneby: Constituents of Fruit Essential Oils and Antimicrobial Activity. *Nat. Prod. Res.* **2018,** *33* (11), 1637–1640.

Henderson, L. *Alien Weeds and Invasive Plants. A Guide to Declare Weeds and Invader in South Africa*, 2nd ed.; Chapman Press South Africa, 2001; pp 136–139.

Holm, L. G.; Pancho, J.; Herberger, J. P.; Plucknett, D. L. *A Geographical Atlas of World Weeds*; Knieger Publishing Company: Florida, 1979.

Joshua, P. E.; Nwodo, O. F. C. Hepatoprotective Effect of Ethanolic Leaf Extrat of *Senna hirsuta (Cassia hirsuta)* Against Carbon Tetrachloride (CCl_4) Intoxication in Rats. J. Pharm. Res. 2010, *3* (2), 310–316.

Madankumar, D.; Pari, S. Preliminary Phytochemical Activity and Quantitative Analysis of *Senna alata* and *Senna hirsuta* Medicinal Tree. *J. Emerg. Technol. Innov. Res.* **2019,** *6* (6), 375–380.

Neuwinger, H. *African Traditional Medicine: A Dictionary of Plant Use and Applications*; Medpharm Scientific: Stuttgart, Germany, 2000.

Oliver, B. *Medicinal Plants in Nigeria.* Nigeria College of Arts. In linking-hub.elsevier.com/retrieve/pi. Feb 23, 2005.

Pacific Island Ecosystems at Risk (PIER). *Senna hirsuta* (L.) H. S. Irwin & Barneby, Fabaceae (Leguminosae): Plant Threats to Pacific Ecosystems. www.hear.org/Pier/species/senna_hirsuta.htm. Institute of Pacific Islands Forestry, Hawaii, USA. Accessed March 2011.

Rahman, M. A.; Rahman, M. A.; Ahme, N. U. Phytochemical and Biological Activities of Ethanolic Extract of *C. hirsuta* Leaves. *Bangladesh J. Sci. Ind. Res.* **2013,** *48* (1), 43–50.

Sangat-Roemantyo, H. *Senna hirsuta* (L.) Irwin and Barneby. In *Plant Resources of South-East Asia No. 11. Auxilliary Plant*; Faridah, H. I., Van der Maesen, L. J. G., Eds.; Prosea Foundation: Bogor, Indonesia, 1997; pp 231–232.

Shalavadi, M. H.; Mangannavar, C. V.; Muchchandi, I. S.; Hulakoti, B. Qualitative and Quantitative Phytochemical Analysis of Seeds *Cassia hirsuta. Asian J. Pharm. Pharmacol.* **2019,** *5* (2), 290–297.

Sofowora, A. *Medicinal Plants and Tradition Medicine in Africa*, 3rd ed.; Spectrum Books Limited: Ibadan, Nigeria, 2008; pp 199–204.

Vellingiri, V.; Aruna, N.; Hans, K. B. Antioxidant Potential and Health Relevant Functionality of Traditionally Processed *Cassia hirsuta* L. Seeds: An Indian Underutilized Food Legume. *Plant Foods Human Nutr.* **2011,** *66* (3), 245–253.

http://www.theplantlist.org/tpl1.1/record/ild-103

https://keys.lucidcentral.org/keys/v3/eafrinet/weeds/key/weeds/Media/Html/Senna_hirsuta (Hairy_Senna).htm.

CHAPTER 29

Bioactive Metabolites and Pharmacological Activities of *Vigna radiata* (L.) Wilczek (Mung Bean)

DESAI KRISHNA, NAINESH R. MODI, and ANJALI SHUKLA

Department of Botany, Bioinformatics and Climate Change Impacts Management, Gujarat University, Ahmedabad, Gujarat 38009, India

Corresponding author. E-mail: nrmodi@gujaratuniversity.ac.in

ABSTRACT

Vigna radiata (L.) Wilczek (Mung bean) is an annual legume crop that, in addition to being highly nutritious, also offers considerable health benefits for pharmacological purposes. Previous reports have demonstrated a balanced nutritional profile of mung beans including proteins, fibres, minerals and carbohydrates along with other non-nutritive metabolites such as flavonoids, phenols, tannins, phytic acids, and hemagglutinins, that contribute to its medicinal properties. It has also been documented that, different parts of the plant have antioxidant, antimicrobial, anti-inflammatory, anti-diabetic, antiproliferative, antihypertensive, antisepsis, tyrosinase inhibition, and myocardial preservation effects. Moreover, it is also known for detoxification, heatstroke elimination, mental refreshment, and gastrointestinal regulation. The present chapter aims to explore the various pharmacological activities of *Vigna radiata* and provide insight into the major bioactive components of the plant. It highlights the need for more studies to focus on comprehending the mechanism of action for such activities to validate the effectiveness of mung bean and its derivatives as therapeutic agents.

Bioactives and Pharmacology of Legumes. T. Pullaiah, PhD (Ed.)

29.1 INTRODUCTION

Pulses are annually grown legume crops used for food and fodder, which produce seeds of varying color, size, and shape. The bean-based diets are drawing attention as a whole food rich in nutrition because dietary habits are evolving globally, and the portion of noncereal foods in the total intake of calories and protein is increasing. Clinical evidence suggests that food derived from pulses has many significant health benefits for fighting a non-contagious diet-related disease like obesity, cardiovascular diseases, diabetes, and cancer (Singh et al., 2016, 2017; Tang et al., 2014). Pulses are a natural source of phytochemicals including nutraceuticals, detoxifying enzymes, and immuno-modulatory compounds that have potent pharmacological activities. Mung bean (*Vigna radiata* (L.) Wilczek) belongs to the group of legume crops, grown during the warmer season in tropical and subtropical conditions (Lambrides and Godwin, 2007; Chauhan and Williams, 2018). Because of the significant nutritional and medicinal values of Mung bean, there is a great deal of interest in the genetic and genomic analysis of this crop, particularly in developing nations where food deprivation is a major issue (Kim et al., 2015). Mung bean is an inexpensive source of macro and micronutrients including high-grade proteins, carbohydrates, iron, potassium, and calcium. It also contains a high level of antioxidant agents that improve cardiovascular and general health. Many studies revealed the powerful anti-microbial, anti-inflammatory, antiproliferative, and antidiabetic activities of mung bean extract. Moreover, it is also known for detoxification, heatstroke elimination, mentality refreshment, and gastrointestinal regulation (Johnson et al., 2019; Tang et al., 2014).

29.2 BIOACTIVE COMPONENTS

Bioactive components are naturally occurring phytochemicals that have health-care benefits. When a plant species is exposed to biotic or abiotic stress, it accumulates more of these compounds as a defensive mechanism (Sehrawat et al., 2020; Shukla and Tyagi, 2017; Xue et al., 2016). Scientific researches confirmed that bioactive components derived from legumes exert potential human health benefits. Mung bean offers a surprisingly balanced nutrition profile that includes proteins, fibers, minerals, and other active metabolites; these components remarkably enhance human health (Mubarak, 2005; Hou et al., 2019; Xie et al., 2019). It consists of high protein ranging from 20 to 24%, by dry weight, of which globulin (60%) and albumin

(25%) are the most important storage proteins (Yi-Shen et al., 2018). These proteins are rich in essential amino acids including aromatic amino acids, isoleucine, leucine, and valine, whereas threonine, lysine, tryptophan, and sulfur-containing amino acids are lacking. Besides that, proteolytic cleavage of protein during germination increases the amino acid content (EI-Adawy et al., 2003; Mubarak, 2005). Mung beans have a much higher carbohydrate content compared to other legumes, among which starch is the most prevalent (Tang et al., 2014). The majority of polysaccharides were reported to exhibit potential antioxidant and immunomodulatory effects (Hou et al., 2019).

Mung beans also contain other non-nutritive components such as flavonoids, phenols, tannins, phytic acids, and hemagglutinin, which possess different pharmacological activities and eliminate toxins (Lin and Li, 1997). The primary phenolic groups in mung beans are flavonoids, phenolic acids, and tannins (Lee et al., 2011; Shi et al., 2016; Sing et al., 2017). Most flavonoids are classified as polyphenol because of polyhydroxy substitution. The most important flavone glycoside, Apigenin 8-C-glucoside (Vitexin) (51.1 mg/g) and Apigenin 6-C-glucoside (Isovitexin) (51.7 mg/g) have been reported to be present in seeds of mung bean (Tang et al., 2014). HPLC analysis of various extracts of mung bean reported the presence of active flavonoids including rutin, kaempferol, robinin, quercetin, isoquercitrin, kaempferol-3-O-glucoside, kaempferol-7-O-rhamnoside, and kaempferol-3-O-rutinoside (Ganesan and Xu, 2018). Phenolic acids are the most important bioactive synthesized by shikimate or pentose phosphate pathways. Capillary gas-liquid chromatography of mung bean sprouts revealed the presence of phenolic acids like catechin, p-coumaric acid, epicatechin, ferulic acid, p-hydroxy benzoic acid, syringic acid, gallic acid, isovitexin, vitexin, protocatechuic acid, quercetin, sinapic acid, caffeic acid, and 5-O-caffeoylquinic acid (Sosulski and Dabrowski, 1984; Silva et al., 2013) (Figure 29.1).

29.3 BIOLOGICAL ACTIVITIES

29.3.1 Antioxidant Activity

The seeds, sprouts, and mung bean hulls contain a large amount of proteins, polysaccharides, and secondary metabolites, which all have potent anti-oxidant activity. These compounds can restrict the release of free radicals by chelating metal ions or inhibiting key enzymes active in a free radical generation. Glycosylated quercetins and kaempferol 3-O- neohesperidoside significantly reduce the metal chelating potential (Reynoso-Camacho et

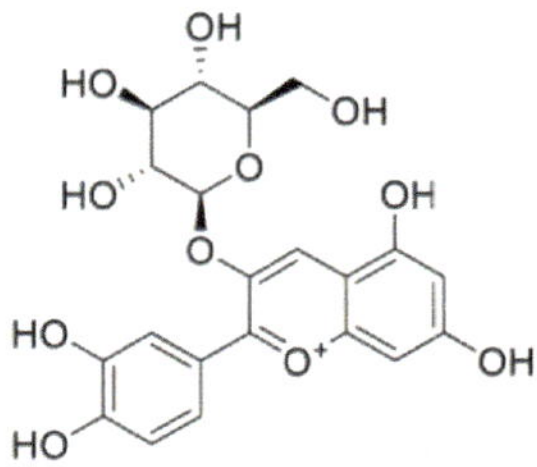

(1)

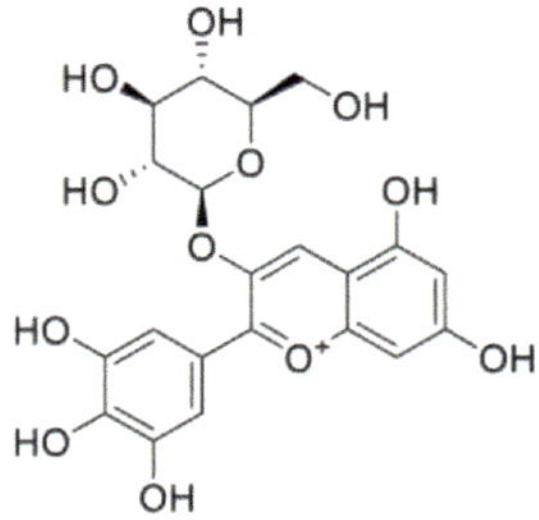

(2)

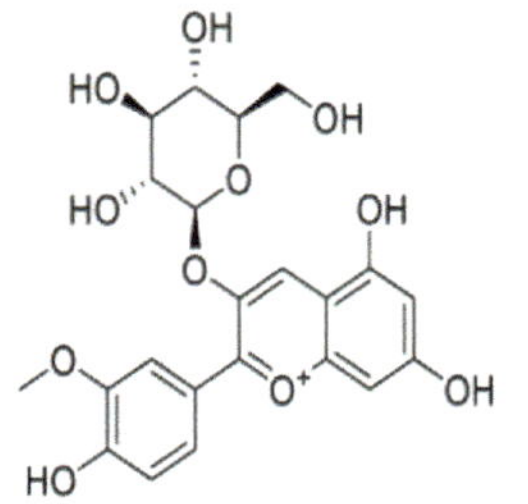

(3)

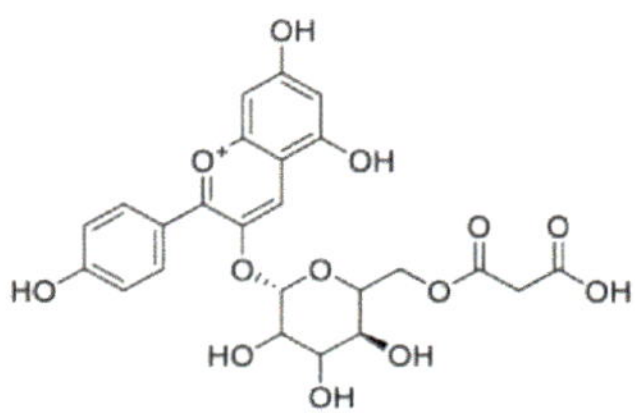

(4)

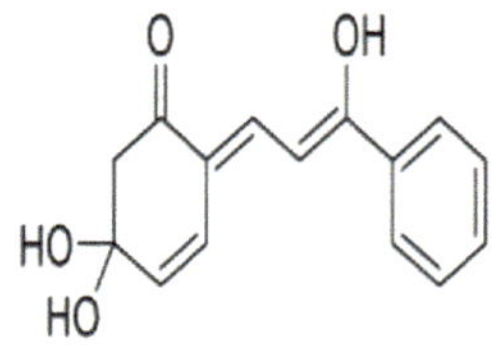

(5)

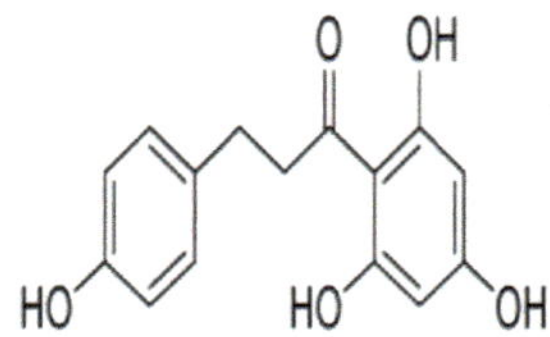

(6)

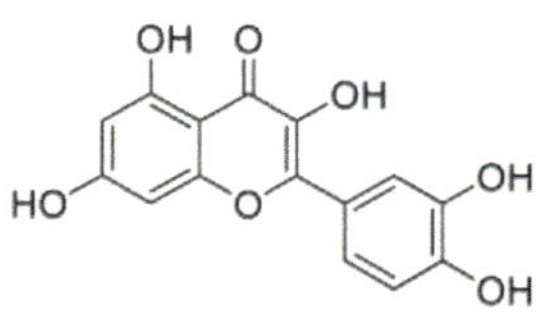

(7)

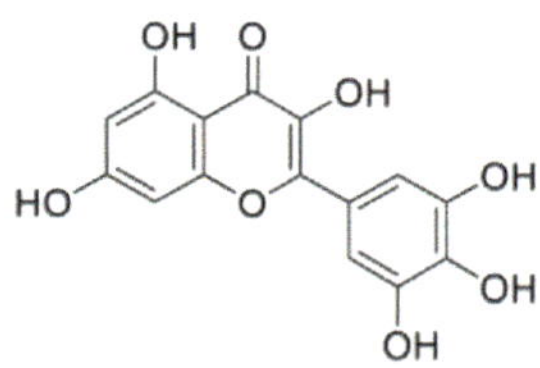

(8)

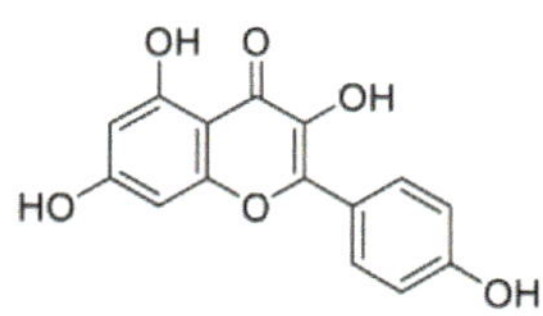

(9)

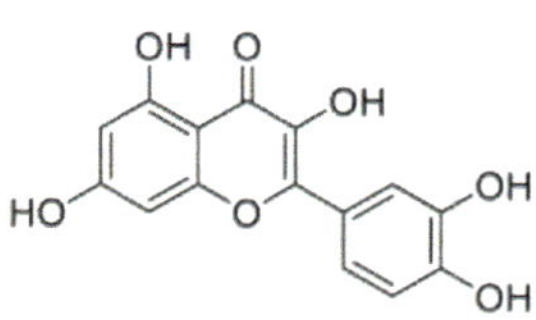

(10)

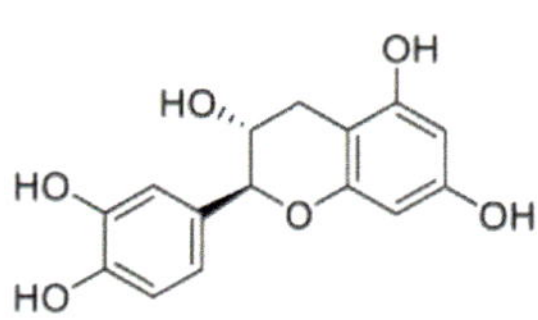

(11)

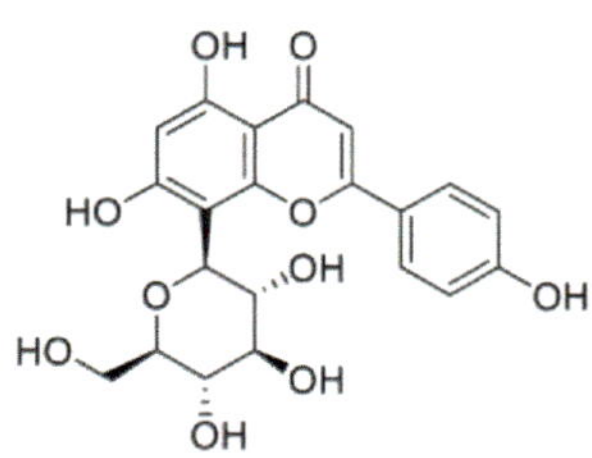

(12)

FIGURE 29.1 *(Continued)*

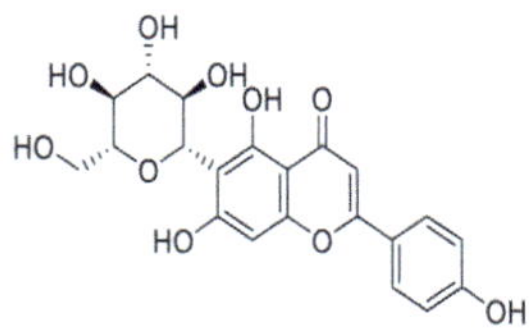

(13)

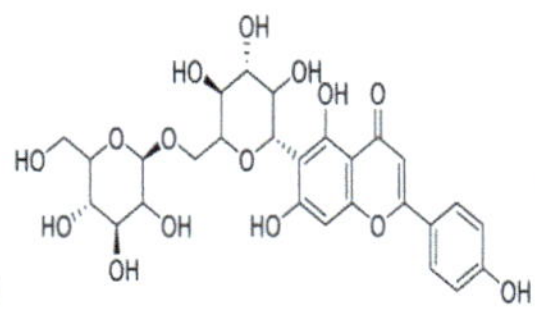

(14)

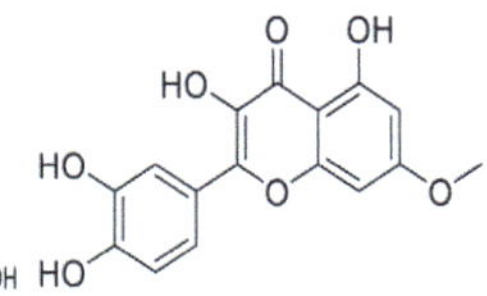

(15)

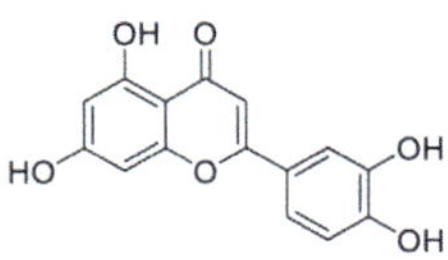

(16)

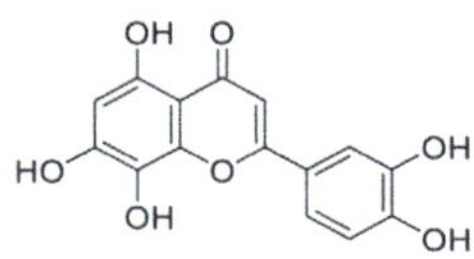

(17)

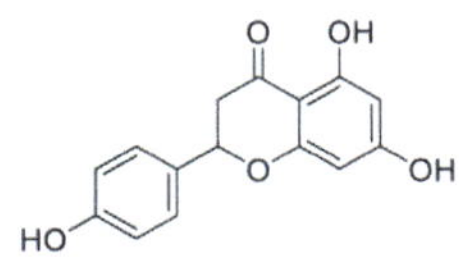

(18)

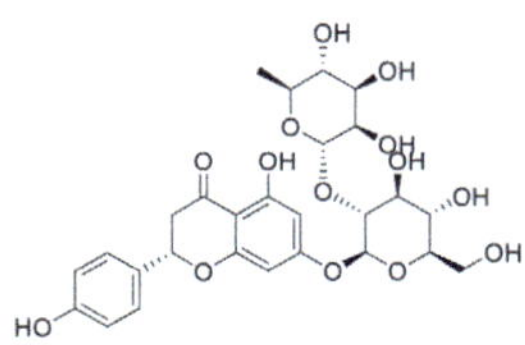

(19)

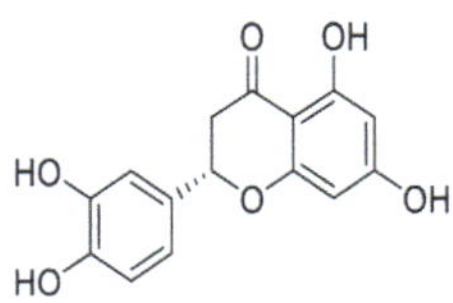

(20)

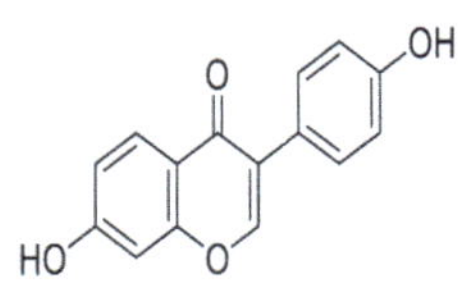

(21)

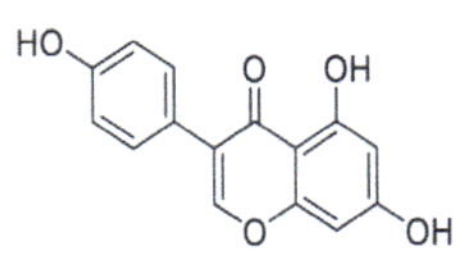

(22)

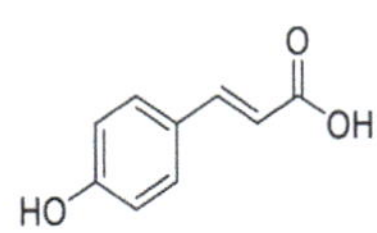

(23)

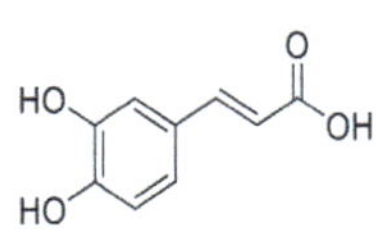

(24)

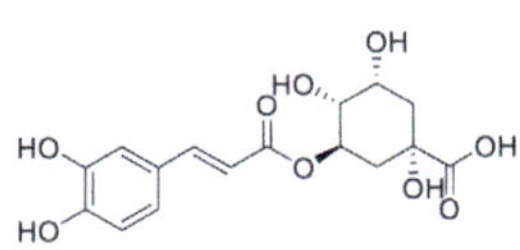

(25)

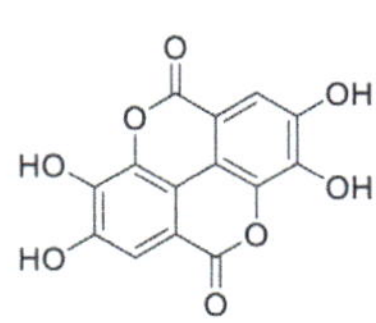

(26)

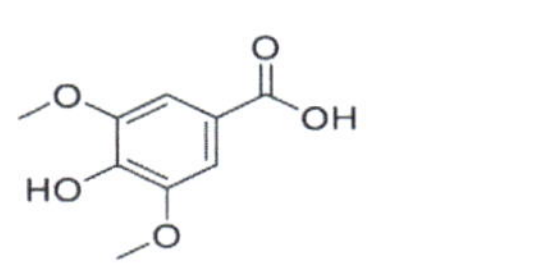

(27)

FIGURE 29.1 Bioactive components of *Vigna radiata* (L.) Wilczek (Mung bean), Cyanidin-3-glucoside (1), Delphinidin 3-O-glucoside (2), Peonidin-3-glucoside (3), Pelargonidin-3-(6"-malonylglucoside) (4), 2′,4,4′-Trihydroxychalcone (5), Phloretin (6), Quercetin (7), Myricetin (8), Kaempferol (9), Kaempferol 3-O rutinoside (10), Catechin (11), Vitexin (12), Isovitexin (13), Isovitexin-6"-O-α-L-glucoside (14), Rhamnetin (15), Luteolin (16), Hypolaetin (17), Naringenin (18), Naringin (19), Eriodictyol (20), Daidzein (21), Genistein (22), p-Coumaric acid (23), Caffeic acid (24), Chlorogenic acid (25), Ellagic acid (26), Syringic acid (27).

al., 2006; Hayat et al., 2014). Many studies reported the strong antioxidant activity of mung bean extracts. According to one of them, the highest free radical scavenging activity of mung bean extracts was observed in ABTS (36.65 ± 0.63 μmol/g), followed by FRAP (31.85 ± 3.03 μmol/g) and DPPH (11.33 ± 0.24 μmol/g) (Tang et al., 2014). Rani et al. (2018) reported a higher phenol content (68.62 mg GAE/g) in methanolic extracts of the seed coat, leading to higher DPPH scavenging potential (0.61 mg/mL). Germination of mung bean sprouts has led to a dramatic increase in the amount of Vitamin C which is 24 times higher than the initial concertation in mung bean seeds, and also an increase in total phenols and flavonoids in a time-dependent manner. As a result, the free radical scavenging activity of sprouts increased by six times compared to mung bean seed (Guo et al., 2012). Another study also reported the escalated level of vitamin C and flavonoid content in euphylla after 6 days of seed germination, with maximum peroxyl radical scavenging activity (Wang et al., 2017). The DPPH radical scavenging activity of various extracts of young mung bean sprouts and seeds ranged from 18.5 to 90.9% and from 13.5 to 24.9%, respectively, of which ethyl acetate fractions showed the highest radical scavenging potential in both sprouts and seeds (Kim et al., 2012). Krishnappa and the team assessed the antioxidant activity of mung bean exudate, husk, and the germinated seed of different stages. Exudate had substantially lower IC_{50} values for ABTS, DPPH, and H_2O_2 radical scavenging activity and higher absorbance for reducing power assay compared to standard antioxidant butylated hydroxyanisole (BHA), which indicate stronger radical scavenging potential of the sample (Krishnappa et al., 2017). Many studies reported the strong antioxidant potential of mung bean isolates. DPPH, ABTS, and FRAP assay was performed to study the antioxidant potential of isolates. Almost all of the studied compounds showed higher ABTS scavenging activity, and moderately strong activity in FRAP assay, while only two of them give better results against DPPH (Bai et al., 2016a).

29.3.2 Antimicrobial Activity

The use of phyto-components as natural antimicrobial agents is becoming more prevalent. Enzymes, polypeptides, and polyphenols derived from mung bean exhibit both antifungal and antimicrobial activities. Antifungal activity is usually performed using the crescent inhibition method, while antimicrobial activity is assessed by the agar well diffusion method or the deferred plate method (Wang et al., 2009a, b). Many studies reported the increase in phenol content and enhancement in defence mechanisms during germination. The screening of mung bean sprouts was carried out against 12 gram-positive and gram-negative bacteria and 12 fungi using the disc diffusion method, followed by the determination of minimum inhibitory concentration. The results of the analysis indicate that the mung bean sprout extract has significant activity against 11 bacteria and 2 fungi. It also showed remarkable antimicrobial potential against multiple drug-resistant bacteria such as *Staphylococcus aureus*, MRSA super bug, multiple drug-resistant *Escherichia coli* O157:H7, *Salmonella typhimurium,* and *Klebsiella pneumoniae*, and pathogenic fungus like *Trichoderma harzianum* and *Trichophyton rubrum*. This study suggests mung bean extract as a key source of novel, inexpensive, and readily accessible antimicrobial agents (Hafidh et al., 2011). Camalxaman and coworkers studied the antibacterial activity of methanol and chloroform extract of mung bean sprouts against gram-negative bacteria (*Escherichia coli, Pseudomonas aeruginosa, Salmonella* spp., and *Klebsiella pneumoniae)* by performing the disc diffusion assay and quantifying minimum inhibitory concentration (MIC). Both the above-mentioned extracts possessed efficacious antibacterial activity against all the gram-negative bacteria excluding *K. pneumoniae,* which was resistant (Camalxaman et al., 2013). Mung bean sprouts also possess effective antiviral and prophylactic activities comparable to Acyclovir against respiratory syncytial virus and Herpes simplex virus. The bioactive components of mung bean sprouts stimulate the release of cytokines in the human body and negate the actions on viral proliferation (Hafidh et al., 2015). Additionally, two proteins isolated from seeds of mung bean, that is, Mungin and Chitinase exhibited potential antifungal activity (Wang et al., 2005).

29.3.3 Anti-Inflammatory Activity

Inflammation was considered the basis of many diseases. A natural source of anti-inflammatory agents is important in preventing such diseases. Mung beans are an important part of Asian cuisine and are also used as folk remedies to treat heatstroke associated with thirst, irritability, and high body temperature, and such health-promoting effects are believed to be associated with inflammatory responses. The anti-inflammatory activity of mung bean extract was assessed by the protease inhibition model. The whole bean showed potential inhibition of protease (IC_{50} value 1.45 mg/mL) (Luo et al., 2016). Ali and coworkers studied the in-vitro and in-vivo anti-inflammatory activity of fermented mung bean, germinated mung bean, and untreated mung bean. The inhibition of nitric oxide (inflammatory mediator) represents the in-vitro study, while the in-vivo study was carried out as inhibition of ear edema and decrement in response to pain stimuli. Based on the outcomes, both germinated mung bean and fermented mung bean possessed efficacious anti-inflammatory activity in a dose-dependent fashion. The outcomes of the in-vitro study revealed that germinated mung bean and fermented mung bean strongly inhibited nitric oxide at 2.5 and 5 mg/mL. Moreover, the in-vivo study revealed that the aqueous extract of germinated mung bean and fermented mung bean reduced arachidonic acid-induced ear edema in mice at a concentration of 1000 mg/kg (Ali et al., 2014).

29.3.4 Antidiabetic Activity

Mung bean extracts have also been tested for their antidiabetic properties. Because of weak digestibility and release of saccharides, beans are known as low glycemic index (GI) food (Chung et al., 2008). Many studies reported that low GI foods help to reduce the complications and prevalence of obesity and diabetes (Sreerama et al., 2012; Mushtaq et al., 2014; Kang et al., 2015). A study conducted in 2012 revealed that fermented and nonfermented mung bean extracts had no substantial hypoglycaemic effect, but there is a noteworthy reduction in blood sugar level of glucose-induced and alloxan-induced hyperglycemic mice (Yeap et al., 2012). Natural resistance against enzyme α-glycosidase may be the primary approach to develop antidiabetic molecules. According to the research, lignans have a more positive effect on the α-glucosidase inhibitor than flavonoids, which was in great agreement with the FRAP analysis method. Furthermore, the presence of glucose molecule at 7-OH or 4′-OH significantly reduces α-glucosidase inhibition between isoflavones skeleton (Bai et al., 2016b). Another study reported considerable aldose reductase inhibition potential of the hull (IC_{50} value 1.45 mg/mL), cotyledon (IC_{50} value 3.51 mg/mL), and whole bean of mung (IC_{50} Value 3.21 mg/mL) (Luo et al., 2016). According to Yao and coworkers, the mung bean sprout (MBS) and mung bean seed coat also exhibited potent antidiabetic activity against type 2 diabetic mice (Male KK-Ay mice). When the extracts of MBS and MBSC were orally administered to male KK-Ay mice for five weeks, the results indicated a considerable increase in glucose tolerance and insulin immunoreactive levels, and also a decrease in blood glucose level, glucagon, plasma C-peptide, cholesterol, triglycerides, and blood urea nitrogen level (Yao et al., 2008).

29.3.5 Anticancer Activity

A novel antitumor and immunostimulatory effects of methanolic extracts of mung bean sprouts were studied in HeLa (cervix adenocarcinoma) and HepG2 (hepatocellular carcinoma) cell lines by assessing anticancer cytokines, immunological cytokines, regulatory genes of the cell cycle, apoptotic gene expression, tumor-suppressing genes, and the percentage of apoptotic cells. These findings strongly suggest that the mung bean is a powerful antitumor and immunostimulatory agent, opening up new avenues for anticancer therapy (Ganesan and Xu, 2018). Kim et al. performed the 3-(4, 5-dimethylthiazol -2-yl)-2, 5-diphenyltetrazolium bromide (MTT) assay to check the antiproliferative properties of different extracts of mung bean seeds and sprouts. The extracts of mung bean sprouts showed more positive effects against Calu-6 (human pulmonary carcinoma) and SNU-601 (human gastric carcinoma) compared to seeds. Also, ethanolic extract of both sprouts and seeds showed the highest activity, followed by n-hexane and butanol (Kim et al., 2012). Another study also reported the anticancer and immunomodulatory potential of mung bean sprouts and investigated the possible mechanism against both human cervical and hepatocarcinoma cancer cells. The findings indicated the MBS extract as a very promising cytotoxic agent in anticancer therapies because it affects both known and unknown targets. These antioxidant activities are driven by more than one factor, which allows MBS extracts to provide highly strong, multi-mechanism, and synergistic anticancer effects (Hafidh et al., 2012).

29.3.6 Anti-Hypertensive Effects

The administration of fractions of raw sprouts, dried sprouts, and enzyme digested sprouts resulted in a considerable reduction of systolic blood pressure of rats. Related findings were achieved in the plasma angiotensin I` converting enzyme (ACE) activity of sprouts. Longer-term intervention studies have been carried out which include treatment with freshly prepared sprout powder, dried sprout powder, and concentrated extracts. The findings suggest that the concentrated extracts were more effective than the dried powders (Hsu et al., 2011; Tang et al., 2014). Angiotensin I-converting enzyme activity of mung bean protein isolates was determined, which were prior hydrolyzed with the commercially available proteases alcalase and neutrase. The non-hydrolyzed proteins did not show any inhibitory activity. The hydrolyzed proteins obtained with alcalase exhibited higher ACE inhibitory activity (IC_{50} value 0.64 mg protein/mL), while hydrolysates obtained with neutrase showed comparatively lower inhibitory potential. The results revealed that the protein isolated from mung bean can be a good source of peptides with ACE inhibitory potential when hydrolyzed with enzyme alcalase (Li et al., 2005).

29.3.7 Antisepsis Effects

The aqueous extract of mung bean coat (MBC) exerts in-vitro and in-vivo antisepsis effects. The inhibition of nucleosomal protein named high mobility group box 1(HMGB1), which is identified as a late mediator of lethal inflammation, causes antisepsis effects. MBC was observed to reduce the LPs-induced generation of HMGB1 and many other chemokines in macrophage cultures in a dose-dependent manner. The survival rate of animals after oral administration of MBC was greatly improved, increasing from 29.4 to 70% (Zhu et al., 2012). Chlorogenic acid has also been found to protect against fatal sepsis by inhibiting sepsis-related late mediators. Furthermore, external administration of chlorogenic acid greatly reduced systematic HMGB1 aggregation in-vivo and avoided mortality rate by sepsis like endotoxemia and microbial sepsis (Lee et al., 2012). The other study reported that MBC extract is very much protective against malignant sepsis by stimulating autophagic degradation of HMGB1 (Zhu et al., 2012).

29.3.8 Myocardial Preservation Effects

An antecedent of vitexin, iso-vitexin, and mung bean sprout extract had a considerable protective effect on ISO-induced myocardial ischemia by clearly inhibiting an increase in serum levels of LDH, CK, and AST (Bai et al., 2016).

29.3.9 Tyrosinase Inhibition Activity

Tyrosinase is a crucial rate-limiting enzyme in the development of melanogenesis. Its abnormal manifestations can lead to various dermatological diseases. According to the previous analysis, the ethanolic extract of mung bean was found to have the highest potential tyrosinase inhibition among

16 legumes (Yao et al., 2011). Methanolic extract and fractions of sprouts exhibited maximum tyrosinase inhibition potential than seeds. Ethyl acetate fraction of sprouts possesses the highest activity, which is followed by butanol and n-hexane (Kim et al., 2012). The proanthocyanidins of mung bean seed were comprised of procyanidins, prodelphinidins, and their rhamnosides. The enzyme dynamic study of these compounds suggests them to be a powerful, reversible, and mixed type of tyrosinase inhibitor. They obstructed enzyme activity by interfering with both the enzyme and substrates (Chai et al., 2018).

KEYWORDS

- ***Vigna radiata***
- **legume crop**
- **bioactive components**
- **pharmacological profile**
- **anti-microbial activity**
- **antihypertensive effects**

REFERENCES

Ali, N. M.; Mohd Yusof, H.; Yeap, S. K.; Ho, W. Y.; Beh, B. K.; Long, K.; Koh, S. P.; Abdullah, M. P.; Alitheen, N. B. Anti-Inflammatory and Antinociceptive Activities of Untreated, Germinated, and Feremented Mung Bean Aqueous Extract. *Evid. Based Complement. Altern. Med.* **2014,** Article ID 350507.

Bai, Y.; Chang, J.; Xu, Y.; Cheng, D.; Liu, H.; Zhao, Y.; Yu, Z. Antioxidant and Myocardial Preservation Activities of Natural Phytochemicals from Mung Bean (*Vigna radiata L.*) Seeds. *J. Agric. Food Chem.* **2016a,** *64* (22), 4648–4655.

Bai, Y.; Xu, Y.; Change, J.; Wang, X.; Zhao, Y.; Yu, Z. Bioactives from Stems and Leaves of Mung Beans (*Vigna radiata L.*). *J. Funct. Foods*. **2016b,** *25*, 314–322.

Camalxaman, S. N.; Zain, Z. M.; Amom, Z.; Mustakim, M.; Mohamed, E.; Rambely, A. S. In Vitro Antimicrobial Activity of *Vigna radiata* (L) Wilzeck Extracts Against Gram Negative Enteric Bacteria. *World Appl. Sci. J.* **2013,** *21* (10), 1490–1494.

Chai, W. M.; Ou-Yang, C.; Huang, Q.; Lin, M. Z.; Wang, Y. X.; Xu, K. L.; Huang, W. Y.; Pang, D. D. Antityrosinase and Antioxidant Properties of Mung Bean Seed Proanthocyanidins: Novel Insights Into The Inhibitory Mechanism. *Food Chem.* **2018,** *260*, 27–36.

Chauhan, Y. S.; Williams, R. Physiological and Agronomic Strategies to Increase Mungbean Yield in Climatically Variable Environments of Northern Australia. *Agronomy*. **2018,** *8* (6), 83.

Chung, H. J.; Liu, Q.; Pauls, K. P.; Fan, M. Z.; Yada, R. In Vitro Starch Digestibility: Expected Glycemic Index and Some Physicochemical Properties of Starch and Flour from Common Bean (*Phaseolus vulgaris* L.) Varieties Grown in Canada. *Food Res. Int.* **2008,** *41* (9), 869–875.

El-Adawy, T. A.; Rahma, E. H.; El-Bedawey, A. A.; El-Beltagy, A. E. Nutritional Potential and Functional Properties of Germinated Mung Bean, Pea and Lentil Seeds. *Plant Foods Hum. Nutr.* **2003,** *58* (3), 1–13.

Ganesan, K.; Xu, B. A Critical Review on Phytochemical Profile and Health Promoting Effects of Mung Bean (*Vigna radiata*). *Food Sci. Hum. Well.* **2018,** *7* (1), 11–33.

Guo, X.; Li, T.; Tang, K.; Liu, R. H. Effect of Germination on Phytochemical Profiles and Antioxidant Activity of Mung Bean Sprouts (*Vigna radiata*). *J. Agric. Food Chem.* **2012,** *60* (44), 11050–11055.

Hafidh, R. R.; Abdulamir, A. S.; Bakar, F. A.; Jalilian, F. A.; Abas, F.; Sekawi, Z. Novel Molecular, Cytotoxical, and Immunological Study on Promising and Selective Anticancer Activity of Mung Bean Sprouts. *BMC Complement. Altern. Med.* **2012,** *12* (1), 1–24.

Hafidh, R. R.; Abdulamir, A. S.; Bakar, F. A.; Sekawi, Z.; Jahansheri, F.; Jalilian, F. A. Novel Antiviral Activity of Mung Bean Sprouts Against Respiratory Syncytial Virus and Herpes Simplex Virus −1: An In Vitro Study on Virally Infected Vero and MRC-5 Cell Lines. *BMC Complement. Altern. Med.* **2015,** *15* (1), 1–16.

Hafidh, R.; Abdulamir, A. S.; Vern, L. S.; Bakar, F. A.; Abas, F.; Jahanshiri, F.; Sekawi, Z. Novel In-Vitro Antimicrobial Activity of *Vigna radiata* (L.) R. Wilczek Against Highly Resistant Bacterial and Fungal Pathogens. *J. Med. Plants Res.* **2011,** *5* (16), 3606–3618.

Hayat, I.; Ahmad, A.; Masud, T.; Ahmed, A.; Bashir, S. Nutritional and Health Perspectives of Beans (*Phaseolus vulgaris* L.): An Overview. *Crit. Rev. Food Sci. Nutr.* **2014,** *54* (5), 580–592.

Hou, D.; Yousaf, L.; Xue, Y.; Hu, J.; Wu, J.; Hu, X.; Feng, N.; Shen, Q. Mung Bean (*Vigna radiata* L.): Bioactive polyphenols, polysaccharides, peptides, and health benefits. *Nutrients*, **2019,** *11* (6), 1238.

Hsu, G. S. W., LU, Y. F.; Chang, S. H.; Hsu, S. Y. Antihypertensive Effect of Mung Bean Sprout Extracts in Spontaneously Hypertensive Rats. *J. Food Biochem.* **2011,** *35* (1), 278–288.

Johnson, J.; Collins, T.; Power, A.; Chandra, S.; Skylas, D.; Portman, D.; Panozzo, J.; Blanchard, C.; Naiker, M. Antioxidative Properties and Macrochemical Composition of Five Commercial Mungbean Varieties in Australia. *Legume Sci.* **2019,** *2* (1), e27.

Kang, I.; Choi, S.; Ha, T. J.; Choi, M.; Wi, H. R.; Lee, B. W.; Lee, M. Effects of Mung Bean (*Vigna radiata* L.) Ethanol Extracts Decrease Proinflammatory Cytokine-Induced Lipogenesis in the KK-Ay Diabese Mouse Model. *J. Med. Food.* **2015,** *18* (8), 841–849.

Kim, D. K.; Jeong, S. C.; Gorinstein, S.; Chon, S. U. Total Polyphenols, Antioxidant and Antiproliferative Activities of Different Extracts in Mungbean Seeds and Sprouts. *Plant Foods Hum. Nutr.* **2012,** *67* (1), 71–75.

Kim, S. K.; Nair, R. M.; Lee, J.; Lee, S. H. Genomic Resources in Mungbean for Future Breeding Programs. *Front. Plant Sci.* **2015,** *6*, 626.

Krishnappa, N. P.; Basha, S. A.; Negi, P. S.; Prasada Rao, U. J. Phenolic Acid Composition, Antioxidant and Antimicrobial Activities of Green Gram (*Vigna radiata*) Exudate, Husk, and Germinated Seed of Different Stages. *J. Food Process. Preserv.* **2017,** *41* (6), e13273.

Lambrides, C. J.; Godwin, I. D. Mungbean. In *Pulses, Sugar and Tuber Crops*; Springer: Berlin, Heidelberg, 2007; pp 69–90.

Lee, C. H.; Yoon, S. J.; Lee, S. M. Chlorogenic Acid Attenuates High Mobility Group Box 1 (HMGB1) and Enhances Host Defence Mechanisms in Murine Sepsis. *Mol. Med.* **2012,** *18* (1), 1437–1448.

Lee, J. H.; Jeon, J. K.; Kim, S. G.; Kim, S. H.; Chun, T.; Imm, J. Y. Comparative Analyses of Total Phenols, Flavonoids, Saponins and Antioxidant Activity in Yellow Soy Beans and Mung Beans. *Int. J. Food Sci. Technol.* **2011,** *46*, 2513–2519.

Li, G. H.; Le, G. W.; Liu, H.; Shi, Y. H. Mung-Bean Protein Hydrolysates Obtained with Alcalase Exhibit Angiotensin I-Converting Enzyme Inhibitory Activity. *Food Sci. Technol. Int.* **2005,** *11* (4), 281–287.

Lin, X. X. L. H.; Li, W. Z. The Research of Mung Bean SOD Oral Liquid. *Food Sci.* **1997,** *18*, 25–26.

Luo, J.; Cai, W.; Wu, T.; Xu, B. Phytochemical Distribution in Hull and Cotyledon of Adzuki Bean (*Vigna angularis* L.) and Mung Bean (*Vigna radiata* L.), and Their Contribution to Antioxidant, Anti-Inflammatory and Anti-Diabetic Activities. *Food Chem.* **2016,** *201*, 350–360.

Mubarak, A. E. Nutritional Composition and Antinutritional Factors of Mung Bean Seeds (*Phaseolus aureus*) as Affected by Some Home Traditional Processes. *Food Chem.* **2005,** *89* (4), 489–495.

Mushtaq, Z.; Imran, M.; Salim-ur-Rehman; Zahoor, T.; Ahmad, R. S.; Arshad, M. U. Biochemical Perspectives of Xylitol Extracted from Indigenous Agricultural By-Product Mung Bean (*Vigna radiata*) Hulls in a Rat Model. *J. Sci. Food Agric.* **2014,** *94* (5), 969–974.

Rani, S.; Lamba, S.; Khabiruddin, M. Interactive Study of Phytochemicals and Their Antioxidant Efficiencies in Mungbean (*Vigna radiata* L.). *J. Pharmacogn. Phytochem.* **2018,** *7* (1), 1739–1744.

Reynoso-Camacho, R.; Ramos-Gomez, M.; Loarca-Pina, G. Bioactive Components in Common Beans (*Phaseolus vulgaris* L.). In *Advances in Agricultural and Food Biotechnology*; Guevara-Gonzalez, R. G., Torres-Pacheco, I., Eds.; 2006; pp 217–236.

Sehrawat, N.; Yadav, M.; Kumar, S.; Upadhyay, S. K.; Singh, M.; Sharma, A. K. Review on Health Promoting Biological Activities of Mungbean: A Potent Functional Food of Medicinal Importance. *Plant Arch.* **2020,** *20* (2), 2969–2975.

Shi, Z.; Yao, Y.; Zhu, Y.; Ren, G. Nutritional Composition and Antioxidant Activity of Twenty Mung Bean Cultivars in China. *Crop J.* **2016,** *4* (5), 398–406.

Shukla, S.; Tyagi, B. Comparative Phytochemical Screening and Analysis of Different *Vigna* Species in Organic Solvents. *Austin. J. Biotechnol. Bioeng.* **2017,** *4* (3), 1084.

Silva, L. R.; Pereira, M. J.; Azevedo, J.; Gonçalves, R. F.; Valentão, P.; de Pinho, P. G.; Andrade, P. B. *Glycine max* (L.) Merr., *Vigna radiata* L. and *Medicago sativa* L. Sprouts: A Natural Source of Bioactive Compounds. *Food Res. Int.* **2013,** *50* (1), 167–175.

Singh, B.; Singh, J. P.; Singh, N.; Kaur, A. Saponins in Pulses and Their Health Promoting Activities: A Review. *Food Chem.* **2017,** *233*, 540–549.

Singh, J.; Kanaujia, R.; Singh, N. P. Pulse Phytonutrients: nutritional and medicinal importance. *J. Pharm. Nutr. Sci.* **2016,** 6, 160–171.

Sosulski, F. W.; Dabrowski, K. J. Composition of Free and Hydrolyzable Phenolic Acids in the Flours and Hulls of Ten Legume Species. *J. Agric. Food Chem.* **1984,** *32* (1), 131–133.

Sreerama, Y. N.; Takahashi, Y.; Yamaki, K. Phenolic Antioxidants in Some *Vigna* Species of Legumes and Their Distinct Inhibitory Effects on—Glucosidase and Pancreatic Lipase Activities. *J. Food Sci.* **2012,** *77*, C927–C933.

Tang, D.; Dong, Y.; Ren, H.; Li, L.; He, C. A Review of Phytochemistry, Metabolite Changes, and Medicinal Uses of the Common Food Mung Bean and Its Sprouts (*Vigna radiata*). *Chem. Cent. J.* **2014,** *8* (1), 1–9.

Wang, J.; Ye, Y.; Li, Q.; Abbasi, A. M.; Guo, X. Assessment of Phytochemicals, Enzymatic and Antioxidant Activities in Germinated Mung Bean (*Vigna radiata* L. Wilezek). *Int. J. Food Sci. Technol.* **2017,** *52* (5), 1276–1282.

Wang, S.; Ng, T. B.; Chen, T.; Lin, D.; Wu, J.; Rao, P.; Ye, X. First Report of a Novel Plant Lysozyme with Both Antifungal and Antibacterial Activities. *Biochem. Biophys. Res. Commun.* **2005,** *327* (3), 820–827.

Wang, S.; Rao, P.; Ye, X. Isolation and Biochemical Characterization of a Novel Leguminous Defence Peptide with Antifungal and Antiproliferative Potency. *Appl. Microbiol. Biotechnol.* **2009a,** *82* (1), 79–86.

Wang, S.; Shao, B.; Fu, H.; Rao, P. Isolation of a Thermostable Legume Chitinase and Study on the Antifungal Activity. *Appl. Microbiol. Biotechnol.* **2009b,** *85* (2), 313–321.

Xie, J.; Du, M.; Shen, M.; Wu, T.; Lin, L. Physico-Chemical Properties, Antioxidant Activities and Angiotensin-I Converting Enzyme Inhibitory of Protein Hydrolysates from Mung Bean (*Vigna radiata*). *Food Chem.* **2019,** *270*, 243–250.

Xue, Z.; Wang, C.; Zhai, L.; Yu, W.; Chang, H.; Kou, X.; Zhou, F. Bioactive Compounds and Antioxidant Activity of Mung Bean (*Vigna radiata* L.), Soybean (*Glycine max* L.) and Black Bean (*Phaseolus vulgaris* L.) During the Germination Process. *Czech J. Food Sci.* **2016,** *34* (1), 68–78.

Yao, Y.; Chen, F.; Wang, M.; Wang, J.; Ren, G. Antidiabetic Activity of Mung Bean Extracts in Diabetic KK-Ay Mice. *J. Agric. Food Chem.* **2008,** *56* (19), 8869–8873.

Yao, Y.; Cheng, X.; Wang, L.; Wang, S.; Ren, G. Biological Potential of Sixteen Legumes in China. *Int. J. Mol. Sci.* **2011,** *12* (10), 7048–7058.

Yeap, S. K.; Mohd Ali, N.; Mohd Yusof, H.; Alitheen, N. B.; Beh, B. K.; Ho, W. Y.; Koh, S. P.; Long, K. Antihyperglycemic Effects of Fermented and Nonfermented Mung Bean Extracts on Alloxan-Induced-Diabetic Mice. *J. Biomed. Biotechnol.* **2012,** *2012*, Article ID 285430.

Yi-Shen, Z.; Shuai, S.; FitzGerald, R. Mung Bean Proteins and Peptides: Nutritional, Functional and Bioactive Properties. *Food Nutr. Res.* **2018,** 62. https://doi.org/10.29219/fnr.v62.1290.

Zhu, S.; Li, W.; Li, J.; Jundoria, A.; Sama, A. E.; Wang, H. It Is Not Just Folklore: The Aqueous Extract of Mung Bean Coat Is Protective Against Sepsis. *Evid. Based Complement. Altern. Med.* **2012,** *2012*, Article ID 498467.

CHAPTER 30

Bioactive Compounds and Pharmacological Activities of *Acacia dealbata* Link and *Acacia mearnsii* De Wild. (Wattle)

YOGESH CHAND YADAV[1*], PANKAJ YADAV[1], PRADEEP KUMAR[2], NEETESH K. JAIN[3], and SUMEET DWIVEDI[4]

[1]*Faculty of Pharmacy, Uttar Pradesh University of Medical Sciences, Saifai, Etawah 206130, Uttar Pradesh, India*

[2]*Chaudhary Sughar Singh College of Pharmacy, Jaswant Nagar Etawah 206130 Uttar Pradesh, India*

[3]*OCPR, Faculty of Pharmacy, Oriental University, Indore, Madhya Pradesh, India*

[4]*UIP, Faculty of Pharmacy, Oriental University, Indore, Madhya Pradesh, India*

**Corresponding author. E-mail: drycy31@gmail.com*

ABSTRACT

Acacia dealbata (Black wattle) is a flowering plant belonging to the legume family Fabaceae. *A. dealbata* absolute oil has been used in the flavor and perfume industries and now its industrial production is estimated about five tons per year worldwide. It consists of many flavonoids such as luteolin derivatives (with 6- or 8-O-substitutions), a dihydroflavonol, quercetin derivatives with 3-O- substitutions and a phenolic acid derivative. Pharmacological activities of *Acacia dealbata* include antimicrobial activity and cytotoxic activity. It has good antioxidant potential to scavenging free radical active species.

Bioactives and Pharmacology of Legumes. T. Pullaiah, PhD (Ed.)

Acacia mearnsii is commonly known as black wattle, late black wattle or green wattle. The bioactives of *A. mearnsii* include leuco-fisetinidin, a flavan-3,4-diol (leucoanthocyanidin) and a monomer of the condensed tannins. The pharmacological activities reported include antiobesity, anti-diabetic, lipase inhibition and glucosidase activities and alleviating itching associated with atopic dermatitis, by preventing skin dryness.

30.1 INTRODUCTION

Acacia dealbata Link belongs to the legume family Fabaceae. It is distributed in Australia, in Mediterranean, warm temperate, and highland tropical landscapes (Kull et al., 2011). *Acacia dealbata* leaves, flowers, bark, and gum have been reported for the treatment of diseases (Sowndharajan et al., 2015).

Acacia mearnsii De Wild. [Syn. *Racosperma mearnsi* (De Wild.) Pedley] is commonly known as black wattle, late black wattle, or green wattle. *A. mearnsii* is a spreading shrub or erect tree that typically grows to a height of 10 m.

30.2 BIOACTIVES

30.2.1 Bioactives of Acacia dealbata

Perriot et al. (2010) reported that *Acacia dealbata* absolute oil has been used in the flavor and perfume industries and now its industrial production is estimated to be about five tons per year worldwide and acacia absolute oil contains straight-chain analogues from C6 to C26 with different functional groups (hydrocarbons, esters, aldehydes, diethyl acetals, alcohols, and ketones) identified in the volatile fraction. These include (Z)-heptadec-8-ene, heptadecane, nonadecane, and palmitic acid that are the most abundant, and 2-phenethyl alcohol, methyl anisate, and ethyl palmitate are present in smaller amounts. The heavier constituents were mainly triterpenoids such as lupenone and lupeol, which were identified as two of the main components. (Z)-Heptadec-8-ene, lupenone, and lupeol were quantified by GC-MS in SIM mode using external standards and represent 6%, 20%, and 7.8% (w/w) of the absolute oil (Perriot et al., 2010).

Many flavonoids in *Acacia dealbata* pollen were determined which include luteolin derivatives (with 6- or 8-O-substitutions), a dihydroflavonol, quercetin derivatives with 3-O-substitutions and a phenolic acid derivative, and two unknown compounds (Anjos et al., 2014).

Other bioactive compounds are also identified predominantly in *A dealbata* such as resorcinol, maculosin and moretenone in leaves, stigmasterol, d-alpha-tocopherol quinone, and lupaninin pods, and methyl p-anisate, p-anisyl alcohol, stigmasterol, and anisalin flowers and barks (Aguilera et al., 2015).

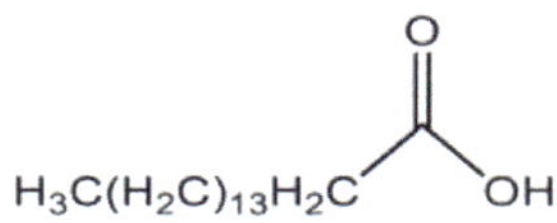

Palmitic acid

Ethyl palmitac

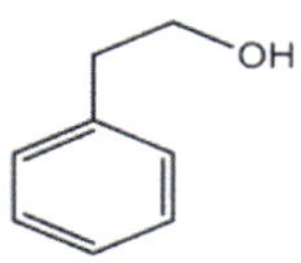

2-phenylalcohol

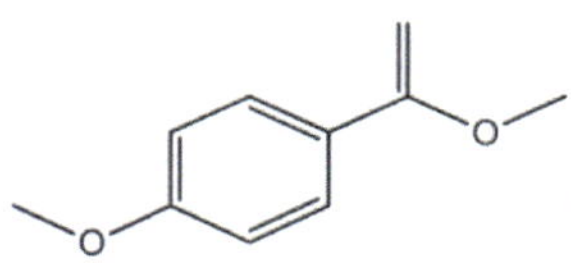

Methyl anisate

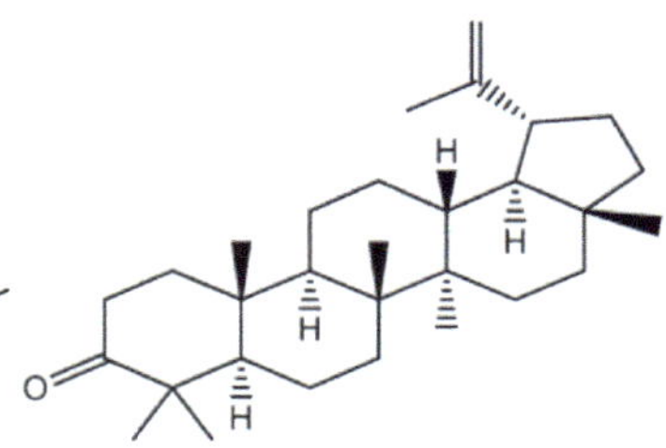

Lupenone

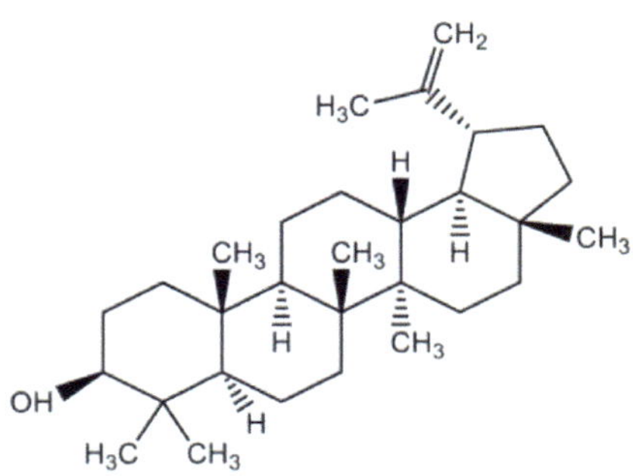

Lupeol

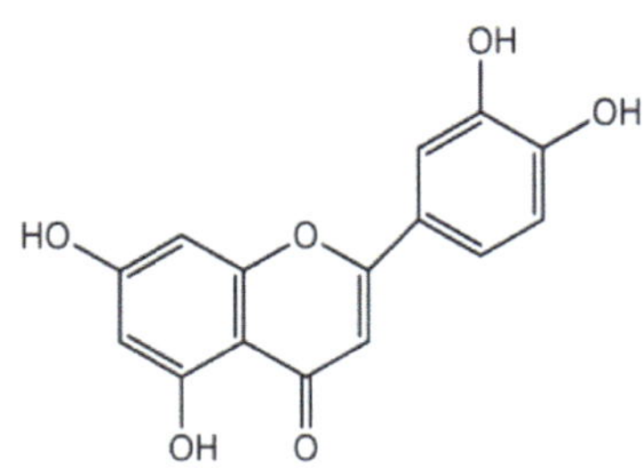

Luteolin

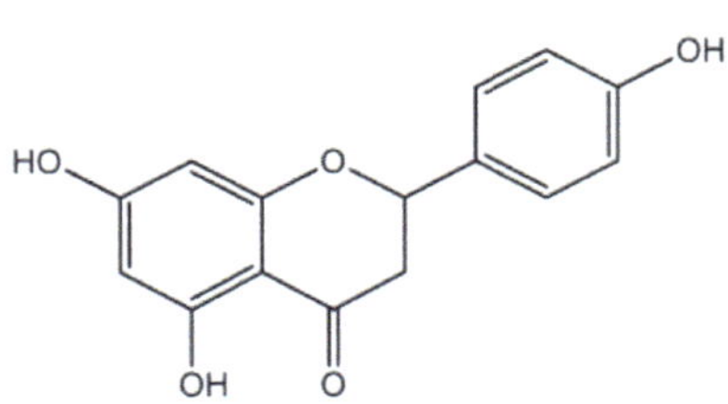

Naringenin

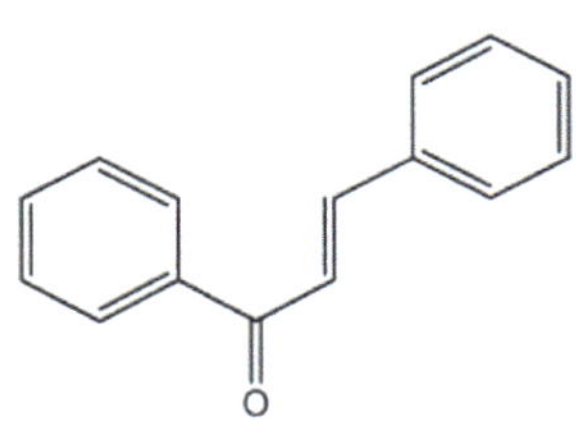

Chalcone

30.2.2 Bioactives of A. mearnsii

Leuco-fisetinidin, a flavan-3,4-diol (leucoanthocyanidin), and a monomer of the condensed tannins called profisetinidins can be extracted from the heartwood of *A. mearnsii* (Roux and Poulus, 1962). *A. mearnsii* bark extract consists of angular oligomers based on catechin and gallocatechin as starter units and fisetinidol and predominantly robinetinidol as extender units (Venter et al., 2012).

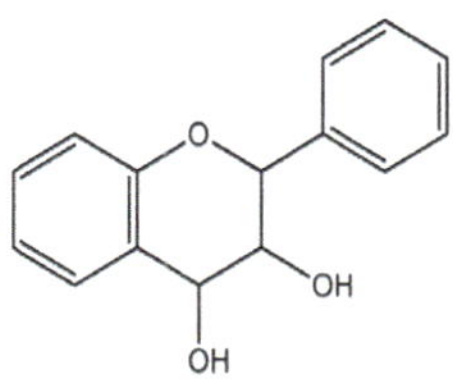

Leucoanthocyanidin

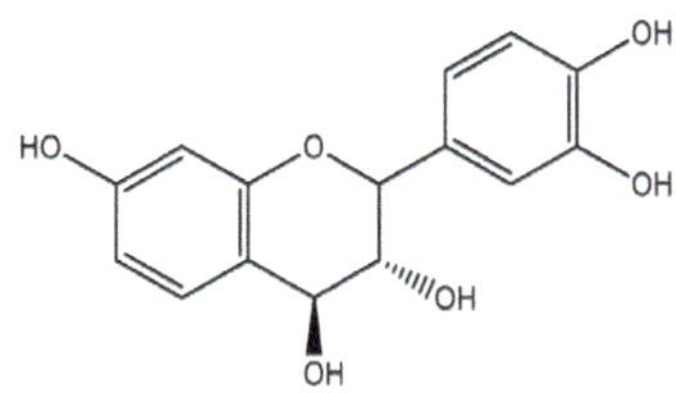

Leuco-fisetinidin

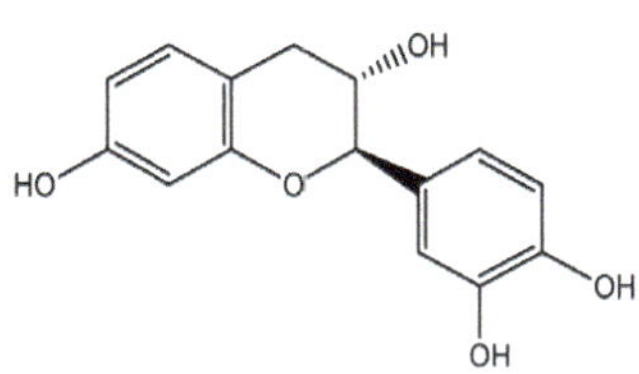

Fisetinidol

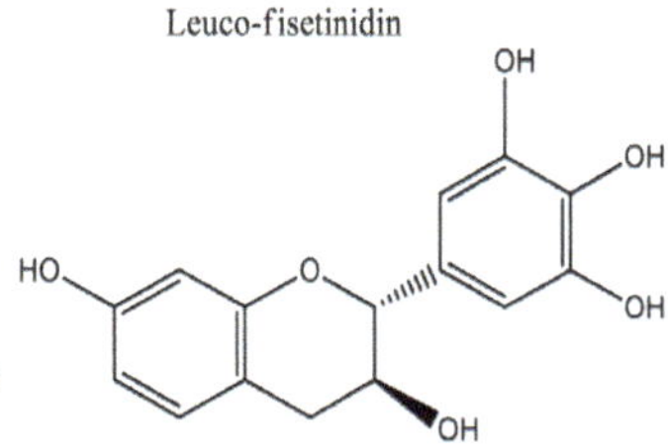

Robinetinidol

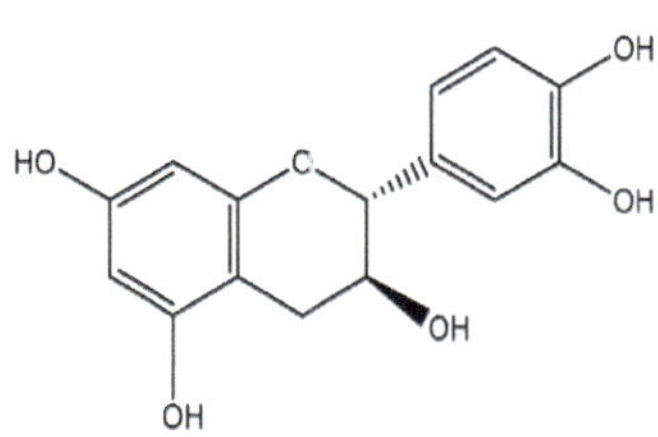

Catechin

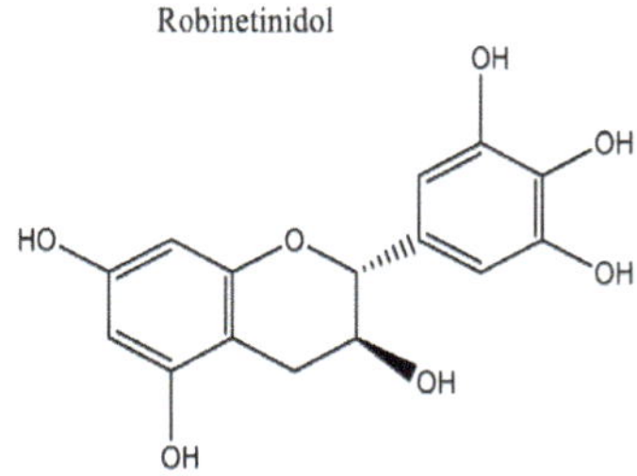

Galocatechin

30.3 PHARMACOLOGY

30.3.1 Pharmacology of Acacia dealbata

30.3.1.1 Antimicrobial Activity

The ethanolic extract of *Acacia dealbata* shows high antimicrobial activity than other species of *Acacia* (Silva et al., 2016). The sapwood, heartwood, and bark of *A. dealbata* extracts were tested for antimicrobial activity by the agar well diffusion method for bacteria and yeast against *Staphylococcus aureus* ATCC 25923, *Escherichia coli* ATCC 25922, *Enterococcus faecalis* ATCC 29212, *Pseudomonas aeruginosa* ATCC 27853, *Salmonella typhimurium* ATCC 14028, *Klebsiella pneumoniae* ATCC 13883, *Proteus mirabilis* ATCC 7002, *Listeria monocytogenes* ATCC 43251, *Candida parapsilosis* ATCC 22019, and *Candida albicans* ATCC 10231. The antimicrobial activity of *A. dealbata* was determined visually and measuring the diameter of clear inhibition zone around agar well shows positive antimicrobial activity. The MIC values for the ethanol extract were 6.25 ±1.07 mg/mL against *S. aureus* and 7.92 ±1.52 mg/mL against *C. albicans* (Yildiz et al., 2018).

30.3.1.2 Antioxidant Activity

Sowndharajan et al. (2015) reported the acetone extract of *Acacia dealbata* bark antioxidant activity. Hydroalcoholic extract of *A. dealbata* contains highest concentration of total phenols and had the greatest antioxidant activity. The antioxidant activity of this extract was performed by two methods, one was DPPH scavenging assay and the second one was β-carotene bleaching test (Luice et al., 2012). The various studies had documented that the extract of wood part (sapwood, heartwood, and bark) in methanol gives highest antioxidant activity (Yildiz et al., 2018).

The acetone and methanol extracts of *A. dealbata* bark showed antioxidant activity measured by using various antioxidant assays such as phosphomolybdenum assay, metal chelating activity, DPPH radical scavenging activity, and β-carotene bleaching assay (Sowndharajan et al., 2013).

30.3.1.3 Cytotoxic Activity

The cytotoxic activity was determined against colon carcinoma (HCT-116), pancreatic adenocarcinoma (PSN1), lung adeno-carcinoma epithelial cell line (A549), and Caucasian human glioblastoma cells (T98G), from the European collection of cell cultures by using *Acacia dealbata* flower extract, which was extracted with ethanol. All the extracts showed similar cytotoxicity against tumoral cell lines. The acidified extract showed highest inhibitory activity against lung A549 cells and colon HTC 116 cells (Casas et al., 2020).

30.3.2 Pharmacology of Acacia mearnsii

Acacia bark extract was evaluated for antiobesity, antidiabetic, lipase inhibition, and glucosidase activities in mouse models. *A. mearnsii* polyphenols demonstrated efficacy for alleviating metabolic syndrome, inhibiting lipase and glucosidase activities (Ikarashi et al., 2011), and alleviating itching associated with atopic dermatitis, by preventing skin dryness (Ikarashi et al., 2012). Although *A. mearnsii* bark have been extensively studied for phytochemical composition and biological activities, minimal research has been conducted on the leaves. Reportedly, leaves accumulate flavonoids including myricitrin, quercitrin, mearnsitrin, catechin, gallocatechin, and leuco-delphinidin tannin (Saayman and Roux, 1965; MacKenzie, 1967, 1969).

Xionga et al. (2016) reported antioxidant activities and anti-inflammatory activity of crude extract and semi-purified fractions in mouse macrophage cell line RAW 264.7.

KEYWORDS

- ***Acacia dealbata***
- ***Acacia mearnsii* antimicrobial activity**
- **cytotoxic activity**
- **antiobesity**
- **anti-diabetic activity**

REFERENCES

Aguilera, N.; Becera, J.; Parada, C. V.; Lorenzo, P.; Gonzalez, L.; Hernandez, V. Effects and Identification of Chemical Compounds Released from the Invasive *Acacia dealbata* Link. *Chem Ecol.* **2015,** *31* (6), 479–493.

Anjos, O.; Amâncio, D.; Serrano, M.; Campos, M. G. Determination of Structural Phenolic Compounds of *Acacia dealbata* Pollen by HPLC/DAD. *Planta Med.* **2014,** *80* (16), 80–82.

Casas, M. P.; Conde, E.; Ribeiro, D.; Fernandes, E.; Domínguez, H.; Torres, M. D. Bioactive Properties of *Acacia dealbata* Flowers Extracts. *Waste Biomass Valorization* **2020,** *11*, 2549–2557.

Ikarashi, N.; Toda, T.; Okaniwa, T.; Ito, K.; Ochiai, W.; Sugiyama, K. Anti-Obesity and Anti-Diabetic Effects of *Acacia* Polyphenol in Obese Diabetic KKAy Mice Fed High-Fat Diet. *Evidence-Based Complem. Altern. Med.* **2011,** 1–10. Article Id. 952031

Ikarashi, N.; Sato, W.; Toda, T.; Ishii, M.; Ochiai, W.; Sugiyama, K. Inhibitory Effect of Polyphenol-Rich Fraction from the Bark of *Acacia mearnsii* on Itching Associated with Allergic Dermatitis. *Evidence-Based Complem. Altern. Med.* **2012,** 1–9. DOI: 10.1155/2012/120389

Kull, C. A.; Shackleton, C. M.; Cunningham, P. J.; Ducatillon, C.; Dufour-Dror, J. M.; Esler, K. J. et al. Adoption, Use and Perception of Australian Acacias Around the World. *Divers. Distrib.* **2011,** *17* (5), 822–836.

Luice, A.; Gil, N.; Amaral, M. E.; Duarte, A. P. Antioxidant Activities of Extracts from *Acacia melanoxylon, Acacia dealbata* and *Olea europaea* and Alkaloids Estimation. *Int. J. Pharm. Pharm. Sci.* **2012,** *4*, 225–231.

MacKenzie, A. M. Mearnsitrin: A New Flavonol Glycoside from the Leaves of *Acacia mearnsii. Tetrahedron Lett.* **1967,** *8*, 2519–2520.

MacKenzie, A. M. The Flavonoids of the Leaves of *Acacia mearnsii. Phytochemistry* **1969,** *8*, 1813–1815.

Perriot, R.; Breme, K.; Meirerhenrich, U. J.; Carenini, E.; Ferrando, G.; Baldovini, N. Chemical Composition of French Mimosa Absolute Oil. *J. Agric. Food Chem.* **2010,** 58(3), 1844–1849.

Roux, D. G.; Paulus, E. Condensed Tannis. 12. Polymeric Leucofisetinidin Tannins from the Heartwood of *Acacia mearnsii. Biochemical J.* **1962,** *82 (2), 320–324.*

Saayman, H. M.; Roux, D. G. The Origins of Tannins and Flavonoids in Black-Wattle Barks and Heartwoods, and Their Associated "Non-Tannin" Components. *Biochem. J.* **1965,** *97*, 794–801.

Silva, E.; Fernandes, S.; Bacelar, E.; Sampaio A. Antimicrobial Activities of Aqueous Ethanolic and Methanolic Leaf Extracts from *Acacia* Species and *Eucalyptus nicholii. Afr. J. Tradit. Complem. Altern Med.* **2016,** *13* (6), 130–134.

Sowndharajan, K.; Hong, S.; Jhoo, J. W.; Kim, S.; Chin, N. L. Effect of Acetone Extract from Stem Bark of *Acacia* Species (*Acacia dealbata, Acacia ferruginea* and *Acacia leucopholoea*) on Antioxidant Enzymes Status in Hydrogen Peroxide- Induced HepG2 Cells. *Saudi J. Bio. Sci.* **2015,** *22*, 685–691.

Sowndharajan, K.; Joseph, J. M.; Manian, S. Antioxidant and Free Radical Scavenging Activities of Indian Acacias: *Acacia leucophloea* (Roxb.) Willd., *Acacia ferruginea* DC., *Acacia dealbata* Link. and *Acacia pennata* (L.) Willd. *Int. J. Food Properties* **2013,** *16*, 1717–1729.

Venter, P. B.; Senekal, N. D.; Kemp, G.; Amra-Jordaan, M.; Khan, P.; Bonnet, S. L.; van der Westhuizen, J. H. Analysis of Commercial Proanthocyanidins. Part 3: The Chemical Composition of Wattle (*Acacia mearnsii*) Bark Extract, *Phytochemistry* **2012,** *83*, 153–167.

Xionga, J.; Grace, M. H.; Esposito, D.; Wang, F.; Lila, M. A. Phytochemical Characterization and Anti-Inflammatory Properties of *Acacia mearnsii* Leaves. *Nat. Prod. Communications* **2016,** *11* (5), 649–653.

Yildiz, S.; Gürgen, A.; Can, Z.; Tabbouche, S. A.;Kilic, A. O. Some Bioactive Properties of *Acacia dealbata* Extract and Other Potential Utilization in Wood Protection. *Drewno* **2018,** *61* (202). DOI: 10. 12841/wood. 1644–3985. 255. 03

CHAPTER 31

An Insight of Phytochemical and Pharmacological Prospective of *Senna auriculata* (L.) Roxb.

LEPAKSHI M. D. BHAKSHU[1*], K. VENKATA RATNAM[2] and R. R. VENKATA RAJU[3]

[1]*Department of Botany, PVKN Government College (A), Chittoor, Andhra Pradesh 517002, India*

[2]*Department of Botany, Rayalaseema University, Kurnool, Kerala 518007, India*

[3]*Department of Botany, Sri Krishnadeveraya University, Ananthapuramu, Andhra Pradesh, 515003, India*

**Corresponding author. E-mail: lmbakshu@pvkngcchittor.ac.in*

ABSTRACT

Senna auriculata (L.) Roxb. (Syn.: *Cassia auriculata* L.) of Fabaceae is a popular and common shrub species widely growing in open places throughout India. It is popular with the regional names like tanner's cassia, tarwar, thangedu, avarthani, avarkin etc., *C. auriculata* is a well-known traditional medicinal plant used in Ayurvedic medicines. All parts such as roots, leaves, flowers, flower buds, stem bark, raw fruits are being used for treating human ailments such as diabetes, rheumatic pains, urinary disorders, fever, constipation etc. The plant is scientifically investigated for phytochemicals as well as pharmacological properties for which it is used. The plant is rich source of natural compounds such as flavonoids, avarasides, emodin, tannins, and anthracene glycosides. The

Bioactives and Pharmacology of Legumes. T. Pullaiah, PhD (Ed.)

present review has disclosed its phyto-constituents characterized so far and the pharmacological properties such as antimicrobial, antidiabetic, anti-inflammatory, anticancer activities etc., of *S. auriculata*. The plant possesses a diversified biological effect to retain the normal health which were discussed herein.

31.1 INTRODUCTION

Senna auriculata (L.) Roxb. (Syn.: *Cassia auriculata* L.) belongs to the family Fabaceae. It is evergreen shrub (1–1.5 m height) with attractive yellow color flowers. Leaves are dull green, alternate stipulate, paripinnately compound, with 16–24 pairs of leaflets, stipules auriculate, compound corymb, flowers bright yellow, irregular, and large (up to 5 cm). Legumes ranges 7.5–11 cm long and 1.5 cm broad, the shape is oblong, obtuse, with the persistent style, pods are thin and flattened, undulated and became brownish when dried, contain 10–20 seeds in each pod. It grows in various parts of India and Sri Lanka, appears commonly in open places, as well as other parts of Asia. *S. auriculata* is popularized with the names viz., tanner's cassia (English), avartaki, pitapuspa, pitakalika, manojyana, carmaranga, pitakala (Sanskrit); tarwar (Hindi), tangedu (Telugu), avarike (Kannada), avarai, avaram (Tamil), mataran tea, tanner's tea (Malaysia).

This is a well-known medicinal plant and all parts were used extensively in traditional systems since ages for curing ailments such as skin infections, diabetes, fever, inflammation, cold, hair cleanser, digestive issues, diarrhoea, worm infestation, improving fertility in women, illness, venereal disease, reducing body heat, vomiting, toothache, leprosy, urinary disorders, rheumatism, conjunctivitis and as tonic etc. (Jha et al., 2013; Guruprasad and Reddy, 2015; Salma et al., 2020). All parts of the plant, namely, leaves are used for bowl movements and constipation and digestive disorders, floral parts are beneficial in diabetes, the seeds are used for eye infections, whereas the stem bark is used in gout, gonorrhea, rheumatic pain, while the root powder is given for fever, urinary diseases etc., in the traditional systems of medicine especially in Ayurveda (Ayyanar and Ignacimuthu, 2008; Anandan et al., 2011; Kainsa et al., 2012; Padmavathi, 2018; Meena et al., 2019; Aye et al., 2019; Salma et al., 2020).

31.2 PHYTOCHEMICAL CONSTITUENTS

Chemical investigation of the bark of *S. auriculata* yielded two triterpenoid glycosides, 3β 24-dihydroxyurs-12-en-28-oic acid-24-*O*-β-D-xylopyranoside and 3β 24-dihydroxyurs-12-en-28-oic acid-3-*O*-β-D-xylopyranoside (Sanghi et al., 2000). A benzocoumarin glycoside, avaraside I, avaraol II, luteolin, kaempferol, quercetin, myricetin, 3-methoxy-luteolin, kaempferol 3-O-beta-D-glucopyranoside, isoquercetin, myricetin-3-O-D-glucopyranoside, kaempferol-3-O-rutinoside, rutin, myricetin 3-O-rutinoside, lanceolatin B, pseudosemiglabrin, (+)-catechin, (+)-gallocatechin, epicatechin, 6-demethoxycapillarisin, 6-demethoxy-7-methylcapillarisin, (2S)-7,4-dihydroxyflavan(4-8)-catechin, (2S) -7, 4-dihydroxyflavan (4-8)–gallocatechin, (2S)-7,4-dihydroxy-flavan (4-8)–epicatechin, (2S) -7,4- dihydroxy- flavan (4-8)–epigallo-catechin, chrysophanol, emodin, physcion, roseoside, bridelionoside-C, benzyl-*O*-D-apiofuranosyl (1,2)-D-glucopyranoside, epigallocatechin, etc., from the leaves (Markham et al., 1978; Nakamura et al., 2014).

Chemical investigation of roots reported for the presence of anthraquinone glycosides such as 1,3-dihydroxy-2-methyl anthraquinone; 1,3,8-trihydroxy–6-methoxy-2-methylanthraquinone; 1,8-dihydroxy-6-methoxy-2-methylanthraquinone-3-*O*-rutinoside, 1,8-dihydroxy-2-methylanthraquinone-3-*O*-rutinoside and a flavone glycoside and 7', 4'-dihydroxyflavone-5-*O*-D-galactopyranoside (Rai and Dasundhi, 1990). Rao et al. (2000) isolated and elucidated Myristyl alcohol, β-sitosterol β-D-glucoside, quercetin 3-*O*-glucoside, rutin, and di-2-ethyl-hexyl-phthalate from the ethanol extract of fresh leaves along with di-2-ethyl- hexyl phthalate. Chalcones such as 3',6'-dihydroxy-4-methoxychalcone, two leuco-anthocyanins, viz. leucocyanidin-3-*O*-L-rhamnopyroside and leucopeonidin-3-*O* L-rhamnopyranoside were isolated from the root bark (Balakrishna et al., 2011).

In the phytochemical analysis of methanol fraction of seed extract of *S. auriculata* fatty acid esters, triterpene, fatty acid, diterpene alcohols, and phytol were identified as the major chemical groups and the structures were elucidated by using GC-MS data, and major components are linoleic acid, n-hexadecanoic acid, 9-octadecenoic acid, E, Z-1,3,12-nona-decatriene, stearic acid, etc. (Raj et al., 2012). The phytochemical analysis showed the presence of alkaloids, flavonoids, carbohydrates, fats, tannins, saponins, terpenoids, gum and mucilage, lignin, fixed oils, and sterols (Raja et al., 2013; Selvi et al., 2014). The chemical constituents of the flower extracts of *S. auriculata* were identified by the gas chromatography coupled mass spectrophotometry (GC-MS) namely, n-dodecane, pentadecane, tetradecane, hexane, eicosane, heptadecane, pentadecane, diethyl phthalate, phthalic acid, 4-(4-methyl-phenoxy phenol, squalene, dibutyl chloro-ethenylsilane and phenol. The qualitatively identified phytoconstituents are flavonoids, terpenoids, tannins, alkaloids, saponinis, glycosides, carbohydrates, coumarins, steroids from the flowers (Soundharajan and Ponnusamy, 2014). Certain benzocoumarin glycosides, cassiaglycoside I, naphthol glycoside, cassiaglycoside II, chromone glycoside, and cassiaglycoside III, phenylethyl glycoside, cassiaglycoside IV were reported from *S. auriculata* seeds (Nakamura et al., 2016).

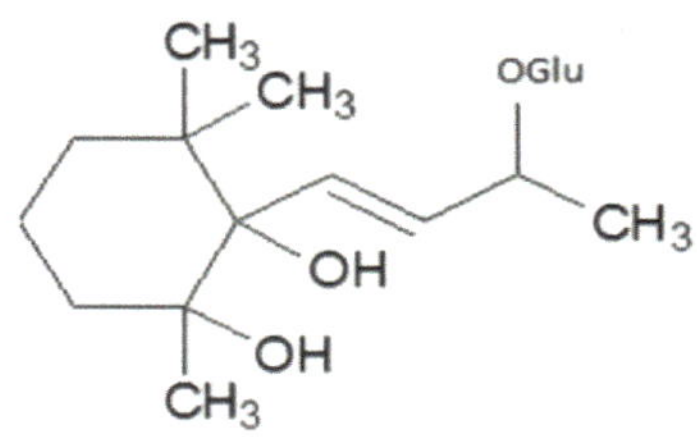

Bridelionoside

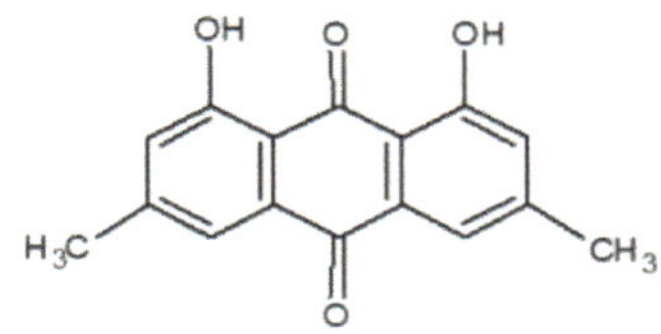

Physcion

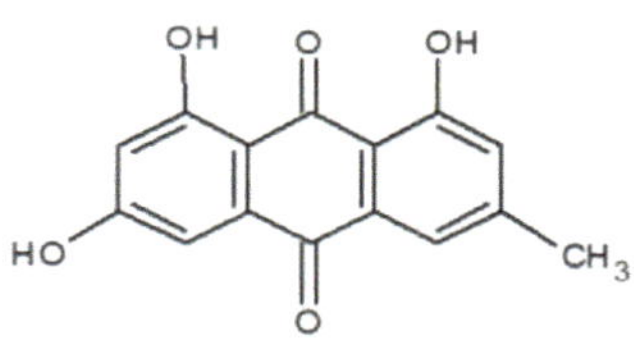

Emodin

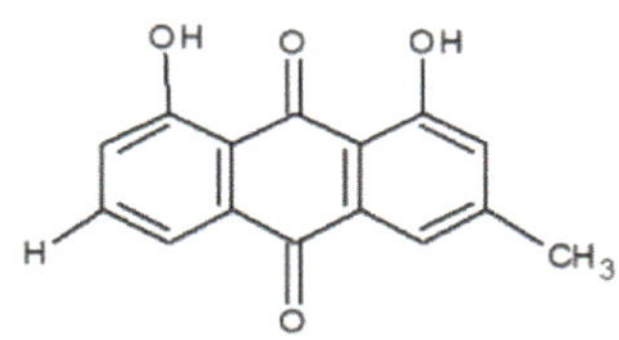

Chrysophenol

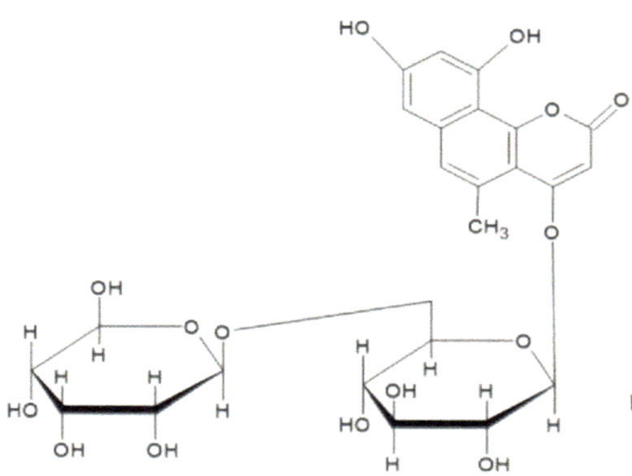

Avaraside-1

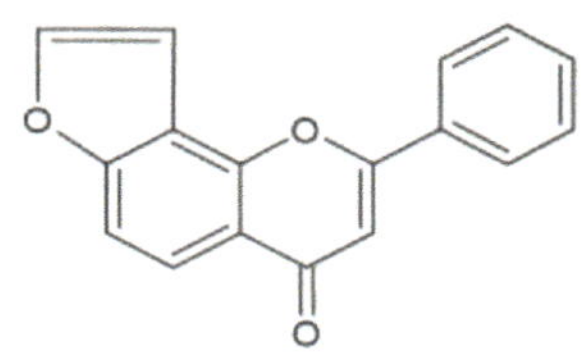

Lanceolatin-B

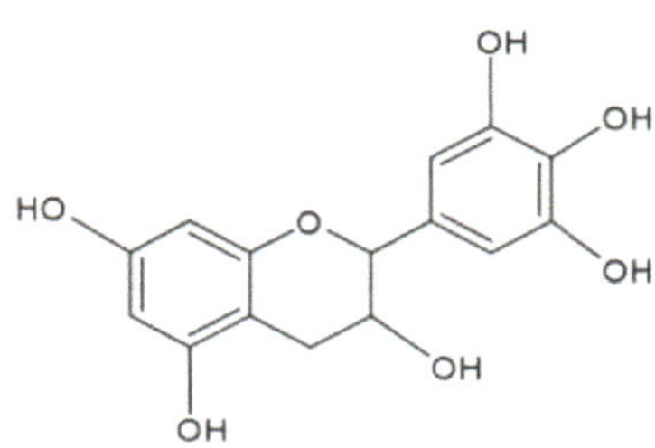

+(-) Gallocatachin

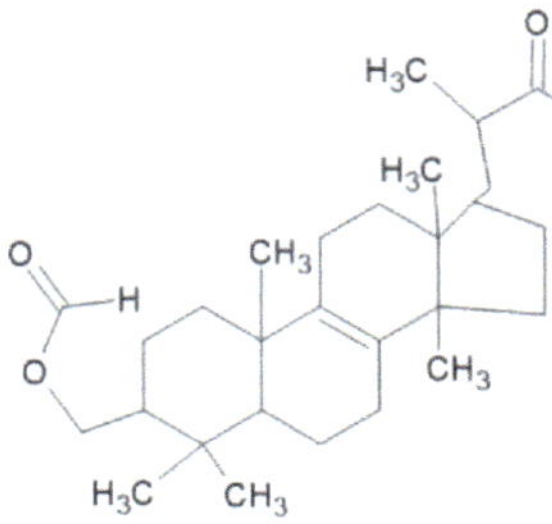

Propanoic acid -2-4,3-acetoxy-4,4, 14 trimethylandrost-8-en17

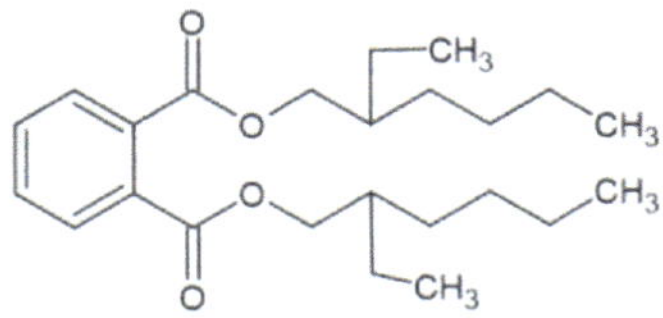

Di-2-ethyl hexl phthalate

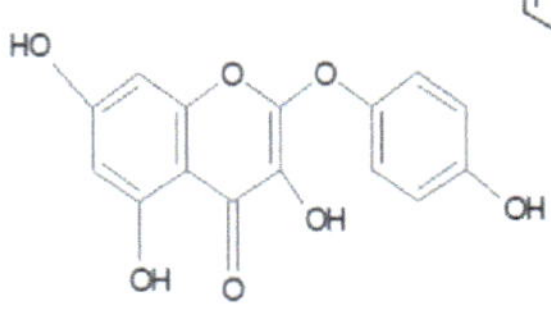

6-demethoxy capillarisin

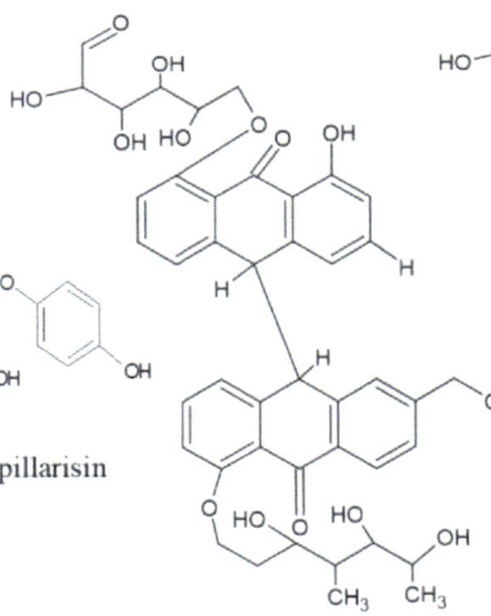

Diantrone – A

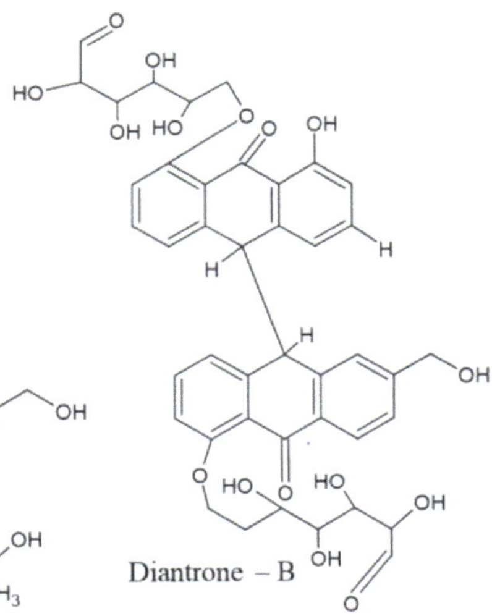

Diantrone – B

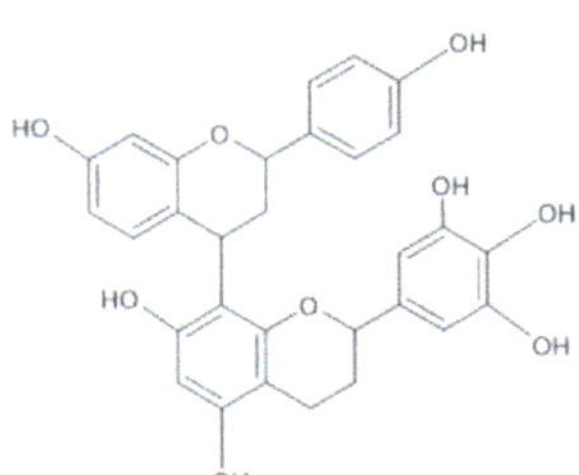

(2S)-7,4-dihydroxy flavan (4β----8) - epigallocatachin

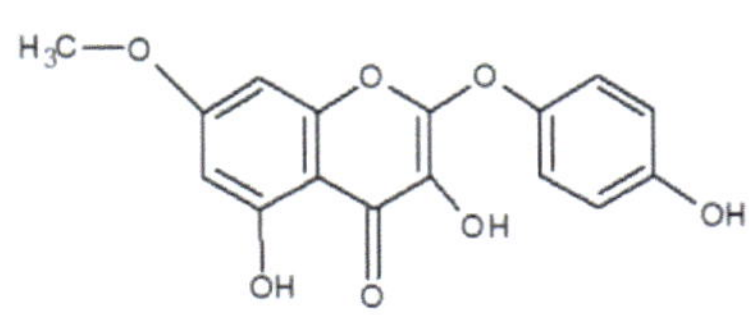

6-dimethoxy-7= methyl capillarisin

Five new diterpene glycosides, auricuosides I–V, a new flavonol glycoside, auricuoflavonoside-I, and a new megastigmane glycoside, auricuomegastigmane-I, were isolated from the seeds of *S. auriculata*, together with 11 known constituents (Zhang et al., 2015). Forty-eight compounds were identified in ethanolic fraction of *S. auriculata* leaf extract analyzed by

GC-MS and reported the compounds such as orcinol, eptacosanol, utanol, dihydro-benzofuran, 3-methyl-acetate, eptacosanol, eptacosanol, gmast-5-en-3-ol, (3β), 2,5,7,8–tetramethyl-2-(4,8,12-trimethyltridecy, benzamide, n-(2-methoxyphenyl)-2-(pyrrolidi) (Gowri et al., 2018).

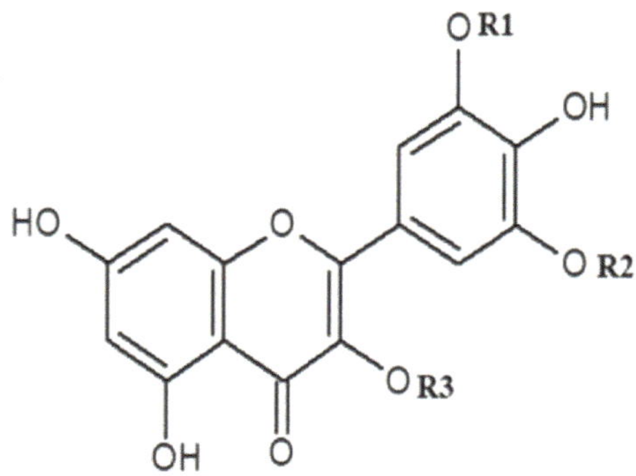

Compound	R1	R2	R3
Luteolin	H	OH	H
Kaempferol	OH	H	H
Quercetin	OH	OH	H
Myricetin	OH	OH	OH
3-methoxy luteolin	OMe	OH	H
Kaempferol 3-O-β-D-Glucopyranoside	OGlc	H	H
Isoquercetin	OGlc	OH	H
Myricetin-3-*O*-β-D-Glucopyranoside	OGlc	OH	OH
Kaempferol-3-*O*-β-rutinoside	ORut	H	H
Rutin	ORut	OH	H
Myricetin-3-O-rutinoside	ORut	OH	OH

31.3 PHARMACOLOGICAL ACTIVITIES

Senna auriculata is being used for long time in various chronic diseases therapeutically and gains importance throughout the globe. All parts of the plant are effective in the curing of infectious or metabolic ailments without having the side effects besides its wide availability (Shiradkar et al., 2011; Guruprasad and Reddy, 2015).

31.3.1 The Antiviral Activity

The aerial parts of *S. auriculata* reported the antiviral activity against Ranikhet disease virus and vaccinia virus (Dhawan et al., 2012). The methanolic extracts of *S. auriculata* flowers were found effective antiviral (in vitro) and cytotoxic when tested on HeLa, Vero, CRFK, and HEL Cell lines. The extract reported a significant antiviral effect on Herpes simplex-1 virus and it is an important source of anti-herpetic [Herpes simplex virus (HSV-1 and II)] and the extract is found as nontoxic at the tested doses (Arthanari et al., 2013).

31.3.2 The Antibacterial/Fungal Activity

The extracts of different parts exhibited potential inhibition effect on the growth of broad spectrum of bacteria and fungal strains were discussed (Table 31.1).

31.3.3 Antidiabetic Activity

All parts of *S. auriculata* exhibited a significant antidiabetic effect at its differential potentiality and its complications. The beneficial role of *S. auriculata* in the management of hyperglycemic conditions and in the maintenance of lipid profile or cardiac supporting effects were given in Table 31.2.

31.3.3.1 Beneficial Role in Biochemical Markers

The bark extracts of *S. auriculata*, (SABE) showed a significant lipid-lowering effect and liver protective activity in STZN provoked diabetes in rats and influenced significantly the serum lipid profile total cholesterol, triglycerides, HDL, LDL, and VLDL, in addition to plasma enzymes such as aspartate aminotransferase, alanine aminotransferase, alkaline phosphatase, and acid phosphatase, total protein, urea, uric acid, and creatinine through oral administration. The test extracts showed positive elevation levels significantly in lipid profiles, plasma and kidney functions, whereas SABME-treated rats were recovered more actively when compared with hexane (SABHE), ethyl acetate (SABEA), and aqueous extracts (SABAL) along with the beneficial effects of hypolipidemic, hepatoprotective and alleviated the renal damage (Daisy et al., 2013).

TABLE 31.1 Antimicrobial Activity of *S. auriculata*

Sl. No.	Particulars of the plant extracts	Effect on the tested microorganisms	References
1	Organic solvent extracts of leaf	Exhibited broad spectrum of antimicrobial effect on *Enterococcus faecalis, Staphylococcus aureus, Escherichia coli, Klebsiella pneumoniae, Proteus vulgaris, Pseudomonas aeruginosa, Salmonella typhi, Salmonella paratyphi, Shigella boydii* and *Vibrio cholera*	Anushia et al. (2009)
2	Flower extracts	Effective on *S. aureus, Enterococcus faecalis, B. subtilis, S. typhi, S. paratyphi* A, *E. coli, P. mirabilis, V. cholerae*, and *Shigella dysenteriae* except *P. aeruginosa* and *K. pneumoniae.*	Maneemegalai and Naveen (2010)
3	Flower extracts with chloroform, ethanol, acetone, methanol, and water	Reported the inhibition of *Bacillus* sp., *Lactobacillus* sp., *Streptococcus* sp., not showing the effect on *Pseudomonas* sp., and *Proteus* sp.,	Muthukumaran et al. (2011); Gaurav et al. (2011); Kanthimathi and Soranam (2014)
4	Methanol extracts of leaf and flowers	Reported antibacterial effect on *B. subtilis, S. epidermidis, E. faecalis, E. coli*	Perumalsamy and Ignacimithu (2000); Duraipandiyan et al. (2006)
5	Leaf extracts	*E. coli, S. typhi, Proteus mirabilis* and *K. pneumoniae* and isolated oleanolic acid as an active principle.	Senthilkumar and Reetha (2011)
6	Methanol fraction of aerial parts	Effective on *P. aeruginosa, A. niger, C. albicans*, and *Penicillium* sp. along with other microorganisms. Identified 1,2-benzenedi carboxylic acid and di-(2-ethylhexyl) phthalate found in fractions from the extract	Senthilrani and Renuka Devi (2014)
7	Chloroform extract of aerial parts	*B. subtilis, S. aureus, P. aeruginosa, E. coli*, and fungal cultures *Candida albicans* and *Aspergillus niger*	Raja et al. (2013)
8	Chloroform, Methanol and water extracts of dried leaf	Effective on the growth of *B. cereus, S. aureus, E. coli, K. pneumoniae, P. aeruginosa*, and *P. mirabilis*	Murugan et al. (2013)

TABLE 31.1 *(Continued)*

Sl. No.	Particulars of the plant extracts	Effect on the tested microorganisms	References
9	Leaves and flower extracts	*S. aureus, E. coli,* and *B. subtilis*	Thambidurai et al. (2010); Subhadradevi et al. (2011)
10	Flavonoid rich fraction of leaf	Effective on *S. aureus*, *E. coli, B. subtilis, P. aeruginosa,* and *P. vulgaris*	Sumathy et al. (2013)
11	Ethanolic extract of flower	Effective on bacteria such as *P. aeruginosa, E. coli, S. aureus, B. subtilis*, and one fungal stain, *C. albicans*	Jayashree et al. (2015); Santosh et al. (2013)
12	Flower extracts	Significant antibacterial effect using petroleum ether and ethanol extracts	Kavimani et al. (2015)
13	Petroleum ether (PE) and ethyl acetate (EA) extracts of leaf	Ethyl acetate extract more effective on *E. coli, P. aeruginosa* than petroleum ether extracts.	Kulkarni et al. (2015); Padmalochana (2018)
14	Aqueous and organic extracts leaf and pod	Significant inhibition on *E. coli*, *P. aeruginosa*, and *S. aureus* and the antifungal activity on *A. niger* and *Trichoderma viride.*	Monisha et al. (2017)
15	Ethanol, methanol and aqueous extracts of dry flower and fresh flowers	*S. aureus, E. faecalis, B. subtilis, S. typhi, S. paratyphi A, E. coli, P. mirabilis, V. cholerae*, and *S. dysenteriae* except *P. aeruginosa* and *K. pneumoniae*	Rajendran and Vasanthi (2017)
16	Leaf extracts from diethyl ether, ethyl acetate, methanol and water.	*S. aureus, E. coli, P. mirabilis, A. flavus*, and *C. albicans*	Pavunraj et al. (2019)

TABLE 31.2 Antidiabetic Effect of *S. auriculata* (Prepared by Authors)

Sl. No.	Details of Extract and dose and mode of administration	Model studied and standards	Findings and Inference	References
1.	SAFWE of 0.15, 0.3, 0.45 gr/kg body weight (bw) given orally for 30 days	STZ induced diabetic rats Glibenclamide	• Significant effect reported along with normalization of BLGL, HG, HK, GPK activities • Lipid peroxides, GSH, SOD, CAT, GPx, and GST normalized at 0.45 g/kg BW	Pari and Latha (2002a, b)
2.	SAFALE at 100 and 200 mg/kg bw for three days	Alloxan induced diabetic rats (Male Wister albino)	• Antihyperglycemic effect in normal rats, glucose uptake, and glycogen deposition suggesting no direct insulin-like effect	Sabu and Subbu Raju (2002)
3.	SAFWE of 0.15, 0.3, 4.5 gr/kg bw given orally for 30 days	STZ induced diabetic rats glibenclamide	• Stimulation of Hepatic carbohydrate meatabolyzing enzymes • Gluconeogenesis path suggested as antidiabetic mechanism	Latha and Pari, (2003a, b)
4.	Methanol extract of leaf	In vitro and in AGH inhibitory actions Acarbose	• Significantly inhibited the activity of AGH in vitro with an IC_{50} value of 0.023 mg/mL • In vivo AGH inhibition is effective at ED_{50} of 4.9 mg/kg	Abesundara et al. (2004)
5.	SAFALE and SAFWE at 250 mg/kg bw given orally in a short-term test for 4 hours	Alloxan induced diabetic rats (Male Wister albinos)	• SAFALE is significantly controlled serum glucose level • SAFWE not significant • No toxicity at 2 gr/kg BW • Suggested that the SAFALE acts through non-insulin-dependent path	Hatapakki et al. (2005)

TABLE 31.2 *(Continued)*

Sl. No.	Details of Extract and dose and mode of administration	Model studied and standards	Findings and Inference	References
6.	SAWP-ALE 400 mg/kg and WE at 250 mg/kg and 500 mg/kg, 28 days.	STZN induced diabetic rats glibenclamide	• Reported hypo-glycemic as well as antihyper glycemic activity	Juvekar and Halade (2006)
7.	Aqueous extracts of roots, leaves, and flowers	Diabetic Wistar male albino rats induced with Alloxan. Tolbutamide used as a reference compound	• Glucose tolerance (acute and chronic) study • Significant reduction in the serum glucose, triglycerides, and cholesterol and enhanced plasma insulin levels • Leaves and flowers extracts are significant than root extracts	Devi et al. (2006)
8.	Thirty days' oral treatment with water-soluble fraction of SAFAL at 0.25 and 0.5 gr/kg BW	Alloxan induced rats glibenclamide as a standard	• Serum glucose controlled and insulin level enhanced in the treated rats • Triglycerides and total cholesterol were effectively controlled. • Marker enzymes of liver toxicity such as serum alanine/aspartate transaminase, and acid/alkaline phosphatase were normalized • Restoration of the liver glycogen and glycogen synthase levels	Hakkim et al. (2007)
9.	n-Butanol and ethyl acetate fractions of hydro-methanolic extract of flowers at 0.20 g/kg bw	Alloxan induced rats. Phenformin as a standard	• Diabetic rats showed substantial weight loss, affecting carbohydrate, lipid, and protein metabolism	Surana et al. (2008)

TABLE 31.2 *(Continued)*

Sl. No.	Details of Extract and dose and mode of administration	Model studied and standards	Findings and Inference	References
			• Tested extracts were effectively restored body weight • n-Butanol fraction is highly effective and • Diabetic key indicators were normalizcd	
10.	SALALE and SAFAL at 120 mg/Kg BW each, oral administration for 15 days	Alloxan induced male Wistar albino rats	• No toxicity noticed during study at dose up to 1000 mg/kg BW • Both extracts exerted antihyper glycemic activity • Improvement of antioxidant enzymes during the treatment • Reported potential in vitro antioxidant effects with the extracts	Kalaivani et al. (2008)
11.	SALWE at 100, 200, and 400 mg/kg doses for 1 day 400 mg/kg continued for 3 weeks	STZ induced mild and severely affected diabetic rats	• 400 mg/kg dosage if is effectively controlled glycemic level after 5 by 13.9% and 17.4% in mild diabetic and severe diabetic rats, respectively. • Significant reduction in fasting blood glucose level and oxidized hemoglobin after three-week treatment. It also exerted antilipidemic effect in both cases in term treatment.	Gupta et al. (2009 a, b, c)
12.	Combined SAFWE and leaf extracts of Bel (*Aegle marmelos)* of 250, 350, and 450 mg/kg bw	STZ induced diabetic rats	• Combined extracts exerted a significant glucose-lowering effect along with enhanced serum insulin and declined lipid.	Sivaraj et al. (2009)

TABLE 31.2 *(Continued)*

Sl. No.	Details of Extract and dose and mode of administration	Model studied and standards	Findings and Inference	References
	(one group fed with 7 days and another group treated for 25 days)		• Restoration of β-cells of pancreas has been reported in histopathological studies at 450 mg/kg bw • Significant normalization effect of the lipid profile.	
13.	SALWE at 400 mg/kg once a day, doses for 3 weeks	STZ induced mild and severely affected diabetic rats	• Significant control reported on FBLG • Prominent elevated levels of insulin in the serum • Improvement in the activities of hexokinase and phospho-fructo-kinase of liver • Declined activities of glucose-6-phosphatase and fructose-1-6-bis-phosphatase • Raise in the glycogen levels of muscle and liver • Histopathological studies reported that improvement in the active β-cells of pancreas • The suggested mechanism is activation of insulin secretion, ameliorating the carbohydrate metabolism, revealing its potential insulinogenic or extra-insulinogenic pathways of controlling type-II diabetes	Gupta et al. (2010)

TABLE 31.2 *(Continued)*

Sl. No.	Details of Extract and dose and mode of administration	Model studied and standards	Findings and Inference	References
14.	Seed extracts obtained from PE (200), Chloroform (300) EA (200), EtOH (200) and aqueous extracts (100) mg/kg bw each investigated for seven days	Alloxan induced diabetic rats, standard is tolbutamide 250 mg/kg bw	• No significant toxic effect up to 5000 mg/kg bw seed extract • The PE and EA extracts controlled FBLG • The restoration of body mass in PE and EA extracts, and the treatment indicated the reversal effect may be due to the modulation of gluconeogenesis and glycogenolysis • The extracts also controlled the oxidative stress • Decrease in the hepatic glycogen reported • Improvement in the lipid profile indicated the insulinotropic mechanism	Aruna and Roopa (2011)
15.	SALALE at 150 mg/kg bw given orally14 days	Alloxan induced diabetic Wistar albino rats	• Elicited significant reduction in the blood glucose, improvement in the lipid profile, creatinine and urea • Raise of serum insulin • Enhanced activity of antioxidant enzymes [CAT, SOD, GSH, GPx]	Shanmugasundaram et al. (2011)
16.	SAFALE 150 and 250 mg/kg bw for 28 days	In vitro inhibition effect on α-amylase (AM) and α-glucosidase (AGA) standard: acarbose	• Significant control on fasting blood glucose and improvement in serum insulin. • Improvement in antioxidant and lipid profile	Jyothi et al., 2012

TABLE 31.2 *(Continued)*

Sl. No.	Details of Extract and dose and mode of administration	Model studied and standards	Findings and Inference	References
		In vivo studies on the STZ induced diabetic rats standard: Glibenclamide	• Exhibited in vitro antioxidant effect on the free radicals • Potential inhibition of the activity of AGA and AM • Phenolics identified in the extracts	
17.	SAFALE at 200 mg and 400 mg/kg BW once a day for 15 days	Male albino Wistar rats induced with STZ-nicotinamide standard: Glibenclamide	• Extracts controlled the blood glucose level to normal within 14 days in treated experimental rats.	Sharada et al. (2012)
18.	SAFALE at 300 mg and 500 mg/kg BW for 4 weeks	Male Wistar Rats tested with diet having high fat after four weeks induced with STZ standard: Glibenclamide	• Estimated the flavonoid (21.1 g), phenol (9.2 g), and alkaloid content (6.66 g) in 100 g of flower extract. • 500 mg/kg BW treatment reported as a significant effect on the normalization of sugar levels • Elevated C-peptide levels as an indication of enhanced insulin secretion • Normalization of serum antioxidant enzymes in the treated rats. • Enhanced glycogen production • Significantly decreased lipid content in the treated rats • Improvement in the lipid profile	Rani et al. (2014)

TABLE 31.2 *(Continued)*

Sl. No.	Details of Extract and dose and mode of administration	Model studied and standards	Findings and Inference	References
19.	Seed extract (400 mg/kg bw) for thirty days	Alloxan-induced diabetic rats and Glyclazide as reference drug	• Normalized blood sugar values • Normalized lipid profile and lipid peroxides antioxidants in plasma and pancreatic tissues	Subramanian et al. (2011)
20.	In vitro studies	Effect on AM/AGH	• Potent inhibition effect demonstrated	Rajendran and Vasanthi (2017)
21.	Using isolated models	Glucose uptake in isolated rat hemidifragm Inhibition of a-amylase	• Demonstrated beneficial effects along with in vitro DPPH radical scavenging.	Muthukumar et al. (2016); Shravan Kumar et al. (2015)
22.	SALWE at 100, 200, 400, and 600 mg/kg BW once a day for 21 days	Mildly induced diabetic (Alloxan-induced diabetic MD) and severely induced diabetic (SD) rabbits	• Significant control of blood glucose with 400 mg/kg BW reported • Elevation of insulin in the serum • Decreased TC and TG and LDL-C level along with enhanced HDL.	Vijayakumar and Vasanthi (2017b)
23.	Flower extracts (250, 350, and 450 mg/kg bw for 21 days) and fractions leading to the isolation of dianthrone A& B as active principle	Alloxan-induced rats standard: Glibenclamide	• No toxicity reported with the extract at 5000 mg/kg bw • Significant control over the blood glucose with the tested extracts • Gluconeogenesis is effectively controlled • Improvement in the lipid profile reported	Khader et al. (2017)
24.	Ethanol extract of flower buds in comparison with flower extracts	STZN induced diabetic rats fed with high fat standard: Metformin	• Flower bud extracts were better in reversal of diabetic symptoms	Nambirajan et al. (2018)

TABLE 31.2 *(Continued)*

Sl. No.	Details of Extract and dose and mode of administration	Model studied and standards	Findings and Inference	References
			• Bud extracts regulated the IRS genes, whereas flower extracts not shown • Exhibited efficient inhibition of AM/AGH • Improvement in antioxidant mechanism • Improvement in organ protection (liver, pancreas, and kidneys) • Expression levels of GRIA2 and IRS2 genes and mRNA were enhanced.	

SAFWE, S. auriculata flower Water Extract, *SAFALE, S. auriculata* flower alcohol extract, *SALALE. S. auriculata* leaf alcohol extract, *SALWE, S. auriculata* leaf Water Extract, *STZ*, Streptozotocin, *HG, BLGC, HK, AGH* (α-glucosidase), *TC*, Total cholesterol; *TG*, Triglyceride; *HDL*, High density lipoprotein; *LDL*, Low density lipoprotein; *VLDL*, Very low-density lipoprotein; *CAT*, catalase; *SOD*, superoxide dismutase; *GSH*, reduced glutathione; *GPx*, Glutathione peroxidase.

The MESAR (protective role of methanolic extract of *S. auriculata* roots) on ethyl alcohol effected hepato-toxicity in rats with ethanol (40%, 4 g/kg bw) was administered for 21 days. The effected rats were treated with MESAR (300 and 600 mg/kg bw) and silymarin (100 mg/kg bw) for 7 days. The rat models were pre-treated with the extract (MESAR), silymarin and isoniazid rifampicin and pyrazinamide were exhibited protective role against the alcohol induced hepato-toxicity with in so days. The protective role of the MESAR has been synchronized as per the standard drugs and shown similar biochemical profile. The serum markers, lipid profile, anti-oxidant enzymes along with total proteins restored in liver in a dose-dependent pattern along with prevented the peroxidation of lipids (Jaydeokar et al., 2014).

31.3.4 PTP 1B Inhibitory Activity

Protein Tyrosine Phosphatase 1B (PTP 1B) plays an essential counteracting role in insulin or leptin regulation and the antogonists or its inhibitors useful in controlling diabetes pathogenesis. Venkatachalam et al. (2013) isolated and characterized, propanoic acid 2-(3-acetoxy-4, 4, 14-trimethylandrost-8-en-17-yl) from *S. auriculata* flowers reported for PTP 1B controlling effect and the possible mechanism based on the molecular docking.

31.3.5 Pharmacokinetic Studies

Puranik et al. (2011) evaluated the hydro-alcoholic extract of defatted seeds using the super critical extraction method and tested on the diabetic Wistar albino rats. The standard biochemical markers such as homocysteine, CKMB, and troponin were evaluated in the studies and reported beneficial effects of the tested extracts in addition to the diabetic biochemical markers indicating its cardioprotective role.

S. auriculata and *Cardiospermum halicacabum* enhanced the steady-state levels of theophylline by 32.5% and 48.2%, respectively, when compared with experimental rats given with theophylline (a Tea component) and found that co-administration prevents the bio-availability of the medicaments in the treatment (Thabrew et al., 2004).

Elango et al. (2015) elucidated that the co-administration of *S. auriculata* extracts with Metformin boosted the antidiabetic effect and reversed to the normal state based on the pharmaco-dynamic and kinetic studies and suggested reducing the doses of metformin when *S. auriculata* is given in diabetes.

31.3.6 In Silico Studies on Signaling Mechanisms of Antidiabetic Effects

Fazlin et al. (2017) determined the antidiabetic effects using an *in silico* and in vivo studies and revealed, *S. auriculata* the strong correlation in the entire cascade of diabetic pathogenesis and inhibition of the patho-biochemical markers and secretion of insulin-signaling levels. The secondary metabolite epigallo-catechin gallate assumed to involve in the glucose transporter expressions along with AKT2 and P13K- γ was also being found in the in vivo experiments.

The antidiabetic efficiency of bud and flower extracts of *S. auriculata* was studied in high fat diet (HFD) and streptozotocin (STZ) induced diabetic rats. The parameters such as feed intake, water intake, and body weight were monitored for 21 days and blood parameters such as insulin, glucose, lipid profile, hepatic function test, renal function test, and oxidative stress markers were analyzed.

31.3.7 Organ Protective Roles

S. auriculata leaf extract significantly reduced lipid peroxidation in tissues and higher levels of both enzymatic and non-enzymatic-mediated antioxidants in alcohol induced liver damage in animal models. The extracts also proved for emollient effect (Kalaivani et al., 2008). The ethanolic extract of the roots of *S. auriculata* exhibited noteworthy nephroprotective activity in cisplatin and gentamicin challenged renal damage in male albino rats (Annie et al., 2005).

31.3.8 Hepatoprotective Activity

The leaf extract of *S. auriculata* (SALE) demonstrated the protective role in rats with ethanol-induced liver damage in animal models for a period of 30 days. *S. auriculata* leaf extract was administered at a dose of 250 mg/kg b.wt. daily in one group and 500 mg/kg b.wt. daily in another group of alcohol-treated rats and was fed with standard pellets and the control rats with isocaloric glucose solution. The SALE significantly altered alcohol-induced elevated levels of phospholipids, cholesterol, and triglyceride levels in the liver, brain, kidney, and intestine of treated rats. Microscopic studies of liver and brain tissue slices of SALE-treated animals showed that the extract markedly reversed fatty changes in liver sections and spongiosis in brain sections (Kumar et al., 2002). Chauhan et al. (2009) experimentally proved hepatoprotective activity of *S. auriculata.*

Hepatoprotective effect of *S. auriculata* leaf extract at a dose of 250 mg kg^{-1} body weight and 500 mg kg^{-1} body weight in ethanol-challenged toxicity for a period of two months was investigated. It revealed that the plant extract significantly lowered the higher levels of liver marker enzymes such as ALT, ALP, and AST, and enhanced the enzymatic and nonenzymatic antioxidant enzymes in rat models (Rajagopala et al., 2003). The hepatoprotective effect of *S. auriculata* leaf extract was tested with two concentrations, that is, 250 mg/kg and 500 mg/kg bodyweight in alcohol-treated rats. Rats treated with the extract exhibited noteworthy increase in body weight, cholesterol levels, triglycerides, and phospholipids. Microscopic study revealed that the extract reversed alcohol induced fatty changed in liver tissue and spongiosis in brain tissues (Kumar et al., 2003).

Jeeva Jothi and Ganapathi (2011) investigated on the methanolic extract of *S. auriculata* leaf and reported the hepatoprotective activity in carbon tetrachloride induced liver toxicity in albino rat models for a period of two months. In vivo study results revealed that the extract significantly decreased LDH, SGPT, SGOT, GGPT, and lipid-peroxidation levels and enhanced GPx, GST, CAT, and SOD enzymes levels in the treated animal group. The extract exhibited concentration-dependent cytotoxic activity in HepG2 cell lines. It further enhanced the antioxidant levels in in vitro studies.

The extract of *S. auriculata* leaves (80% aqueous acetone) showed a protective effect on D-galactosamine-induced cytotoxicity in primary cultured mouse hepatocytes and isolated a new benzo coumarin glycoside, avaraoside I and a new flavanol dimer, avaraol I, together with 29 known constituents. Among the compounds, pseudo-semiglabrin, (2S)-7,4'-dihydroxyflavan(4β→8)-catechin, and (2S)-7,4'-dihydroxyflavan(4β→8)-gallocatechin displayed hepatoprotective effects equivalent to silyb in a standard hepatoprotective agent (Nakamura et al., 2014).

31.3.9 Antipyretic Activity

The antipyretic effect of both leaves and flower extracts of *S. auriculata* exhibited a significant inhibition on oral administration. Moreover, the flower extract showed good results for antipyretic activity within a single day which indicated its efficacy (Rao and Vedavathy, 1991).

31.3.10 Anti-Inflammatory Activity

The flower of *S. auriculata* was found to contain a flavonol glycoside 5-*O*-methylquercetin 7-*O*-glucoside. The 50% acetone extract of the flower of *S. auriculata* showed marked anti-inflammatory activity (56%) in carrageenan-induced edema in rats (Manogaran and Sulochana, 2004). The anti-inflammatory activity of aqueous, methanolic, ethyl acetate, and hydroalcoholic extracts of *S. auriculata* leaves was investigated in carrageenan-induced rat paw edema. Among all extracts methanolic extract showed maximum anti-inflammatory potential. Healthy Wistar albino rats (150–180 g) in groups of six each were treated with vehicle, aqueous, methanolic hydroalcoholic extracts at 250 and 500 mg/kg and indomethacin (10 mg/kg) one hour prior to carrageenan injection. 0.1 mL of 1% carrageenan was injected into the sub-plantar region of hind paw of the rats. At the same time, the control group was administered orally with 1 mL/kg of vehicle and indomethacin 10 mg/kg. The aqueous extract of *S. auriculata* 250 and 500 mg/kg showed significant activity at 6 h with percentage inhibition of 31.06 and 30.62%, respectively. The EtOAc (ethyl acetate) extract of *S. auriculata* at 250 and 500 mg/kg shows significant activity in second phase of inflammation induced by carrageenan with percentage inhibition of 34.16 and 30.79%, respectively. The hydroalcoholic extract of *S. auriculata* 250 and 500 mg/kg shows significant activity at 6 h with percentage inhibition of 23.73 and 30.95%, respectively. Similarly, methanolic extract showed inhibition of paw edema in second phase with percentage inhibition of 37% at 250 mg/kg and 31.63% at 500 mg/kg. The standard indomethacin showed significant activity from 3 h onward, maximum at 6 h with inhibition of 42.56% (Mali et al., 2012).

Ethanolic extract of *S. auriculata* flowers showed noteworthy anti-inflammatory activity against the human red blood cell (HRBC) membrane stabilization model. It showed the strong inhibitory activity in membrane stabilization at 100 μg/mL concentration as 29.15±1.75%, for 200 μg/mL as 49.09±1.72%, for 400 μg/mL as 62.47±1.49%, for 600 μg/mL as

79.05±1.09%, and for 800 μg/mL as 86.42±1.94%, respectively. The inhibition of albumin denaturation activity showed the percentage of inhibition of membrane stabilization for 100 μg/mL as 29.15±1.75%, for 200 μg/mL as 49.18±1.87%, for 400 μg/mL as 58.26±1.58%, for 600 μg/mL as 85.16±0.49, and for 800 μg/mL as 89.24±1.70% indicating its protective role (Muruganantham et al., 2015).

31.3.11 Anticlastogenic Activity

Animals treated with ethyl acetate and flavonoids rich extracts of *S. auriculata* root at 100 mg/kg and 200 mg/kg demonstrated noteworthy protection against cyclophosphamide-induced chromosomal aberration (Deshpande et al., 2013).

31.3.12 Antimutagenic Activity

Antimutagenic capacity of flavonoid-rich extract and ethyl acetate extracts of *S. auriculata* was evaluated (Deshpande et al., 2013). It revealed that flavonoids-rich extract significantly attenuated cyclophoshamide-induced genotoxicity in bone marrow and ethyl acetate extract significantly reversed cyclophosphamide-induced chromosomal aberrations.

31.3.13 Anticancer Activity

The in vitro anticancer effect of SALE was evaluated in MCF-7 and Hep-2 cell lines. SALE strongly suppressed the division of MCF-7 and Hep-2 cell lines in a concentration-dependent manner. The extract showed its mechanism of action by decreasing anti-apoptotic proteins such as Bcl-2 and by enhancing the expression of pro-apoptotic Bax protein. In addition to suppression of MCF-7 and Hep-2 cells the extract strongly inhibited colon cancer cell lines (Prasanna et al., 2009). But, Esakkirajan et al. (2014) evaluated anticancer activity of isolated compound 4-(2,5 dichlorobenzyl)-2,3,4,5,6,7 hexahydro7(4methoxyphenyl) benzo[h][1,4,7] triazecin8(1H)-one from *S. auriculata* against HCT15 cell lines. The study results revealed that the compound exhibited noteworthy cytotoxic activity in HCT-15 cell lines.

Further, SALE reported for the anticancer activity against HepG2 (liver cancer) cell lines with potential suppression of the cell viability up to 50% at

200 μg/mL against both cell lines. Plant extracts showed a more significant activity as compared with the positive control. The extract showed significant inhibition in the cell viability in a dose-dependent manner. The treatment with ethanol extract against HepG2 cell lines significantly decreased the viability of cells at 200 g/mL when compared with other extracts. The cells were contacted with 50, 100, 150, 200, 250, and 300 μg/mL of each extract exhibiting a decreased number of cell viability. The results indicate that SALE has an anticancer activity in HepG2 cell lines (Padmalochana, 2018).

The ethanolic extract of *S. auriculata* was used to check whether it had the capability of inducing cytotoxic effects on the MCF-7 cell lines. In the MTT assay method both the cell lines were used by using the concentration range of 7.8–2000 μg/mL and the IC_{50} concentration was determined. The IC_{50} concentration of MCF-7 cell line was 84.56 μg/mL. The anticancer activity of MCF-7 cell line with the silver nanoparticles of *S. auriculata* leaf extracts showed a very good inhibition effect as compared with standard (Nawaz et al., 2020). The zinc oxide nanoparticles (ZnONPs) were synthesized using an aqueous extract of *S. auriculata* leaves (SAE) at room temperature using standardized protocols. The ZnONPs enhanced the tumoricidal activity of SAE in MCF-7 breast cancer cell lines (Prasad et al., 2020).

31.3.14 Antilipidemic Activity

Antihyperlipidemic efficiency of *S. auriculata* flower ethanolic extract was studied in oleic acid-induced hyperlipidemia in yeast cells. It revealed that the extract significantly lowered oleic acid-induced higher levels of sterol esters, triglycerides. These results were confirmed with mRNA expression and confocal microscopic studies (Vijayakumar et al., 2017a). The results were compared with standard drug Atorvastatin. Similarly, antihyperlipidemic efficiency of ethanolic extract obtained from *S. auriculata* flowers was investigated against Triton-induced hyperlipidemia in rat models (Vijayaraj et al., 2011). The extract significantly reversed higher levels of lipid markers to normal levels.

The antilipase activity of the ethanol extract prepared using aerial parts of *S. auriculata* and purified phenolic compounds (5 Nos) was studied. The crude extract displayed noteworthy inhibitory activity against pancreatic lipase. Among the tested five components kaempferol-3-*O*-rutinoside showed potent antilipase activity than standard compound Orlistat. The remaining components exhibited very weak activity (Habtemariam, 2012).

The antihyperlipidemic and antioxidative effect of ethanolic extract *S. auriculata* flower on hyperlipidemic rats in triton WR 1339 induced hyperlipidemia at three different concentrations for 14 days of experimental schedule. Animals treated with ethanolic extract showed significantly reduced levels of lipid markers, lipid peroxidation, and elevated antioxidant enzymes levels. Effective changes were observed at 450 mg/kg-BW of Et-SAF for two weeks and it was comparable to the standard drug lovastatin (Vijayaraj et al., 2013). The Et-SAF depicted antihyperlipidemic effects in the budding yeast cells as a model and hyperlipidemic conditions were induced with oleic acid exerting an increased level of lipid markers. The antihyperlipidemic effect has been conformed with the Et-SAF without any side effects (Vijayakumar et al., 2017a).

31.3.15 Anthelmintic Activity

Anthelmintic activity of aqueous extract prepared from *S. auriculata* leaves was studied against earthworms, tapeworms, and roundworms at different concentrations ranging from 10 mg to 50 mg/mL. The tested extract at higher concentrations exhibited noteworthy anthelmintic activity. Piperazine citrate was used as a standard drug (Wadekar et al., 2011; Gaikwad et al., 2011; Joy et al., 2012; Kulkarni et al., 2015). In another experiment anthelmintic activity of three organic solvent extracts, that is, ethanol, petroleum ether, and ethyl acetate and aqueous extracts of *S. auriculata* leaves were investigated against earthworms at three concentrations. The aqueous extract demonstrated to possess dose-dependent anthelmintic (Sushma et al., 2012).

31.3.16 Antioxidant Studies

Agarwal (2007) reported antioxidant potential of *S. auriculata* seedlings against UV-B-induced oxidative stress. It revealed that the extracts showed notable reduction in UV-B-induced oxidative stress by enhancing the levels of antioxidant enzymes. Methanol extract of fruits showed the highest antioxidant activity, while the methanol extract of leaves and flowers possessed the lowest antioxidant activity (Thambidurai et al., 2010). In another experiment methanol and ethanol extracts prepared from *S. auriculata* flowers showed potent antioxidant capacity in different antioxidant models (Kumaran and Karunakaran 2007; Jeeva Jothi and Ganapathi, 2011).

Antioxidant properties of *S. auriculata* flowers, leaves, and roots by hydroxyl radical scavenging activity and improved N, N-dimethyl-p phenylenediamine methods were studied. The methanolic extract of the plant parts was evaluated by two improved in-vitro antioxidant models. The flowers showed prominent amount of activity in comparison to leaves and roots, thereby suggesting the probable antioxidant activity which could be attributed to the content of flavonoids noticed in preliminary phytochemical evaluation (Doshi et al., 2011).

The sequential extraction of powder of *S. auriculata* flower was with petroleum ether, ethanol, and methanol using soxhlet extractor and it was subjected to antioxidant activity and qualitative phytochemical screening to detect different types of phytoconstituents. The DPPH radical scavenging activity of the extracts was increased with the increasing concentration, the reducing power of extracts was carried out with ascorbic acid as a standard. The methanolic extract of *S. auriculata* exhibited higher scavenging and reducing power than the other extracts (Elayarani et al., 2011). The synthesis and characterization of silver nanoparticles using *S. auriculata* flower evaluated their antioxidant activity (Velavan et al., 2012). Studies on the in vitro antioxidant properties were reported by many workers using various parts and the extracts exhibited strong to moderate antioxidant activity (Pari and Ramalingam, 2006; Gupta et al., 2009b; Devi et al., 2011; Elayarani et al., 2011; Velavan et al., 2012; Raghavendra et al., 2013; Sachin and Amit, 2014; Purushotham et al., 2014; Yadao et al., 2015; Ashok et al., 2015). The antioxidant potential of the plant extract was evaluated with the help of DPPH radical scavenging assay, nitric oxide scavenging assay, and reducing power assay. The significant correlation was observed in many studies between the phenolics contents and antioxidant activity. Flowers, leaves, and seeds of *S. auriculata* were screened for the quantification of phenolics and flavonoids for which the activities were attributed (Doshi et al., 2015).

The ethanolic extract prepared using flowers of *S. auriculata* demonstrated notable antioxidant activity than DPPH assay. The result showed the percentage of cytotoxicity for 1000 μg/mL as 61.15%, for 500 μg/mL as 59.03%, for 125 μg/mL as 48.31%, and for 31.25 μg/mL as 29.12%, respectively, when compared with ABTS assay activity. The sample possesses ABTS assay activity with the percentage of cytotoxicity for 1000 μg/mL as 72.81%, for 500 μg/mL as 59.58%, for 125 μg/mL as 46.53%, and for 31.25 μg/mL as 23.12%. The total antioxidant activity of the plant is 372.1 μg/mL (Muruganantham et al., 2015). The antioxidant activity of flowers, leaves, and seeds of *S. auriculata* was studied using different well-known and standard in vitro antioxidant models. The antioxidant studies indicated that the extracts of *S. auriculata* were significant (Doshi et al., 2015; Monisha et al., 2017). Padmalochana (2018) reported that the antioxidant efficacy of DPPH scavenging capacity of methanolic and ethyl acetate extracts of *S. auriculata* were found to be higher and comparative with standard ascorbic acid at concentration of 2–10 μg/mL. The antioxidant potential ethanol and water extracts of four *Senna (=Cassia)* species were evaluated by in vitro antioxidant assay models and compared with standard antioxidant component ascorbic acid. The extracts of *S. auriculata* have been found to possess highest activity among the four tested *Senna* species (Kolar et al., 2018).

31.3.17 Carbohydrate Hydrolyzing Enzyme Inhibitor Property

Antidiabetic efficiency of chloroform, methnanol, ethyl acetate, petroleum ether, and water extracts of three plants, that is, *S. auriculata* (SA), *Delonix regia* (DR), and *Vinca rosea* (VR) was evaluated using α-glucosidase/amylase inhibitory assays. Among the tested extracts methanol exhibited promising activity with IC_{50} values of 58.52, 83.46, and 77.41 μg/mL for *S. auriculata, D. regia*, and *V. rosea*, respectively. Chromatographic analysis of the active extract possesses good amount of phenolic and flavonoids and potent antioxidant activity in tested in vitro methods (Yadao et al., 2015).

31.3.18 Aphrodisiac Property

The studies related to the male potential using hydro-alcoholic extracts of *S. auriculata* flower examined on the male Wistar albino rat's sexual behavior using Sildenafil as a positive control. These findings demonstrated the aphrodisiac potential of *S. auriculata* flower in vivo found as an effective sexual stimulating agent (HariPriya et al., 2018).

31.3.19 The Immunomodulatory Activity

After oral administration of SALME to the healthy rats, Chakraborthy (2009) assessed the immune-modulatory effects using antibody-titer and foot pad swelling as immune responses to the antigenic effect by sheep RBCs and by neutrophil adhesion test and found a significant increase in adhesion of neutrophils and delayed-type hypersensitivity reaction, without effecting the humoral response to sheep RBCs, suggested its potential the cellular immunity.

In addition, the extracts rich in polyphenolics were demonstrated and showed improved levels of T and B cells, splenocytes in LPS-induced or resting rats. It correlated with elevated levels of CD4+, CD8+, CD4+, CD25+ regulatory cells which is acted through the antioxidative mechanism in the neutrophils (Cini et al., 2011).

31.3.20 Laxative Activity

The ethanol extract of pods of *S. auriculata* reported at different doses for laxative activity in rats and compared with standard senna and the laxative activity showed as significant at the dose of 200 mg/kg (Suresh et al., 2007).

31.3.21 Cardioprotective Effect

Protective effect of *S. auriculata* flower extract was tested in isoproterenol-induced myocardial infarction in male albino rats. Pretreatment of the aqueous extract of *S. auriculata* significantly reduced isoproterenol-induced elevated levels of lipid markers to normal level. The results clearly demonstrated that *S. auriculata* flowers have potent cardioprotective effect (Manimegalai and Venkatalakshmi, 2019).

31.3.22 Antivenom Studies

The protective efficacy of *S. auriculata* against saw-scaled viper venom induced toxicity. *S. auriculata* leaf methanol extract (SAME) significantly inhibited enzymatic activities of venom proteases ($96 \pm 1\%$), PLA2($45 \pm 5\%$) and hyaluronidases (100%) in vitro studies. In vivo studies showed that the extract reduced venom-induced toxic effects to normal levels (Nanjaraj et al., 2015).

31.3.23 Effect on Melanogenesis

The methanol extract of *S. auriculata* seeds found to reduce melanin production in theophylline stimulated B16 melanoma 4A5 cells. The two components characterized from methanol extract, that is, auriculataosides A and B exhibited melanin inhibitory activity in the tested cell lines models (Wang et al., 2019).

31.3.24 Antinociceptive Effect

The effect of *S. auriculata* aqueous extract on nociception, experimental diabetes, and hyperlipidemia in mice and rats was assessed and reported when administered orally in different doses on nociception in mice proved for anti-nociceptive activity in mice (George et al., 2007).

KEYWORDS

- ***Senna auriculata***
- ***Cassia auriculata* traditional uses**
- **flavonoids**
- **anthracene glycosides**
- **emodin**
- **avarasides**

REFERENCES

Abesundara, K. J.; Matsui, T.; Matsumoto, K. α-Glucosidase Inhibitory Activity of Some Sri Lanka Plant Extracts, One of Which, *Cassia auriculata*, Exerts a Strong Anti-Hyperglycemic Effect in Rats Comparable to the Therapeutic Drug Acarbose. *J. Agric. Food Chem.* **2004,** *52* (9), 2541–2545.

Agarwal, S. Increased Antioxidant Activity in *Cassia* Seedlings Under UV-B Radiation. *Biol. Plant.* **2007,** *51*, 157–160.

Anandan, A.; Eswaran, R.; Doss, A.; Sangeetha, G.; Anand, S. P. A. Chemical Compounds Investigation of *Cassia auriculata* Leaves Folklore Medicinal Plant. *Bull. Environ. Pharmacol. Life Sci.* **2011,** *1*, 20–23.

Annie, S; Rajagopal, P. L.; Malini, S. Effect of *Cassia auriculata* Linn. Root Extract on Cisplatin and Gentamicin-Induced Renal Injury. *Phytomedicine* **2005,** *12*, 555–560.

Anushia, C.; Sampathkumar, P.; Ramkumar, L. Antibacterial and Antioxidant Activities in *Cassia auriculata. Glob. J. Pharmacol.* **2009,** *3* (3), 127–130.

Arthanari, S.; Vanitha, J.; Krishnaswami, V.; Renukadevi, P.; Deivasigamani, K.; De, C. E. Antiviral and Cytotoxic Screening of Methanolic Extract of *Cassia auriculata* Flowers in HeLa, Vero, CRFK and HEL Cell Lines. *Drug Invent. Today* **2013,** *5*, 28–31.

Aruna, P.; Roopa, K. Evaluation of Antidiabetic Activity of *Cassia auriculata* Linn. Seeds for Alloxan Induced Diabetes in Rats. *J. Pharma. Res. Opin.* **2011,** *1* (1), 30–33.

Ashok, J. P.; Harish, P. H.; Prasad, W. V.; Ashok, W. P. Comparative Assessment of Antioxidant Potential of *Cassia auriculata* (Linn.) Flower, Leaf and Seed Methanolic Extracts. *Intern. J. Pharm. Pharma. Sci.* **2015,** *7* (9), 381–385.

Aye, M. M.; Aung, H. T.; Sein, M. M.; Armijos, C. A Review on the Phytochemistry, Medicinal Properties and Pharmacological Activities of 15 Selected Myanmar Medicinal Plants. *Molecules* **2019,** *24*, 293.

Ayyanar, M.; Ignacimuthu, S. Pharmacological Actions of *Cassia auriculata* L. and *Cissus quadrangularis* Wall: A Short Review. *J. Pharmacol Toxicol.* **2008,** 3, 213–221.

Balakrishna, R. B.; Patel, J. R.; Prabhakaran, V. *Cassia auriculata*—A Phytopharmacological Review. *J. Adv. Drug Res.* **2011,** *1*, 46–57.

Chakraborthy, G.S, Evaluation of Immunomodulatory Activity of *Cassia auriculata* Linn. *J. Herbal Med. Toxicol.* **2009,** *3*, 111–113.

Chauhan, K. N.; Patel, M. B.; Valera, H. R.; Patil, S. D.; Surana, S. J. Hepatoprotective Activity of Flowers of *Cassia auriculata* Against Paracetamol Induced Liver Injury. *J. Nat. Remed.* **2009,** *9*, 85–90.

Cini, J. M.; Sandrasaigaran, P.; Tong, C. K.; Adam, A.; Ramasamy, R. Immunomodulatory Activity of Polyphenols Derived from *Cassia auriculata* Flowers in Aged Rats. *Cell Immunol.* **2011,** *271*, 474–479.

Daisy, P.; Feril, G.; Jeeva, K. Hypolipidemic and Hepatoprotective Effects of *Cassia auriculata* Linn Bark Extracts on Streptozotocin Induced Diabetics in Male Wistar Albino Rats. *Asian J. Pharma. Clin. Res.* **2013,** *6* (2), 43–48.

Deshpande, S. S.; Kewatkar, S.; Paithankar, V. V. Anticlastogenic Activity of Flavonoid Rich Extract of *Cassia auriculata* Linn. on Experimental Animal. *Indian J. Pharmacol.* **2013,** *45* (2), 184–186.

Devi, P. U.; Selvi, P. S.; Suja, S.; Selvam, K.; Chinnaswamy, P. Antidiabetic and Hypolipidemic Effect of *Cassia auriculata* in Alloxan Induced Diabetic Rats. *Int. J. Pharmacol.* **2006,** *2*, 601–607.

Devi, V. S.; Ushanandhini, R.; Kuppusamy, A.; Muthuswamy, U.; Andichettiar, T. S.; Puliyath, J. Antioxidant Activity of *Cassia auriculata* Linn. Flowers. *Pharmacol. Online* **2011,** *2*, 490–498.

Dhawan, B N. Anti-Viral Activity of Indian Plants. *Proc. Natl. Acad. Sci. India, Sect. B* **2012,** *82* (1), 209–224.

Doshi, G. M.; Aggarwal, G. V.; Shanbagh, P. P.; Shastri, K. V.; Agarwal, O. K.; Bhalerao, A. B.; Desai, S. K. Studies on *Cassia auriculata* Plant Parts by Improved In Vitro Antioxidant Models. *J. Pharma Biomed. Sci.* **2011,** *10*, 1–2.

Doshi, G. M.; Sandhya, D. K.; Shahare, M. D.; Aggarwal, G. V.; Pillai, P. G. Comparative Antioxidant Studies of *Cassia auriculata* Plant Parts. *Intern. J. Appl. Biol. Pharma. Sci.* **2015,** *7* (9), 381–385.

Duraipandiyan, V.; Ayyanar, M.; Ignacimuthu, S. Antimicrobial Activity of Some Ethnomedicinal Plants Used by Paliyar Tribe from Tamil Nadu, India. *BMC Complement. Altern. Med.* **2006,** *6*, 35–35.

Elango, H.; Ponnusankar, S.; Sundaram, S. Assessment of Pharmacodynamic and Pharmacokinetic Interaction of Aqueous Extract of *Cassia auriculata* L. and Metformin in Rats. *Pharmacogn. Mag.* **2015,** *11* (Suppl 3), S423–S426.

Elayarani, M.; Shanmuganathan, P.; Muthukumaran, P. In Vitro Anti-Oxidant Activity of the Various Extracts of *Cassia auriculata* L. Flower by UV Spectrophotometer. *Asian J. Pharm. Technol.* **2011,** *1*, 70–72.

Esakkirajan, M.; Prabhu, N. M.; Manikandan, R. et al. Apoptosis Mediated Anti-Proliferative Effect of Compound Isolated from *Cassia auriculata* Leaves Against Human Colon Cancer Cell Line. *Spectrochim. Acta Mol. Biomol. Spectrosc.* **2014,** *127*, 484–489.

Fazlin, M. F.; Cini, M. J.; Arunkumar, K.; Hamse, Y. M.; Rajesh, R.; Aishah, A.; Andreas, B. Understanding the Mode of Action of *Cassia auriculata* via In Silico and In Vivo Studies Towards Validating It as a Long-Term Therapy for Type II Diabetes. *J. Ethnopharmacol.* **2017,** *197*, 61–72.

Gaikwad, S. A.; Kale, A. A.; Jadhav, B. G.; Deshpande, N. R.; Salvekar, J. P. Anthelmintic Activity of *Cassia auriculata* L. Extracts—In Vivo Study. *J. Nat. Prod. Plant Resour.* **2011,** *1*, 62–66.

Gaurav, M.; Doshi, S.; Supriya, S.; Gayatri, V. A.; Preeja, P. P.; Abhijeet, B.; Bhalerao, S.; Desai, K. Antibacterial Potential of *Cassia auriculata* Flowers. *J. Microbiol. Biotechnol. Res.* **2011,** *1*, 15–19.

George, M.; Joseph, L.; Ramaswamy. Effect of *Cassia auriculata* Extract on Nociception, Experimental Diabetes and Hyperlipidemia in Mice and Rats. *Highland Med. Res. J.* **2007,** 5 (2), 11–19.

Gowri, R.; Durgadevi, K.; Mini, S. T.; Ramamurthy, V. Phytochemical Profiling of Ethanolic Leaves Extract of *Cassia auriculata. Intern. J. Pharm. Biol. Sci.* **2018,** *8* (4), 1177–1183.

Gupta, S.; Sharma, S. B.; Bansal, S. K.; Prabhu, K. M. Antihyperglycemic and Hypolipidemic Activity of Aqueous Extract of *Cassia auriculata* L. Leaves in Experimental Diabetes. *J. Ethnopharmacol.* **2009a,** *123* (3), 499–503.

Gupta, S.; Sharma, S. B.; Prabhu, K. M. Ameliorative Effect of *Cassia auriculata* L. Leaf Extract on Glycaemic Control and Atherogenic Lipid Status in Alloxan-Induced Diabetic Rabbits. *Indian J. Exp. Biol.* **2009c,** *47* (12), 974–980.

Gupta, S.; Sharma, S. B.; Prabhu, K. M.; Bansal, S. K. Protective Role of *Cassia auriculata* Leaf Extract on Hyperglycemia-Induced Oxidative Stress and Its Safety Evaluation. *Indian J. Biochem. Biophys.* **2009b,** *46*, 371–377.

Gupta, S.; Sharma, S. B.; Usha, R. S.; Bansal, S. K.; Prabhu, K. M. Elucidation of Mechanism of Action of *Cassia auriculata* Leaf Extract for Its Antidiabetic Activity in Streptozotocin Induced Diabetic Rats. *J. Med. Food* **2010,** 13 (3), 528–534.

Guruprasad, C. N.; Reddy, K. R. C. A Phytopharmacological Review of Plant—*Cassia auriculata. Intern. J. Pharma. Biol. Arch.* **2015,** *6*, 1–9.

Habtemariam, S. Antihyperlipidemic Components of *Cassia auriculata* Aerial Parts: Identification Through In Vitro Studies. *Phytother. Res.* **2012,** *27 (1)*, 152–155.

Hakkim, F. L.; Girija, S.; Kumar, R. S.; Jalaludeen. Effect of Aqueous and Ethanol Extracts of *Cassia auriculata* L. Flowers on Diabetes Using Alloxan Induced Diabetic Rats. *Int. J. Diab. Metabol.* **2007,** *15*, 100–106.

Haripriya, V. M.; Dhamotharan, K.; Shukla, S. K.; Suvekbala, V.; Ragupathy, L.; Kumaran, A. Aphrodisiac Properties of Hydro-Alcoholic Extract of *Cassia auriculata* Flower in Male Rats. *Andrologia* **2018,** *51* (2), e13180, 1–9.

Hatapakki, B. C.; Suresh, H. M.; Bhoomannavar, V.; Shivkumar, S. I. Effect of *Cassia auriculata* Linn. Flowers Against Alloxan-Induced Diabetes in Rats. *J. Nat. Remedies* **2005,** *5*, 132–136.

Jayashree, V.; Anju, K. C.; Ragavendran, M. P.; Ravichandiran, V. In Vitro Antimicrobial Activity Using Ethanolic Extract of Flower and Stem Extract of *Cassia auriculata* Linn. *Res. J. Pharm. Technol.* **2015,** *8* (7), 901–905.

Jaydeokar, A. V.; Bandawane, D. D.; Bibave, K. H.; Patil, T. V. Hepatoprotective Potential of *Cassia auriculata* Roots on Ethanol and Antitubercular Drug-Induced Hepatotoxicity in Experimental Models. *Pharm. Biol.* **2014,** *52*, 344–355.

Jeeva Jothi, D.; Ganapathy, M. Hepatoprotective Effect of *Cassia auriculata* L. Leaf Extract on Carbon Tetrachloride Intoxicated Liver Damage in Wister Albino Rats. *Asian J. Biochem.* **2011,** *6* (1), 104–112.

Jha, A.; Sharma, G. K.; Elango, K. Evaluation of Pharmacodynamic and Pharmacokinetic Interactions of *Cassia auriculata* with Metformin in Streptozotocin Induced Diabetic Rats. *Indo Am. J. Pharma. Res.* **2013,** *3,* 1578–1584.

Joy, P. V.; Raj, J. M. P.; Yesu, J. R. Medicinal Values of Avaram (*Cassia auriculata*) Linn-A Review. *Intern. J. Curr. Pharma. Res.* **2012,** *4* (3), 1–3.

Juvekar, A. R.; Halade, G. V. Hypoglycemic Activity of *Cassia auriculata* in Neonatal Streptozotocin-Induced Non-Insulin Dependent Diabetes Mellitus in Rats. *J. Nat. Remedies* **2006,** *6* (1),14–18.

Jyothi, S. G.; Chavan, C. S.; Somashekaraiah, B. V. In Vitro and In Vivo Antioxidant and Antidiabetic Efficacy of *Cassia auriculata* L. Flowers. *Global J. Pharmacol.* **2012,** *6*, 33–40.

Kainsa, S.; Kumar, P.; Rani, P. Pharmacological Potentials of *Cassia auriculata* and *Cassia fistula* plants: A Review. *Pak. J. Biol. Sci.* **2012,** *15* (9), 408–417.

Kalaivani, A.; Umamaheswari, A.; Vinayagam, A.; Kalaivani, K. Anti-Hyperglycemic and Antioxidant Properties of *Cassia auriculata* Leaves and Flowers on Alloxan Induced Diabetic Rats. *Pharmacol. Online* **2008,** *1*, 204–217.

Kanthimathi, M.; Soranam, R. Phytochemical Screening and In Vitro Antibacterial Potential of *Cassia auriculata* Linn. Flowers Against Pathogenic Bacteria. *Intern. Res. J. Pharma. Biosci.* **2014,** *1* (1), 45–56.

Kavimani, V.; Ramadevi, A.; Kannan, K.; Gnanavel, S. Antibacterial Activity of *Cassia auriculata* Linn. *J. Chem. Pharmaceut. Res.* **2015,** *7* (9), 479–485.

Khader, S. Z. A.; Syed, Z. A.; S, Balasubramanian, S. K.; Arunachalam, T. K.; Kannappan, G.; Mahboob, M. R.; Ponnusamy, P.; Ramesh, K. Modulatory Effect of Dianthrone Rich Alcoholic Flower Extract of *Cassia auriculata* L. on Experimental Diabetes. *Integr. Med. Res.* **2017,** *6* (2), 131–140.

Kolar, F. R.; Gogi, C. L.; Khudavand, M. M.; Choudhari, M. S.; Patil, S. B. Phytochemical and Antioxidant Properties of Some *Cassia* Species. *Nat. Prod. Res.* **2018,** *32* (11), 1324–1328.

Kulkarni, A.; Govindappa, M.; Channabasava, C.; Ramachandra, Y.; Koka, P. Phytochemical Analysis of *Cassia species* and It's In Vitro Antimicrobial, Antioxidant and Anti-Inflammatory Activities. *Adv. Med. Plant Res.* **2015,** *3*, 8–17.

Kumar, R. S.; Manickam, P.; Periyasamy, V.; Namasivayam, N. Activity of *Cassia auriculata* Leaf Extract in Rats with Alcoholic Liver Injury. *J. Nutr. Biochem.* **2003,** *14* (8), 452–458.

Kumar, R. S.; Ponmozhi, M.; Viswanathan, P.; Nalini, N. Effect of *Cassia auriculata* Leaf Extract on Lipids in Rats with Alcoholic Liver Injury. *Asia Pac. J. Clin. Nutr.* **2002,** *11* (2), 157–163.

Kumaran, A.; Karunakaran, R. J. Antioxidant Activity of *Cassia auriculata* Flowers. *Fitoterapia* **2007,** *78*, 46–47.

Latha, M.; Pari, L. Anti-Hyperglycaemic Effect of *Cassia auriculata* in Experimental Diabetes and Its Effects on Key Metabolic Enzymes Involved in Carbohydrate Metabolism. *Clin. Exp. Pharmacol. Physiol.* **2003a,** *30* (1–2), 38–43.

Latha, M.; Pari, L. Preventive Effects of *Cassia auriculata* L. Flowers on Brain Lipid Peroxidation in Rats Treated with Streptozotocin. *Mol. Cell. Biochem.* **2003b,** *243*, 23–28.

Mali, A. A.; Hivrale, M. G.; Bandawane, D. D.; Chaudhari, P. D. Study of Anti-Inflammatory Activity of *Cassia auriculata* Linn. Leaves in Wistar Rats. *Indian Drugs* **2012,** *49*, XI.

Maneemegalai, S.; Naveen, T. Evaluation of Antibacterial Activity of Flower Extracts of *Cassia auriculata* L. *Ethnobot. Leaflets* **2010,** *14*, 182–192.

Manimegalai, S.; Venkatalakshmi, P. Cardioprotective Effect of *Cassia auriculata* Linn. Petal Extract on Isoproterenol Induced Myocardial Infarction in Male Albino Rats. *Intern. J. Pharma. Sci. Res.* **2019,** *29*, 848–852.

Manogaran, S.; Sulochana, N. Anti-Inflammatory Activity of *Cassia auriculata. Ancient Sci Life* **2004,** *26* (2), 65–67.

Markham, K. R.; Ternai, B.; Stanley, R.; Geiger, H.; Mabry, T. Carbon-13 NMR Studies of Flavonoids-III: Naturally Occurring Flavonoid Glycosides and Their Acylated Derivatives. *Tetrahedron* **1978,** *34* (9), 1389–1397.

Meena, V. Himangshu B.; Parveen R. *Cassia auriculata*: A Healing Herb for All Remedy. *J. Pharmacogn. Phytochem.* **2019,** *8* (3), 4093–4097

Monisha, M.: Sowmiya, M.: Ragunathan, R.; Jesteena J. Extraction of Bio Active Compounds from *Cassia auriculata* Pods and Leaves and Its Medicinal Uses. *Int. J. Curr. Microbiol. App. Sci.* **2017,** *6* (8), 425–434.

Murugan, T.; Wins, J. A.; Murugan, M. Antimicrobial Activity and Phytochemical Constituents of Leaf Extracts of *Cassia auriculata. Indian J. Pharm. Sci.* **2013,** *75* (1), 122–125.

Muruganantham, N.; Solomon, S.; Senthamilselvi, M. M. Anti-Oxidant and Anti-Inflammatory Activity of *Cassia auriculata* (Flowers). *British J. Pharma. Res.* **2015,** *8* (1), 1–9.

Muthukumar A.; Mohan, S.; R. Sundharaganapathy, R.; Nagaraja G. P. Anti-Diabetic Activity of *Cassia auriculata* Flowers in α- amylase Inhibition and Glucose Uptake by Isolated Rat Hemi-Diaphragm. *Der Pharmacia Lettre* **2016,** *8* (16), 101–105.

Muthukumaran, P.; Elayarani, M.; Shanmuganathan, P.; Cholarajan, A. Antimicrobial Activities of *Cassia auriculata* L and *Morinda tinctoria* Roxb. *Intern. J. Res. Pure Appl. Microbiol.* **2011,** *1*, 9–12.

Nakamura, C; Yasumoto, E.; Nakano, K.; Nakayachi, T.; Hashimoto, K.; Kusama, K.; Fukuda, M.; Sakashita, H.; Shirahata, A.; Sakagami, H. Changes in Intracellular Concentrations of Polyamines During Apoptosis of HL-60 cells. *Anticancer Res.* **2003,** *23* (6C), 4797–803.

Nakamura, S.; Xu, F.; Ninomiya, K.; Nakashima, S.; Oda, Y.; Morikawa, T.; Muraoka, O.; Yoshikawa, M.; Matsuda, H. Chemical Structures and Hepatoprotective Effects of Constituents from *Cassia auriculata* Leaves. *Chem. Pharm. Bull.* **2014,** *62*, 1026–1031.

Nakamura, S.; Zhang, Y.; Nakashima, S.; Oda, Y.; Wang, T.; Yoshikawa, M.; Matsuda, H. Structures of Aromatic Glycosides from the Seeds of *Cassia auriculata. Chem. Pharm. Bull.* **2016,** *64*, 970–974.

Nambirajan, G.; Karunanidhi, K.; Ganesan, A.; Rajendran, R.; Kandasamy, R.; Elangovan, A.; Thilagar, S. Evaluation of Antidiabetic Activity of Bud and Flower of Avaram Senna (*Cassia auriculata* L.) in High Fat Diet and Streptozotocin Induced Diabetic Rats. *Biomed. Pharmacother.* **2018,** *108*, 1495–1506.

Nanjaraj, U. A. N.; Yariswamy, M.; Joshi, V.; Suvilesh, K. N.; Sumanth, M. S.; Das, D.; Nataraju, A.; Vishwanath, B. S. Local and Systemic Toxicity of *Echis carinatus* Venom: Neutralization by *Cassia auriculata* L. Leaf methanol extract. *J. Nat. Med.* **2015,** *69* (1), 111–122.

Nawaz, M. P.; Afroos, A. B.; Mohamed S. R.; Palanivelu, M.; Ayeshamariam, A. Anticancer Activity of Silver Nanoparticle by Using *Cassia auriculata* Extract. *Eur. J. Med. Plants* **2020,** *31* (2), 1–9.

Padmalochana, K. Antibacterial, Anticancer and Antioxidant Activity of *Cassia auriculata* Leaves Methanolic, Petroleum Ether and Ethyl Acetate Extracts. *Int. J. Pharma. Res. Health Sci.* **2018,** *6* (3), 2610–2614.

Padmavathi, V. Phyto-Chemical Constituents and Their Biological Aspects of *Senna auriculata*. *Indo Am. J. Pharma. Sci.* **2018,** *5* (7), 6470–6475.

Pari, L.; Latha, M. Antidiabetic Activity of *Cassia auriculata* Flowers: Effect on Lipid Peroxidation in Streptozotocin Diabetic Rats. *Pharm. Biol.* **2002b,** *40* (7), 512–517.

Pari, L.; Latha, M. Effect of *Cassia auriculata* Flowers on Blood Sugar Levels, Serum and Tissue Lipids in Streptozotocin Diabetic Rats. *Singapore Med. J.* **2002a,** *43* (12), 617–621.

Pari, L.; Ramalingam, S. Role of Diasulin, an Herbal Formulation on Antioxidant Status in Chemical Induced Diabetes. *Int. J. Pharmacol.* **2006,** *2*, 110–115.

Pavunraj, M.; Rajeshkumar, S.; L.; Bhuvana, L.; Babujanarthanam, R. Preparation of *Cassia auriculata* Plant Extracts Using Different Solvents and Its Antibacterial and antifungal Activity Against Clinical Pathogens. *Drug Invent. Today* **2019,** *11* (1), 142–146.

Perumalsamy, R.; Ignacimuthu, S. Antibacterial Activity of Some Folklore Medicinal Plants Used by Tribals in Western Ghats of India. *J. Ethnopharmacol.* **2000,** *69*, 63–71.

Prasad, K. S.; Prasad, S. K.; Ansari, M. A.; Mohammad, A. A.; Mohammad, N. A.; Sami A.; Chandrashekar, S.; Mahadevamurthy, M.; Veena M. A.; Chandan, S. Tumoricidal and Bactericidal Properties of ZnONPs Synthesized Using *Cassia auriculata* Leaf Extract. *Biomolecules* **2020,** *10* (7), 982.

Prasanna, R.; Harish, C. C.; Pichai, R.; Sakthisekaran, D.; Gunasekaran, P. Anti-Cancer Effect of *Cassia auriculata* Leaf Extract In Vitro Through Cell Cycle Arrest and Induction of Apoptosis in Human Breast and Larynx Cancer Cell Lines. *Cell Biol. Intern.* **2009,** *33*, 127–134.

Puranik, A. S.; Halade, G.; Kumar, S; Mogre, R.; Apte, K.; Vaidya, A. D.; Patwardhan, B. *Cassia auriculata*: Aspects of Safety Pharmacology and Drug Interaction. *Evid. Based Complement. Alternat. Med.* **2011,** *6*, 915240.

Purushotham, K. N.; Annegowda, H. V.; Sathish, N. K.; Ramesh, B.; Mansor, S. M. Evaluation of Phenolic Content and Antioxidant Potency in Various Parts of *Cassia auriculata* L.: A Traditionally Valued Plant. *Pak. J. Biol. Sci.* **2014,** *17*, 41–48.

Raghavendra, M.; Reddy, A. M.; Yadav, P. R.; Raju, A. S.; Kumar, L. S. Comparative Studies on the In Vitro Antioxidant Properties of Methanolic Leafy Extracts from Six Edible Leafy Vegetables of India. *Asian J. Pharm. Clin. Res.* **2013,** *6*, 96–99.

Rai, K. N.; Dasundhi, R. A. A New Flavone Glycoside from the Roots of *Cassia auriculata*. *J. Bangladesh Acad. Sci.* **1990,** *14*, 57–61.

Raj, J. Y.; Peter, M.; Joy, V.; Yesu, J.; Paul, J. Chemical Compounds Investigation of *Cassia auriculata* Seeds: A Potential Folklore Medicinal Plant. *J. Asian Sci.* **2012,** *2*,187–192.

Raja, D. K.; Jeganathan, N. S.; Rajappan, M. In Vitro Antimicrobial Activity and Phytochemical Analysis of *Cassia auriculata* Linn. *Intern. Curr. Pharma. J.* **2013,** *2* (6), 105–108.

Rajagopala, S. K.; Manikama, P.; Periyasamyb, V.; Namasivayam, N. Activity of *Cassia auriculata* Leaf Extract in Rats with Alcoholic Liver Injury. *J. Nutr. Biochem.* **2003,** *14*, 452e8.

Rajendran, V.; Vasanthi, N. Lipid Lowering Effect of *Cassia auriculata* Flower Extract in Hyperlipidemic Yeast and Mammalian System. Ph.D. Thesis. Bharathidasan University, Department of Biochemistry, Tiruchirappalli, Tamil Nadu, India, 2017.

Rani, S. S.; Kumar, J. S.; Tharaheswari, M.; Subhashree, S. *Cassia auriculata* Flower Extract Articulates Its Antidiabetic Effects by Regulating Antioxidant Levels in Plasma, Liver and Pancreas in T2DM Rats. 2014. https://www.semanticscholar.org/paper/

Cassia-auriculata-Flower-Extract-Articulate-its-by-Rani-Kumar/01ae7748e3d08b0e4d396af3a7669817f40a851c.

Rao, G. N.; Kumar, P. M.; Dhandapania, V. S.; Ramakrishnaa, T.; Toshimitsu, H. Constituents of *Cassia auriculata*. *Fitoterapia* **2000,** *71*, 82–83.

Rao, K. N.; Vedavathy, S. Antipyretic Activity of Six Indigenous Medicinal Plants of Tirumula Hills, *J. Ethnopharmacol.* **1991,** *33*, 193–196.

Sabu, M. C.; Subbu Raju, T. Effect of *Cassia auriculata* Linn. on Serum Glucose Level, Glucose Utilization by Isolated Rat Hemidiaphragm. *J. Ethnopharmacol.* **2002,** *80* (2–3), 203–206.

Sachin, C.; Amit, K. J. Phytochemical Analysis and Assessment of In-Vitro Anthelmintic Activity of *Cassia auriculata* Linn Leaves. *J. Am. Clin. Therapeut.* **2014,** *2*, 161–167.

Salma, B.; Muthukumar, S. P.; Avinasha, S.; Manjula, S. N. Review on Ethnobotany, Phytochemistry, and Pharmacological Properties of *Cassia auriculata*. *Pharm. Pharmacol. Int. J.* **2020,** *8* (2), 106–111.

Sanghi, R.; Tripathi, K.; Sharma, J. P. New Triterpenoid, O-beta-D-xylopyranosides from *Cassia auriculata*. *Indian J. Chem. Sect. B, Org. Including Medicinal* **2000,** *39* (6), 477e9.

Santosh, C.; Idress, H. A.; Reddy, R. R.; Madan, M. M. Phytochemical Analysis, Antimicrobial and Antioxidant Activity of *Cassia auriculata*. *Int. J. Bioassays* **2013,** *2* (10), 1411–1414.

Selvi, V.; Isaivani, I.; Gomathi, R.; Karpagam, S. Phytochemical Screening of Crude Extract of Leaf and Flowers of *Cassia fistula* and *Cassia auriculata*. *Mintage J. Pharma. Med. Sci.* **2014,** *3* (4), 7–8.

Senthilkumar, P. K.; Reetha, D. Isolation and Identification of Antibacterial Compound from the Leaves of *Cassia auriculata*. *Eur. Rev. Med. Pharmacol. Sci.* **2011,** *15* (9), 1034–1038.

Senthilrani, S.; Renuka Devi, P. Biological Activities and Chemical Composition of *Cassia auriculata*. *Asian J. Biochem.* **2014,** *9*, 195–202.

Shanmugasundaram, R.; Devi, K. V.; Soris, T. P.; Maruthupandian, A.; Mohan, V. R. Antidiabetic, Anti-Hyperlipidaemic and Antioxidant Activity of *Senna auriculata* (L.) Roxb. Leaves in Alloxan Induced Diabetic Rats. *Int. J. Pharm. Tech. Res.* **2011,** *3* (2), 747–756.

Sharada, N.; Venkateswarlu, G.; Manoranjan, S.; Someshwar, K.; Noorunnisa, M. B.; Rao, K. V. Salubrious Effect of Ethanolic Extract of *Cassia auriculata* Linn. in Streptazotocin-Nicotinamide Induced Diabetes in Rat Model. *Asian J. Pharm. Tech.* **2012,** *2* (3), 104–106.

Shiradkar, M.; Pawankumar, G.; Shah, K. Pharmacological Evaluation of *Cassia auriculata* Bark Extract. *Intern. J. Pharma. Bio. Sci.* **2011,** *2*, 758–766.

Shravan Kumar, N.; Tulasi, P.; Errabelli S. In Vitro Anti-Diabetic Activity of Seed Extracts of *Cassia auriculata* and *Cassia angustifolia*. *Eur. J. Exp. Biol.* **2015,** *5* (5), 12–17.

Sivaraj, A.; Devi, K.; Palani, S.; Kumar, P. V.; Kumar, B. S.; David, E. Antihyperglycemic and Antihyperlipidemic Effect of Combined Plant Extract of *Cassia auriculata* and *Aegle marmelos* in Streptozotocin-Induced Diabetic Albino Rats. *Int. J. Pharm. Tech. Res.* **2009,** *1*, 1010–1016.

Soundharajan, S. R.; Ponnusamy, R. D. Chemical Composition and Characterization Studies of *Cassia auriculata* Flower Extract. *Intern. J. Pharm. Pharma. Sci.* **2014,** *6*, 327–330.

Subhadradevi, V.; Asok Kumar, K.; Umamaheswari, M.; Sivashanmugam, A. T.; Ushanandhini, J. R.; Jagannath, P. Antimicrobial Activity of Leaves and Flowers of *Cassia auriculata* L. *Bangladesh J. Sci. Ind. Res.* **2011,** *46*, 513–518.

Subramanian, S.; Uma, S. K.; Sriram, P. G. Biochemical Evaluation of Antidiabetic, Antilipidemic and Antioxidant Nature of *Cassia auriculata* Seeds Studied in Alloxan-Induced Experimental Diabetes in Rats. *Intern. J. Pharma. Sci. Rev. Res.* **2011,** *11* (2), 137–144.

Sumathy, R.; Sankaranarayanan; S.; Bama, P.; Ramachandran, J.; Vijayalakshmi, M.; Deecaraman, M. Antibacterial and Antioxidant Activity of Flavonoid Rich Fraction from the Petals of *Cassia auriculata* an In Vitro Study. *Int. J. Pharm. Pharm. Sci.* **2013,** *5* (3), 492–497.

Surana, S. J.; Gokhale, S. B.; Jadhav, R. B.; Sawant, R. L.; Jyoti, B. W. Antihyperglycemic Activity of Various Fractions of *Cassia auriculata* Linn. in Alloxan Diabetic Rats. *Indian J. Pharm. Sci.* **2008,** 70 (2), 227–229.

Suresh, H. M.; Hatapakki, B. S.; Shivakumar, S. I.; Hallikeri, C. S.; Swamy, B. M.; Chandur, V. K. Laxative Activity of *Cassia auriculata* Pods in Rats. *J. Nat. Rem.* **2007,** *7* (1), 150–154.

Sushma, K.; Praveen, K.; Dahiya, R. S. Investigation of In Vitro Anthelmintic Activity of *Cassia auriculata* Leaves. *J. Nat. Prod. Plant Resour.* **2012,** *2* (4), 460–464.

Thabrew, M. I.; Munasinghe, T. M.; Senarath, S.; Yapa, R. M. Effects of *Cassia auriculata* and *Cardiospermum halicacabum* Teas on the Steady State Blood Levels of Theophylline in Rats. *Drug Metabol. Drug Interact.* **2004,** *20* (4), 263–272.

Thambidurai, M.; Rajesh, P.; Balamurugan, B.; Kannan, V. R. J. In Vitro Antioxidant and Antimicrobial of Study on *Cassia auriculata* L. *J. Int. Biol. Sci.* **2010,** *2*, 1–7.

Velavan, S.; Arivoli, P.; Mahadevan, K. Biological Reduction of Silver Nanoparticles Using *Cassia auriculata* Flower Extract and Evaluation of Their In Vitro Antioxidant Activities. *Nanosci. Nanotechnol. Int. J.* **2012,** *2*, 30–35.

Venkatachalam, M.; Singaravelu, G.; Govindaraju, K.; Jong, S. A. PTP-1B Inhibitory Action of a Phytochemical Propanoic Acid, 2- (3-acetoxy-4,4,14 Trimethylandrost-8-en-17-yl). *Curr. Sci.* **2013,** *105* (6), 827–831.

Vijayakumar, R.; Anilkumar, K.; Vasanthi, N. Antihyperlipidemic Activity of *Cassia auriculata* Flower Extract in Oleic Acid Induced Hyperlipidemia in *Saccharomyces cerevisiae*. *J. Food Sci. Technol.* **2017a,** *54* (9), 2965–2972.

Vijayakumar, R.; Vasanthi, N. *Cassia auriculata* Flower Extract Attenuates Hyperlipidemia in Male Wistar Rats by Regulating the Hepatic Cholesterol Metabolism. *Biomed. Pharmacother. Biomed. Pharmacotherapie* **2017b,** *95*, 394–401.

Vijayaraj, P.; Muthukumar, K.; Sabarirajan, J.; Nachiappan, V. Antihyperlipidemic Activity of *Cassia auriculata* Flowers in Triton WR 1339 Induced Hyperlipidemic Rats. *Exp. Toxicol. Pathol.* **2013,** *65* (1–2), 135–141.

Vijayaraj, P.; Muthukumar, K.; Sabarirajan, J.; Nachiappan, V. Evaluation of Antihyperlipidemic Activity of Ethanolic Extract of *Cassia auriculata* Flowers. *Indian J. Biochem. Biophys.* **2011,** *48* (1), 54–58.

Wadekar, J.; Sawant, R.; Honde, B. Anthelmintic and Antibacterial Potential of *Cassia auriculata* Roots. *Int. J. Pharm. Front. Res.* **2011,** *1*, 93–98.

Wang, W.; Zhang, Y.; Nakashima, S.; Nakamura, S.; Wang, T.; Yoshikawa, M.; Matsuda, H. Inhibition of Melanin Production by Anthracenone Dimer Glycosides Isolated from *Cassia auriculata* Seeds. *J. Nat. Med.* **2019,** *73*, 439–449.

Yadao, N.; Charles, L. P.; Kokati, V. B. R. Carbohydrate Hydrolyzing Enzyme Inhibitor Property, Antioxidant and Phytochemical Analysis of *Cassia auriculata*, *Delonix regia* and *Vinca rosea* Linn: An In Vitro Study. *J. Appl. Pharma. Sci.* **2015,** *5* (05), 18–27.

Zhang, Y.; Nakamura, S.; Nakashima, S.; Wang, T.; Yoshikawa, M.; Matsuda, H. Chemical Structures of Constituents from the Seeds of *Cassia auriculata*. *Tetrahedron* **2015,** *71* (38), 6727–6732.

CHAPTER 32

Phytochemical and Pharmacological Profile of *Clitoria ternatea* L.

AJAY NEERAJ*, R. Y. HIRANMAI, SUPRIYA VAISH, and SUNIL SONI

School of Environment and Sustainable Development, Central University of Gujarat, Gandhinagar 382030 Gujarat, India

**Corresponding author. E-mail: ajayneeraj.hansraj92@gmail.com*

ABSTRACT

Clitoria ternatea L., known as *Aprajita* in Hindi and Butterfly pea in English, is a herbal medicinal plant mostly found in Asia. This plant has also some traditional value as memory enhancer, antidepressant agent etc. It contains a wide variety of secondary metabolites, including triterpenoids, flavonoids, flavanol glycosides, anthocyanins, steroids, alkaloids and glucosides. Many pharmacological characteristics are found in its extracts, including antibacterial and analgesic properties; diuretic; anaesthetic; insecticide; antidiabetic; and platelet aggregation-inhibiting capabilities. This review is an effort to study the phytochemical and pharmacological investigations of this plant, coupled with an essential appraisal of its future.

32.1 INTRODUCTION

Clitoria ternatea L., also known as butterfly pea or blue pea, is a well-known plant of the Fabaceae family that is used in traditional medicine. Shoots, roots, leaves, and seeds possess memory enhancer, anxiolytic, anti-stressor, anticonvulsant, nootropic, antidepressant, tranquillizing properties and is also a sedative agent (Mukherjee et al., 2008). The plant originated in

Bioactives and Pharmacology of Legumes. T. Pullaiah, PhD (Ed.)

tropical Asia and later spread across Africa and distributed across Angola, Australia, Benin, Brazil, Central Africa, China, Colombia, Cuba, Ecuador, Egypt, Ethiopia, Fiji, Gambia, Ghana, Guinea, Gulf of Guinea, Madagascar, Maldives, Mexico, Myanmar, Namibia, Pakistan, Philippines, Nepal, Taiwan, Tanzania, Nigeria, Saudi Arabia, Somalia, Sudan, Uganda, Venezuela, Vietnam, and Zambia.

32.2 PHYTOCHEMICALS

C. ternatea leaves contain phytochemicals, namely, kaempferol-3-O-rhamnosyl-(1,6)-glucoside, kaempferol-3-O-rhamnosyl-(1,6)-galactoside, and kaempferol-3-O-rhamlnosyl-(1,2) Ochalmnosyl-(1,2)-O-[rhamnosyl-(1,6)]-glucoside (Silalahi, 2020). The flowers of this plant possess tannin, phlobatannin, carbohydrates, cardiac glycosides, saponin, flavonoid, flavanol, glycosides, proteins, alkaloids, anthraquinones, triterpenoid, phenol, anthocyanins, stigmast-4-ene-3,6-dione, essential oils, and steroids (Kamilla et al., 2009b). Fresh flower extract exhibited a higher concentration of anthocyanin compared with dried flower powder. The oxidation of anthocyanin is observed, while the flowers were boiled (Purwaniati and Yuliantini, 2020). Kazuma et al. (2003a, b) reported that the glycoside flavanol from flowers are kaempferol 3-O-(2″-O-a-rhamnosyl-6″-O-malonyl)-β-glucoside, quercetin 3-O-(2″-O-a-rhamnosyl-6″-Omalonil)-β-glucoside, and myricetin3-O-(2″, 6″-di-O-a-rhamnosyl) β-glucoside. Delphinidin derivatives such as ternatin D3, ternatin B3, rutin, quercetin 3-O-dirhamnoside, and manghaslin quercetin 3-[2G]-rhamnosylrutinoside are generated by the herb in addition to ternatin B2, ternatin B4, ternatin C2, and ternate D1 (Nair et al. 2015).

C. ternatea seeds contains glycosides, alkaloids, tannins, saponins, carbohydrate sterols, proteins, sugars, phenolic compounds, and flavonoids (Kalyan et al. 2011) and antifungal proteins (Ajesh and Sreejith, 2014). Vasisht et al. (2016) reported presence of norneolignan compounds clitorieno-lactones A–C, triterpenoid, and taraxerol in the root. Clitorieno lactones A and B are memory enhancers that also suppress acetylcholinesterase (ALT).

Palmitic, linoleic, oleic, stearic, and linolenic acids are some of the fatty acids in addition to anthoxanthin glucoside, cinnamic acid, finotin, a very simple small protein, water-soluble mucilage, delphinidin 3, 3', 5'-triglu-coside, and beta-sitosterol found in the seeds (Kelemu et al., 2004). Total phenolics, flavonoids, and total anthocyanins were reported by Chayarata-nasin et al. (2015) in dried flower extracts. The flowers contained flavonol glycosides. 3-O- (2″-O-alpharhamnosyl- 6″-O-malonyl)-beta-glucoside, 3-O-(6″-O-alpha-rhamnosyl-6″-O-malonyl)-betaglucoside and 3-O-(2″,6″-di-O-alpharhamnosyl)- beta-glucoside of kaemferol, quercetin, and myricetin were isolated from the petals. 3-O-(2″-O-a-rahmnosyl-6″-Omalonyl)- b-glucoside, 3-O-b-glucoside,3-O-(2″-O-a-rahmnosyl)-b-glucoside, Delphinidin glycosides of delphinidin, and eight anthocyanins (ternatins C1, C2, C3, C4, C5, and D3, and preternatins A3 and C4) were also isolated from the flowers (Terahara et al., 1996; Kogawa et al., 2007).

Delphinidin 3-O-(2″-O-alpha-rhamnosyl-6″-O-malonyl)-beta-glucoside, together with three other anthocyanins, was recently isolated from the petals of a mauve line. The whole blue petal contained lines ternatins, a group of 15 (poly)acylated delphinidin glucosides (Kazuma et al., 2003a, b).

32.3 PHARMACOLOGY

C. ternatea possess antioxidant, antimicrobial, anti-inflammatory, anticancer, cardiovascular, central nervous, infectious, hypolipidemic, and analgesic properties (Quazi and Yogekar, 2020).

32.3.1 Antifungal Activity

The existence of a small molecular weight cystein rich finotin, protein, from plants, and a flavanoid in the leaf extract are correlated to antifungal activity of *C. ternatea* (Kelemu et al. 2004). The seeds of *C. ternatea* revealed an antifungal protein with a molecular mass of 14.3 kDa broad-spectrum activities that prevent fungicide like *C. laurentii, C. albidus, Candida albicans, Cryptococcus neoformans*, and *C. parapsilosis*. Its inhibitory activity was also on mycelial growth of *Candida albicans, C. parapsilosis, Curvularia* sp., *Alternaria* sp., *Cladosporium* sp., *A. niger, A. fumigatus, Aspergillus flavus, Rhizopus* sp., and *Sclerotium* sp. (Ajesh and Sreejith 2014). The leaf extract of *C. ternatea* also showed a significant activity against *A. niger* (Kamilla et al., 2009a). Cytoplasm in fungal in the SEM (scanning electron microscopy) observation showed hyphae and wall reduced and became slightly thinner, disoriented, and resulted in the disruption of cell wall after *C. ternatea* extract treatment. Conidiophore changes were also visualized in *A. niger* treated with *C. ternatea* leaf extract. A single protein (finotin) derived from seeds of *Clitoria ternatea* showed broad spectrum inhibitory effects on the growth rate of various notable fungal pathogens of plants (*Fusarium solani, Rhizoctonia solani, Colletotrichum lindemuthianum, Lasiodiplodia theobromae, Pyricularia grisea, Bipolaris oryzae*, and *Colletotrichum gloeosporioides)* (Kelemu et al. 2004).

32.3.2 Antibacterial Activity

Leaves, stems, flowers, and roots are used for in vitro antibacterial activity. 12 bacterial and 2 yeast species were inhibited with the methanol extracts of *C. ternatea* (Kamilla et al., 2009b). *C. ternatea* seed protein inhibits the growth of microbes such as *Micrococcus luteus* (Ajesh and Sreejith 2014). Various extracts of *C. ternatea* inhibit growths of *Klebsiella pneumoniae, Escherichia coli, Bacillus subtilis, Pseudomonas aeruginosa, Aeromonas hydrophila, Aeromonas formicans,* and *Streptococcus agalactiae*. The leaf has potent antibacterial activity against *E. coli*, *V. cholerae*, *S. aureus*. The antibacterial efficacy of the leaf extract was higher than that of the root extract (Mazumder et al., 2007). Extracts prepared from seeds and leaf derived calli of *C. ternatea* showed maximum zone of inhibition against *E. coli.* The callus extract had the greatest inhibition zone against *Salmonella typhi.* Extracts from the leaves of *C. ternatea* were investigated against *Staphylococcus aureus*, *Bacillus cereus*, *Proteus vulgaris*, *Klebsiella pneumoniae,* and *Salmonella typhi* using organic solvents. Antibacterial activity against the investigated microbial pathogens was effective (Anand et al., 2011).

The antibacterial activity of alcoholic and aqueous extracts from in vitro produced calli was investigated using the agar well diffusion technique against Gram-negative bacteria. Aqueous extracts showed antibacterial activity against *Streptococcus pyogenes, Bacillus subtilis*, and *Bacillus cereus.* The extracts also showed antibacterial activity against *Salmonella* spp. and *Shigella dysenteriae*, which cause enteric fever (Shahid et al., 2009). The leaf was shown to have high antibacterial action against the bacteria *Escherichia coli*, which cause dysentery, and *Staphylococcus aureus*, which causes fever. The antibacterial activity of the leaf extract was stronger than the root extract. Mhaskar et al. (2010) investigation that the crude extract from *Clitoria ternatea* seeds showed a maximum zone (22 ± 0.5mm) of inhibition against *Escherichia coli* at 0.75 mg concentration and a minimum zone (14 ± 1.0mm) of inhibition against *Micrococcus flavus* at 0.75 mg concentration. *Salmonella typhi* showed the highest zones of inhibition (16 ± 2 mm), whereas *Escherichia coli* and *Staphylococcus aureus* exhibited the lowest (12 ± 1mm and 12 ± 0.9 mm, respectively).

32.3.3 Anti-Inflammatory, Antipyretic, and Analgesic Effects

Dose of oral treatment *C. ternatea* roots of methanol extract decreased both carrageenan-induced paw edema, and acetate acid-induced vascular permeability in experimental rats and also reduced yeast-induced pyrexia in rats (Devi et al., 2003). *C. ternatea* floral extract was beneficially inflammatory and analgesic in rats (carrageenan paw edema) and mice (hot plate experiment) (Shyamkumar and Ishwar, 2012). In animals experimental both leaf and root *C. ternatea* extracts exhibited a strong antinociceptive activity in mice. According to the formalin test results, the antinociceptive effect of extracts can be mediated at both central and peripherals levels. Furthermore, *C. ternatea* root extract mediated antinociceptive activity centrally at supraspinal and spinal levels in hot plate and tail-flick studies, but antinociceptive activity centrally only at the supraspinal level mediated by the *C. ternatea* leaf extract. The opioid receptors are thought to be involved in the antinociceptive action of both *C. ternatea* root extract according to the scientist (Kamilla et al., 2014). Parimaladevi et al. (2004) investigated the anti-pyretic potential of a methanol extract of blue-flowered variety of *C. ternatea* root (MECTR) on normal body temperature and yeast-induced pyrexia in albino rats. After 19 h of subcutaneous injection, yeast suspension increased rectal temperature. The extract induced a significant reduction in normal body temperature and yeast-provoked high temperature in a dose-dependent manner at dosages of 200, 300, and 400 mg/kg.

32.3.4 Antihelmintic and Insecticidal Effects

The Indian earthworm *Pheritima posthuma* is paralyzed in 15–20 min and dies in 28–30 min after being exposed to an ethanol extract of *C. ternatea* (100mg/ml) (Shekhawat and Vijayvergia, 2011). Antihelmintic efficacy of ethanol extracts of *C. ternatea* leaves, stems, flowers, and roots was tested on adult Indian earthworms *Pheretima posthuma.* According to the findings the plant of *C. ternatea* roots required less time to paralyze and kill earthworms. The roots were then extracted in a series of solvents and examined for antihelmintic activity and were found to be effective (Nirmal et al., 2008). Khadatkar et al. (2008) reported the antihelmintic activity of the *C. ternatea* plant and when compared to a standard reference, the crude alcoholic extract of CT and its ethyl acetate and methanol fractions considerably displayed paralysis and also induced worm mortality, especially at higher concentrations of 50 mg/ ml. citrate of piperazine. The extract from *C.*

ternatea leaves, roots, flowers, and seeds was effective against the larvae of all three species—*Aedes stephensi, Aedes aegypti*, and *C. quinquefasciatus* (Mathew et al., 2009).

32.3.5 Antioxidant Activity

Oxidative stress is a significant contributor to the development of many degenerative and chronic diseases. The petals of *C. ternatea* have been shown to have antioxidant properties (Quazi and Yogekar 2020). The results of Jacob and Latha (2013) indicated that the methanol extract of *C. ternatea* is a potential antioxidant source that could be used to cure and control diseases caused by free radicals. The 1,1-diphenyl-2-picryl-hydrazyl (DPPH) radical scavenging technique revealed that various solvent extracts of *C. ternatea* leaf had potent in vitro free radical scavenging activity.

Patil and Patil (2011a) reported that extracts of roots of blue and white-flowered *C. ternatea* (CT) greatly inhibited the DPPH free radical. The activity of *C. ternatea* leaves, as antioxidant blue and white flowers, was notable, with the sample from the blue flower-bearing plant showing stronger scavenging activity (Sivaprabha et al., 2008). The *C. ternatea* flower petal extract was significant antioxidant activity and protective ability. *C. ternatea* extract remarkably protected erythrocytes, protein carbonyl group formation, reduced membrane lipid peroxidation, and prevented the reduction of glutathione concentration in AAPH-induced oxidation of erythrocytes (Phrueksanan et al., 2014). The yeast cell was used to investigate the antioxidant properties and apoptotic research of *C. ternatea* leaves. *C. ternatea* leaf extracts successfully reduced the level of DNA damage (Balakrishnan et al., 2013. In Thailand, flower extracts from *Clitoria ternatea* (butterfly pea) are used in cosmetics and the chemical properties of the flowers additionally suggest that flower extract has antioxidant properties. *Clitoria ternatea* aqueous extracts were shown to have stronger antioxidant activity than ethanol extracts (Kamkaen and Wilkinson, 2009). The amounts of enzymatic and nonenzymatic antioxidants in aqueous leaf extracts of *Clitoria ternatea* were used to assess their antioxidant capacity. Different assays, such as the Ferric reducing power assay (FRAP), reducing activity test, were used to measure in vitro antioxidant capability. The results were comparable to standard antioxidants such as butylated hydroxyl toluene (BHT), ascorbic acid, and rutin in the diphenypicrylhydrazyl (DPPH) test and hydroxyl radical scavenging efficacy (Rao et al., 2009). The seeds of *C.ternatea* varieties with domesticated ether, chloroform, and methanol were studied for their in-vitro antioxidant capacity. Pet ethers from seeds of *C. ternatea* chloroform and methanol extract of a white-flowered variety of *C. ternatea* significantly inhibited DPPH. The methanol extract from the seeds of the white flowering variety of *C. ternatea* showed more significant antioxidant activity than the blue flowering variety (Patil and Patil, 2011b).

32.3.6 Hepatoprotective Activity

In the paracetamol-induced liver damage on mice, *C. ternatea* leaf extract had shown a significant hepatoprotectivity and antioxidative function. Analysis of the levels of alanine aminotransferases enzymes, aspartate aminotransferases, and bilirubin, as well as various other histopathological analyses, were used to ascertain its activities. The studies revealed that mice paracetamol-induced liver toxicity tests *C. ternatea* leaf had significantly lowered alanine aminotransferase, AST, and bilirubin levels than given paracetamol (Nithianantham et al., 2011). Leaf extracts of *C. ternatea* have also shown its ameorilating effects against adverse histopathological changes (Chakraborthy et al., 2018).

32.3.7 Antidiabetic Activity

CH_3-OH, H_2O, petroleum ether, and chloroform extracts of *C. ternatea* leaves were tested for acute and subacute hypoglycemic effects in streptozotocin-induced diabetic rats *C. ternatea* extract dramatically lowered blood glucose levels. Study showed that the acute effects of the methanol extract with a concentration of 200 and 400 mg/kg produced a very comparable impact, whereas 200 mg/kg exhibited a fine drop in blood glucose level in the early stage of 30 min (Abhishek et al., 2015).

The pancreatic totipotency capacity at different concentrations of an ethanol extract of *C. ternatea* aerial part was investigated by Verma et al. (2013). In streptozotocin-induced diabetic mice, the antidiabetic and anti-hyperlipidemic potential was assessed and linked with antioxidant activity in vivo and in vitro. The ethanol extract and butanol soluble fraction showed the most substantial pancreatic regeneration, antidiabetic, and antihyperlipidemic activities.

In alloxan-induced diabetes in rats, the hypo-glycemic effects of an aqueous extract of *C. ternatea* foliars and flowers were estimated. Aqueous extracts of *C. ternatea* foliar and flowers reduced serum glucose levels, glycosylated hemoglobin, and glucose-6-phosphatase activity, but it increased the levels of insulin in blood, hepatocites, and skeletal-muscle-glycogen, and the activity of glucokinase (Daisy et al., 2009).

32.3.8 Wound-Healing Effect

Excision, incision, and dead-space wound-healing effects *C. ternatea* seed and root extracts were investigated on mice models. Oral as well as topical as an ointment, *C. ternatea* seed, and root extracts greatly enhanced wound healing in excision, incision, and dead-space mice models. These results were like those shown with cotrimoxazole ointment. *C. ternatea* was shown to affect all the three different wound-healing phases: inflammatory stage, proliferation condition, and remodeling, according to the findings (Solanki and Jain 2012).

32.3.9 Antihistaminic and Antiasthmatic Effect

Taur and Patil (2009) investigated the antiasthmatic activity of CH_3-OH and C_6H_6 extracts of *C. ternatea.* However, in milk-induced eosinophilia, ethanol extract inhibited eosinophilia in a dose-dependent response, but benzene extract did not suppress eosinophilia. *C. ternatea* root extract reduces milk-induced leucocytosis and eosinophilia in mice, protects against egg albumin-induced mast cell degranulation in rats, and inhibits the region of blue dye leakage in passive cutaneous anaphylaxis in rats (Taur and Patil, 2011). *C. ternatea* ethanolic extract protected rats from histamine-induced bronchoconstriction and reduced inflammatory cell infiltration in the airway and inhibited the induction of histamine-like mediators from mast cells through stabilizing it (Chauhan et al., 2012).

32.3.10 Central Nervous Effect

C. ternatea root extract was known to cause long-term changes in the brain, which were closely linked to improved learning ability (Rai et al., 2002). *C. ternatea* extracts were found to also have significant nootropic, anxiolytic, antidepressant, and antistress characteristics. The use of nootropic medications enhances mental performance, learning, and memory (Gupta et al., 2010). In a study, Rai et al. (2000a) employed open field behavior test, spontaneous alternation test, rewarded alternation test, and passive avoidance test to assess the effect of CT aqueous root extract on learning and memory in 7-day-old rat pups. The findings of this investigation revealed that oral administration of CT roots extract at various dosages greatly improved rat memory.

Jain et al. (2003) studied on the CNS, the spectrum of action of *Clitoria ternatea* (CT) methanolic extract. The CT's impact on cognition behavior, anxiety, sadness, stress, and convulsions generated by pentylenetetrazol (PTZ), as well as maximal electroshock, was investigated. Sedative effect of *C. ternatea* is utilized to treat conditions like syncope, vertigo, and brain weakness. The effects of *C. ternatea* on cognition, perplexity, depression, stress, and convulsions were investigated by Sethiya (2009). The methanolic extract of *C. ternatea* was shown to have nootropic, anxiolytic, antidepressant, anticonvulsant, and antistress efficacy. *C. ternatea* seeds and leaves are believed to be brain tonic and improve memory and intellect. *C. ternatea*'s activity in Alzheimer's disease was investigated by Shahnas and Akhila (2014). In the study, two groups of mice, newly-born and young-adult mice, were treated with a concentration of 100 mg/kg *C. ternatea* aqueous root extract (CTR) for a period of 30 days, showing significant increase in acetylcholine (ACh) levels in their hippocampi compared to their respective age-matched control rats. Increased ACh levels in their hippocampus region may be the reason for neurochemical basis for their enhanced memory and cognitive capacity (Rai et al., 2002). The alcoholic extracts of *C. ternatea's* aerial part and roots increased rat brain acetylcholine levels and activity of acetylcholinesterase (Taranalli and Cheeramkuzhy, 2000). (Z)-9,17-octadecadienal and n-hexadecenoic acids stopped the monoamine oxidase activity and were potentially against antidepressant, anxiety, and cognitive illness in Alzheimer's and Parkinson's diseases (Margret et al., 2015).

Rai et al. (2001) demonstrated that a dose of 100 mg/kg of aqueous root extract significantly enhanced acetylcholine(Ach) levels in the hippocampus of newly born rats having a range of 52.79 ± 12.36 to 68.83 ± 9.87 nmol/g tissue, whereas that of young adult rats had a range of 33.9 ± 6.22 to 52.79 ± 12.36 nmol/g tissue. This extract also increased dendritic arborization in brain structures such as hippocampus neurons and the amygdala, resulting in an increase in protein production such as acetylcholinesterase (Rai et al., 2001). Rai et al. (2000b) reported that CT aqueous root extract administration increases amygdaloid neuron dendritic arborization as well as hippocampal CA3 neuron dendritic arborization. Enhanced dendritic arborization of amygdaloidal nerve cells correlates with improved passive avoidance learning and memory in *C. ternatea*-treated rats (Rai et al., 2005).

32.3.11 Diuretic and Anti-Urolithiasis Effect

To analyze the inhibitory effect of in vitro calcium oxalate crystal formation (a typical major constituent of stones in the urinary system), the titrimetric approach was used by different *C. ternatea* extracts. The inhibitory efficacy of *C. ternatea* alcoholic extract was found to be equivalent to Cystone (a medicine for treating renal stones through dissolution). In vitro, the alcoholic extract of *C. ternatea* foliars inhibited calcium oxalate crystallization more effectively than cystone (Quazi et al., 2014).

32.3.12 Hypolipidemic Effect

The aqua-alcoholic extract of *C. ternatea* roots and seeds decreased STC (serum total cholesterol), triglycerides, very-low-density lipoprotein cholesterol, and low-density lipoprotein cholesterol significantly. In diet-induced hyperlipidemic rats, the atherogenic index and HDL/LDL ratio were both found normal after the treatment (Solanki and Jain, 2010a).

32.3.13 Antiulcer Effect

In model rats with experimentally induced ulcer, the antiulcer capacity of hydrous and ethanolic extracts of entire plant of *C. ternatea* was assessed. Regarding ulcer induction, various parameters such as volume of GAS (gastric acid secretion), pH (potential of hydrogen), total acidity, ulcer index, and various other antioxidative parameters were measured and these parameters were then compared with extracts group, standard group, and control groups. In pylorus ligation and indomethacin caused ulceration, on which the high dose of the alcoholic extract showed significant antiulcer efficacy (Rai et al., 2015).

32.3.14 Anticancer Effect

A crude methanol extract of *C. ternatea* foliars, seedlings, and bark of stems exhibited significant cytotoxic activity in a brine shrimp lethality bioassay test. The cytotoxic effect of crude CH_3-OH extract and fraction CH_3-OH of leaves was found very promising (Rahman et al., 2006). The in vitro cytotoxic and antioxidant activities of an ethanolic extract of *C. ternatea* were investigated by Ramaswamy et al. (2011). In the trypan blue dye exclusion method using DLA cell lines, the extract showed potent cytotoxic activity. Petroleum ether and ethanolic flower extracts of *C. ternatea* had a strong dose-dependent cytotoxic effect on cells (Kumar and Bhat, 2011).

32.3.15 Anesthetic Activity Studies

Kulkarni et al. (1988) investigated the anesthetic activity of an alcoholic extract of *C. ternatea* aerial parts using ocular anesthetic in some rodents and plexus anesthetic in frogs. In induced anesthesia, the results were nearly as reliable as xylocaine.

32.3.16 Effect on General Behavior

In a neuropharmacological-effect study on rats and mice, an ethanolic extract of the root of *C. ternatea* (CTEE) was tested for a variety of behavior, including general behavior, exploratory behavior, muscle relaxant activity, and phenobarbitone-induced sleep duration. Additionally, CTEE significantly increased with the period of phenobarbitone-induced sleep (Boominathan et al., 2003).

32.3.17 Immunomodulatory Effect

The immunomodulatory impact of the alcoholic extract of CT root and the hydroalcoholic extract of CT seed was observed, which might be attributed to reduced immune cell sensitization, immune cell presentation, and phagocytosis (Solanki and Jain, 2010b).

In a male albino rat model, alcoholic extracts of *C. ternatea* seed and root revealed significant immunosuppressive action. Immunoinhibition may be helped by the antioxidant and anti-inflammatory properties of plants. The presence of flavonoid and phenolic chemicals may be responsible for the immunomodulatory effect (Daisy et al., 2004). The immunostimulatory properties of hydrous extracts of *C. ternatea* leaf and flower were evaluated by giving alloxan-induced diabetic mice aqueous extract of *C. ternatea* for 60 days, which dramatically reduced blood glucose and cholesterol levels. In treated mice, total white blood cells, red blood cells, T-lymphocytes, and B-lymphocytes all increased significantly, but monocytes and eosinophils decreased. These results also imply that the plant extracts are having immunomodulatory properties that help the immune system function better (Daisy et al., 2004). In rabbits, the anthocyanin ternatin D1 extracted from the petals of *C. ternatea* inhibited platelet aggregation in vitro. It is because collagen and ADP-induced platelet aggregation are significantly inhibited (Honda et al., 1991).

KEYWORDS

- ***C. ternatea***
- **medicinal plant**
- **antimicrobial**
- **anticancer**
- **antipyretic**
- **anti-inflammatory**

REFERENCES

Abhishek, S.; Pankaj, M.; Vikas, S. Hypoglycemic Effects of *Clitoria ternatea* (Linn) Leaves Extract. *J. Pharmacol. Toxicol. Studies* **2013,** *1* (1), 4–7.

Abhishek, S.; Vikas, S.; Minu, K.; Pankaj, M. Comparative Hypoglycemic Effects of Different Extract of *Clitoria ternatea* Leaves on Rats. *IOSR J. Pharm. Biol. Sci.* **2015,** *10*, 60–65.

Ajesh, K.; Sreejith, K. A Novel Antifungal Protein with Lysozyme-Like Activity from Seeds of *Clitoria ternatea. Appl. Biochem. Biotechnol.* **2014,** *173* (3), 682–693.

Anand, S. P.; Doss, A.; Nandagopalan, V. Antibacterial Studies on Leaves of *Clitoria ternatea* Linn. A High Potential Medicinal Plant. *Intern. J. Appl. Biol. Pharmaceut. Technol.* **2011,** *2* (3), 453–456.

Balakrishnan, B.; Ayyavoo, J.; Sadayan, P.; Abimannan, A. Evaluation of Antioxidant Activity of *Clitoria ternatea* and *Alternanthera sessilis* Plant Extracts Using Model System for Yeast Cells. *Afr. J. Basic Appl. Sci.* **2013,** *5* (3), 134–138.

Boominathan, R.; Parimaladevi, B.; Mandal, S. C. Studies on Neuropharmacological Effects of *C. ternatea* Linn. Root Extracts in Rats and Mice. *Nat. Prod. Sci.* **2003,** *9* (4), 260–263.

Chakraborthy, G. S.; Kumar, V.; Gupta, S.; Kumar, A.; Gautam, N.; Kumari, L. Phytochemical and Pharmacological Aspects of *Clitoria ternatea*—A Review. *J. Appl. Pharma. Sci. Res.* **2018,** *1*, 3–9.

Chauhan, N.; Rajvaidhya, S.; Dubey, B. K. Antiasthmatic Effect of Roots of *Clitorea ternatea* Linn. *Intern. J. Pharma. Sci. Res.* **2012,** *3* (4), 1076–1079.

Chayaratanasin, P.; Barbieri, M. A.; Suanpairintr, N.; Adisakwattana, S. Inhibitory Effect of *Clitoria ternatea* Flower Petal Extract on Fructose-Induced Protein Glycation and Oxidation-Dependent Damages to Albumin In Vitro. *BMC Complem. Altern. Med.* **2015,** *15* (1), 1–9.

Daisy, P.; Priya, N.; Rajathi, M. Immunomodulatory Activity of *Eugenia jambolana, Clitoria ternatea* and *Phyllanthus emblica* on Alloxan-Induced Diabetic Rats. *J. Exp. Zool.* **2004,** *7* (2), 269–278.

Daisy, P.; Santosh, K.; Rajathi. M. Antihyperglycemic and Antihyperlipidemic Effect of *C. ternatea* Linn. in Alloxan–Induced Diabetic Rats. *Afr. J. Microbiol. Res.* **2009,** *3* (5), 287–291.

Devi, B. P.; Boominathan, R.; Mandal, S. C. Anti-Inflammatory, Analgesic and Antipyretic Properties of *Clitoria ternatea* Root. *Fitoterapia* **2003,** *74* (4), 345–349.

Gupta, G. K.; Chahal, J.; Bhatia, M. *Clitoria ternatea* (L.): Old and New Aspects. *J. Pharm. Res.* **2010,** *3* (11), 2610–2614.

Honda, T.; Saito, N.; Kusano, T.; Ishsone, H.; Funayama, N.; Kubota, T.; Araogi, S. Isolation of Anthocyanins (Ternatin A1, A2, B1, B2, D1and D2) from *Clitoria ternatea* cv. (Double Blue) Having Blood Platelet Aggregation-Inhibiting and Vascular Smooth Muscle Relaxing Activities. Japan Kokai Tokyo Koha, 1991; p 7.

Jacob, L.; Latha, M. S. In Vitro Antioxidant Activity of *Clitoria ternatea* Linn. *Intern. J. Res. Phytochem. Pharmacol.* **2013,** *3* (1), 35–39.

Jain, N. N.; Ohal, C. C.; Shroff, S. K.; Bhutada, R. H.; Somani, R. S.; Kasture, V. S.; Kasture, S. B. *Clitoria ternatea* and the CNS. *Pharmacol. Biochem. Behavior.* **2003,** *75* (3), 529–536.

Kalyan, B. V.; Kothandam, H.; Palaniyappan, V.; Praveen, A. R. Hypoglycaemic Activity of Seed Extract of *Clitoria ternatea* Linn in Streptozotocin-Induced Diabetic Rats. *Pharmacogn. J.* **2011,** *3* (19), 45–47.

Kamilla, L.; Mansor, S. M.; Ramanathan, S.; Sasidharan, S. Effects of *Clitoria ternatea* Leaf Extract on Growth and Morphogenesis of *Aspergillus niger*. *Microsc. Microanalysis* **2009a,** *15* (4), 366–372.

Kamilla, L.; Mansor, S. M.; Ramanathan, S.; Sasidharan, S. Antimicrobial Activity of *Clitoria ternatea* (L.) Extracts. *Pharmacol. Online.* **2009b,** *1*, 731–738.

Kamilla, L.; Ramanathan, S.; Sasidharan, S.; Mansor, S. M. Evaluation of Antinociceptive Effect of Methanolic Leaf and Root Extracts of *Clitoria ternatea* Linn. in Rats. *Indian J. Pharmacol.* **2014,** *46* (5), 515–520.

Kamkaen, N.; Wilkinson, J. M. The Antioxidant Activity of *Clitoria ternatea* Flower Petal Extracts and Eye Gel. *Phytother. Res.* **2009,** *23* (11), 1624–1625.

Kazuma, K.; Noda, N.; Suzuki, M. Flavonoid Composition Related to Petal Color in Different Lines of *Clitoria ternatea. Phytochemistry* **2003a,** *64* (6), 1133–1139.

Kazuma, K.; Noda, N.; Suzuki, M. Malonylated Flavonol Glycosides from the Petals of *Clitoria ternatea. Phytochemistry* **2003b,** *62* (2), 229–237.

Kelemu, S.; Cardona, C.; Segura, G. Antimicrobial and Insecticidal Protein Isolated from Seeds of *Clitoria ternatea*, a Tropical Forage Legume. *Plant Physiol. Biochem.* **2004,** *42* (11), 867–873.

Kogawa, K.; Kazuma, K.; Kato, N.; Noda, N.; Suzuki, M. Biosynthesis of Malonylated Flavonoid Glycosides on the Basis of Malonyltransferase Activity in the Petals of *Clitoria ternatea*. *J. Plant Physiol.* **2007,** *164* (7), 886–894.

Khadatkar, S. N.; Manwar, J. V.; Bhajipale, N. S. In-Vitro Antihelmintic Activity of Root of *Clitoria ternatea* Linn. *Pharmacogn. Netw. Worldwide* **2008,** *4* (13), 148–150.

Kulkarni, C.; Pattanshetty, J. R.; Amruthraj, G. Effect of Alcoholic Extract of *Clitoria ternatea* L. on Central Nervous in Rodents. *Indian J. Exp. Biol.* **1988,** *26* (12), 957–960.

Kumar, B. S.; Bhat, K. I. In-Vitro Cytotoxic Activity Studies of *Clitoria ternatea* Linn Flower Extracts. *Intern. J. Pharmaceut. Sci. Rev. Res.* **2011,** *6* (2), 120–121.

Macedo, M. L. R.; Xavier-Filho, J. Purification and Partial Characterization of Trypsin Inhibitors from Seeds of *Clitoria ternatea*. *J. Sci. Food Agric.* **1992,** *58* (1), 55–58.

Margret, A. A.; Begum, T. N.; Parthasarathy, S.; Suvaithenamudhan, S. A Strategy to Employ *Clitoria ternatea* as a Prospective Brain Drug Confronting Monoamine Oxidase (Mao) Against Neurodegenerative Diseases and Depression. *Nat. Prod. Bioprospecting* **2015,** *5* (6), 293–306.

Mathew, N.; Anitha, M. G.; Bala, T. S. L.; Sivakumar, S. M.; Narmadha, R.; Kalyanasundaram, M. Larvicidal Activity of *Saraca indica*, *Nyctanthes arbor-tristis*, and *Clitoria ternatea* Extracts Against Three Mosquito Vector Species. *Parasitol. Res.* **2009,** *104* (5), 1017–1025.

Mazumder, A.; Roy, P.; Mazumder, R. In Vitro Antibacterial Activity of Leaf and Root Extracts of *Clitoria ternatea* Linn. *Ethiopian Pharmaceut. J.* **2007,** *25* (2), 145–150.

Mehla, J.; Pahuja, M.; Gupta, P.; Dethe, S.; Agarwal, A.; Gupta, Y. K. *Clitoria ternatea* Ameliorated the Intracerebroventricularly Injected Streptozotocin Induced Cognitive Impairment in Rats: Behavioral and Biochemical Evidence. *Psychopharmacol.* **2013,** *230* (4), 589–605.

Mhaskar, A. V.; Prakesh, K.; Vishwakarma, K. S. Maheshwari VL, Callus Induction and Antimicrobial Activity of Seed and Callus Extracts of *Clitoria ternatea* L. *Curr Trends Biotechnol. Pharam.* **2010,** *3* (4), 561–567.

Morris, J. B. Characterization of Butterfly Pea (*Clitoria ternatea* L.) Accessions for Morphology, Phenology, Reproduction and Potential Nutraceutical, Pharmaceutical Trait Utilization. *Genet. Resour. Crop Evol.* **2009,** *56* (3), 421–427.

Mukherjee, P. K.; Kumar, V.; Kumar, N. S.; Heinrich, M. The Ayurvedic Medicine *Clitoria ternatea*—From Traditional Use to Scientific Assessment. *J. Ethnopharmacol.* **2008,** *120* (3), 291–301.

Nair, V.; Bang, W. Y.; Schreckinger, E.; Andarwulan, N.; Cisneros-Zevallos, L. Protective Role of Ternatin Anthocyanins and Quercetin Glycosides from Butterfly Pea (*Clitoria ternatea* Leguminosae) Blue Flower Petals Against Lipopolysaccharide (LPS)-Induced Inflammation in Macrophage Cells. *J. Agric. Food Chem.* **2015,** *63* (28), 6355–6365.

Neda, G. D.; Rabeta, M. S.; Ong, M. T. Chemical Composition and Anti-Proliferative Properties of Flowers of *Clitoria ternatea*. *Intern. Food Res. J.* **2013,** *20* (3), 1229–1234.

Nirmal, S. A.; Bhalke, R. D.; Jadhav, R. S.; Tambe, V. D. Anthelmintic Activity of *Clitoria ternatea*. *Pharmacol. Online* **2008,** *1*, 114–119.

Nithianantham, K.; Shyamala, M.; Chen, Y.; Latha, L. Y.; Jothy, S. L.; Sasidharan, S. Hepatoprotective Potential of *Clitoria ternatea* Leaf Extract Against Paracetamol Induced Damage in Mice. *Molecules* **2011,** *16* (12), 10134–10145.

Patil, A. P.; Patil, V. R. Comparative Evaluation of In Vitro Antioxidant Activity of Root of Blue and White Flowered Varieties of *Clitoria ternatea* Linn. *Intern. J. Pharmacol.* **2011a,** *7* (4), 485–491.

Parimaladevi, B.; Boominthan, R.; Mandal, S. C. Evaluation of Antipyretic Potential of *C. ternatea* L Extract in Rats. *Phytomedicine.* **2004,** *11* (4), 323–326.

Patil, A. P.; Patil, V. R. Evaluation of In Vitro Antioxidant Activity of Seeds of Blue and White Flowered Varieties of *Clitoria ternatea* Linn. *Intern. J. Pharm. Pharmaceut. Sci.* **2011b,** *3* (4), 330–336.

Phrueksanan, W.; Yibchok-anun, S.; Adisakwattana, S. Protection of *Clitoria ternatea* Flower Petal Extract Against Free Radical-Induced Hemolysis and Oxidative Damage in Canine Erythrocytes. *Res. Vet. Sci.* **2014,** *97* (2), 357–363.

Piala, J. J.; Madissoo, H.; Rubin, B. Diuretic Activity of Roots of *Clitoria ternatea* L. in Dogs. *Experientia* **1962,** *18* (2), 89–89.

Purwaniati, A. A. R.; Yuliantini, A. Analysis of Total Anthocyanin Content in Telang Flowers Preparations (*Clitoria ternatea*) with pH Differential Method Using Visible Spectrophotometry. *Farmagazine* **2020,** *7* (1), 18–23.

Quazi, S.; Rathore, P.; Sharma, A.; Sharma, P.; Panchariya, N.; Sharma, S. Inhibition of Calcium Oxalate Crystallization In Vitro by *Clitoria ternatea* Root. *Indian J. Drugs* **2014,** *2* (1), *24* 25.

Quazi, S.; Yogekar, T. A Review on Ethanopharmacological Importance of *Clitoria ternatea*. *Scholars Acad. J. Biosci.* **2020,** *8* (3), 63–67.

Rahman, A. S.; Saha, R.; Talukder, N.; Khaleque, S. M. A.; Ali, H. A. Bioactivity Guided Cytotoxic Activity of *Clitoria ternatea* Utilizing Brine Shrimp Lethality Bioassay. *Bangladesh J. Physiol. Pharmacol.* **2006,** *22* (1), 18–21.

Rai, K. S.; Murthy, K. D.; Karanth, K. S.; Nalini, K.; Rao, M. S.; Srinivasan, K. K. *Clitoria ternatea* Root Extract Enhances Acetylcholine Content in Rat Hippocampus. *Fitoterapia* **2002,** *73* (7–8), 685–689.

Rai, K. S.; Murthy, KD.; Rao, MS.; Karanth, K. S. Altered Dendritic Arborization of Amygdala Neurons in Young Adult Rats Orally Intubated with *Clitoria ternatea* Aqueous Root Extract. *Phytother. Res.* **2005,** *19* (7), 592–598.

Rai, K. S.; Rao, M. S.; Karanth, K. *Clitoria ternatea* Enhances Learning and Memory- An Experimental Study on Rats. International Congress on frontiers in Pharmacology and Therapeutics in 21st Century. New Delhi. *India. Indian J. Pharmacol.* **2000a,** *32*, 150.

Rai, K. S.; Murthy, K. D.; Karanth, K. S.; Rao, M. S. *Clitoria ternatea* (Linn) Root Extract Treatment During Growth Spurt Period Enhances Learning and Memory in Rats. *Indian J. Physiol. Pharmacol.* **2001,** *45* (3), 305–313.

Rai, K. S.; Murthy, K. D.; Rao, M. S.; Karanth, K. S. *Clitoria ternatea* (Linn) Root Extract Treatment in Rats During Growth Spurt Period Affects Dendritic Morphology of Hippocampal CA3 Neurons, Third Congress of Federation of Indian Physiological Societies (FIPS). Abstract no. 4.5, Calcutta, India, 2000b.

Rai, S. S.; Banik, A.; Singh, A.; Singh, M. Evaluation of Anti-Ulcer Activity of Aqueous and Ethanolic Extract of Whole Plant of *Clitoria ternatea* in Albino Wistar Rats. *Intern. J. Pharmaceut. Sci. Drug Res.* **2015,** *7* (1), 33–39.

Ramaswamy, V.; Varghese, N.; Simon, A. An Investigation on Cytotoxic and Antioxidant Properties of *Clitoria ternatea* L. *Intern. J. Drug Discov.* **2011,** *3* (1), 74–77.

Rao, D. B.; Kiran, C. R.; Madhavi, Y.; Rao, P. K.; Rao, T. R. Evaluation of Antioxidant Potential of a *Clitoria ternatea* L. and *Eclipta prostrata* L. *Indian J. Biochem. Biophys.* **2009,** *46*, 247–252.

Sethiya, N. K.; Nahata, A.; Mishra, S. H.; Dixit, V. K. An Update on Shankhpushpi, a Cognition Boosting Ayurvedic Medicine. *Zhong Xi Yi Jie He Xue Bao* **2009,** *7* (11), 1001–1022.

Shahnas, N.; Akhila, S. Phytochemical, In Vitro and Silico Evaluation on *Clitoria ternatea* for Alzheimer's Disease. *Pharma Tutor.* **2014,** *2* (9), 135–149.

Shahid, M.; Shahid, A.; Anis, M. Antibacterial Potential of the Extracts Derived from Leaves of Medicinal Plants *Pterocarpus marsupium* Roxb., *Clitoria ternatea*, Korean Medical Database. *Oriental Pharm. Exp. Med.* **2009,** 174–181.

Shekhawat, N.; Vijayvergia, R. Anthelmintic Activity of Extracts of Some Medicinal Plants. *Intern. J. Computation. Sci. Math.* **2011,** *3* (2), 183–187.

Shyamkumar, I. B.; Ishwar, B. Anti-Inflammatory, Analgesic, and Phytochemical Studies of *Clitoria ternatea* Linn Flower Extract. *Intern. Res. J. Pharm.* **2012,** *3* (3), 208–210.

Silalahi, M. *Clitoria ternatea* L. Fabaceae. Ethnobotany of the Mountain Regions of Southeast Asia, 2020; pp 1–7.

Sivaprabha, J.; Supriya, J.; Sumathi, S.; Padma, P. R.; Nirmaladevi, R.; Radha, P. A Study on the Levels of Nonenzymic Antioxidants in the Leaves and Flowers of *Clitoria ternatea*. *Ancient Sci. Life*. **2008,** *27* (4), 28–32.

Solanki, Y. B.; Jain, S. M. Antihyperlipidemic Activity of *Clitoria ternatea* and *Vigna mungo* in Rats. *Pharmaceut. Biol.* **2010a,** *48* (8), 915–923.

Solanki, Y. B.; Jain, S. M. Immunomodulatory Activity of Ayurvedic Plant Aparajita (*Clitoria ternatea* L.) in Male Albino Rats. *Global J. Sci. Front. Res.* **2010b,** *10* (3), 2–8.

Solanki, Y. B.; Jain, S. M. Wound Healing Activity of *Clitoria ternatea* L. in Experimental Animal Models. *Pharmacologia* **2012,** *3* (6), 160–168.

Taranalli, A. D.; Cheeramkuzhy, T. C. Influence of *Clitoria ternatea* Extracts on Memory and Central Cholinergic Activity in Rats. *Pharma. Biol.* **2000,** *38* (1), 51–56.

Taur, D. J.; Patil, R. Y. Effect of *Clitoria ternatea* Seeds Extract on Milk-Induced Leucocytosis and Eosinophilia in Mice. *J. Pharm. Res.* **2009,** *2* (12), 1839–1841.

Taur, D. J.; Patil, R. Y. Evaluation of Antiasthmatic Activity of *Clitoria ternatea* L. Roots. *J. Ethnopharmacol.* **2011,** *136* (2), 374–376.

Terahara, N.; Oda, M.; Matsui, T.; Osajima, Y.; Saito, N.; Toki, K.; Honda, T. Five New Anthocyanins, Ternatins A3, B4, B3, B2, and D2, from *Clitoria ternatea* Flowers. *J. Nat. Prod.* **1996,** *59* (2), 139–144.

Vasisht, K.; Dhobi, M.; Khullar, S.; Mandal, S. K.; Karan, M. Norneolignans from the Roots of *Clitoria ternatea* L. *Tetrahedron Lett.* **2016,** *57* (16), 1758–1762.

Verma, P. R.; Itankar, P. R.; Arora, S. K. Evaluation of Antidiabetic Antihyperlipidemic and Pancreatic Regeneration, Potential of Aerial Parts of *Clitoria ternatea*. *Revista Brasileira de Farmacognosia* **2013,** *23* (5), 819–829.

CHAPTER 33

Secondary Compounds Profile and Bioactive Properties of *Sesbania sesban* (L.) Merr.

J. M. SASIKUMAR[1*], SIBBALA SUBRAMANYAM[2], K. N. JAYAVEERA[3], NAFYAD I. BATU[4], and MESERET C. EGIGU[1]

[1]*School of Biological Sciences and Biotechnology, College of Natural and Computational Sciences, Haramaya University, Haramaya, P.O. Box 138, Ethiopia*

[2]*Department of Pharmaceutical Chemistry, College of Health and Medical Sciences, Haramaya University, Haramaya, Ethiopia*

[3]*Department of Chemistry, Jawaharlal Nehru Technological University, Anantapur, Andhra Pradesh 515002, India*

[4]*Department of Biology, Bonga University, Bonga, P.O. Box 334, Ethiopia*

[*]*Corresponding author. E-mail: jmsashikumar@gmail.com*

ABSTRACT

Sesbania sesban (L.) Merr. (Fabaceae) is used in ethnomedicine as a remedy for various ailments. The major objective of this review is to discuss the information on traditional medicinal uses, secondary metabolites composition and bioactive studies on *S. sesban*. The outcome of the literature survey showed that *S. sesban* is known to contain flavonols, glycosides, proteins, tannins, triterpenoids, saponins, sterols and steroids. Pharmacological investigations based on *in vitro* and *in vivo* studies of *S. sesban* revealed that it possessed antimicrobial, antioxidant, larvicidal, cytotoxic, anti-inflammatory,

Bioactives and Pharmacology of Legumes. T. Pullaiah, PhD (Ed.)

antinociceptive, anthelmintic, antidiabetic, antifertility, molluscicidal and spermicidal activities.

33.1 INTRODUCTION

Sesbania sesban (L.) Merr., belonging to family Fabaceae, is commonly known as jayant (Bengali); sesbania (English); jainti, jait, rawasan (Hindi); champai, chithagathi, karunchembai (Tamil); sesaban (Arabic). It has synonyms as *Sesbania aegyptiaca* Poir., *Aeschynomene sesban* L., *Coronilla sesban* Willd., and *Emerus sesban* (L.) Kuntze. It is an 8-m-tall shrub and short-lived, with 2–18 cm long, pinnately compound leaves and with 6–27 pairs of linear-oblong leaflets with the length of 26 × 5 mm. Raceme of this plant comprises 2–20 flowers and the yellow corolla possesses streaks of purple or brown. The plant contains sub-cylindrical pods that are straight or slightly curved with a length of 30 cm and a width of 5 mm with 10–50 seeds. The origins of *S. sesban* are not known. It is broadly distributed and cultivated through tropical Africa and Asian nations (Gomase et al., 2012; Humperys, 1995; Mani et al., 2011).

In ethnomedicine, the poultice of leaves of *S. sesban* helps in suppuration of sores and swellings and absorption of rheumatic inflammations. Juice from fresh leaves is found to possess anthelmintic properties (Anonymous, 1980). In the folk-lore medicines, the juice of leaves is used in scorpion stings. Leaves are useful in diabetes, colic, skin diseases and found to have application in Vicharchika that is an Eczema-like skin disease. Seeds are used as stimulant, astringent, emmenagogue, and also utilized in diarrhea and excessive flow of menstruation, and the seeds are known for reducing spleen enlargement (Nadkarni, 1982; Rastogi and Mehrotra, 1993; Yusuf et al., 1994). Barks are used in the treatment of ulcers, leucorrhoea, vitiated conditions of pitta, anemia, bronchitis, tumors, dysentery, inflammations, cirrhosis of the liver, and hypertension (Warrier et al., 1996–1997). The plant is known for its medicinal applications as antifertility, anthelmintic, astringent, anti-inflammatory, antimicrobial, carminative, demulcent, and laxative. It is also found as a medicine for the treatment of fever and ulcers (Sheik et al., 2012).

33.2 PHYTOCHEMICAL CONSTITUENTS

The ethanolic extract of *S. sesban* was analyzed by GC-MS, which revealed the presence of three compounds such as squalene, hexatriacontane, and lauroyl peroxide. *S. sesban* was found to comprise amino acids, carbohydrates, fat, glycosides, kampferol, proteins, tannins, triterpenoids, saponins, sterol, steroids, and vitamins (Chatterjee and Pakrashi, 1992). The flowers comprise cyanidin and delphinidin glucosides. Pollen and pollen tubes of the plant comprise alpha-ketoglutaric, pyruvic and oxaloacetic acids. Earlier phytochemical investigations of the plant were directed to the extraction of amino acids, fatty acids, stigmasta-5, 24(28)-diene-3-ol-3-0-β-D-galactopyranoside, and oleanolic acid (Gupta and Grasdalen, 1989). Some kinds of lignins with guaiacyl and syringyl and, P-hydroxyphenylpropane building units and also kaempferol disaccharide, an anticancer principle were isolated (Upadhyaya and Singh, 1991). The leaves of *S. sesban* were found to contain triterpenes (Tokayo, 2000). The flowers of *S. sesban* have been reported to have cyanidin and delphinidin glycosides and flavonols. The seeds of this plant were found to contain saponins, palmitic acid, stearic acid, lignoceric acid, oleic acid, and linolenic acid. The leaves of *S. sesban* comprise kaempferol (Kinghorn, 2001). Kaempferol and β-sitosterol were isolated from the stem extracts of *S. sesban* (Singh et al., 2017a) (Figure 33.1).

33.3 BIOACTIVE STUDIES

33.3.1 Antimicrobial Activity

Leaf extracts of *S. aegyptiaca* (Synonym of *S. sesban*) were evaluated for antimicrobial activities by Sasikumar et al. (2005). In this study, crude petroleum ether, benzene, chloroform, ethyl acetate, methanol, and water extracts were tested against *Escherichia coli* MTCC 443, *Staphylococcus aureus* MTCC 737, *Salmonella typhi* MTCC 734, and *Pseudomonas aeruginosa* MTCC 741 at 10, 5, 2.5 mg/mL concentrations using the disc diffusion method and minimum inhibitory concentration (MIC). The extracts were also tested against fungal pathogens such as *Mucor* sp., *Aspergillus niger*, *A. fumigatus*, and *Alternaria alternata*. All the extracts showed appreciable activity against the tested pathogens. *P. aeruginosa* was found to be most susceptible to the plant extracts among the tested pathogens (MIC = 0.125 mg/mL).

In another study, antibacterial and antifungal activity of leaves of *S. sesban* was investigated by Hossain et al. (2007). The n-hexane, chloroform, and carbon tetrachloride fractions of methanol extract of the leaves were subjected for antibacterial and antifungal activity by using the disc diffusion technique. The fractions exhibited inhibitory activity against the tested gram-negative and gram-positive bacteria. Also, a strong inhibition against *Aspergillus niger* was established by the extracts.

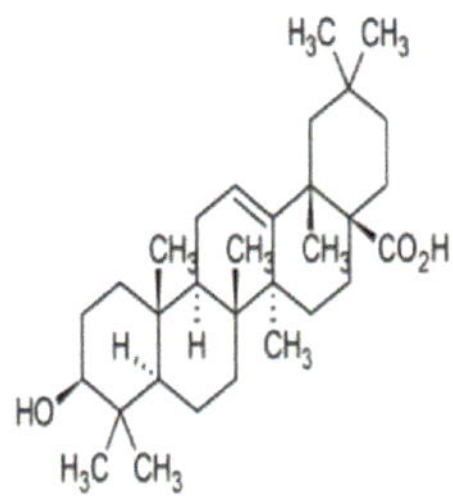

Oleanolic acid

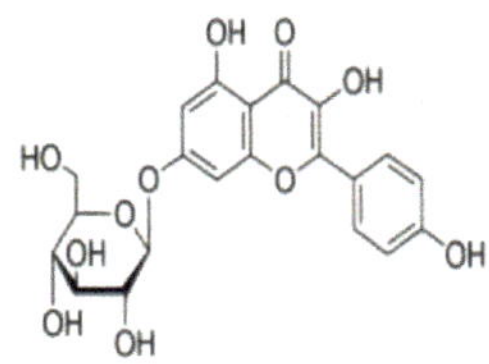

Kaemferol-3-rutinoside

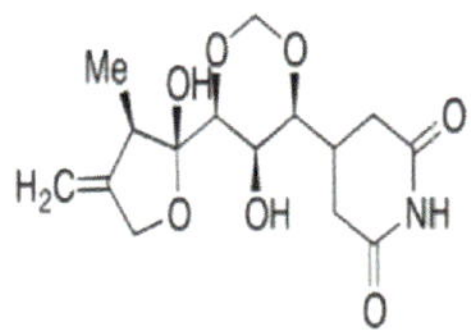

Sesbanimide

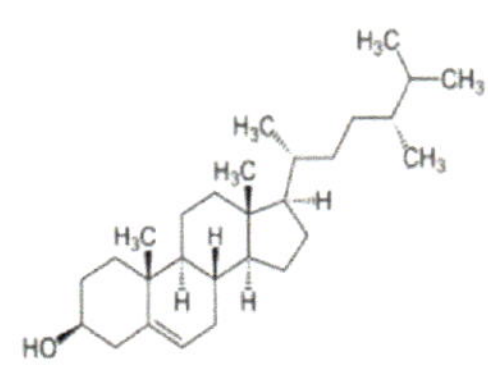

Campesterol

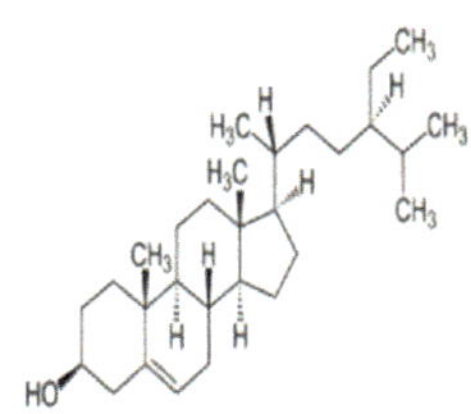

B-Sitosterol

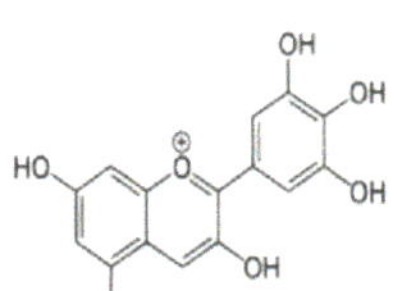

Delphinidin

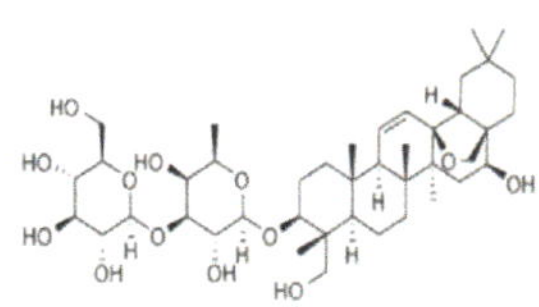

Kaikosaponin

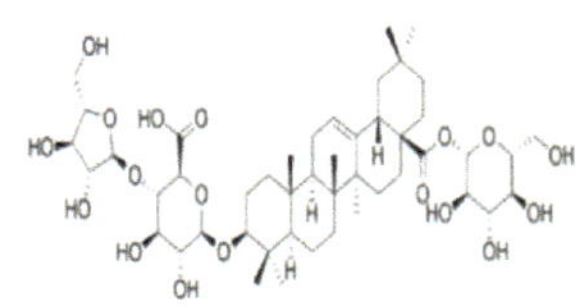

Chikusetsu saponin IV

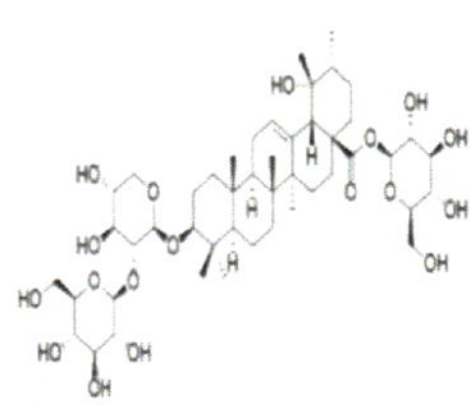

Ilexoside VIII

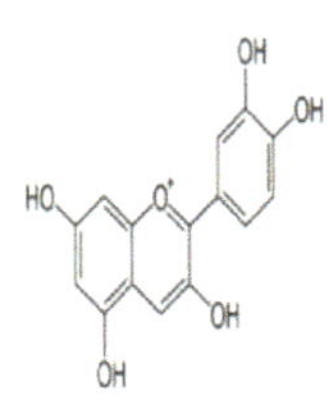

Cyanidine

FIGURE 33.1 Phytochemical constituents of *S. sesban.*

The antimicrobial activity of the methanol stem extract was examined against bacterial and fungal pathogens (Mythili and Ravindhran, 2012). A strong antibacterial activity was displayed against *Erwinia amylovora* with an inhibition zone of 17.25 mm and *Escherichia coli* with 16 mm at the concentration of 250 µg/mL. A complete inhibition was observed against the fungi *Fusarium oxysporum* and *Curvularia lunata*.

A study was undertaken to evaluate antimicrobial activity of the diethyl ether, chloroform, and ethanol extracts of the bark of *S. sesban*. The activity was performed against gram-positive (five), gram-negative (nine) bacteria, and fungi (seven) using disc diffusion and broth macro dilution methods. The antimicrobial activity was found to be strong against the tested bacterial and fungal pathogens. The minimum inhibitory concentration (MIC) values of the examined bark extracts were effective against tested pathogens (Ahmed et al., 2013).

Leaves of *S. sesban* were investigated for antimicrobial activity of their solvent extracts against bacterial pathogens (gram negative) such as *Escherichia coli*, *Klebsiella pneumoniae*, and *Pseudomonas aeruginosa*. Among the extracts tested, ethanol extract displayed the maximum activity. Zone of inhibition caused by ethanol extract against *Escherichia coli*, *Klebsiella pneumoniae*, and *Pseudomonas aeruginosa* at 50, 100, 150 µL concentrations was 25%, 18.8%, and 24.5% higher, respectively, than produced by the methanol and aqueous extracts (Nirosha et al., 2019).

33.3.2 Antioxidant Activity

Ethanolic extract of *S. sesban* was subjected for DPPH and nitric oxide scavenging activities (Mani et al., 2011). The DPPH scavenging activity was measured toward DPPH-free radical using the Blois method. The NO radical scavenging activity was determined against sodium nitroprusside generated nitric oxide and quantified by the Greiss reaction. The scavenging activities were found to be concentration dependent.

Earlier, anthocyanin was extracted from *S. sesban* colored flower petals (Kathiresh et al., 2011). Solvents such as acidified methanol and methanol were used to isolate anthocyanins from *S. sesban* flower petals. Initially, the crude extracts of the flower petals were subjected for phytochemical studies to extract anthocyanins, total flavonoids, and total phenols. The antioxidant properties of anthocyanins and the extracts of *S. sesban* flower petals were analyzed. The antioxidant activity of the *S. sesban* flower petals of the acidified methanol extract exhibited strong quenching activity (84%) at the concentration of 1 mg/mL.

Antioxidant activity of *S. sesban* leaves extract was studied using the DPPH (2,2-diphenyl-1-picrylhydrazyl) quenching test and correlation was measured with its total phenolic, flavonoid, and carotenoid content (Fitriansyah et al., 2017). Hexane, ethyl acetate, and ethanol extracts of leaves possessed DPPH scavenging activity (IC_{50} < 50 μg/mL). The ethanol extract displayed the highest of total phenolics (5.18 g GAE/100 g) and highest total flavonoid composition (4.56 g QE/100 g), while the highest content of total carotenoid (4.56 g BE/100 g) was shown by the hexane extract. Total phenolic content in *S. sesban* leaf extracts showed significantly negative correlation with their DPPH scavenging activities.

33.3.3 Larvicidal Activity

In a study, petroleum ether and ethanolic extracts of *S. sesban* were screened for their larvicidal activity toward *Culex quinquefasciatus* at 2, 4, 8, 10, 15, 20% concentrations and the highest mortality was seen after 72 h in the petroleum ether extract. Low level of mortality was observed in the ethanolic extract after 48 h (Venkatesh and Arunprasath, 2019).

33.3.4 Cytotoxic Activity

Ahmed et al. (2013) conducted brine shrimp lethality bioassay to study cytotoxic activity of ethanol, diethyl ether, and chloroform extracts of *S. sesban* bark. The results showed that the LC_{50} values of ethanol, diethyl ether, and chloroform extracts of *S. sesban* bark were found to be 1280, 640, and 320 μg/mL, respectively.

33.3.5 Anti-Inflammatory Activity

Dande et al. (2010) tested the crude saponins extract of *S. sesban* to evaluate the topical anti-inflammatory activity against rat paw edema-induced carrageenan by preparing the gel formulation in Wistar albino rats. The study animals received two doses of crude saponin gel at the concentrations of 1% and 2% w/w respectively and diclofenac sodium gel (1% w/w) was used as a standard. The gel formulation from the crude extract at 2% w/w exhibited a more significant anti-inflammatory activity than control group and the results were equivalent to the activity displayed by the standard.

Patil et al. (2010) subjected petroleum ether, chloroform, and methanol extracts of *S. sesban* bark for anti-inflammatory activity at the concentration of 300 mg/kg b.w. p.o. The adjuvant-carrageenan induced rat paw swellings of primary and secondary levels, and splenomegaly was effectively reduced after prophylactic administration of extracts of bark of *S. sesban*.

Earlier, a research was performed to investigate anti-inflammatory potential of petroleum ether, chloroform, and methanol extracts and three constituents were separated from the leaves of *S. sesban*. Among the extracts studied, the methanol extract displayed significant anti-inflammatory activity by decreasing carrageenin-induced rat paw edema. The constituent no. 2 separated from the methanol extract also exhibited a significant anti-inflammatory activity (Sajid et al., 2012).

Anti-inflammatory activity of saponins from *S. sesban* leaves on inflammations (chronic and acute) in carrageenan-induced edema was evaluated (Tatiya et al., 2013). Rats (male Wistar) and mice (Swiss albino) were utilized for the experiment. The paw volume was examined after the treatment. It was found that the carrageenan-induced edema was significantly reduced by the crude saponins and the activity was analogous to that of reference indomethacin.

Subramanian and Kalava (2014) investigated the aqueous extract of *S. sesban* seeds for the anti-inflammatory potential by inducing paw edema by carrageenan in rats. The anti-inflammatory activity was measured using paw thickness for the assessment of potency of the extract and the inhibition percentage of the paw edema was also calculated. The protein and nitric oxide levels were assessed in the serum. Histopathological analysis was also carried out in the hind paw. The extract administration reduced the paw thickness of treatment group animals significantly. Upon inflammation, the protein level was found to be decayed and the levels of nitric oxide were found to be raised. The histopathological examination showed the activity by the plant extract in the tissues of investigational animals post inflammation.

Singh et al. (2017b) used the stabilization of the human RBC membrane method and the rat paw edema method for investigating in vitro and in vivo anti-inflammatory activity of the extracts of *S. sesban* root and stem. The effectiveness of the extracts of root and stem of *S. sesban* was compared with diclofenac sodium (10 mg/kg/b.w.). The extracts of stem and roots at the concentrations of 100, 200, and 300 mg/kg were found to have maximum potency by reducing rat paw edema. The extracts also presented the most significant activity anti-inflammatory properties on membrane stabilization action on human RBC membrane.

33.3.6 Antinociceptive Activity

Petroleum ether, chloroform, ethyl acetate, ethanol, and water extracts (50 and 100 mg/kg, ip) of *S. sesban* wood were investigated for their antinociceptive activity using the hot plate method and writhing induced by the acetic acid method in mice. Among the extracts tested, petroleum ether, chloroform, ethyl acetate extracts showed significant reduction in writhing caused by acetic acid which is comparable with the control group (Nirmal et al., 2012).

33.3.7 Anthelmintic Activity

The ethanolic extract (10, 20, 30, 40, and 50 mg/mL) of *S. sesban* leaves was evaluated for anthelmintic activity of using *Pheretima posthuma* (earth worm) as test worms (Rajan et al., 2013). Piperazine citrate was taken as a standard reference. The extracts of the plant displayed a significant anthelmintic activity by causing paralysis followed by the death of *P. posthuma* worms, but the activity was not significant as exhibited by the standard reference.

In an earlier study, investigation was performed to substantiate in vitro anthelmintic activity of hydroethanolic and aqueous extract of *S. sesban* leaf against *Moneizia expansa* and *Paramphistomes* using the petri-dish method (Limsay et al., 2014). For *M. expansa*, the hydroethanolic extract was found to cause cessation of motility (paralysis) at 5 and 10 mg/mL concentrations after the exposure for 4:42 and 3:35 h, whereas the complete cessation of motility (death) was noticed after the exposure for 5:55 and 5:16 h respectively. Paralysis was produced by the aqueous leaf extract after 5:15 and 4:58 h and death was found after 7:56 and 7:20 h of exposure at the doses of 5 and 10 mg/mL respectively. In the case of *Paramphistomes*, the hydroethanolic extracts of leaf caused paralysis after 2:25 and 2:05 h and death after 5:05 and 4:58 h of exposure. The aqueous extract of leaf caused paralysis after 3:25 and 2:50 h and death was observed after 5:38 and 5:22 h of contact at the concentrations of 5 and 10 mg/mL respectively.

The in vitro and in vivo anthelmintic effectiveness of the methanol extract of *S. sesban* var. *bicolor* leaves was examined using *Hymenolepis diminuta*-rat (cestode) and *Syphacia obvelata*-mice (nematode) as test parasites (Soren et al., 2020). The standard drugs used were albendazole and praziquantel. The extract at the dose of 30 mg/mL concentration, *H. diminuta* and *S. obvelata* exposed mortality at 0.81 ± 0.01 and 15.17 ± 0.05 h, respectively. The in vivo findings revealed that the extract displayed a higher cestocidal effect in a concentration-dependent fashion, while 400 mg/kg treatment with the methanolic extract produced 65.10% decrease in eggs per gram (EPG) of faeces and 56% drop in worm counts. *S. obvelata*-infected mice exposed to the same dose displayed 34.32% and 47.08% decrease in EPG and worm counts at necropsy, respectively.

33.3.8 Antidiabetic Activity

The renal protective activity of *S. sesban* leaf extract (250 and 500 mg/kg/day) was evaluated in diabetic rats induced by streptozotocin (Pandhare et al., 2010). The rats induced with diabetes displayed excessive hyperglycemia with noticeable surge in albuminuria and proteinuria. The decrease in albuminuria, proteinuria, lipid, and glycated hemoglobin (HbA1c) content in rats with diabetes was observed. Histopathological investigations on diabetic rats' kidney revealed severe upsurge in cells of mesangial and glomeruli matrix with hyaline thickening of wall of arteiole, resulted from selective proteinuria and albuminuria. The changes were inhibited by the aqueous extract of the plant. The findings of the study revealed that *S. sesban* was able to reduce progress of nephropathy in streptozotocin-induced diabetic rats.

The aqueous extract of *S. sesban* leaves was evaluated for its antidiabetic activity in diabetic rats induced by streptozotocin (Pandhare et al., 2011). The normal and streptozotocin-induced diabetic rats received the aqueous extract at 250 and 500 mg/kg body weight (b.w.) concentrations per day for 30 days. The fasting blood glucose, the level of serum insulin, and biochemical parameters such as glycosylated hemoglobin, total cholesterol, triglycerides, high density lipoproteins, and low-density lipoproteins were assessed and the above-mentioned parameters were compared with reference glibenclamide at the dose of 0.25 mg/kg b.w. The results showed significant progress in the body weight, liver glycogen, serum insulin, and high density lipoproteins levels and decline in glycosylated hemoglobin, blood glucose, total cholesterol, and serum triglycerides when related to the standard drug.

Pandhare et al. (2012) investigated the anti-hyperglycemic properties of the petroleum ether extract of *S. sesban* roots. The extract was administered orally at different doses (250, 500, and 1000 mg/kg) to normal and streptozotocin (STZ)-induced type-2 diabetic mice. The blood glucose (fasting), biochemical markers in serum, body weight change, weight of internal organs, food and water intake, and glycogen level in liver were examined. All the doses of the extract produced a noticeable reduction of fasting glucose in streptozotocin-induced diabetic mice. The plant extract reduced the cholesterol level, triglyceride level, urea and creatinine levels and amplified the insulin, high-density lipoproteins cholesterol, and the level of total protein. Decline in body weight and level of glycogen brought by streptozotocin was reestablished. Rise in water intake and food intake induced by STZ was reduced.

Hypoglycemic properties of petroleum ether extract of *S. sesban* roots in streptozotocin-induced diabetes in mice were investigated by Aggarwal and Gupta (2012). The activity of three doses of the extracts at concentrations of 250, 500, and 1000 mg/kg was evaluated in streptozotocin-induced diabetic mice during treatment for 15 days. Everyday treatment of petroleum ether extract resulted in dosage-dependent reduction (50 %) in Fasting Blood Glucose (FBG) levels. There was a significant drop in glucose level at doses of 500 and 1000 mg/kg on the 15th day.

33.3.9 Antifertility Activity

An investigation reported the antifertility activity of different doses of *Sesbania sesban* seed powder. They were found to hinder the function of ovary, change the structure of uterine, and avert the implantation, and hence, control the fertility ability of female albino rats. The extracts of *S. sesban* root displayed that oleanolic acid 3-β-D glucuronide possessed spermicidal activity (Singh, 1990).

33.3.10 Stimulatory Effect on Central Nervous System

Extracts obtained from *S. sesban* bark were evaluated for their stimulating effects on central nervous system in albino mice. The standard reference drug used was Caffeine. The crude extracts and caffeine showed a significant stimulatory effect on the central nervous system when compared to control group (Naik et al., 2011).

33.3.11 Molluscicidal Activity

Methanol extract of the *S. sesban* plant was investigated for the biological and physiological parameters of *Bulinus truncatus* snails (Hasheesh et al., 2011). Myracidia of *Schistosoma haematobium* were used in bioassay and infection assessments. The methanolic extracts of *S. sesban* displayed toxic effects against the snails *B. truncatus*.

33.3.12 Spermicidal Activity

Oleanolic acid 3-beta-D-glucuronide, an active principle isolated from the plant *S. sesban*, was investigated for the spermicidal effect (Das et al., 2011). There was a concentration-dependent increase in sperm immobilization with an increase in the concentration of OAG. The minimum concentration of OAG that instigated 100% immobilization of sperm within 20s with no restoration of motility after following 60 min of incubation in Baker's buffer at 37 °C was considered to be the Minimum Effective Concentration (MEC).

KEYWORDS

- **bioactivity**
- **ethnomedicine**
- **pharmacology**
- **phytochemistry**
- ***Sesbania sesban***

REFERENCES

Aggarwal, N.; Gupta, P. Effect of Petroleum Ether Extract of *Sesbania sesban* (Merr.) Roots in Streptozotocin (STZ) Induced Diabetes in Mice. *Asian Pac. J. Trop. Biomed.* **2012,** *2* (3), S1254–S1260.

Ahmed, A.; Md. Sariful Islam H.; Dey, S. K.; Hira, A.; Hossain, Md. H. Phytochemical Screening, Antimicrobial and Cytotoxic Activity of Different Fractions of *Sesbania sesban* Bark. *Intern. J. Basic Med. Sci. Pharm.* **2013,** *3* (1), 2049–2063.

Anonymous. *The Wealth of India (Raw Materials)*, Vol. 9; Council of Scientific and Industrial Research Publication: New Delhi; 1980, pp 295–298.

Chatterjee, A.; S. Pakrashi, S. *The Treatise of Indian Medicinal Plants*, Vol. 2; Deep Publication: New Delhi; 1992, 121pp.

Dande, P. R.; Talekar, V. S.; Chakraborthy, G. S. Evaluation of Crude Saponins Extract from Leaves of *Sesbania sesban* (L.) Merr. for Topical Anti-Inflammatory Activity. *Int. J. Res. Pharm. Sci.* **2010,** *1* (3), 296–299.

Das, N.; Chandran, P.; Chakraborty, S. Potent Spermicidal Effect of Oleanolic Acid 3-Beta-D-Glucuronide, an Active Principle Isolated from the Plant *Sesbania sesban* Merrill. *Contraception* **2011,** *83* (2), 167–175.

Gomase, P.; Gomase, P.; Anjum, S.; Shakil, S.; Shahnavaj, K. M. *Sesbania sesban* Linn: A Review on Its Ethnobotany, Phytochemical and Pharmacological Profile. *Asian J. Biomed. Pharm. Sci.* **2012,** *2*, 11–14.

Fitriansyah, S. N.; Fidrianny, I.; Ruslan, K. Correlation of Total Phenolic, Flavonoid and Carotenoid Content of *Sesbania sesban* (L.) Merr. Leaves Extract with DPPH Scavenging Activities. *Int. J. Pharmacogn. Phytochem. Res.* **2017,** *9* (1), 89–94.

Gupta, A. K.; Grasdalen, H. NMR Studies of Composition and Side-Chain Arrangement in *Sesbania aegyptiaca* Seed Galactomannan. *Carbohydr. Res.* **1989,** *81*, 239–244.

Hasheesh, W. S.; Mohamed, R. T.; El-Monem, S. A. Biological and Physiological Parameters of *Bulinus truncatus* Snails Exposed to Methanol Extract of the Plant *Sesbania sesban* plant. *Adv. Biol. Chem.* **2011,** *1* (03), 65.

Hossain, M. A.; Rahman, M. S.; Chowdhury, A. S.; Rashid, M. A. Bioactivities of *Sesbania sesban* Extractives. *Dhaka Univ. J. Pharm. Sci.* **2007,** *6* (1), 61–63.

Humperys, L. R. Agriculture: Legumes for the Tropics *Sesbania sesban*; 5th ed.; Patridge I. J.; 1995.

Kathiresh, M.; Suganya P.; Saravanakumar, M. Antioxidant Effect of *Sesbania sesban* Flower Extract. *Int. J. Pharm. Sci.* **2011,** *3* (2), 1307–1312.

Kinghorn, A. D. Pharmacognosy in the 21st Century. *J. Pharm. Pharmacol.* **2001,** *53*, 135–148.

Limsay, R. P.; Jangde, C. R.; Jahan, A.; Swati, U. P. *In Vitro* Anthelmintic Activity of Hydroethanolic and Aqueous of *Sesbania sesban* Pers. Leaf Extract Against *Moneizia expansa* and *Paramphistomes*. *Int. J. Pharm. Pharm. Sci.* **2014,** *6* (2), 504–505.

Mani, R. P.; Pandey, A.; Shambaditya, G.; Tripathi P., Kumudhavalli, V.; Singh A. P. Phytochemical Screening and *In-Vitro* Evaluation of Antioxidant Activity and Antimicrobial Activity of the Leaves of *Sesbania sesban* (L.) Merr. *Free Radic. Antioxid.* **2011,** *1* (3), 66–69.

Mythili, T.; Ravindhran, R. Phytochemical Screening and Antimicrobial Activity of *Sesbania sesban* (L.) Merr. *Asian J. Pharm. Clin. Res.* **2012,** *5* (4), 18–23.

Nadkarni, K. M. *Indian Materia Medica,* 3rd ed., Vol. I; Popular Prakashan: Bombay; 1982, 1130pp.

Nirmal, S. A.; Bairagi, J. H.; Patil, A. N.; Pal, S. C.; Upasani, C. D.; Mandal, S. C. Antinociceptive Activity of *Sesbania sesban* (Linn) Wood Extracts, a Preliminary Study. *Indian J. Exp. Biol.* **2012,** *50*, 61–64.

Nirosha, M.; Pavani, K.; Priya, A. K.; Haseena, U.; Jyothsna, S. Antimicrobial Activity of Organic Leaf Extract of *Sesbania sesban* Against Gram Negative Pathogenic Bacteria. *Int. J. Life Sci. Pharm. Res.* **2019,** *9* (2), 31–38.

Naik, N.; Tare, H. L.; Sherikar, A. K.; Deore, S. R.; Dama, G. Y. Central Nervous System Stimulant Effect of Extracts Obtained from the Barks of *Sesbania sesban*. *Int. J. Institutional Pharm. Life Sci.* **2011,** *1* (1), 77–92.

Pandhare, R. B.; Sangameswaran, B.; Mohite, P. B.; Khanage, S. G. Aqueous Extracts of the Leaves of *Sesbania sesban* Reduces Development of Diabetic Nephropathy in Streptozotocin-Induced Diabetic Rat. *Bangladesh J. Pharmacol.* **2010,** *5* (2), 103–106.

Pandhare, R. B.; Sangameswaran, B.; Mohite, P. B.; Khanage, S. G. Antidiabetic Activity of Aqueous Leaves Extract of *Sesbania sesban* (L.) Merr. in Streptozotocin Induced Diabetic Rats. *Avicenna J. Med. Biotechnol.* **2011,** *3* (1), 37–43.

Pandhare, R. B.; Sangameswaran, B.; Mohite, P. B.; Khanage, S. G. Attenuating Effect of *Sesbania sesban* (L.) Merr. Extract on Neuropathic Pain in Streptozotocin-Induced Diabetic Rats: An Evidence of Neuroprotective Effects. *Phytopharmacology* **2012,** *2* (1), 190–201.

Patil, R. B.; Nanjwade, B. K.; Manvi, F. V. Effect of *Sesbania grandiflora* and *Sesbania sesban* Bark on Carrageenan Induced Acute Inflammation and Adjuvant-Induced Arthritis in Rats. *Int. J. Pharm. Sci.* **2010,** *1* (1), 75–89.

Rajan, M. A; Cassiano, M.; Prinkesh Fanasia, P.; Koshy, S.; Gopikrishna. Evaluation of *In-Vitro* Anthelmintic Activity of Ethanolic and Aqueous Extract of *Sesbania sesban* Leaves. *Intern. J. Pharm. Chem. Sci.* **2013,** *2* (4), 1803–1805.

Rastogi, R. P.; Mehrotra, B. N. *Compendium of Indian Medicinal Plants,* 1st ed., Vol. II; Central Drug Research Institute: Lucknow; 1993, 62pp.

Sajid, S.; Vijay, T. P; Rageeb, Md.; Usman, Md. Anti-Inflammatory Activity of *Sesbania sesban* (L) Merr. *Int. Res. J. Pharm.* **2012,** *3* (1), 176–180.

Sasikumar, J. M.; Palaniswamy, M.; Doss, A.; Doss, P. A.; Geetha, V. Antibacterial and Antigungal Activities of *Sesbania aegyptiaca* Pers. Leaves. *Indian J. Pharm. Sci.* **2005,** *67* (5), 765–767.

Singh, B.; Sharma, R.; Kalia, A. N. Antiinflammatory Potential of Root and Stem Extracts of *Sesbania sesban. Ann. Plant Sci.* **2017a,** *6* (11), 1790–1793.

Singh, B.; Sharma, R.; Kalia, A. N.; Kumar, V. Isolation and Identification of Chemical Compounds from Stem and Roots of *Sesbania sesban. Ann. Plant Sci.* **2017b,** *6* (10), 1718–1719.

Singh S. P. Fertility Control of Female Through *Sesbania sesban* Seeds. *J. Res. Educ. Indian Med.* **1990,** *9* (4), 227–232.

Soren, A. D.; Chen, R. P.; Arun K. Y. *In Vitro* and *In Vivo* Anthelmintic Study of *Sesbania sesban* var. *bicolor*, a Traditionally Used Medicinal Plant of Santhal Tribe in Assam, India. *J. Parasit. Dis.* **2020**.

Subramanian, K.; Kalava, S. V. Anti-Inflammatory Potential of Aqueous Extract of *Sesbania sesban* (L.) Merr. *Int. J. Pharm. Pharm. Sci.* **2014,** *6* (2), 301–304.

Tatiya, A. U.; Dande, P. R.; Mutha, R. E.; Surana, S. J. Effect of Saponins from of *Sesbania sesban* L. (Merr.) on Acute and Chronic Inflammation in Experimental Induced Animals. *J. Biol. Sci.* **2013,** *13* (3), 123–130.

Tokayo, U. Triterpene Saponins from Leaves of *Sesbania sesban. Nat. Med.* **2000,** *54* (6), 350.

Upadhyaya, J. S.; Singh, S. P. Chromatographic Studies of Oxidation Products of Lignin from *Sesbania sesban. Cell. Chem. Technol.* **1991,** *25*, 219–226.

Venkatesh, G.; Arunprasath, A. Evaluation of Mosquito Larvicidal Potential Against *Culex quinquefasciatus* and Phytochemical Profiling in *Sesbania sesban* Merr. *Int. J. Entomol. Res.* **2019,** *4* (3), 35–38.

Warrier, P. K.; Nambiar, V. P. K.; Ramankutty, C. (eds.). *Indian Medicinal Plants a Compendium of 500 Species*, Vol. V; Orient Longman Ltd: Madras; 1996–1997.

Yusuf, M.; Chowdhury, J. U.; Wahab, M. A.; Begum, J. *Medicinal Plants of Bangladesh*; BCSIR: Dhaka, Bangladesh; 1994.

CHAPTER 34

A Review on Bioactives and Pharmacological Activities of *Senna occidentalis* L.

K. S. SHANTHI SREE[1*], A. SUVARNA LATHA[1],
P. LAKSHMI PADMAVATHI[1], D. BHARATHI[1], and B. KAVITHA[2]

[1]*Department of Biosciences and Sericulture, Sri Padmavati Mahila Visvavidyalayam, Tirupati, Andhra Pradesh 517502, India*

[2]*Department of Botany, Rayalaseema University, Kurnool, Andhra Pradesh 518007, India*

[*]*Corresponding author. E-mail: shanthi4botany@gmail.com*

ABSTRACT

Senna occidentalis is an Ayurvedic medicinal plant that has been used in many traditional medicines to treat various diseases. This plant has been known to possess pharmacological activities like antimicrobial, anti-inflammatory, antipyretic, antioxidant, anti-diabetic, hepatoprotective activity, antitrypanosomal activity, wound healing and sun protective effects. Chemical compounds like achrosin, aloeemodin, emodin, anthraquinones, anthrones, apigenin, aurantiobtusin, campesterol, cassiollin, chryso-obtusin, chrysophanic acid, chrysarobin, chrysophanol, chrysoeriol etc. have been isolated from this plant. The present review summarizes the information related to the bioactive compounds and Pharmacognostical activities of *Senna occidentalis* plant extracts.

Bioactives and Pharmacology of Legumes. T. Pullaiah, PhD (Ed.)

34.1 INTRODUCTION

Senna occidentalis L. belongs to the family Fabaceae. It is commonly called as coffee senna. Synonyms of this species include *Cassia occidentalis* L., *Cassia caroliniana* Walter, *Cassia ciliata* Raf., *Cassia falcata* L., *Cassia foetida* Pers., *Cassia macradenia* Collad., *Cassia obliquifolia* Schrank, *Cassia planisiliqua* Burm.f., *Cassia planisiliqua* L., *Cassia plumieri* DC., *Ditramexa occidentalis* Britton & Rose, and *Ditremexa occidentalis* (L.) Britton & Wilson (Lavanya, 2014).

It is a plant that grows in wastelands and on roadsides (Francis, 2002). *Senna occidentalis* was mentioned as "Kasamarda" in various nighantus like Rajnighantu, Dhanwantari, and Bhavaprakasa (Warrier et al., 1994). It is widely distributed from Himalayas to the West Bengal, South India, Myanmar, and Sri Lanka and found all over India, at an altitude of 1500 mt (Nadkarni, 2005). *S. occcidentalis* grows in tropical and subtropical regions, which includes United States (Texas to Iowa eastward), Asia, Africa, and Australia (Khare, 2008; Liogier, 1988).

The *S. occidentalis* is known by different common names and they vary from one area to the other, such as Barkichakor, Chilmile, Tulo tapre, and Panwar in Nepal (Hanelt and Mansfeld, 1986); Bricho, Brusca, Frijolillo, Guanina, Dirjinni, and Fihaari in Spain and Somalia; Moshabela moha, Tsinyembane, and Umnwanda nyoka in South Africa (Huang and Williams, 1999); and Msalafu and Mrambazi in Kenya (Mahanthesh and Jalalpure, 2016). In English, it is called as Negro coffee, Kasaumdi in Hindi, Kasamardah in Sanskrit, and Kasinda in Telugu (Warrier et al., 1994).

The plant *S. occidentalis* is a diffuse offensively odorous undersharb with subglabrous branches and three to five pairs of leaflets, inflorescence axillary corymb, and terminal panicle. Flowers are yellow in color, have stamens numbering 10, which are free and not equal in size. Seven stamens are perfect and they are reduced into staminodes. Pods, which are transversely septate, contain 20–30 seeds (Warrier et al., 1994).

This plant is widely used as a substitute for coffee by local people. A coffee like beverage is prepared from seeds, which helps in the treatment of asthma, and the infusion of flower is used in treating bronchitis (Chukwujekwu et al., 2006). Leaves can be used in treating ailments like gonorrhea, fevers, edema, and urinary tract disorders (Emmanuel et al., 2010). The roots are used as febrifuge, diuretic, and tonic. Roots are also used in menstrual problems, anaemia, tuberculosis, and liver problems. It can also be used as a tonic for general weakness and illness (Fidèle et al., 2015).

In Ayurveda, extract taken from *S. occidentalis* is used to cure eye inflammation. In Jamaica, it is used as folk medicine for treating dysentery, diarrhoea, constipation, fever, cancer, eczema, and venereal diseases. It is the main ingredient of medicines like Liv. 52 (Fidèle et al., 2017) and a polyherbal formulation called Herbolax, which is used in treating constipation (Garba et al., 2015). "Bonnisan," a metabolic corrective for new-borns and infants is also made from *S. occidentalis* (0.5 mg/5 mL), which gives immediate relief from the disturbances caused by gastric wind (Ginde et al., 1970).

34.2 BIOACTIVES

The phytochemicals which have been isolated from *S.occidentalis* include aloe-emodin, achrosin, emodin, islandicine, kaempferol, obtusifolin, obtusin, physcion (Kanakam et al., 2013), apigenin, anthraquinones, aurantiobtusin, cassiollin, campesterol, chrysoobtusin, chrysarobin, chrysophanol, chrysophanic acid, (Huang and Williams, 1999), chrysoeriol, funiculosin (Khare, 2008), quercetin, rhein, rhamnosides, rubrofusarin, sitosterols, tannins, xanthorine (Kitanaka and Takido, 1992), lignoceric acid, Ferol, linoleic acid, linolenic acid, mannopyranosyl, mannitol, matteucinol, and oleic acid (Chukwujekwu et al., 2006).

Phytochemical reports on *S. occidentalis* reveal that according to climate change the quantity and nature of phytochemicals may vary. For example, root bark, stem, and leaves of the plant from Ivory Coast (Africa) contain low quantity of saponins and has no quinones, sterols, alkaloids, triterpines, tannins, and flavonoids.

Ararabinol has been isolated from the whole plant of *S. occidentalis* (Sastry et al., 2011). Compounds like α-L-arbinopyranoside and β-D xylopyranoside (Yadava and Satnami, 2011), and two pigments like 7-methyl physcin and 7-methyltorosachrysone have been reported (Takahashi et al., 1976). In a further investigation, new flavonoid compound cassioccidentalis A, B, and C (Kitanaka and Takido, 1992) and isolation of cassiollin from *S. occidentalis* have been reported (Ginde et al., 1970) (Table 34.1).

S. occidentalis contains chemical compounds like aloe-amine, anthrone, kaempferein, isorhein, 1,8-dihydroxyl-2-methyl anthraquinone, 1,4,5-trihydroxy-3-methyl-7-methoxy anthraquinone, achrosine, xanthorin, aurantiobtusin, galactopyranosyl, helminthosporin, and islandicin (Vijayalakshmi et al., 2013; Yadav et al., 2010; Ginde et al., 1970; Oliver, 1986; Shah and Shinde, 1969; Jawahar and

TABLE 34.1 Chemical Structures of Reported Phytochemicals of *S. occidentalis.*

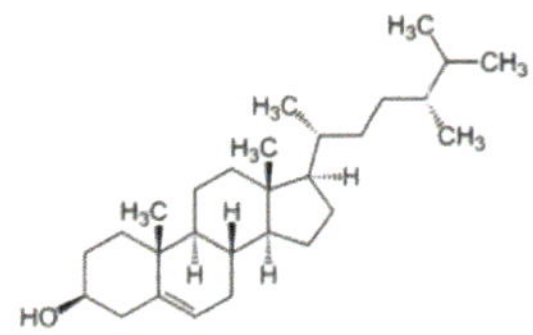 Campsterol	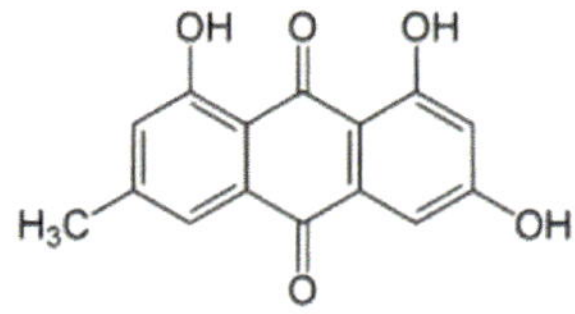 Emodin	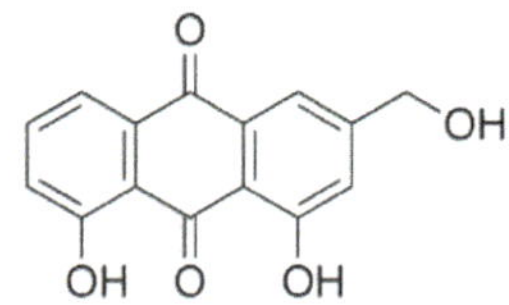 Aloe-emodin
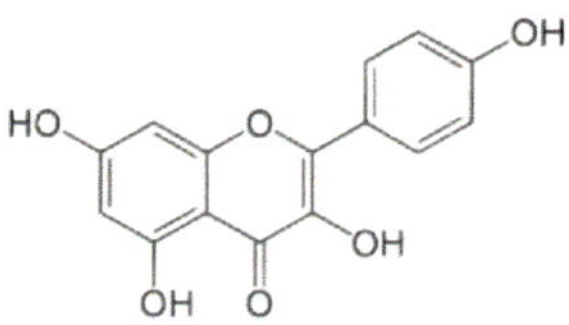 kaempferol	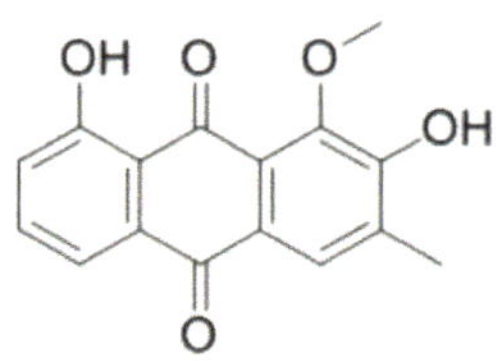 obtusifolin	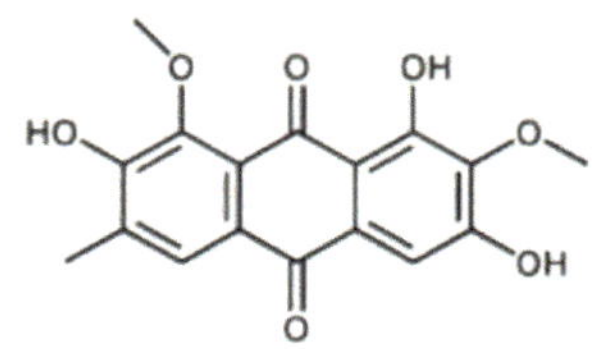 obtusin
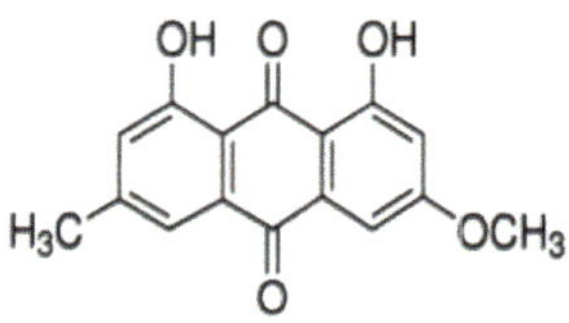 physcion	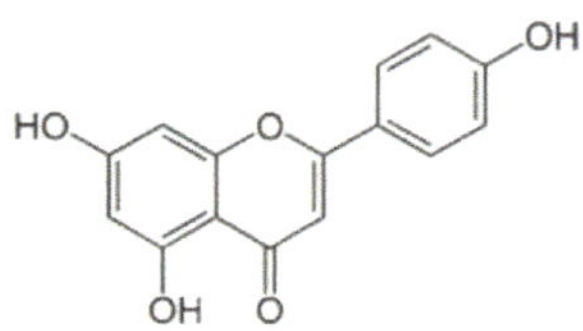 Apigenin	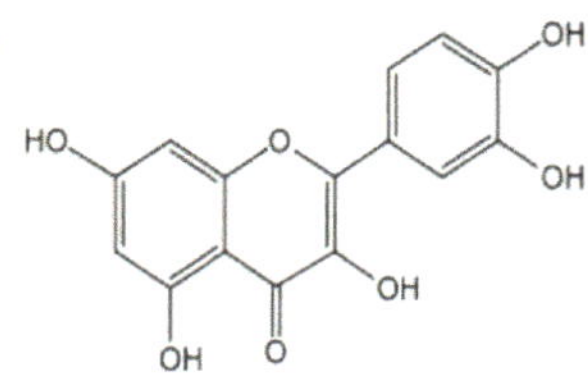 Quercetin
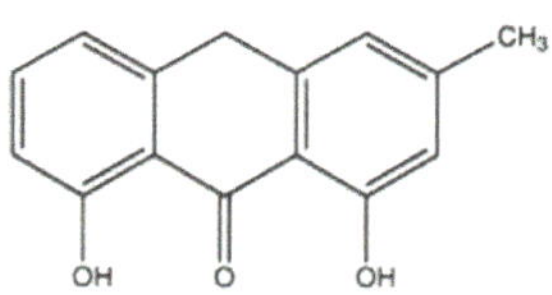 chrysarobin	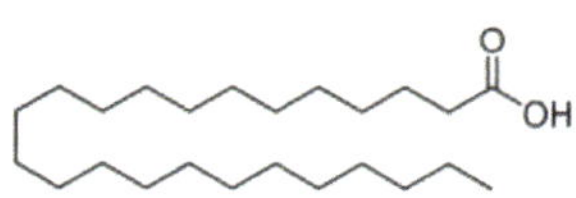 lignoceric acid	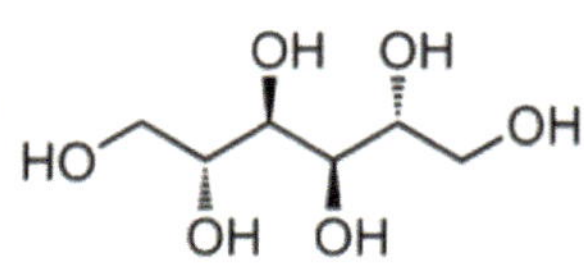 Mannitol
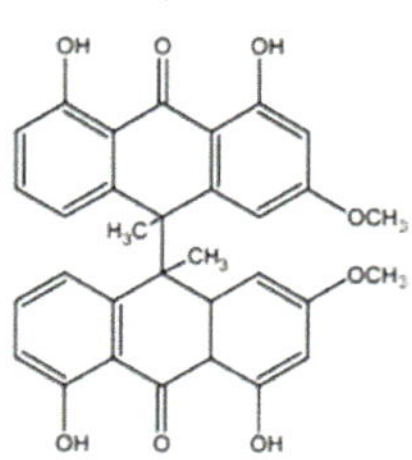 4, 4 5, 5 tetrahydroxy 2, 2, methoxy 9, 9 bianthraquinone	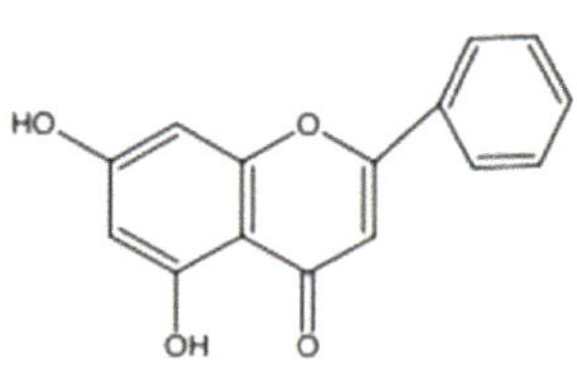 α-L arabino pyranoside	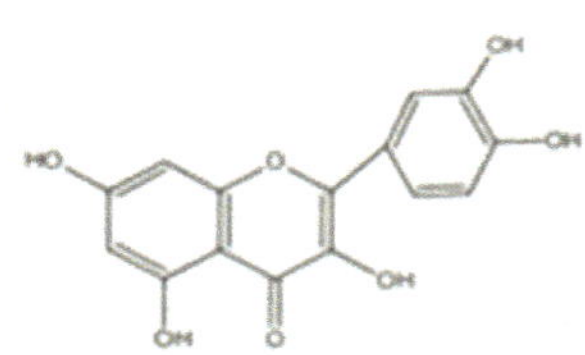 β-D xylopyranoside

TABLE 34.1 *(Continued)*

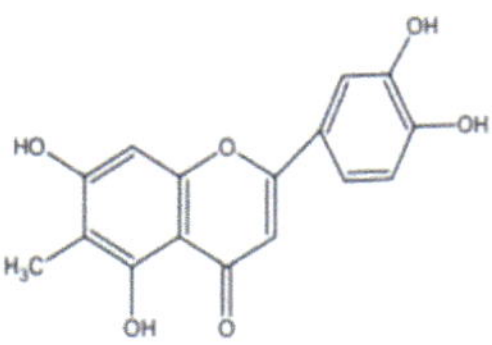	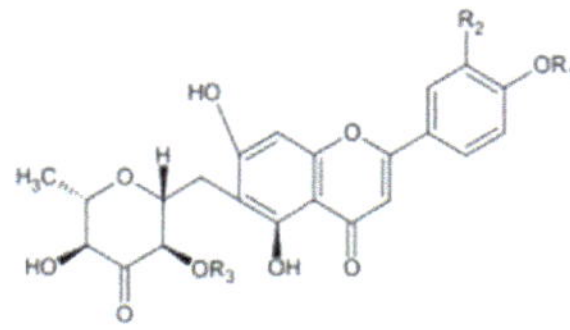	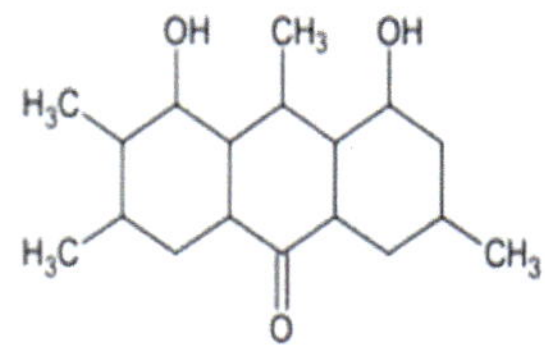
β-D galactopyranoside	Cassiaoccidentalis A, B and C	7-methyl physcin
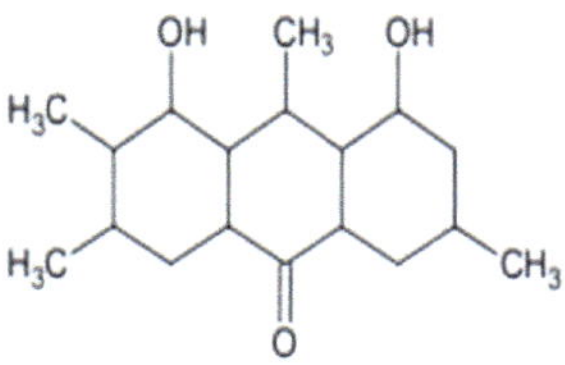	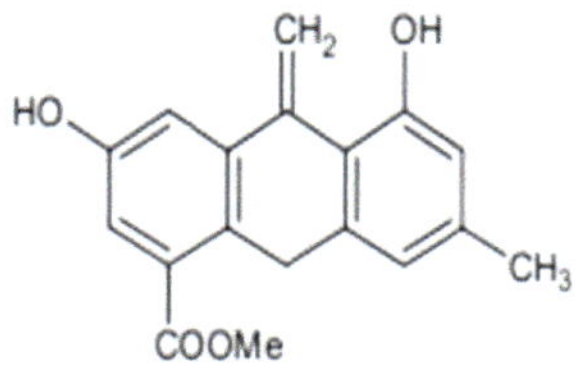	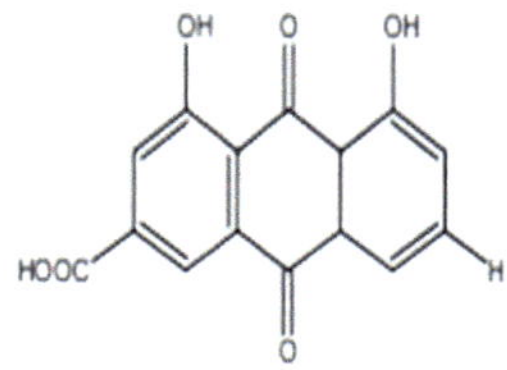
7-methyltorosachrysone	Cassiollin	Rhein
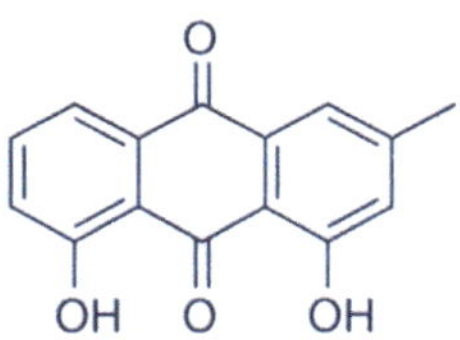	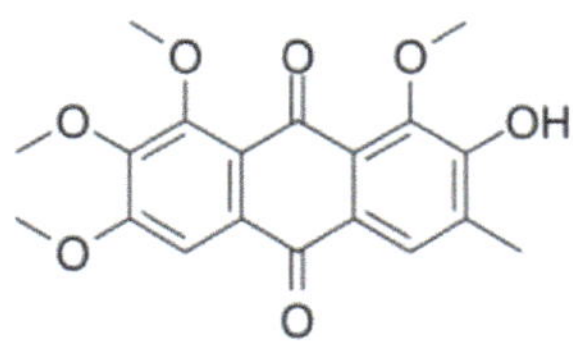	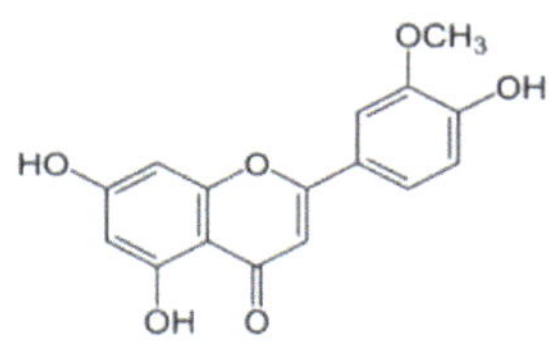
Chrysophanic acid	Chrysoobtusin	Chrysoeriol
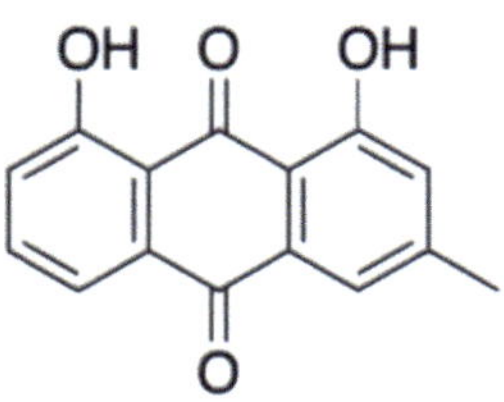	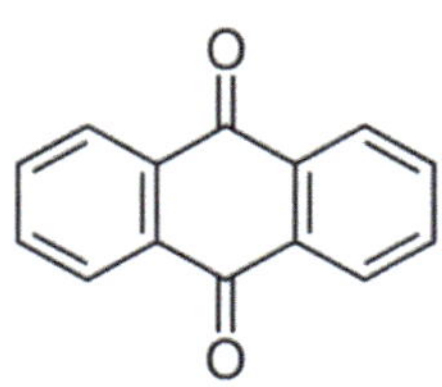	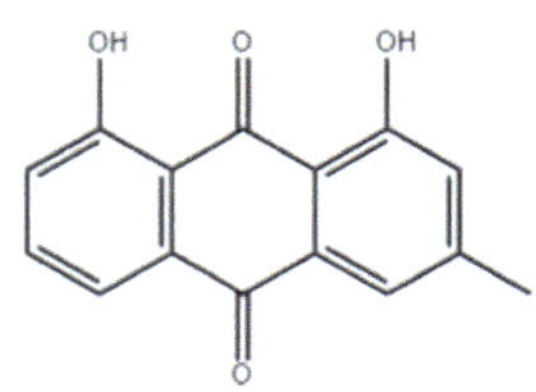
Chrysophanol	Anthraquinone	Chrysophanic acid
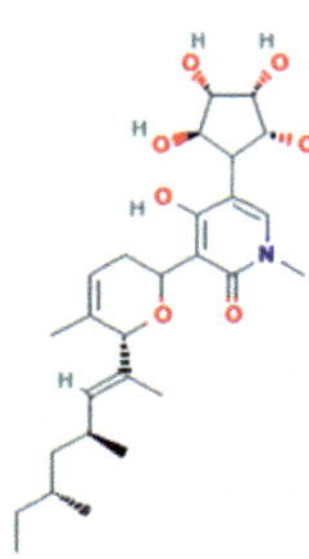	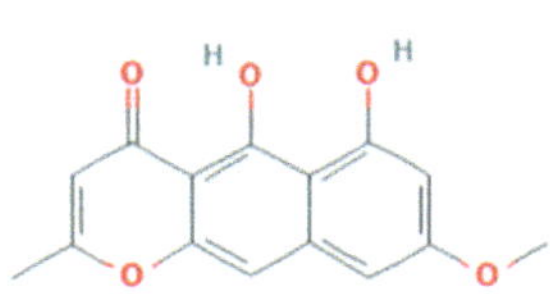	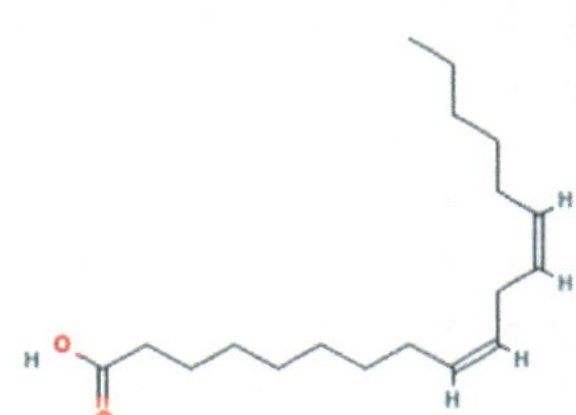
Funiculosin	Rubrofusarin	Linoleic acid

TABLE 34.1 *(Continued)*

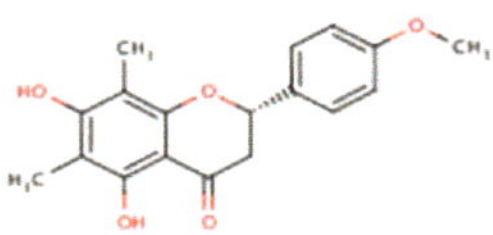	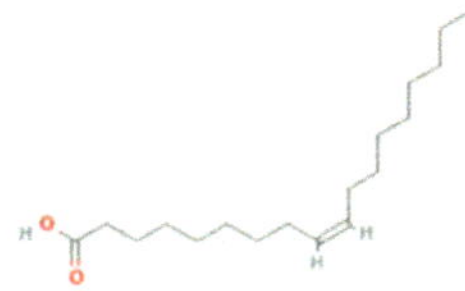	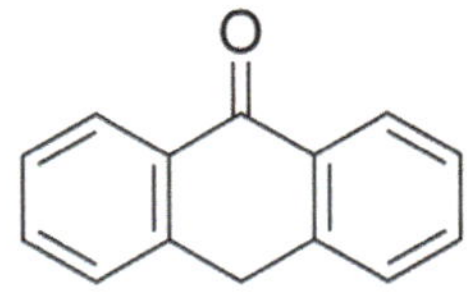
Matteucinol	Oleic acid	Anthrone
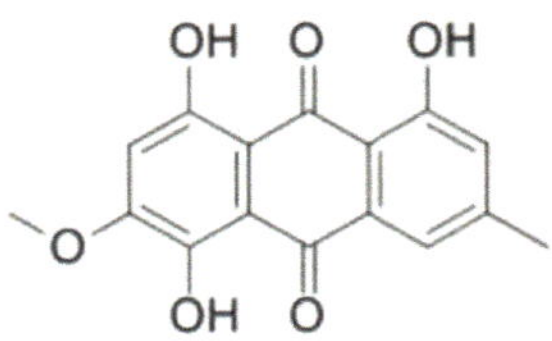	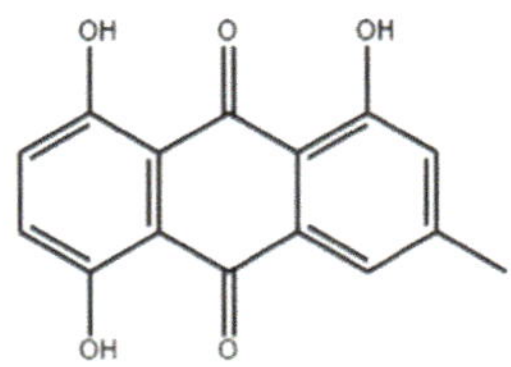	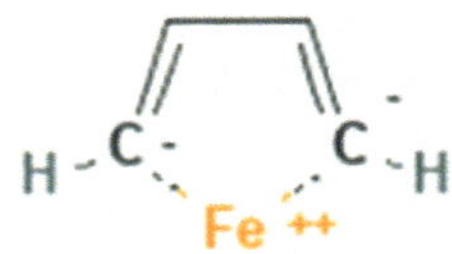
Xanthorin	Helminthosporin	Ferol

Gupta, 1974; Tsutomu et al., 1999). 5,7-dihydroxyflavone-5-O--d-xylopyranosyl-7-O--l-rhamnopyranosyl-(13)-O--l-arabinopyranoside, 3,5,7,3′,4′-pentahydroxy flavone-3-O--l-rhamnopyranosyl-7-O--d-glucopyranosyl-(13)-O--d-xylopyranoside, and 5,7,3′,4′-tetrahydroxy-6-methoxyflavone-5-O--l-arabinopyranosyl-(14)-O--l-rhamnopyranosyl-(13)-O--d-galactopyranoside (Yadava and Satnami, 2011) are the three new compounds that have been isolated from the seeds of *S. occidentalis*.

1,3-dihydroxy-6,7,8-trimethoxy-2-methylanthraquinone 3-*O*-α-rhamnopyranosyl (1→6)-β-glucopyranosyl (1→6)-β-galactopyranoside and 1-hydroxy-3,6,7,8-tetramethoxy-2-methylanthraquinone 1-*O*-α-rhamnopyranosyl (1→6)-β-glucopyranosyl (1→6)-β-galactopyranoside are the two new anthraquinone glycosides that have been isolated from the leaves of *S. occidentalis*. Glycosides like 3,2′-dihydroxy-7,8,4′–trimethoxy-flavone-5-*O*-{β-D-glucopyranosyl (1→2)}-β-D-galacto-pyranoside and apigenin-7-*O*-β-D-allopyranoside have been isolated from *S. occidentalis* (Chauhan et al., 2011; Hatano et al., 1999).

34.3 PHARMACOLOGICAL ACTIVITIES

34.3.1 Antimicrobial Activity

Leaf extracts of *S. occidentalis* shows high antimicrobial activity on *E. coli* at the concentration of 900–1000 mg. Hexane extract was most effective on *E. coli* at concentrations between 500 and 1000 mg compared to other organisms like *Pseudomonas multocida, Salmonella typhi, S. typhimurium, S. pyogenes,* and *S. pneumoniae*, which were not showing any antimicrobial activity (Saganuwan and Gulumbe, 2006). All the extracts were effective when compared with standard drugs like greseofulvine and nystatin (Vipul and Anjan, 2011).

The antimicrobial activity of the organic and aqueous leaf extracts of *S. occidentalis* were tested using disc diffusion assay against six pathogenic bacteria like *P. aeruginosa, Klebsiella pneumoniae, Proteus mirabilis, E. coli, Staphylococcus aureus,* and *S. epidermidis*, and one fungal strain, that is, *C. albicans*. Significant antibacterial activity was observed in methanol and aqueous extracts against tested microbes. *P. aeruginosa* showing 18 mm zone of inhibition in aqueous extract was most the susceptible microorganism, followed by *P. mirabilis* showing 15 mm of zone of inhibition and 8 mm of zone of inhibition by *Candida albicans* in methanol extract (Vedpriya et al., 2010).

In further studies, the antibacterial activity of methanol leaf extract of *S. occidentalis* was screened against *K. pneumoniae, E. coli, C. albicans, S. aureus, P. aeruginosa*, and *S. typhi*. The study was carried by agar well diffusion method using various concentrations of the methanol extract from 80 to 360 mg/mL. Serial dilution method was employed to determine minimum inhibitory concentration (MIC), while minimum bactericidal concentration (MBC) was studied by plating various dilutions of the extract. The results show the antibacterial activity of the extract with MIC ranging from 160 to 280 mg/mL and the minimum bacterial concentration (MBC) ranging from 160 to 320 mg/mL. *S. aureus* and *P. aeruginosa* were sensitive when compared with the *K. pneumonia*, which shows some resistance compared to other organisms tested (Tsado et al., 2016).

34.3.2 Anti-Inflammatory and Antipyretic Effects

The cold extraction of the leaves of *S. occidentalis* was done by using equal volumes of ethyl acetate, petroleum ether, and methanol. The extract was used for pharmacological screening by using rat paw edema model, which has been induced by carrageenan. At the end of 4 h, 36.68% inhibition of edema was observed at 400 mg/kg dose of the extract (Kanakam et al., 2013). The anti-inflammatory activity of *S. occidentalis* leaf powder was carried out by carrageenan-induced rat paw edema in male albino rats. *S. occidentalis* extract was effective at a dose of 2000 mg/kg (Sadique et al., 1987). The antipyretic activity of leaf extracts was tested using yeast-induced pyrexia method in rats. The pyrexia induced by yeast was reduced significantly by ethanol and water extracts (Sini et al., 2011).

34.3.3 Antioxidant Activity

Methanolic leaf extracts of stems, leaves, and seeds of *S. occidentalis* were used to test the antioxidant activity by using in-vitro nitric oxide scavenging activity, hydroxyl radical scavenging activity, reducing power, and superoxide radical scavenging activity. Highest hydroxyl radical, superoxide radical scavenging activity was observed in methanol extract of the seeds compared to the leaf and stem extracts. Highest metal chelating and nitric oxide radical scavenging activities were observed in leaves and stems when compared with seeds (Arya and Yadav, 2011).

The ethanolic extracts from *S. occidentalis* were tested against CCl_4-induced oxidative stress in wistar albino rats (Rani et al., 2010). The administered CCl_4 increased the concentration of lipid peroxides and the activities of enzymatic and nonenzymatic antioxidants have been decreased. Almost a normal antioxidant status was maintained after pretreatment with ethanolic extract, which significantly prevented the alterations induced by CCl_4. In CCl_4-induced rats, the enzyme activity has been reduced whereas the ethanolic-extract treated rats showed increase in enzyme activity. This study proves the potency of ethanolic extract in recovering CCl_4-induced oxidative stress (Kumar and Abbulu, 2011).

34.3.4 Anti-Diabetic Activity

In a further investigation, by treating with aqueous extract of *S. occidentalis*, a significant decrease in the blood glucose levels were observed in normal and Alloxan-induced diabetic rats (Verma et al., 2010). Blood glucose levels in diabetic mice have been significantly reduced when compared with normal rats, by the administration of leaf extract at different doses (Onakpa and Ajagbonna, 2012). Root extract of *S. occidentalis* at 400 mg/kg bwt was given to animal group and result did not differ much from that of the metformin treated animals. This study proves the usage of *S. occidentalis* in traditional medicine for the treatment of diabetes (Garba et al., 2015). Further, the methanolic fraction of leaves was tested in diabetic rats induced by streptozotocin. Histopathological study revealed that pancreatic tissue has been protected by the methanolic extract from streptozotocin-induced damage (Emmanuel et al., 2010).

34.3.5 Hepatoprotective Activity

S. occidentalis leaf aqueous and aqueous-ethanolic extract (50% v/v) was studied on rat liver damage induced by paracetamol and ethyl alcohol. The leaf extract produced significant protection by restoring the function of liver (Sastry et al., 2011; Jafri et al., 1999). Chrysophanol, which has been isolated from *S. occidentalis* at 50 mg/kg bw, and a fraction of methanol (MF) at 200 mg/kg bw were given to rats, which are induced with hepatotoxicity for 7 days. Oral dosage of chrysophanol and methanol fraction significantly reduced the values of CAT, SOP, GP, GSH, Vit-C, and Vit-E. The results show that *S. occidentalis* has significant hepatoprotective action against hepatic damage induced by paracetamol (Rani et al., 2010).

34.3.6 Antitrypanosomal Activity

Similarly in another study, the ethanolic extract of *S. occidentalis* leaf was tested for in-vitro antitrypanosomal activity, which exhibited in-vitro activity on *Trypanosoma brucei* by eliminating the parasites movement within 10 min by post-incubating with 6.66 mg/mL of the extract (Ibrahim et al., 2010).

34.3.7 Wound Healing and Sun Protective Effects

The crude methanolic leaf extract of *S. occidentalis* and chrysophanol isolated from the plant were tested for wound-healing property by using excision, incision, and dead space wound models. Chrysophanol was found to have significant wound-healing property than methanolic extract. Histopathological observation on granulation tissue revealed increased collagenation when compared with the control group of animals (Koche et al., 2010). *S. occidentalis* flowers were tested for the sun protection factor (SPF). It was observed that *S. occidentalis* has high SPF along with anti-oxidant and antibacterial activities. The results showed that flowers can be used against UV-radiation problems as an antimelanocyte agent (Jayanthy and Shafna, 2012).

34.3.8 Acute Toxicity Study

The acute toxicity study of the hydroalcoholic extract of *S. occidentalis* leaf was carried out in male and female wistar rats by giving 0.625, 1.25, 2.5, and 5.0 g/kg (oral dose) to study the general behavior. Mortality rate was recorded for up to 14 days. Subacute toxicity study of *S. occidentalis* was done by giving gauage at the doses of 0.10, 0.50, or 2.5 g/kg/day for 30 days. There was no hazardous symptoms or death reported in the acute toxicity test having LD50 higher than 5 g/kg. There were no changes in the macroscopical study of the organs (Silva et al., 2011).

34.4 CONCLUSION

This review throws light on the enormous biological potential of *S. occidentalis* and it is strongly believed that this plant can be used in the preparation of medicines, by using the information presented on the bioactive compounds and various biological properties of the extracts. Based on the various ecological locations and seasons, the plant may show variation in the efficacy of the medicine.

KEYWORDS

- **traditional medicine**
- **anti-inflammatory**
- **antipyretic**
- **hepatoprotective activity**
- **wound healing**

REFERENCES

Arya, V.; Yadav, J. P. Antioxidant Properties of Methanol Extracts of Leaves, Seeds and Stem of *Cassia occidentalis*. *Res. J. Med. Plants*. **2011,** *5* (5), 547–556.

Chauhan, D.; Chauhan, J. S.; Siddiqui, I. R.; Singh, J. Two New Anthraquinone Glycosides from the Leaves of *Cassia occidentalis*. *Indian J. Chem. Sect. B*. **2011,** *40B*, 860–863.

Chukwujekwu, J. C.; Coombes, P. H.; Mulholland, D. A.; Staden, J. Emodin, an Antibacterial Anthraquinone from the Roots of *Cassia occidentalis*. *S. Afr. J. Bot.* **2006,** *72*, 295–297.

Emmanuel, S.; Rani, M. S.; Sreekanth, M. R. Antidiabetic Activity of *Cassia occidentalis* Linn. in Streptozotocin-Induced Diabetic Rats: A Dose Dependent Study. *Int. J. Pharma Bio Sci.* **2010,** *1* (4), 14–25.

Fidèle, N.; Barama J.; Emmanuel, T.; Théophile, D. Hypolipidemic, Antioxidant and Antiatherosclerogenic Effect of Aqueous Extract Leaves of *Cassia occidentalis* Linn (Caesalpiniaceae) in Diet-Induced Hypercholesterolemic Rats. *BMC Complement Altern. Med.* **2017,** *17* (1), 1–11.

Fidèle, N.; Joseph, B.; David, R. K. A.; Paul, F. S. E. Théophile, D. Diuretic and Antioxidant Activities of the Aqueous Extract of Leaves of *Cassia occidentalis* (Linn.) in Rats. *Asian Pac. J. Trop. Med.* **2015,** *8* (9), 685–693.

Garba, R.; Saidu, A. N.; Adeyemi, H.; Muhammad, R. Y. Effect of Methanolic Extract of *Cassia occidentalis* L. Root Bark on Body Weight and Selected Biochemical Parameters in Alloxan Induced Diabetic Rats. *Br. J. Pharmacol. Toxicol.* **2015,** *6* (2), 39–49.

Ginde, B.; Hosangadi, B. D.; Kudav, N. A.; Nayak, K. V.; Kulkarni, A. B. Chemical Investigations on *Cassia occidentalis* Linn. Part 1. Isolation and Structure of Cassiollin, a New Xanthone. *J. Chem. Soc.* **1970,** *9*, 1285–1289.

Gowrisri, M.; Kotagiri, S.; Vrushabendra, S. B. M.; Archana, S. P.; Vishwanath K. M. Anti-Oxidant and Nephroprotective Activities of *Cassia occidentalis* Leaf Extract Against Gentamicin Induced Nephrotoxicity in Rats. *Res. J. Pharm. Biol. Chem. Sci.* **2012,** *3* (3), 684–694.

Hanelt, P.; Mansfeld, R. Büttner Mansfeld's. In *Encyclopedia of Agricultural and Horticultural Crops*, Vol. 4; Springer: Berlin, 1986; pp 563–564.

Hatano, T.; Mizuta, S.; Ito, H.; Yoshida, T. C-Glycosidic Flavonoids from *Cassia occidentalis*. *Phytochemistry*. **1999,** *52* (7), 1379–1383.

Huang, K. C.; Williams, W. M. *The Pharmacology of Chinese Herbs*; CRC Press: Boca Raton, 1999; p 84.

Ibrahim, M. A.; Aliyu, A. B.; Sallau, M.; Bashir, I. Y.; Umar, T. S. *Senna occidentalis* Leaf Extract Possesses Antitrypanosomal Activity and Ameliorates the Trypanosome-Induced Anemia and Organ Damage. *Pharmacognosy Res.* **2010,** *2* (3), 175–180.

Jafri, M. A.; Jalis Subhani, M.; Javed, K.; Singh, S. Hepatoprotective Activity of Leaves of *Cassia occidentalis* Against Paracetamol and Ethyl Alcohol Intoxication in Rats. *J Ethnopharmacol.* **1999,** *66* (3), 355–361.

Jawahar, L.; Gupta, P. C. Two New Anthraquinones from the Seeds of *Cassia occidentalis* Linn. *Cell Mol. Life Sci.* **1974,** *30* (8), 850–851.

Jayanthy, V.; Shafna, A. Use of Flowers as Antimelanocyte Agent Against UV Radiation Effects. *Am. J. Bio-Pharm. Biochem. Life Sci.* **2012,** *1* (Suppl 1), A68.

Kanakam, V.; Kalakota, C.; Srisailam, K.; Naikal, M. A.; Swathi, S.; Prameela, S. Analgesic and Anti-Inflammatory Activities of the Extract of *Cassia occidentalis* Linn. Animal Model. *Int. J. Res. Pharm. Chem.* **2013,** *3* (4), 759–762.

Khare, C. P. *Indian Medicinal Plants. An Illustration Dictionary*; Springer Science and Business Media: New York, 2008; p 129.

Kitanaka, S.; Takido, M. Studies on the Constituents of the Leaves of *Cassia torosa* Cav. Part III: The Structures of Two New Flavone Glycosides. *Chem. Pharm. Bull.* **1992,** *40* (1), 249–251.

Koche, D. K.; Shirsat, R. P.; Bhadange, D. G.; Kamble, L. H. *In Vitro* Antioxidant and Radical Scavenging Activity of *Cassia occidentalis* L. Leaf Extracts. *Deccan Curr. Sci.* **2010,** *3* (II), 204–208.

Kumar, A. R.; Abbulu, K. Antioxidant Activity of Ethanolic Extract of *Cassia occidentalis* Against Carbon Tetrachloride-Induced Oxidative Stress in Wistar Rats. *Int. J. Chem. Sci.* **2011,** *9* (1), 378–386.

Lavanya, R. Screening of *Cassia occidentalis* L. Leaf and Seed Extracts for Antitubercular Activity. *Glob. J. Res. Anal.* **2014,** *3* (11), 609–610.

Liogier, H. A. Descriptive Flora of Puerto Rico and Adjacent Islands. In *Spermatophyla*, Vol. 2; Editorial de la Universidad dePuerto Rico: Rio Piedras, PR, 1988; p 481.

Mahanthesh, M. C.; Jalalpure, S. S. Pharmacognostical Assessment and Anticonvulsant Activity of Whole Plant of *Cassia occidentalis* Linn. *Int. J. Pharmacogn. Phytochem. Res.* **2016,** *8* (9), 1444–1457.

Nadkarni, K. M. *Indian Materia Medica*; Popular Prakashan Pvt. Ltd.: Mumbai, 2005; p I: 289.

Oliver, B. B.; *Medicinal Plants in Tropical West Africa*; Cambridge University Press: Cambridge, 1986; p 148.

Onakpa, M. M.; Ajagbonna, O. P. Antidiabetic Potentials of *Cassia occidentalis* Leaf Extract on Alloxan Induced Diabetic Albino Mice. *Int. J. Pharmtech Res.* **2012,** *4* (4), 1766–1769.

Rani, A. S.; Emmanuel, S.; Sreekanth, M.; Ignacimuthu, S. Evaluation of In Vivo Antioxidant and Hepatoprotective Activity of *Cassia occidentalis* Linn Against Paracetamol-Induced Liver Toxicity in Rats. *Int. J. Pharm. Pharm. Sci.* **2010,** *2* (3), 67–70.

Sadique, J.; Chandra, T.; Thenmozhi, V.; Elango, V. Biochemical Modes of Action of *Cassia occidentalis* and *Cardiospermum halicacabum* in Inflammation. *J. Ethnopharmacol.* **1987,** *19* (2), 201–212.

Saganuwan, A. S.; Gulumbe, M. L. Evaluation of In Vitro Antimicrobial Activities and Phytochemical Constituents of *C. occidentalis*. *Anim. Res. Int.* **2006,** *3* (3), 566–569.

Sastry, A. V. S.; Girija, V. S.; Appalanaidu, B. K.; Srinivas, Annapurna, A. Chemical and Pharmacological Evaluation of Aqueous Extract of Seeds of *Cassia occidentalis*. *J. Chem. Pharm. Res.* **2011,** *3* (2), 566–575.

Shah, C. S.; Shinde, M. V. Phytochemical Studies of Seeds of *Cassia tora* L. and *C. occidentalis* L. *Indian J. Pharm.* **1969,** *32*, 70.

Silva, M. G.; Aragao, T. P.; Vasconcelos, C. F.; Ferreira, P. A.; Andrade, B. A.; Costa, I. M.; Costa-Silva, J. H.; Wanderley, A. G.; Lafayette SS. Acute and Subacute Toxicity of *Cassia occidentalis* L. Stem and Leaf in Wistar Rats. *J. Ethnopharmacol.* **2011,** *136* (2), 341–346.

Sini, K. R.; Sinha, B. N.; Karpakavalli, M.; Sangeetha, P. T. Analgesic and Antipyretic Activity of *Cassia occidentalis* Linn. *Annals Biol. Res.* **2011,** *2* (1), 195–200.

Takahashi, S.; Takido, M.; Sankawa, U.; Shibata, S. Studies on the Constituents of the Seeds of *Cassia obtusifolia*: The Structures of Two New Lactones, Isotoralactone and Cassialactone. *Phytochemistry* **1976,** *20* (8), 1951–1953.

Tsado, N. A.; Lawal, B.; Kontagora, G. N.; Muhammad, B. M.; Yahaya, M. A.; Gboke, J. A.; Muhammad, U. A.; Hassan, M. K. Antioxidants and Antimicrobial Activities of Methanol Leaf Extract of *Senna occidentalis*. *J. Adv. Med. Pharm. Sci.* **2016,** *8* (2), 1–7.

Tsutomu, H.; Seiki, M.; Hideyuki, I.; Takashi, Y. C-Glycosidic Flavonoids from *Cassia occidentalis*. *Phytochem.* **1999,** *52* (7), 1379–1383.

Vedpriya, A.; Sanjay, Y.; Sandeep, K.; Yadav J. P. Antimicrobial Activity of *Cassia occidentalis* Leaf Against Various Human Pathogenic Microbes. *Life Sci. Med. Res.* **2010,** *9* (1), 1–12.

Verma, L.; Khatri, A.; Kaushik, B.; Patil, K. U.; Pawar, S. R. Antidiabetic Activity of *Cassia occidentalis* (Linn) in Normal and Alloxan-Induced Diabetic Rats. *Indian J. Pharmacol.* **2010,** *42* (4), 224–228.

Vijayalakshmi, S.; Ranjitha, J.; Devi Rajeswari, V.; Bhagiyalakshmi, M. Pharmacological Profile of *Cassia occidentalis* – A Review. *Int. J. Pharm. Pharm. Sci.* **2013,** *5* (3), 29–33.

Vipul, S. D.; Anjan, K. V. Antifungal Activity of Crude Extract of *Cassis occidentalis*. *Int. J. Res. Phytochem. Pharm.* **2011,** *1* (2), 36.

Warrier, P. K.; Nambiar, V. P. K.; Ramakutty, V. (eds.) *Indian Medicinal Plants–A Compandium of 500 Species*, Vol. 1; Orient Longman: Chennai, 1994; pp 342–343.

Yadav, J. P.; Arya, V.; Yadav, S.; Panghal, M.; Kumar, S.; Dhankhar, S. *Cassia occidentalis* L.: A Review on Its Ethnobotany, Phytochemical and Pharmacological Profile. *Fitoterapia* **2010,** *81* (4), 223–230.

Yadava, R. N.; Satnami, D. K. Chemical Constituents from *Cassia occidentalis* Linn. *Indian J. Chem. Sect. B.* **2011,** *50B*, 1112–1118.

CHAPTER 35

Indigofera tinctoria L., Phytochemical and Pharmacological Aspects

THADIYAN PARAMBIL IJINU[1,2*], RAGESH RAVEENDRAN NAIR[3], VASANTHA KAVUNKAL HRIDYA[4], VALPASSERI PURAKKAT AKHILESH[5], and PALPU PUSHPANGADAN[1]

[1]*Amity Institute for Herbal and Biotech Products Development, Thiruvananthapuram 695005, Kerala, India*

[2]*Naturæ Scientific, Kerala University Business Innovation and Incubation Centre, Karyavattom Campus, Thiruvananthapuram 695581, Kerala, India*

[3]*Department of Botany, NSS College Nilamel, Kollam 691535, Kerala, India*

[4]*Chemical Sciences and Technology Division, CSIR-National Institute for Interdisciplinary Science and Technology, Thiruvananthapuram 695019, Kerala, India*

[4]*Amplicon BioLabs, Kinfra Techno-Industrial Park, Malappuram 673635, Kerala, India*

[*]*Corresponding author (E-mail: ijinutp@gmail.com)*

ABSTRACT

Indigofera tinctoria is an important medicinal plant, distributed throughout the tropics and subtropics of Asia, Africa, and America, and is widely cultivated in India. It belongs to the family Fabaceae. *I. tinctoria* has been cultivated and highly valued for centuries as a main source of indigo dye. It is widely used in Ayurveda and traditional Chinese medicine for the treatment

Bioactives and Pharmacology of Legumes. T. Pullaiah, PhD (Ed.)

of nervous disorders, bronchitis, hepatic diseases, ulcers, and haemorrhoids. *I. tinctoria* contains indigotin, indirubin, rotenoids, and isothiocyanates. The major bioactive compounds isolated from *I. tinctoria* include furano-flavones such as pseudo-semiglabrin, semiglabrin, and glabratephrin, as well as kaempferol-4′,7-dirhamnoside, a rare flavonol glycoside. It possesses anti-oxidant, hepatoprotective, neuroprotective, anticancer, immunomodulatory, antiarthritic, antidyslipidemic, antibacterial, and anthelmintic properties.

35.1 INTRODUCTION

Indigofera tinctoria L. (Syn.: *Anila tinctoria* (L.) Kuntze, *Indigofera bergii* Vatke, *I. cinerascens* DC., *I. houer* Forssk., *I. indica* Lam., *I. oligophylla* Baker, *I. orthocarpa* (DC.) O.Berg & C.F.Schmidt, *I. sumatrana* Gaertn., *I. tinctoria* Blanco, and *I. tulearensis* Drake) belongs to the family Fabaceae. Vernacular names of this plant include gouli, lil, neel, nil-ka-jhar, nil-ka-per, nili (Hindi), akika, anjanakesika, asita, bhadra, bharavahi, charatika, dola, dronika, duli, dulika, gandhapushpa, gramina, kala, klitakika, kutsala, mahabala, mandhuparnika, meghavarna, mochakrishna, neela, neelinee, nila, nilam, nilapushpika, nilapuspa, nilavrikshaha, nili, nilika, nilini, ranga-patri, rangapushpi, ranjani, ranjini, sarada, shodhini, shyamalika, sriphali, sthiraranga, tuli, tuni, tuttha, vijaya, vrintika, vyanjankeshi (Sanskrit), amari, avari, madhuparnika, nilam, nili, ranjani, neela amari (Malayalam), acitai, acitam, acotaki, acotam, alampokki, alampokkicceti, alattatai, ancanam, arrippuracatani, arrumarokamnikki, attippuracatani, attipurashadam, avari, aviri, avuri, avuriyilai, cacayapakam, caiver, calikkanni, cantatakacceti, cantatakam, cantatam, caracam, chamundi, cimmai, cirupairikacceti, citavuri, ilancini, irancani, iranci, irasani, kalakacacceti, kalakacam, kalakantaki, kannenci, kantaputpai, karuntolicceti, kasturinilicceti, katura-komatai, kicapatikacceti, kilocakacceti, kurankavacceti, kurankavam, kurcalai, madubarunigai, matupanni, matuparinikai, matuparunikai, matupa-runikaicceti, mekattoniyan, mekavaranai, mekavaranaicceti, mekavarnanai, mocaicceti, neela kashunji, neeli, neelum, nilakkali, nilam, nili avuri, nili-maram, nilicceti, niraimani, pattakari, pentipekam, stirarankam, tamaraik-kanni, tanmavaipatacceti, tanmavaipatam, tarutittam, tolicceti, ulakankattal, ulakankattanceti, ulankaranai, vakkiyecam, vellaikkirikarni (Tamil), ajara, anjooraneeli, bangaali neeli, hennu neeli, kare neeli, mahaaneeli, neeli gida, ollenili, rajani (Kannada), aviri, konda nili, nili, nili-chettu, syaama (Telugu), gulee, neela, neeli, nili (Marathi), ni la (Tibetan), habb-ul-neel, nila (Urdu), ausarahe-nil, darakhte-nil, nil, nilah (Persian), nabatun-nilaj, neyleh, nilaj (Arabic), and black henna, indigo plant (English).

I. tinctoria is a perennial plant with 1–2 m height. Its leaves are compound and imparipinnate; leaflets 1–5 cm long and 0.3–1.2 cm wide, oblong or oblanceolate with short mucronate tip; pale green to greenish-black. The stem is woody, hard, slender, cylindrical, 0.1–1.5 cm in diameter, surface, smooth, with lenticels present; and yellowish-green to grayish brown in color. Its flowers are numerous in nearly sessile spicate racemes and hairy outside. Its fruit pods are nearly cylindrical straight or slightly curved; apiculate, 2–3.2 cm long and 0.15–0.2 cm in diameter, having 8–12 seeds; and is smooth and brown to dark brown. Seeds are somewhat quadrangular, smooth, and yellowish-brown to greenish-brown in color (Sasidharan, 2004, 2011). *I. tinctoria* had been cultivated and highly valued for centuries as a main source of indigo dye. It is distributed throughout the tropics and subtropics of Asia, Africa, and America, and is widely cultivated in India. *I. tinctoria* is widely used in Ayurveda and traditional Chinese medicine for the treatment of nervous disorders, bronchitis, hepatic diseases, ulcers, and hemorrhoids (Anand et al., 1979, 1981; Satyavati et al., 1987). The roots, stems, and leaves of *I. tinctoria* are bitter, thermogenic, gastroprotective, promote hair growth, cardioprotective, hepatoprotective, neuroprotective, nephroprotective, and diuretic. Plant parts are also used against chronic bronchitis, asthma, and skin diseases. Its root decoction is given in renal calculus and juice of the young branches along with honey is used to treat aphthae (mouth ulcer) in children (Pushpangadan et al., 2016).

35.2 PHYTOCHEMICAL CONSTITUENTS

I. tinctoria contains indigotin, indirubin, rotenoids, and isothiocyanates. Indigo is one of the world's oldest dyes, derived from *Indigofera* spp., and it is a derivative of the colorless glucosides of the enol form of indoxyl, e.g., indican (indoxyl-D-glucoside) (Ensley et al., 1983). The dye was obtained from the leaves of *I. tinctoria* by a fermentation process, which has probably not changed for over 3000 years (Clark et al., 1993). Perkin and Bloxam (1907) isolated indigo and indirubin from this plant, and also, it comprises isatan B (indoxyl-β-ketogluconate) as a major precursor, and indican as minor (Epstein et al., 1967; Maier et al., 1990). Kamal and Mangla (1992) isolated six different rotenoids compounds, namely, tephrosin, rotenone, deguelin, rotenol, dehydrodeguelin, and sumatrol.

Singh et al. (2006) isolated an active fraction indigotin from the dried aerial part of *I. tinctoria* and later a compound, trans-tetracos-15-enoic acid was isolated from indigotin. Narendar et al. (2006) have isolated three furano-flavones, namely, pseudo-semiglabrin, semiglabrin, and glabratephrin and a rare flavonol glycoside, kaempferol-4′,7-dirhamnoside from the aerial part of *I. tinctoria*. The extraction of the indigo dyes from the fermentation of *I. tinctoria* in water, and separation of indican, isatin, isatan B, indoxyl, and two main constituents of blue and red color indigo are reported (Laitonjam and Wangkheirakpam, 2011).

Degani et al. (2015) carried out the gas chromatography-mass spectrometry-based optimization for indigotin and indirubin in the extracts of *I. tinctoria*. The percentage of seed oil obtained was 2.66%, and it contains triacylglycerides, glyco- and phospholipids, free fatty acids, and fatty-acid esters of cyclic alcohols (Yuldasheva et al., 2016). Sharma and Singh (2017) isolated a new isothiocyanate-derived compound, 1-[1,2-diisothiocyanato-2-(3-isothiocyanato-2,2-dimethyl-propylsulphanyl)-ethoxy]-3-isothiocyanato-2,2-dimethyl-propane from the aerial parts.

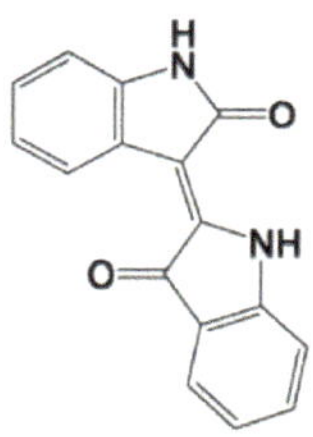

Indirubin

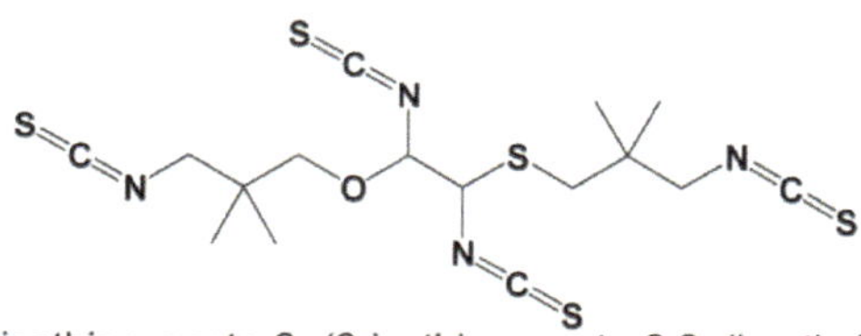

1-[1,2-Diisothiocyanato-2- (3-isothiocyanato-2,2-dimethyl-propylsulphanyl)-ethoxy]- 3-isothiocyanato-2,2-dimethyl-propane

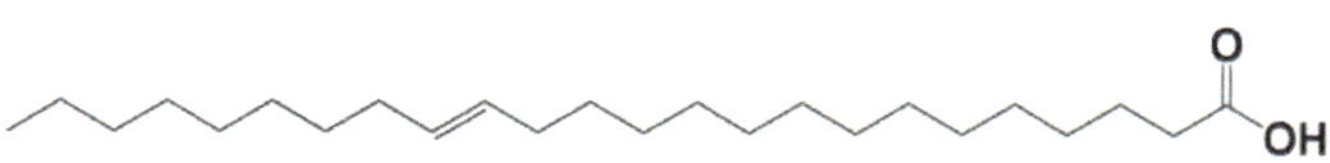

Trans-tetracos-15-enoic acid

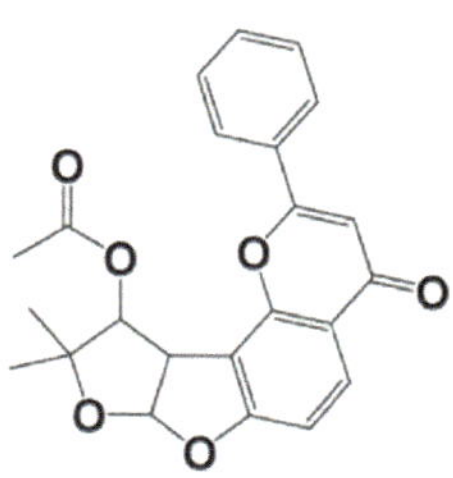

Semiglabrin

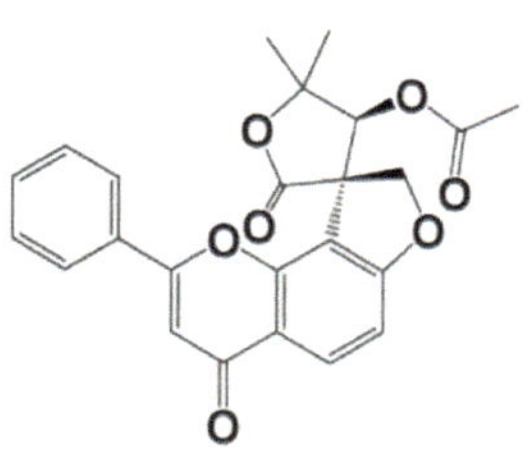

Glabratephrin

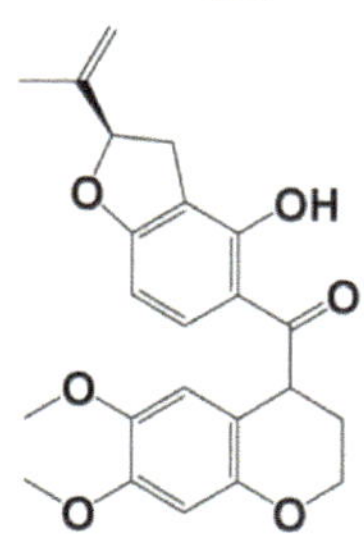

Rotenol

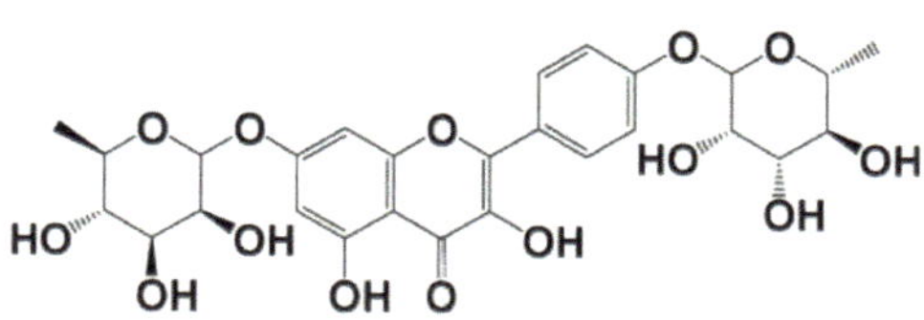

Kaempferol-4',7-dirhamnoside

35.3 PHARMACOLOGICAL STUDIES

35.3.1 Antioxidant Activity

Sharma and Agarwal (2015) found that the 2,2-diphenyl-1-picrylhydrazyl radical scavenging IC_{50} value of methanol and aqueous alcoholic extracts of *I. tinctoria* were 0.709 and 0.881 mg/mL, respectively. The ferric-reducing antioxidant power value for methanol and aqueous alcoholic extracts were 10.807 and 18.256 mM Fe (II) ions/mg of plant extract, respectively. Total antioxidant capacity for methanol and aqueous alcoholic extracts were 622.7 and 722.7 mg/g, respectively. Hydroethanolic extract of *I. tinctoria* showed good antioxidant activity in 2,2-diphenyl-1-picrylhydrazyl, metal chelation, and hydroxyl radical-scavenging studies with IC_{50} values 829, 659, and 26.7 μg/mL respectively. Quantitative measurements showed the abundance of phenolics and flavonoids, 212.9 mg gallic acid equivalent, and 149.7 mg rutin/g of plant extract, respectively (Singh et al., 2015). The aqueous extract of *I. tinctoria* was found to have 2,2-diphenyl-1-picrylhydrazyl (52.08%), nitric oxide (23.12%), and superoxide (26.79%) scavenging activities at the concentration of 250 μg/mL (Srinivasan et al., 2016).

35.3.2 Hepatoprotective Activity

Trans-tetracos-15-enoic acid isolated from *I. tinctoria* showed significant protection against carbon tetrachloride- and paracetamol-induced hepatotoxicity in experimental animal models. The ED_{50} value calculated is 25–45 mg/kg, p.o, and is almost comparable to silymarin and N-acetylcysteine (Singh et al., 2006). Chitra et al. (2003) found that *I. tincoria* leaf extract showed hepatoprotective activity against isoniazid (100 mg/kg)-induced hepatic damage in Wistar rats in a dose-dependent (5 and 10 mL/kg) manner. The extract normalized the alterations in liver enzymes and serum parameters. Alcoholic extract of *I. tinctoria* pretreatment for 21 days showed protective effect against D-galactosamine administered liver toxicity male albino rats. The extract at a dose of 500 mg/kg showed considerable protection by improving the liver antioxiadant potential (Sreepriya et al., 2001). Hydroethanolic extract of *I. tinctoria* containing isothiocyanate derivative and 1-[1,2-diisothiocyanato-2-(3-isothiocyanato-2,2-dimethyl-propylsulfanyl)-ethoxy]-3-isothiocyanato-2,2-dimethyl-propane showed significant liver protective activity by reducing the levels of aspartate aminotransferase, alanine aminotransferase, alkaline phosphatase, and bilirubin, in N-nitrosopyrrolidine followed by carbon tetrachloride-treated mice (Singh and Sharma, 2019). The methanol extract of *I. tinctoria* (50 mg/kg) significantly reduced the serum glutamate pyruvate transaminase and serum glutamate oxaloacetate transaminase levels in the carbon tetrachloride (1.25 mL/kg)- treated liver toxicity in rats (Ravi et al., 2004). *I. tinctoria* aqueous extract (250 and 500 mg/kg) once daily for 10 days showed protective effect against paracetamol-induced liver toxicity in rats, evidenced from serum biochemical and histopathological analysis (Felicia and Muthulingam, 2012).

35.3.3 Immunomodulatory Activity

I. tinctoria aqueous extract at a dose of 200 mg/kg b.w. showed significant stimulation in immune response in Wistar albino rats against normal and chronic noise (100 dB for 1 h, 20 days)-induced stress conditions. The extract significantly prevented the reduction in macrophage phagocytosis, antibody secretion by spleen cells, humoral immune response, lymphocyte proliferation, cell toxicity, expression of tumor necrosis factor alpha, granzyme B, and perforin expression in splenic natural killer cells against noise-induced stress (Boothapandi and Ramanibai, 2017).

35.3.4 Anti-arthritic Activity

The anti-arthritic potential of *I. tinctoria* leaf petroleum ether and ethanol extracts (100, 200, and 300 mg/kg) was carried out by inducing Freunds Complete Adjuvant at the tibiotorsal joint of rats. Both the extracts showed anti-arthritic activity, by significant changes in the behavioral activity, decreased vascular permeability, histamine level, interleukin and tumor necrosis factor, and proliferation of spleenocytes dose-dependently (Katti et al., 2015).

35.3.5 Neuroprotective Activity

Kopalli et al. (2012) found that a compound named SF-6 isolated from *I. tinctoria* showed dose-dependent (1, 5, and 10 μg/mL) protective activity against α-synuclein, hydrogen peroxide-induced cell toxicity, and effectively scavenged the reactive oxygen species in SH-SY5Y cells. SF-6 attenuated the contralateral rotational asymmetry observed by apomorphine challenge in 6-hydroxydopamine-induced mice. Furthermore, SF-6 corrected the behavioral impairments assessed by rotarod test, Y-maze, and passive avoidance tests, and was shown to be more effective than the standard drug deprenyl. In pentylenetetrazole-induced male albino rats, the ethanolic extract of *I. tinctoria* considerably delayed the onset of convulsions and decreased the duration of seizures in a dose-dependent manner. A significant reduction in the duration of hind limb tonic extension and increase in brain gamma-aminobutyric acid levels at higher doses (500 and 1000 mg/kg) were also observed. The rotarod test, pentobarbital-induced sleep time, locomotor activity, and haloperidol-induced catalepsy did not show any indications of neurotoxicity (Garbhapu et al., 2011). The severity of status epilepticus in male Wistar albino rats was considerably decreased after oral administration of ethanol extract of *I. tinctoria* at doses of 500 and 1000 mg/kg 24 h after injection of lithium chloride at a dose of 3 mEq/kg, i.p. (Asuntha et al., 2010). The aqueous extract of *I. tinctoria* (5, 10, and 20 μg/mL) significantly reversed the scopolamine-induced cognitive deficits in experimental amnesic mice, dose-dependently. It further scavenged the lipid peroxide, superoxide, and hydroxyl free radicals with IC_{50} values of 7.28, 5.25, and 7.62 μg/mL, respectively (Kim et al., 2016).

35.3.6 Anticancer Activity

The flavonoid fraction separated from methanolic extract of *I. tinctoria* aerial parts showed antiproliferative effect in A-549 cell line. The isolated fraction reduced A-549 proliferation by inhibiting cell cycle progression in the G0/G1 phase and induced apoptosis (Kameswaran and Ramanibai, 2008). Renukadevi and Suhani (2011) observed that the increased concentration of *I. tinctoria* leaf extract decreased the viability of NCI-H69 (lung cancer) cell line. Aobchey et al. (2007) found that indirubin isolated from *I. tinctoria* at a dose of 30 mM showed significant inhibition (42%) in cell growth of MCF-7 cells within 24 h of treatment. Chrysin isolated from the *I. tinctoria* leaves inhibited the proliferation of human epidermoid carcinoma (A431) cells with IC_{50} value of 23.52 μg/mL. DNA fragmentation analysis indicated that the chrysin-treated A431 cells displayed the fragmented DNA. Chrysin-induced remarkable cell death through apoptosis (87.65%) and arrested the cell cycle at Sub G1 phase. In dose-dependent manner, it also increased the expression of cytochrome c, caspase 9, and caspase 3, and reduced the inhibitor of caspase-activated DNase expression (Boothapandi and Ramanibai, 2018). Ravichandran and Ramanibai (2008) found that the flavanoidal fraction of *I. tinctoria* (100 mg/kg) showed significant improvement in benzo (a) pyrene (BP)-induced lung tumor in Swiss albino mice by improving the cellular antioxidants like superoxide dismutase, glutathione peroxidase, glutathione reductase, and catalase and reduced glutathione.

35.3.7 Antidyslipidemic Activity

The alcoholic extract of *I. tinctoria* and its serial chloroform, butanol, and water fractions were investigated for antidyslipidemic activity in dyslipidemic hamsters fed with high fat diet (Puri et al., 2007). The chloroform fraction at a dose of 250 mg/kg significantly reduced plasma triglycerides by 52%, total cholesterol by 29%, glycerol by 24%, free fatty acids by 14%, and increased high-density lipoproteins by 9%, and high-density lipoprotein/total cholesterol ratio to 52%. Narender et al. (2006) found that *I. tinctoria* containing diastereomeric flavonoid mixture of compounds pseudo-semiglabrin, semiglabrin at a ratio of 80:20 significantly decreased the plasma triglycerides by 60%, total cholesterol 19%, glycerol 13%, and free fatty acid 25% accompanied with increase in high-density lipoproteins-cholesterol by 8% and high-density lipoprotein/total cholesterol ratio 36% in fat diet-fed hamsters at the dose of 50 mg/kg body weight and the compound glabratephrin showed moderate antidyslipidemic activity.

35.3.8 Antibacterial Activity

Indican and sumatrol isolated from methanol extract of *I. tinctoria* demonstrated significant inhibition against *Escherichia coli*, *Bacillus cereus*, and *Klebsiella pneumoniae* (Sindhu and Mathew, 2012).

35.3.9 Anthelmintic Activity

In-vitro and in-vivo anthelmintic potential of *I. tinctoria* against sheep gastrointestinal nematodes was studied by Meenakshisundaram et al. (2016). The ethanolic extract demonstrated significant inhibition on in-vitro egg hatching at concentrations of 40 and 80 mg/mL. At doses of 125, 250, and 500 mg/kg, the extract decreased the fecal egg count by 30.82–47.78% in vivo.

35.3.10 Toxicity Study

Kumar et al. (2009) found that the ehanolic extract (90%) of *I. tinctoria* showed no signs of abnormalities or any mortality for the 15 days of study up to a dose of 2000 mg/kg dose in acute toxicity study.

ACKNOWLEDGMENTS

The authors express their sincere thanks to Dr. Ashok K. Chauhan, Founder President, Ritnand Balved Education Foundation (RBEF) and Amity Group of Institutions, and Dr. Atul Chauhan, Chancellor, Amity University Uttar Pradesh (AUUP) for facilitating this work. Thadiyan Parambil Ijinu has received Young Scientist Fellowship from the Department of Science and Technology, Government of India (SP/YO/413/2018).

KEYWORDS

- **traditional medicinal use**
- **indigotin**
- **indirubin**
- **antioxidant activity**
- **hepatoprotective**
- **activity**
- **neuroprotective activity**

REFERENCES

Anand, K. K.; Chand, D.; Ghatak, B. J. Protective Effect of Alcoholic Extract of *Indigofera tinctoria* Linn. in Experimental Liver Injury. *Indian J. Exp. Biol.* **1979,** *17*, 685–687.

Anand, K. K.; Chand, D.; Ghatak, B. J. R.; Arya, R. K. Histological Evidence of Protection by *Indigofera tinctoria* Linn. Against Carbon Tetrachloride Induced Hepatotoxicity-an Experimental Study. *Indian J. Exp. Biol.* **1981,** *19*, 298–300.

Aobchey, P.; Sinchaikul, S.; Phutrakul, S.; Chen, S. T. Simple Purification of Indirubin from *Indigofera tinctoria* Linn. and Inhibitory Effect on MCF-7 Human Breast Cancer Cells. *Chiang Mai J. Sci.* **2007,** *34*, 329–337.

Asuntha, G.; Prasannaraju, Y.; Prasad, K. V. S. R. G. Effect of Ethanol Extract of *Indigofera tinctoria* Linn (fabaceae) on Lithium/Pilocarpine-Induced Status Epilepticus and Oxidative Stress in Wistar Rats. *Trop. J. Pharm. Res.* **2010,** *9*, 149–156.

Boothapandi, M.; Ramanibai, R. Antiproliferative Activity of Chrysin (5, 7-Dihydroxyflavone) from *Indigofera tinctoria* on Human Epidermoid Carcinoma (A431) Cells. *Eur. J. Integr. Med.* **2018,** *24*, 71–78.

Boothapandi, M.; Ramanibai, R. *In Vivo* Immunoprotective Role of *Indigofera tinctoria* and *Scoparia dulcis* Aqueous Extracts Against Chronic Noise Stress Induced Immune Abnormalities in Wistar Albino Rats. *Toxicol. Rep.* **2017,** *4*, 484–493.

Chitra, M.; Muthusudha, N.; Sasikala, R. Bioefficiency of *Indigogera tinctoria* Linn. on Isoniazid Induced Hepatotoxicity in Albinorats. *Anc. Sci. Life.* **2003,** *23*, 79–89.

Clark, R. J. H.; Cooksey, C. J.; Daniels, M. A. M.; Withnall, R. Indigo, Woad, and Tyrian Purple: Important Vat Dyes from Antiquity to Present. *Endeavour* **1993,** *17*, 191–199.

Degani, L.; Riedo, C.; Chiantore, O. Identification of Natural Indigo in Historical Textiles by GC-MS. *Anal. Bioanal. Chem.* **2015,** *407*, 1695–1704.

Ensley, B. D.; Ratzkin, B. J.; Osslund, T. D.; Simon, M. J.; Wackett, L. P.; Gibson, D. T. Expression of Naphthalene Oxidation Genes in *Escherichia coli* Results in Biosynthesis of Indigo. *Science* **1983,** *222*, 167–169.

Epstein, E.; Nabors, M. W.; Stowe, B. B. Origin of Indigo of Woad. *Nature* **1967,** *216*, 547–549.

Felicia, F. A.; Muthulingam, M. Antihepatotoxic Efficacy of Methanolic Extract of *Indigofera tinctoria* (Linn.) on Paracetamol-Induced Liver Damage in Rats. *J. Nat. Prod. Plant Resour.* **2012,** *2*, 244–250.

Garbhapu, A.; Yalavarthi, P.; Koganti, P. Effect of Ethanolic Extract of *Indigofera tinctoria* on Chemically-Induced Seizures and Brain GABA Levels in Albino Rats. *Iran J. Basic Med. Sci.* **2011,** *14*, 318–326.

Kamal, R.; Mangla, M. *In Vivo* and *In Vitro* Production of Histamine from *Indigofera tinctoria*. *Indian Drugs* **1992,** *29*, 179

Kameswaran, T. R.; Ramanibai, R. The antiproliferative Activity of Flavanoidal Fraction of *Indigofera tinctoria* Is Through Cell Cycle Arrest and Apoptotic Pathway in A-549 Cells. *J. Biol. Sci.* **2008,** *8*, 584–590.

Katti, H. R.; Chandrasekhar, I.; Ramkishan, A.; Chandrashekhar, V. M.; Reddy, Y. S. K.; Muchandi, I. S.; Raghavendra, H. L. Anti-Arthritic Activity of *Indigofera tinctoria* L on Adjuvant Induced Arthritis. *Sci. Technol. Arts Res. J.* **2015,** *4*, 155–168

Kim, J. B.; Kopalli, S. R.; Koppula, S. *Indigofera tinctoria* Linn (fabaceae) Attenuates Cognitive and Behavioral Deficits in Scopolamine-Induced Amnesic Mice. *Trop. J. Pharm. Res.* **2016,** *15*, 773–779.

Kopalli, S. R.; Koppula, S.; Shin, K. Y.; Noh, S. J.; Jin, Q.; Hwang, B. Y.; Suh, Y. H. SF-6 Attenuates 6-Hydroxydopamine-Induced Neurotoxicity: An *In Vitro* and *In Vivo* Investigation in Experimental Models of Parkinson's Disease. *J. Ethnopharmacol.* **2012,** *143*, 686–694.

Kumar, A. S.; Gandhimathi, R.; Lakshmi, S. M.; Nair, R.; Kumar, C. K. A. Evaluation of the Antinociceptive Properties from *Indigofera tinctoria* Leaves Extracts. *J. Pharm. Sci. Res.* **2009,** *1*, 31–37.

Laitonjam, W. S.; Wangkheirakpam, S. D. Comparative Study of the Major Components of the Indigo Dye Obtained from *Strobilanthes flaccidifolius* Nees. and *Indigofera tinctoria* Linn. *Int. J. Plant Physiol. Biochem.* **2011,** *3*, 108–116.

Maier, W.; Schumann, B.; Groger, D. Biosynthesis of Indoxyl in *Isatis tinctoria* and *Polygonum tinctorium*. *Phytochemistry* **1990,** *29*, 817–819.

Meenakshisundaram, A.; Harikrishnan, T. J.; Anna, T. Anthelmintic Activity of *Indigofera tinctoria* Against Gastrointestinal Nematodes of Sheep. *Vet. World* **2016,** *9*, 101–106.

Narender, T.; Khaliq, T.; Puri, A.; Chander, R. Antidyslipidemic Activity of Furanoflavonoids Isolated from *Indigofera tinctoria, Bioorg. Med. Chem.* **2006,** *16*, 3411–3414.

Perkin, A. G.; Bloxam, W. P. XXX-Some Constituents of Natural Indigo. Part I. *J. Chem. Soc. Trans.* **1907,** *91*, 279–288.

Puri, A.; Khaliq, T.; Rajendran, S. M.; Bhatia, G.; Chandra, R.; Narender, T. Antidyslipidemic Activity of *Indigofera tinctoria. J. Herb Pharmacother.* **2007,** *7*, 59–64.

Pushpangadan, P.; George, V., Sreedevi, P.; Ijinu, T. P.; Anzar, S.; Bincy, A. J. *Indigofera tinctoria* L. In *Plants for Health and Nutritional Security*; Amity Institute for Herbal and Biotech Products Development: Thiruvananthapuram, India; 2016, pp 276–279,

Ravi, R.; Venkatanarayanan, R.; Binokingsly, R.; Hemalatha, R.; Suriyaprakash, T. N. K.; Sivakumar, S. M.; Kumar, S. S. Anti-Hepatotoxic Activity of *Indigofera tinctoria. Nig. J. Nat. Prod. Med.* **2004,** *8*, 43–44.

Ravichandran, K.; Ramanibai, K. Protective Effect of Flavanoidal Fraction of *Indigofera tinctoria* on Benzo a Pyrene Induced Lung Carcinogenicity in Swiss Albino Mice. *Int. J. Cancer Res.* **2008,** *4*, 71–80.

Renukadevi, K. P.; Suhani, S. S. Determination of Antibacterial, Antioxidant and Cytotoxicity Effect of *Indigofera tinctoria* on Lung Cancer Cell Line NCI-h69. *Int. J. Pharmacol.* **2011,** *7*, 356–362.

Sasidharan, N. *Biodiversity Documentation for Kerala, Part 6: Flowering Plants*; Kerala Forest Research Institute: Kerala, India; 2004.

Sasidharan, N. *Flowering Plants of Kerala*, Ver. 2.0 (DVD), Serial Number 698329520; Kerala Forest Research Institute: Thrissur, Kerala, India; 2011.

Satyavati, G. V.; Gupta, A.; Tandon, N. *Medicinal Plants of India*; Indian Council for Medical Research: New Delhi, India; 1987, p 138.

Sharma, V.; Agarwal, A. Physicochemical and Antioxidant Assays of Methanol and Hydromethanol Extract of Aerial Parts of *Indigofera tinctoria* Linn. *Indian J. Pharm. Sci.* **2015,** *77*, 729–734.

Sharma, V.; Singh, R. Isolation and Structural Elucidation of An Isothiocyanate Compound from *Indigofera tinctoria* Linn. Extract. *Curr. Sci.* **2017,** *113*, 941–946.

Sindhu, K. K.; Mathew, M. M. Isolation and Identification of Bioactive Compounds from Leaf Extracts and Leaf Callus of *Indigofera tinctoria. J. Trop. Med. Plants*. **2012,** *13*, 1–10.

Singh, B.; Chandan, B. K.; Sharma, N.; Bhardwaj, V.; Satti, N. K.; Gupta, V. N.; Gupta, B. D.; Suri, K. A.; Suri, O. P. Isolation, Structure Elucidation and *In Vivo* Hepatoprotective Potential of Trans-Tetracos-15-Enoic Acid from *Indigofera tinctoria* Linn. *Phytother. Res.* **2006,** *20*, 831–839.

Singh, R.; Sharma, S.; Sharma, V. Comparative and Quantitative Analysis of Antioxidant and Scavenging Potential of *Indigofera tinctoria* Linn. Extracts. *J. Integr. Med.* **2015,** *13*, 269–278.

Singh, R.; Sharma, V. Anti-Hepatotoxic Potential of *Indigofera tinctoria* and Its Isolated Isothiocyanate Compound 'ITC-1' Against NPYR-CCl_4 Intoxicated Mice. *Toxicol. Int.* **2019,** *26*.

Sreepriya, M.; Devaki, T.; Balakrishna, K.; Apparanantham, T. Effect of *Indigofera tinctoria* Linn. on Liver Antioxidant Defense System During D-Galactosamine/Endotoxin-Induced Acute Hepatitis in Rodents. *Indian J. Exp. Biol.* **2001,** *39*, 181–184.

Srinivasan, S.; Wankhar, W.; Rathinasamy, S.; Rajan, R. Free Radical Scavenging Potential and HPTLC Analysis of *Indigofera tinctoria* Linn. (Fabaceae). *J. Pharm. Anal.* **2016,** *6*, 125–131.

Yuldasheva, N. K.; Ul'chenko, N. T.; Glushenkova, A. I.; Ergashev, A. Lipids from Seeds of *Indigofera tinctoria*. *Chem. Nat. Compd.* **2016,** *52*, 32–34.

CHAPTER 36

Bioactives and Pharmacology of *Pterolobium hexapetalum* (Roth) Santapau & Wagh

B. KAVITHA[1*] and N. YASODAMMA[2]

[1]*Department of Botany, Rayalaseema University, Kurnool, Andhra Pradesh 518007, India*

[2]*Department of Botany, Sri Venkateswara University, Tirupati, Andhra Pradesh, India*

[*]*Corresponding author. E-mail: kavithab15@gmail.com*

ABSTRACT

Pterolobium hexapetalum is an extensive, armed straggling, spiny shrub and one of the important herbal plant used by the Chenchu tribes of Nallamalai region against chest pain, fever, cold, cough, tooth ache, dog bite, diarrhoea, ulcer, jaundice, skin disorders, constipation, piles and venereal diseases. It possesses high quantities of phytoconstituents in leaf, stem bark, flower and fruit extracts like flavonoides, alkaloids, phenols, glycosides, saponins, steroids, tannins and quinines. *P. hexapetalum* extracts also proved as effective antimicrobial, antiulcerous, antidiarrhoel, antipyretic, anticancer, antioxidant, wound-healing, ovicidal, larvicidal, and pupicidal activity. This review provides exhaustive information about the pharmacological and bioactives investigations of *P. hexapetalum* till date.

Bioactives and Pharmacology of Legumes. T. Pullaiah, PhD (Ed.)

36.1 INTRODUCTION

Pterolobium hexapetalum (Roth) Santapau & Wagh is a member of Fabaceae/Leguminosae and subfamily Caesalpinioideae, with the synonyms *P. indicum* A. Rich., *P. lacerans* auct. non R.Br, *Caesalpinia lacerans* Roxb., *C. ligulata* DC., *Cantuffa lacerans* (Roxb.) Taub., *C. hexapetala* (Roth) Kuntze., and *Reichardia hexapetala* Roth. It is commonly known as Indian red wing, Campa sieke (or) Bhoca, (Telugu), Baadu bakka, Baadu bakkanamullu, Bada bakka, Kabali mullu, (Kannada), Karu indu, Wale kadudu, Errachikai, Yerrachiku, Yerraseengai (Tamil), Endam (Malayalam), and Kokkichedi, Vaelipparuthi (Irula) (https://indiabiodiversity.org/species/show/32301).

It is very common in all dry deciduous forests, from Godavari southwards to the lower Pulneys, in open places over bushes and on small tress or as pioneer plants in open lands, limited to Peninsular India. It is a native to Myanmar, Bangladesh, Bhutan, and Southern parts of India.

P. hexapetalum is an extensive, armed straggling shrub, with long arching branches, dark pink and prickly; glabrous, leaves 8–10 cm long, alternate and bipinnately compound; consists of 8–10 pairs leaflets, small ovate—oblong, chartaceous, glabrous, base cuneate to truncate, margin entire, apex obtuse, or retuse. Below the rachis, there are pairs of recurved thorns, or solitary straight spines. Flowers in axillary or terminal racemes are yellowish white, sessile, and 0.6–1.0 cm long. This plant is also known as traveler's terror as the rachis bears bent hair pin-like thorns that tears off the skin, and it has pedicellate, bracts subulate, caduceous, lobes 5, bracteoles 0; calyx tube short, cup shaped, persistent, lobes 5 imbricate, the lowest large; petals 5 unequal, clawed, as long as the calyx lobes; stamens 10 free, all fertile, the filaments villous, and anthers uniform. Ovary is sessile, 1-ovuled style subulate, stigma terminal, and dilated. Pods 3–4 × 1–1.2 cm are oblong reddish indehiscent samaroids, 1-seeded ending in an oblong or falcate wing. Seeds are large compressed solitary, base flattened, and obovoid (Pullaiah, 2006; Sharmila et al., 2020).

Young pods and seed paste are used in diarrhea, constipation, and piles (Pullaiah, 2006). Venereal diseases are cured with crushed flowers mixed with goat milk and taken orally. Decoction of leaves is given orally for pregnant women during delivery to minimize delivery pain; stem bark powder is used as tooth powder and also to treat tooth ache (Duraipandiyan et al., 2006). Leaf paste with white egg black gram and onion is applied for bone fracture. During chest pain and vomit, stem and root bark powder is given once in 2 days for 1–3 months. Against jaundice, leaf juice and flowers are taken during night for 1 month. During delivery pain, leaf decoction with asafetida is taken in the morning times once in 2 days (Samuel and Andrews, 2010). Stem bark decoction in case of whooping cough of infants and bark extract in case of dyspepsia in cattle are given (Sankara Rao, 2010).

36.2 BIOACTIVES

The phytochemical qualitative screening of *P. hexapetalum* in different parts with various extracts revealed the presence of carbohydrates, amino acids, proteins, terpenoids, alkaloids, quinines, anthoquinones, coumarins, phenols, flavonoids, tannins, lignins, steroids, glycosides, fixed oils, and saponins in almost all the extracts (Kavitha et al., 2012; Yasodamma et al., 2014b; Kumar and Mohana, 2014; Ganesh et al., 2014; Kokila et al., 2016; Saranya and Elumalai, 2018; Sharmila et al., 2020).

Qualitative analysis of phenolic compounds in the leaf identified protocatechuic acid, homo-protocatechuic acid, p-hydroxy benzoic acid, m-hydroxy benzoic acid, trans-p-coumaric acid, vanillic acid, salicylic acid, syringic acid, and phloretic acid; in stem bark, phloroglucinol, chlorogenic acid, p-hydroxy benzoic acid, cis-p-coumaric acid, m-hydroxy benzoic acid, melilotic acid, cis-ferulic acid, and salicylic acid; in flowers, iso-chlorogenic acid, neochlorogenic acid, cis p-coumaric acid, trans-p-coumaric acid, o-coumaric acid, vanillic acid, trans-ferulic acid, salicylic acid, coumarin, and phloretic acid; and in fruit, neo-chlorogenic acid, trans p-coumaric acid, chlorogenic acid, m-hydroxy benzoic acid, o-coumaric acid, ferulic acid, phloretic acid, scopoletin, o-pyro catechuic acid, cis-ferulic acid, and one unidentified compound (Kavitha and Yasodamma, 2016).

Qualitative analysis of flavonoids in the leaf shows the presence of myricetin, quercetin, and vitexin; in stem bark, luteolin and vitexin; in flower, myricetin, kaempferol, and orientin; and in fruit, myricetin, quercetin, luteolin, and vitexin. Myricetin is common in leaf, flower, and fruit (Kavitha and Yasodamma, 2016).

The main anthocyanidin compounds represented in leaf include cyanidin, peonidin, and rosinidin; in stem bark, cyanidin and luteolinidin; in flower, delphidin, petunidin, cyanidin, and malvidin; and only one luteolinidin in fruit (Kavitha and Yasodamma, 2016).

The HPLC chromatograms of *P. hexapetalum* methanolic leaf extracts showed active compounds like catechin, caffeic acid, rutin, quercetin, naringenin, and cinnamic acid (Sathyanarayanan et al., 2017).

Phenolic compounds in *Pterolobium hexapetalum*.

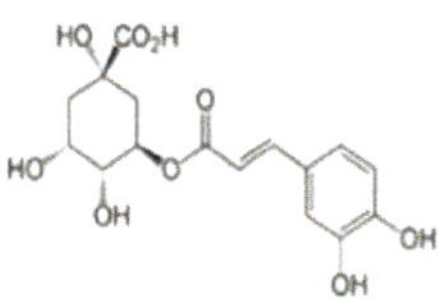

Chlorogenic Acid

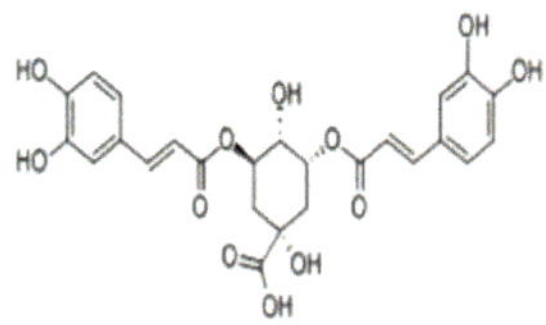

Iso-chlorogenic Acid

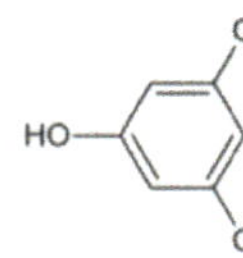

Phloroglucinol

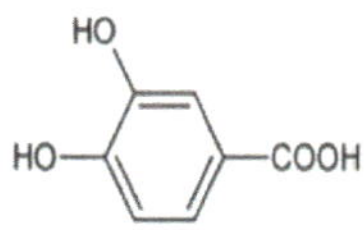

Protocatechuic Acid

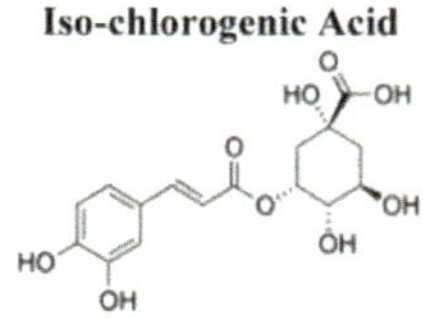

Neo-Chlorogenic Acid

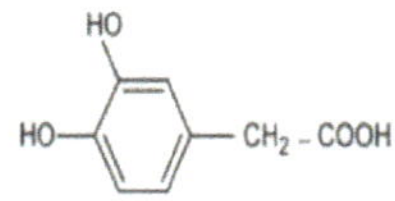

Homo Protocatechuic Acid

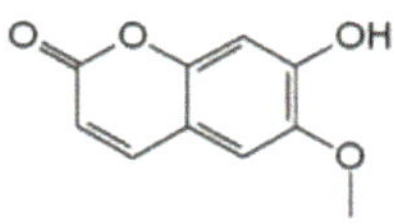

Scopoletin

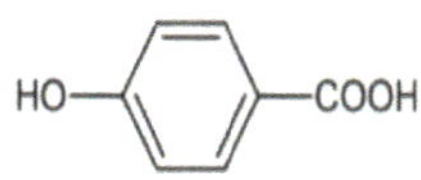

p-Hydroxy Benzoic Acid

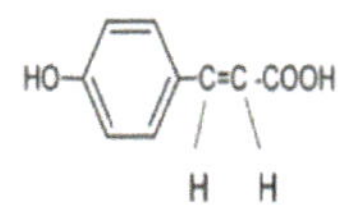

Cisp-Coumaric Acid

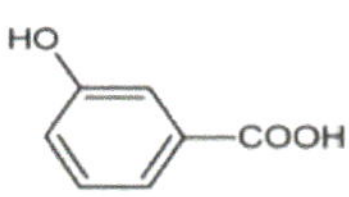

m-Hydroxy Benzoic Acid

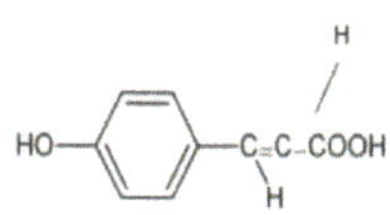

Trans–p-Coumaric Acid

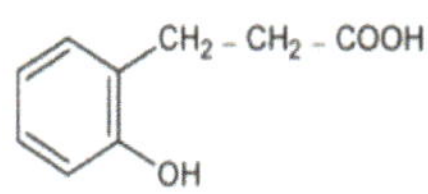

Melilotic Acid

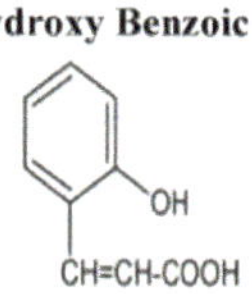

o-Coumaric Acid

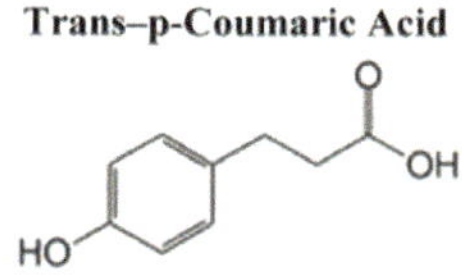

Phloretic Acid

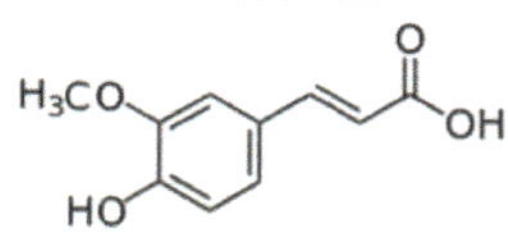

Ferulic Acid

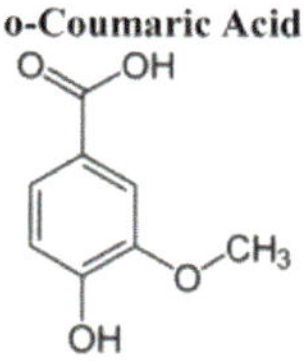

Vanillic Acid

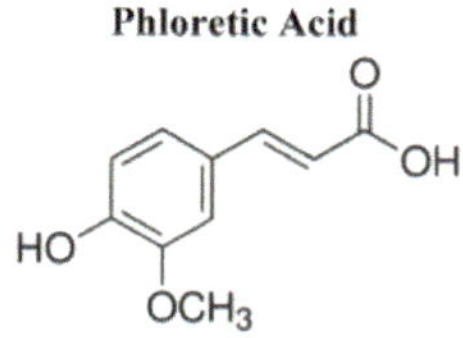

Trans-Ferulic Acid

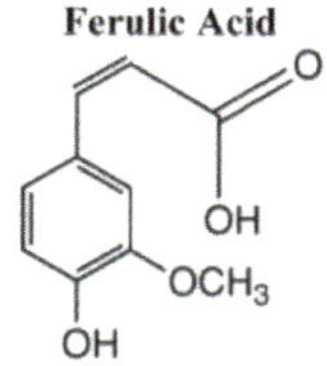

Cis-Ferulic Acid

Syringic Acid

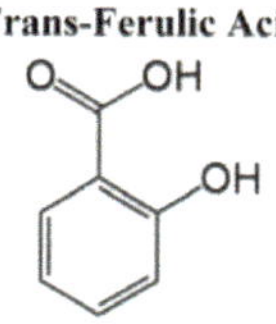

Salicyclic Acid

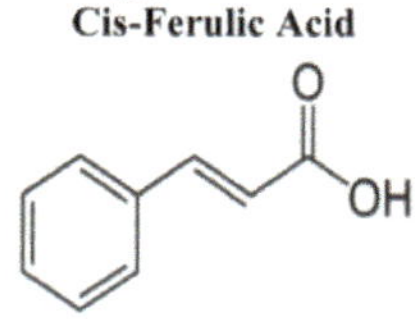

Cinnamic Acid

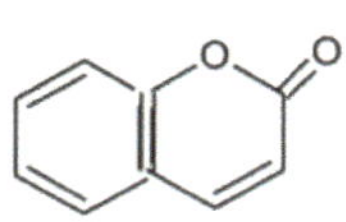

Coumarin

Flavonoids in *Pterolobium hexapetalum*.

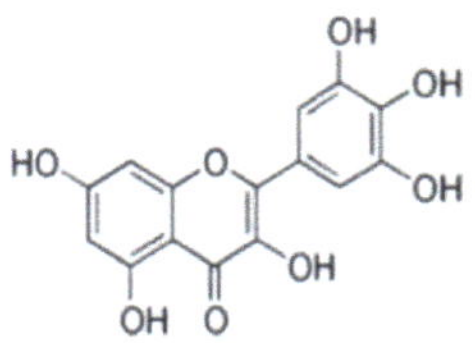

Myricetin

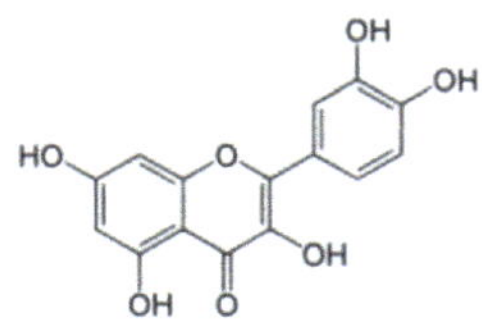

Quercetin

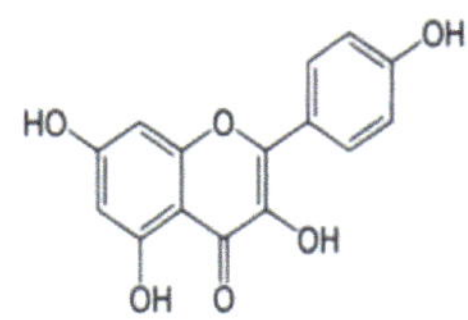

Kaempferol

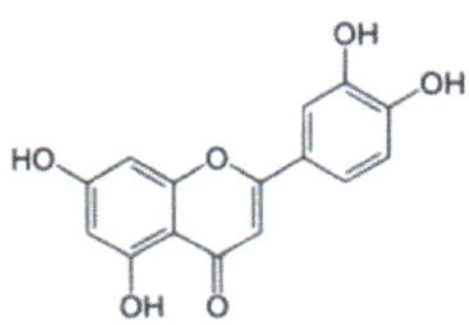

Luteolin

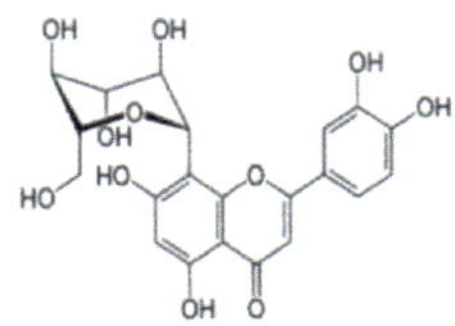

Orientin

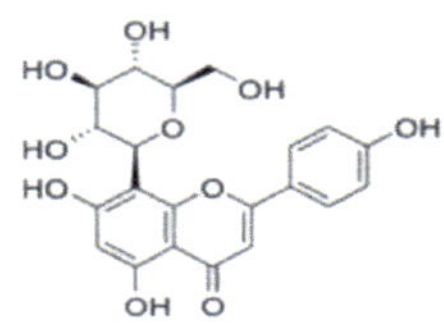

Vitexin

Anthocynadins in *Pterolobium hexapetalum*.

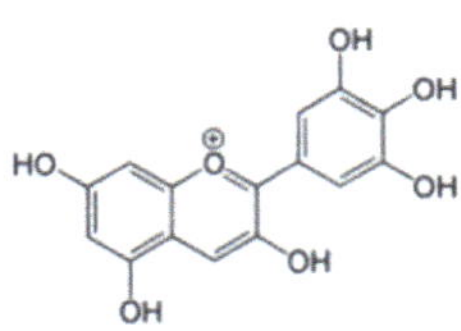

Delphidin

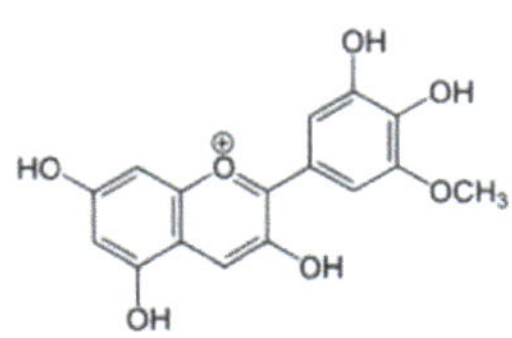

Petunidin

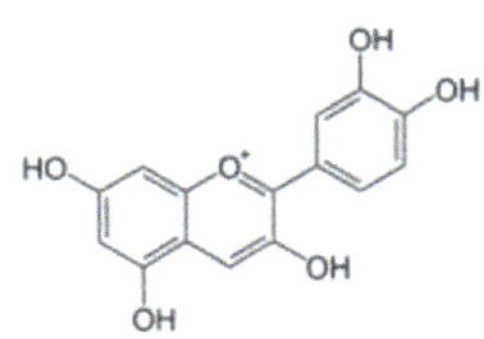

Cyanidin

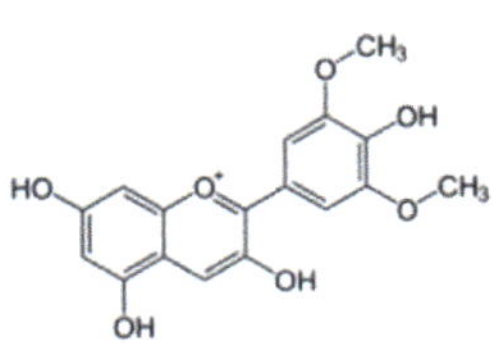

Malvidin

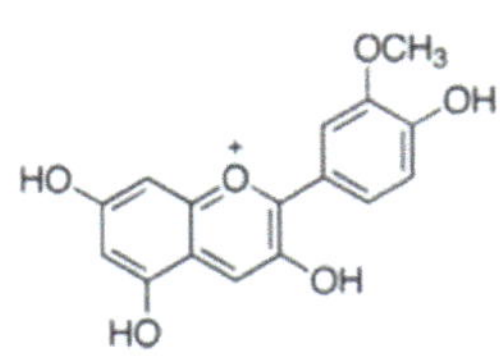

Peonidin

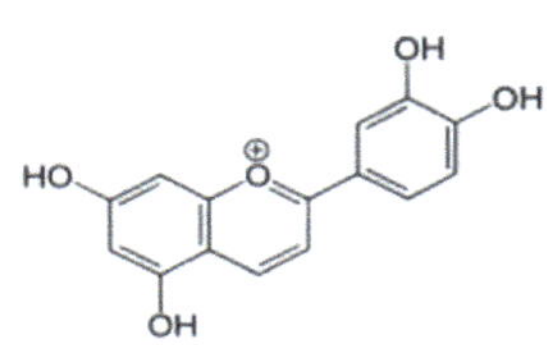

Luteolinidin

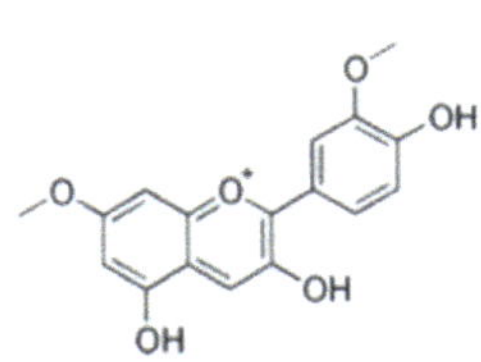

Rosinidin

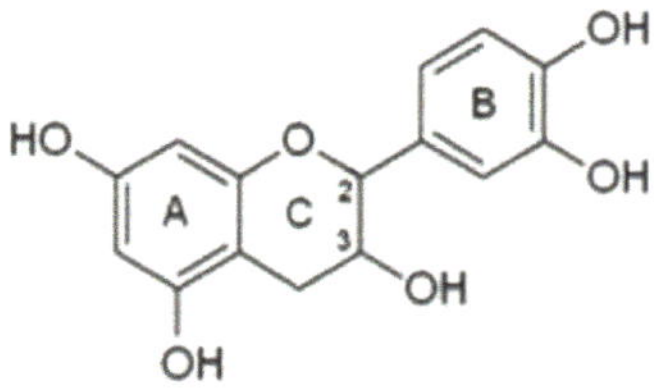

Catechin

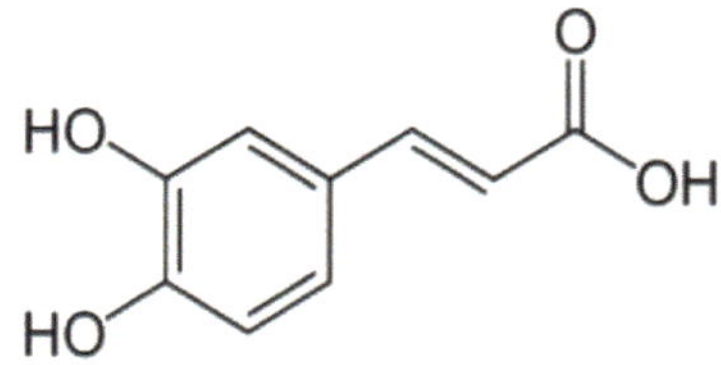

Caffeic Acid

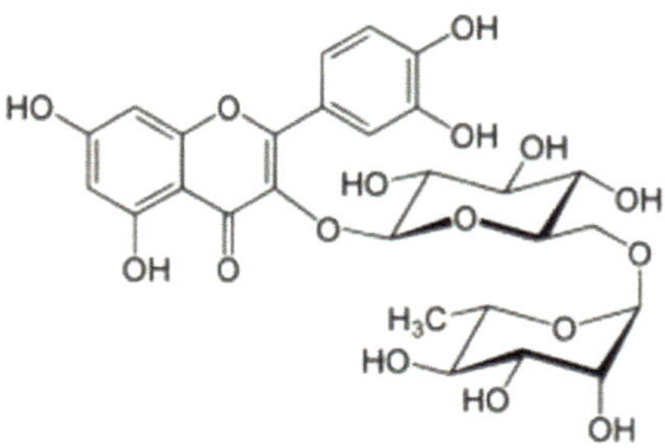

Rutin

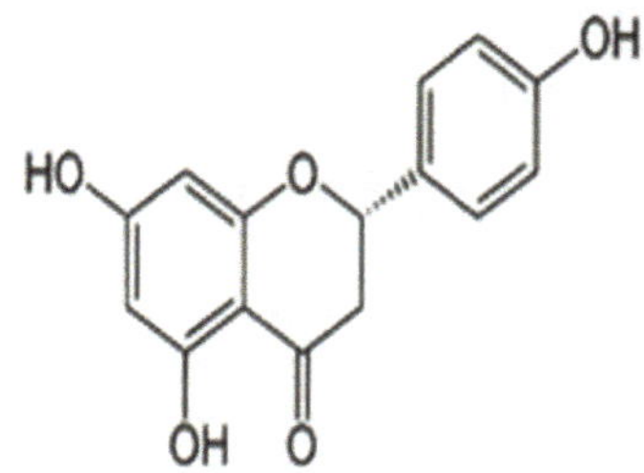

Naringenin

Many useful bioactive compounds have been identified by GC-MS analysis of *P. hexapetalum* seed methanol extract. The major compounds present in the seed extract are 4H-Pyran-4-one; 2,3-dihydro-3,5-dihydroxy-6- methyl-; 1,2,5-oxadiazol-3 carboxamide; 4,4′-azobis-; 2,2′- dioxide; bicyclo[2.2.1] heptan-2-one; 4,7,7-trimethyl-; semicarbazone; 1,2,3-benzenetriol; 2-methyl-4,5-tetramethylene-5-ethyl-2-oxazoline; a-D-glucopyranoside; a-D-glucopyranosyl; 9-octadecynoic acid; 9,12,15-octadecatrienoic acid (ZZZ)-; pyridine-3-carboxamide; 1,2-dihydro-4,6-dimethyl-; acetohydrazide; 2-(3-hydroxy-2-pentylcyclopentyl)-; 7-methyl-Z-tetradecan-1-ol acetate; methyl ester (Z,Z,Z)-; E-11-Hexadecenoic acid; ethyl ester; vitamin E; Z,Z,Z–1,4,6,9-nonadecatetraene; and 2–(7-heptadecynyloxy)tetra hydro-2H–Pyran (Kokila et al., 2016).

The GC-MS studies of *P. hexapetalum* aerial parts ethanolic extract showed the presence of aromatic alcohols, aromatic acid, acid anhydride, esters, amino acids, trihydric alcohol, selenium group, benzene derivative, bicyclic sesquiterpene, alkene, cyclohexane derivative, aromatic carboxylic acid, saturated fatty acid, aromatic hydrocarbon, and amide group. Further characterization studies also showed 19 chemical compounds, they are phenol (CAS), benzenesulfonic acid, 4-hydroxy-(CAS), cis-aconiti anhydride, 1-butanol, 3-methyl-, formate (CAS), L-serine, O-(phenylmethyl)-(CAS), 1,2,3-benzenetriol, á-selinene (CAS), benzene, 1- (1,5-dimethyl-4-hexenyl)-4-methyl- (CAS), (-)-caryophyllene oxide, 12-oxabicyclo [9.1.0] dodeca-3,7-diene, 1,5,5,8-tetramethyl-[1R- (1R*,3E,7E,11R*)]-, mome inositol, hexadecanoic acid (CAS), 2-hexadecen-1-ol, 3,7,11,15-tetramethyl-, [R-[R*,R*-(E)]]-(CAS), 9,12-octadecadienoic acid (Z,Z)- (CAS), hexadecadienoic acid, methyl ester (CAS), 24,25-dihydroxycholecalciferol, 4-hydroxymethyl [2.2.2] paracyclophane, 9-octadecenamide, and squalene (Sharmila et al., 2020).

36.3 PHARMACOLOGICAL USES

36.3.1 Antimicrobial

The in-vitro antimicrobial activity of *P. hexapetalum* aerial parts of different extracts against clinical isolates of *Bacillus subtilis, Escherichia coli, Pseudomonas aeruginosa, Lactobacillus, Staphylococcus aureus, Aspergillus niger*, and *Candida albicans* proved more efficient by all extracts than control drugs Ampicillin, Gentamycin for bacteria, and Nystatin for fungi. Among all extracts, aqueous, methanol, and acetone are the most effective on all organisms. The minimum inhibitory concentrations against selected isolates, the fruit extracts, are effective ranging between 0.312 and 0.625 mg/mL on all organisms compared to that of standard drugs (Kavitha et al., 2012, 2013; Yasodamma et al., 2014b; Kokila et al., 2016). Further studies have been carried out on the biosynthesized copper oxide nanoparticles (CuO NPs) using aqueous leaf extract and it was found to possess effective antibacterial activity on *S. aureus, B. subtilis*, and *E. coli* as compared to the standard drug Gentamycin. Green-synthesized CuO NPs has a promising potential role in the biomedical applications (Nagaraj et al., 2019).

36.3.2 Antioxidant Activity

The antioxidant activity of *P. hexapetalum* stem bark, leaf, flower, fruit, and seed with aqueous, petroleum ether, ethyl acetate, hexane, acetone, and methanol extracts showed equally effective action to that of the standard drug in concentration-dependent manner. Flower aqueous extract at 50 μg/mL showed effective inhibition with 74.70% of free radical scavenging activity, which is approximately equal to that of standard ascorbic acid 86.53%. The IC_{50} values ranged between 3.40 and 215.43 μg/mL. The aqueous extract of flower and fruit showed very high reducing ability when compared to methanol extract. The EC_{50} values of leaf, stem bark, flower and fruit aqueous, and methanol extracts ranges between 18 and 24 μg/mL (Yasodamma et al., 2014a; Kokila et al., 2016; Sathyanarayanan et al., 2017).

36.3.3 Acute Toxicity Study (LD_{50})

The acute toxicity study of *P. hexapetalum* leaf and fruit aqueous found that methanol extracts were devoid of mortality of the animals; in addition, no toxic symptoms were observed and also food and water intake were not affected during the study period. So, these extracts did not show any significant toxicity on Wister albino rats. Hence, 3500 mg/kg was considered as LD_{50} cut-off value. So, the doses selected for experiment as per OECD guidelines was 423 and was fixed up to a maximum of 140 mg/kg (1/25th of 3500 mg/kg) (Kavitha et al., 2014a,b).

36.3.4 Antidiarrhoeal Activity

The antidiarrhoel activity of *P. hexapetalum* leaf and fruit methanol and aqueous extracts on castor oil-induced diarrheal rats at the doses of 50 and 100 mg/kg b.wt reduced the total number of faces as well as delayed the onset of diarrhea in a dose-dependent manner in comparison to control drug Atropine. Enteropooling activity was also very effectively reduced with fruit aqueous extracts at 100 mg/kg b.wt. as 1.4 mL and 60.45% of inhibition than 1.64 mL and 53.67% with leaf methanol extracts to that of the control drug Atropine with 1.68 mL and 52.54% of intestinal fluid inhibition (Kavitha et al., 2014a).

36.3.5 Anti-Ulcer Activity

The anti-ulcer activity of *P. hexapetalum* leaf and fruit methanol and aqueous extracts at 50 and 100 mg/kg b.wt on pyloric ligation-induced ulcer model showed significant reduction in ulcer index between 2.34 and 3.98 mL, ulcer protection between 68.26% and 81.33%, gastric volume, free acidity, and total acidity as compared to the control group Omeprazole-treated rats. Histopathological observation of pyloric ligation gastric lesions in ulcer control group showed extensive damage to the gastric mucosa, oedema, and leucocytes infiltration of the submucosal layer. Rats that received pretreatment with leaf, fruit methanol, and aqueous extracts had comparatively better protection on the gastric mucosa and seen reduction in ulcer area, reduced or absence of submucosal oedema, and leucocytes infiltration. The leaf and fruit extracts has been shown to exert the cytoprotective effect in a dose-dependent manner (Kavitha et al., 2014b).

36.3.6 Antipyretic Activity

The in-vitro antipyretic activity of *P. hexapetalum* methanol and aqueous stem bark extracts against yeast-induced pyrexia in rats showed potential antipyretic activity. It was observed that methanol extract at a dose of 400 mg/kg body weight significantly elevated the body temperature of rabbit, showed maximum antipyretic activity than aqueous extract, and their effects are comparable to that of standard antipyretc drug paracetamol (Kavitha et al., 2017).

36.3.7 Anticancer Activity

Anticancer activity on human breast cancer cells with striking inhibitions, cell death, and the apoptotic activity of MDA-MB-231 cancer cell lines with *P. hexapetalum* aqueous leaf extract and with the biosynthesized copper oxide nanoparticles was reported. It was found that MDA-MB-231 cells proliferation was potentially inhibited by CuO NPs with an IC_{50} value of 30 μg/mL of the concentration. In contrast, anticancer efficiency of CuO NP through inducing oxidative cell damage and apoptosis by ROS generation in MDA-MB-231 breast cancer cells also causes cell death (Nagaraj et al., 2019).

36.3.8 Ovicidal, Larvicidal, and Pupicidal: Dengue Vector

The ovicidal, larvicidal, and pupicidal activity of *P. hexapetalum* leaf ethanol, hexane, and dichloromethane extracts work against the dengue vector *Aedes aegypti.* The ethanol extract at the time period of 72 h, at 800 mg/L showed the highest ovicidal, larvicidal, and pupicidal activity. Whereas, the dichloromethane and hexane extracts showed 83.8% and 85.6% of ovicidal activity at 72 h (Saranya and Elumalai, 2018).

36.3.9 Wound-Healing Activity

Wound-healing activity of *P. hexapetalum* leaf methanol, aqueous, petroleum ether, and ethyl acetate extracts on excision, incision, and dead space wound models showed the methanol extract exhibited 95% maximum wound contraction percentage, whereas, other extracts like petroleum ether exhibited 82%, aqueous 86%, and ethyl acetate 88%. Highest wound contraction percentage and lowest epithelialization period observed 16.33 days of 5% methanolic extract-treated group compared to control group and it was confirmed by histopathology results and scoring analysis. Maximum tensile strain 110.69% of incision and highest hydroxyproline 16.28 mg/g content of dead space wound model are comparable with standard 5% Neomycin (Sathyanarayanan et al., 2017).

KEYWORDS

- ***Pterolobium hexapetalum***
- **bioactive phytoconstituents**
- **pharmacological**
- **acute toxicity**
- **in vitro studies**

REFERENCES

Duraipandiyan, V.; Ayyanar, M.; Ignacimuthu, S. Antimicrobial Activity of Some Ethnomedicinal Plants Used by Paliyar Tribe from Tamil Nadu, India. *BMC Complement. Altern. Med.* **2006,** *6* (1), 1–7.

Ganesh, P.; Nagendrababu, G.; Sudarsanam, G. Pharmacognostical and Analytical Studies of Leaves of *Pterolobium hexapetalum* (Roth) Santapau & Wagh. *J. Pharm. Res.* **2014,** *8* (4), 541–547.

Kavitha, B.; Yasodamma, N. Pharmacognostic Studies of *Pterolobium hexapetalum* (Roth) Sant. and Wagh. *World J. Pharm. Res.* **2016,** *5* (5), 991–1017.

Kavitha, B.; Yasodamma, N.; Alekhya, C. In-Vitro Antibacterial and Phytochemical Studies of *Pterolobium hexapetalum* (Roth.) Sant. and Wagh. A Medicinal Plant of Nallamalai Hills, Andhra Pradesh. *J. Pharm. Res.* **2012,** *5* (5), 2750–2757.

Kavitha, B.; Yasodamma, N.; Alekhya, C. Antifungal Activity of *Pterolobium hexapetalum* (Roth.) Sant. and Wagh. *Indo Am. J. Pharm. Res.* **2013,** *3* (10), 8408–8414.

Kavitha, B.; Yasodamma, N.; Alekhya, C. Antidiarrhoeal Activity of *Pterolobium hexapetalum* (Roth.) Sant. and Wagh. Leaf and Fruit Extracts on Castor Oil Induced Diarrhea. *Indo Am. J. Pharm. Res.* **2014a,** *4* (08), 3355–3359.

Kavitha, B.; Yasodamma, N.; Chaithra, D. Antiulcer Activity of *Pterolobium hexapetalum* Leaf and Fruit Extracts on Pyloric Ligated Rats. *Indo Am. J. Pharm. Res.* **2014b,** *4* (1), 212–219.

Kavitha, B.; Yasodamma, N.; Reddy, E. S. Antipyretic Activity of *Pterolobium hexapetalum* (Roth.) Sant. and Wagh. Stem Bark Extracts. *J. Adv. Pharm. Educ. Res.* **2017,** *7* (3), 253–255.

Kokila, K.; Elavarasan, N.; Sujatha, V. Antioxidant, Antimicrobial and GC–MS Analysis of *Pterolobium hexapetalum* Seed Extracts. *Appl. Sci. Rep.* **2016,** *16* (3), 141–149.

Kumar, C. P.; Mohana, K. N. Phytochemical Screening and Corrosion Inhibitive Behavior of *Pterolobium hexapetalum* and *Celosia argentea* Plant Extracts on Mild Steel in Industrial Water Medium. *Egypt. J. Pet.* **2014,** *23* (2), 201–211.

Nagaraj, E.; Karuppannan, K.; Shanmugam, P.; Venugopal, S. Exploration of Bio-Synthesized Copper Oxide Nanoparticles Using *Pterolobium hexapetalum* Leaf Extract by Photocatalytic Activity and Biological Evaluations. *J. Clust. Sci.* **2019,** *30* (4), 1157–1168.

Pullaiah, T. *Encyclopaedia of World Medicinal Plants*, Vol. 4; Regency Publications: New Delhi; 2006, p 1638.

Samuel, J. K.; Andrews, B. Traditional Medicinal Plant Wealth of Pachalur and Periyur Hamlets Dindigul District, Tamil Nadu. *Indian J. Trad. Knowl.* **2010,** *9* (2), 264–270.

Sankara Rao, M. *Ethnobotanical Studies of Srikakulam District, A. P., India.* Ph.D. Thesis, Andhra University, 2010.

Saranya, J.; Elumalai, K. Ovicidal, Larvicidal and Pupicidal Activity of *Pterolobium hexapetalum* (Roth) Santapau & Wagh Extracts Against the Dengue Vector, *Aedes aegypti* (Linn.) (Diptera: Culicidae). *J. Emerg. Technol. Innov. Res.* **2018,** *5* (6), 716–733.

Sathyanarayanan, S.; Muniyandi, K.; George, E.; Sivaraj, D.; Sasidharan, S. P.; Thangaraj, P. Chemical Profiling of *Pterolobium hexapetalum* Leaves by HPLC Analysis and Its Productive Wound Healing Activities in Rats. *Biomed. Pharmacother.* **2017,** *95*, 287–297.

Sharmila, S.; Dhiva, S. M.; Akilandeswari, D.; Mownika, S.; Ramya, E. K. Pharmacognostic and Analyticalassessment for *Pterolobium hexapetalum* (Roth.) Santapau and Wagh. – A Dynamic Folklore Therapeutic Plant. *Int. J. Res. Pharm. Sci.* **2020,** *11* (3), 3338–3348.

Yasodamma, N.; Kavitha, B.; Alekhya, C. Antioxidant Activity of *Pterolobium hexapetalum* (Roth) Sant. and Wagh. *Int. J. Pharm. Pharm. Sci.* **2014a,** *6*, 143–147.

Yasodamma, N.; Kavitha, B.; Alekhya, C. Phytochemical Screening and Antibacterial Activity of *Pterolobium hexapetalum* (Roth) Sant. and Wagh Flower and Fruit Extracts. *Wkly. Sci. Res. J.* **2014b,** *2* (15), 2321–7871. https://indiabiodiversity.org/species/show/32301

CHAPTER 37

Bioactives and Pharmacology of *Aeschynomene aspera* L. and *Aeschynomene indica* L.

B. KAVITHA[1*] and N. YASODAMMA[2]

[1]*Department of Botany, Rayalaseema University, Kurnool 518007, Andhra Pradesh, India*

[2]*Department of Botany, Sri Venkateswara University, Tirupati 517502, India*

[*]*Corresponding author. E-mail: kavithab15@gmail.com*

ABSTRACT

The genus *Aeschynome* (Fabaceae) represents one of the important medicinal plant genera regarding its chemical constituents and pharmacological activities. *Aeschynome aspera* is a shrub and *Aeschynomene indica* is an erect shrubby or annual herbaceous plant. They are commonly used to treat biliary calculi, spermicidal, antidote to snake bite, to relieve cold, cough and fever, mosquito repellent, jaundice, leprosy relieve painful micturition, antidote to snake bite and to break uric acid calculi. In this review, we surveyed the latest findings on the bioactivities of different extracts of *A. aspera* and *A. indica* and isolated phytochemicals. Major groups of phytochemicals include phenols, flavonoids, and anthocyanidins, isolated from aerial parts of *A. aspera* and *A. indica. A. aspera* and *A. indica* and its bioactive compounds possess outstanding pharmacological properties, especially as antioxidant, antibacterial, antifungal, antihelmintic, hepatoprotective, antidiarhoeal, and wound healing, drugs, in addition to its and analgesic properties.

Bioactives and Pharmacology of Legumes. T. Pullaiah, PhD (Ed.)

37.1 INTRODUCTION

The genus *Aeschynome* belongs to the family Fabaceae. *Aeschynomene aspera* L. is an erect swampy species, with stout glabrous stems, full of white pith, subshrub 1.5–4 m long; leaves 9 × 1.5 cm; leaflets 30–35 pairs; and oblong. Flowers are yellow, in terminal or axillary racemes. Pods are 5.5 × 0.8 cm; 3–6 seeded; jointed, smooth, or echinate; indented along both margins; and seeds round and light brown. *A. aspera* is believed to have originated in Sub-Saharan Africa between Senegal and Sudan. It is widely distributed in the lands of western, central, north-eastern, and southern Africa. In 1986, it was introduced into the Philippines and then it has been grown across South and South-East Asia (AICAF, 1997), distributed around Bangladesh, Cambodia, East Himalaya, India, Java, Laos, Malaya, Myanmar, Nepal, Vietnam, Sri Lanka, Pakistan, Thailand, and the west Himalaya (https://indiabiodiversity.org/species/show/279283). Synonyms of this speces are *A. aquatica* Roxb. ex Steud., *A. lagenaria* Lour., *A. surattensis* Wight & Arn., *A. trachyloba* Miq., and *Hedysarum lagenarium* (Lour) Roxb) (http://www.theplantlist.org/tpl/record/ild-46362). Common names for *A. aspera* include Niti jiligu (Telugu), Sola (Odia), Shola, Kath shola (Bengali), Sola pith plant, pith plant (English), Laugauni (Hindi), Kadessum, Neerkadasam, Ponguchedi (Malayalam), or Attuneddi, Netti (Tamil).

Aeschynomene indica L. is a slender, branched annual herb, about 90 cm tall, identified by its erect stems, and leaves 3.8–5 cm long with 20 or 30 pairs of tiny, close-set, leaves 10 cm long; leaflets 41–61 pairs; and oblong. Flowers are in axillary or terminal racemes, sometimes solitary; peduncle glabrous; calyx tube glabrous; corolla yellow to flame-colored; pods 3–4 cm long; joints with 8–10 seeds; smooth or fine papillose on the face; and seeds round and blackish brown. This species prefers wet conditions and is often found along the borders of ditches or pools, or in wet cultivated land. It is also found in wet open places, sandy areas, and along roadsides. It is distributed within the tropics and sub-tropics, in regions with or without a pronounced dry season, at altitudes from 0 to 1000 m (Aruna, 2010). Synonyms of this species are *A. cachemiriana* Cambess., *A. diffusa* Willd., *A. glaberrima* Poir., *A. macropoda* DC., *A. montana* Span., *A. oligantha* Baker., *A. pumila* L., *A. punctata* Steud., *A. quadrata* Schum. & Thonn., *A. roxburghii* Spreng., *A. subviscosa* DC., *A. viscidula* Willd., *Hedysarum neli-tali* Roxb., and *Smithia aspera* Roxb. (https://indiabiodiversity.org/species/show/279284).

In India, it is vernacularly known as Indian jointvetch, Didhen, Phulan, Chhuimui, and Laugauni (Hindi); Nalabi (Marathi); Sola (Nepali); Chatai, kitai, Kitaichchi, Netti, and Takkaippuntu (Tamil); Neli-tali, Nelitali, and Nellittali (Malayalam); Jeeluga, Tella jeeluga, and Bendu (Telugu); Bendu-kasa and Bendu kasa (Kannada); Kath shola (Bengali); Surlo (Oriya); Kuhila (Assamese); Virginische Schampflanze (Germany); Dinding, Gedeyan, Katisan, Lorotis, and Peupeuteuyan; Tis (Indonesia); Piñata modesta bastarda (Italy); Kusanemu (Japan); Makhiyang lalaki (Philippines); Sano haag kai (Thailand); Angiquinho, Maricazinho, Papquinha, and Pinheirinho (Brazilian Portuguese) (*Aeschynomene indica*. Germplasm Resources Information Network (GRIN)), Budda pea, Curly indigo, Joint vetch, Northern jointvetch, and Sensitive Malayan vetch (English); Anil rizado (Spanish); and Eschynomene (French) (https://www.cabi.org/ISC/datasheet/3451).

This species has a wide range of geographical distribution but its native location is not clear. The species has been introduced widely covering many parts of the Africa, Australia, Asia, and Southeastern United States. The natural habitat of this species made it to inhabit many islands including the Pacific, the Caribbean, and the Indian Ocean, as well as Fiji, the Society Islands, Micronesia, Puerto Rico, Mauritius, and Réunion (*Aeschynomene indica*. Germplasm Resources Information Network (GRIN); United States Department of Agriculture (USDA)). It is also introduced in South America (*Aeschynomene indica*. Pacific Island Ecosystems at Risk (PIER) USFS).

The species of *A. aspera* has been traditionally used as a curative agent in colic, jaundice, and poisoning (Nadkarni, 2003). In Ayurveda system, the aerial parts of the plant have been used to cure different ailments such as to relieve painful micturition and to break uric acid calculi (Rajan and Chezhiyan, 2002), spermicidal (Asolkar et al., 2010), antidote to snake bite, mosquito repellent (Burkill, 1995), to relieve cold, and cough and fever. Also, the young shoot powder is used to increase semen consistency (Panda and Misra, 2011). The paste prepared from the roots is used to treat mumps (Rajendar, 2010). In Siddha medical system, the species is called Aatrunetti, where the leaves are used to cure pains and swellings (Yoganarasimhan, 2001). The root is used to treat jaundice by the Oraon tribe (Pal and Jain, 1998). The Kani tribe uses it as an antidote against snake bites (Ayyanar and Ignacimuthu, 2005).

Ayurvedic Charak samhita explores *A. indica* as Ashanibhed and is used to treat biliary calculi; and is used as spermicidal (Gillett et al., 1971). Oraon tribes use the root as a cure for jaundice (Pal and Jain, 1998); its leaves are edible and are used in the category of genital stimulant or depressant (Burkill, 1985). In Ayurveda and Siddha, this species is called as Kidaichi (Sadai poondu) where the leaves are used for treating leprosy (Jha and Goel, 2004). The Chenchu tribe of Nallamala hills uses this plant for curing kidney troubles (Sambamurthy, 2005).

The plant also is used as charcoal for gun powder (Gillett et al., 1971). It is used as a poison or repellent, fodder, for composting, manuring, hunting and fishing apparatus, household, and domestic and personnel items (Burkill, 1985). In Bengal, *A. indica* finds an important place where quite a good number of art forms are prepared and also exported. Locally known as sola, it is utilized for decorating the idols of Goddesses Durga and Kali. Sola items, like mukut, are worn by both the bride and the groom in Bengal (Jha and Goel, 2004).

37.2 BIOACTIVES

Methanolic extracts of leaves, flowers, and fruits, through phytochemical screening, showed that flavonoids, phenols, and terpenoids are common in all parts, and that anthocyanidins are present in the flowers of *A. aspera*. and *A. indica*. Alkaloids, steroids, and anthraquinones are totally absent in all parts. The flavones are mostly present in all parts of *A. aspera* and *A. indica*. Phenols and terpenoids are exclusively present in all parts of the two species; tannins are absent in *A. indica* and found in *A. aspera* flowers and fruits. Lignins are absent in *A. aspera* and found in *A. indica* flowers and fruits. Indoles are absent in *A. aspera* and present in the leaves of *A. indica*. Glycosides are absent in *A. aspera* and present in the fruits of *A. indica* (Aruna, 2010, 2012a).

Large number of compounds of various phytogroups has been isolated from aerial parts of *A. aspera* and *A. indica*. Major classes include phenols, flavonoids, and anthocyanidins. The phenolic compounds, namely, caffeic acid, phloroglucinol, α-resorcyclic acid, β-resorcyclic acid, neo-chlorogenic acid, protocatechuic acid, homo-protocatechuic acid, trans-p-coumaric acid, cis-ferulic acid, cis-sinapic acid, trans-sinapic acid, cinnamic acid, gentisic acid, m-hydroxy benzoic acid, vanillic acid, syringic acid, melilotic acid, salicylic acid, iso-chlorogenic acid, p-hydroxy benzoic acid, coumarin, aesculetin, scopoletin, trans-sinapic acid, phloretic acid, and o-coumaric acid have been identified. Apart from the 26 compounds, 10 unidentified compounds were also reported. Among them, five are in *A. aspera* and five are in *A. indica* leaves, flowers, and fruits parts. The common compound in all parts is salicylic acid. The common compound in leaves and flowers are coumarin and cis-ferulic acid. The compounds specific in flowers and fruits are homo-protocatechuic acid, 0-coumaric acid, p-hydroxy benzoic acid, and trans-sinapic acid. Phloretic acid, α-resorcyclic acid, m-hydroxy benzoic acid, aesculetin, vanillic acid, melilotic acid, and syringic acid are the compounds found only in leaves. Chlorogenic acid, phloroglucinol, trans-p-coumaric acid, and cis-sinapic acid are found in flowers only. Caffeic acid and cinnamic acid are present only in fruits (Aruna, 2010, 2012b).

Phenolic compounds in *Aeschynome aspera* and *A. indica*.

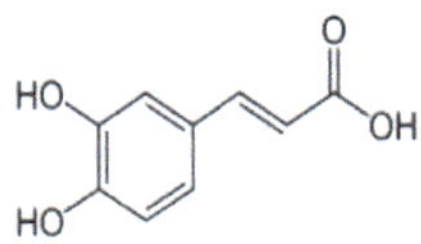

Caffeic Acid

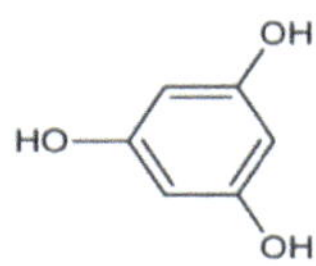

Phloroglucinol

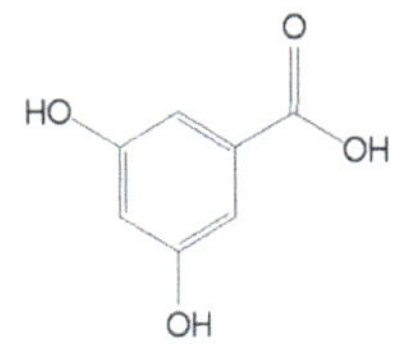

α-Resorcyclic Acid

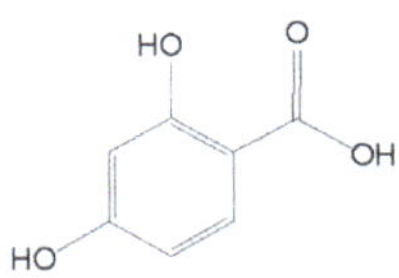

β-resorcylic Acid

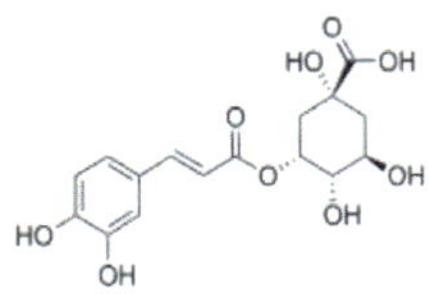

Neo-Chlorogenic Acid

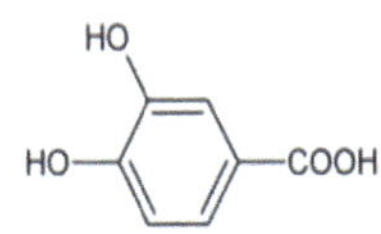

Protocatechuic Acid

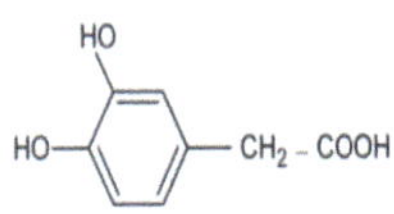

Homo-Protocatechuic Acid

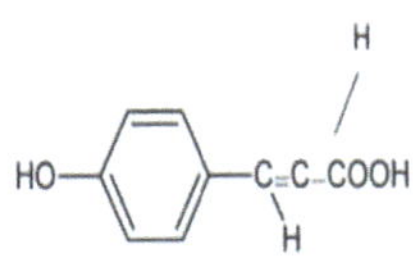

Trans-p-Coumaric Acid

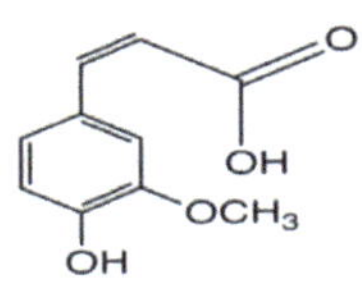

Cis-Ferulic Acid

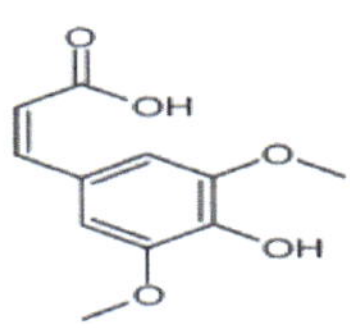

Cis-Sinapic Acid

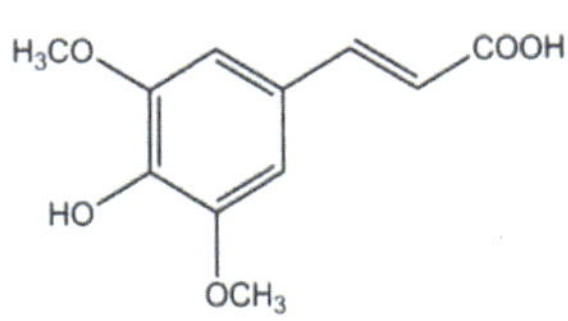

Trans-Sinapic Acid

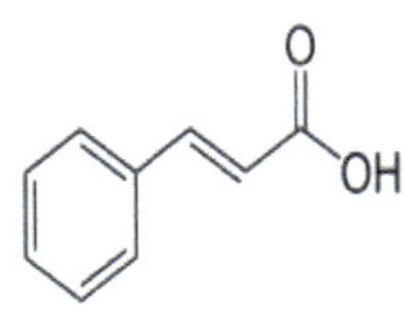

Cinnamic Acid

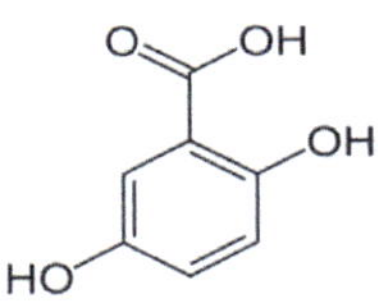

Gentisic Acid

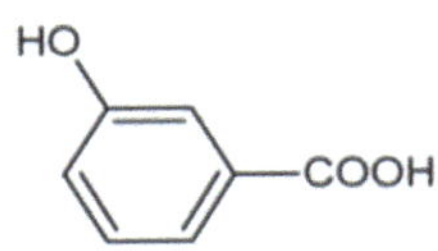

m-Hydroxy Benzoic Acid

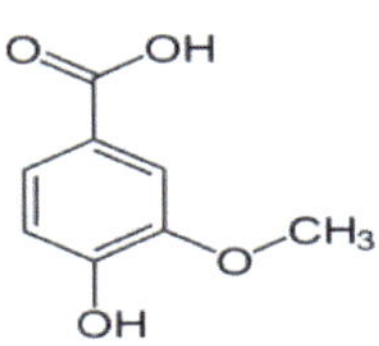

Vanillic Acid

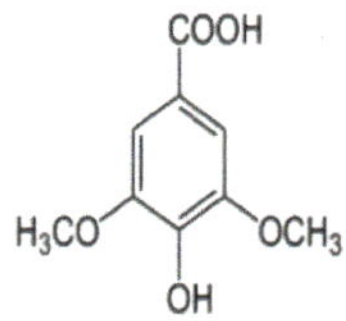

Syringic Acid

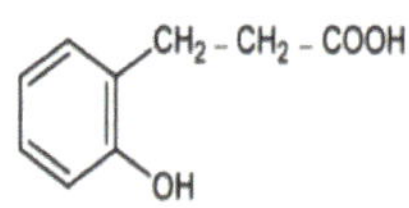

Melilotic Acid

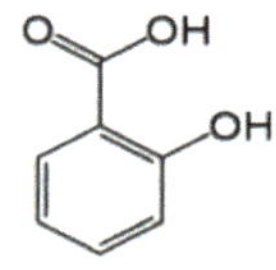

Salicylic Acid

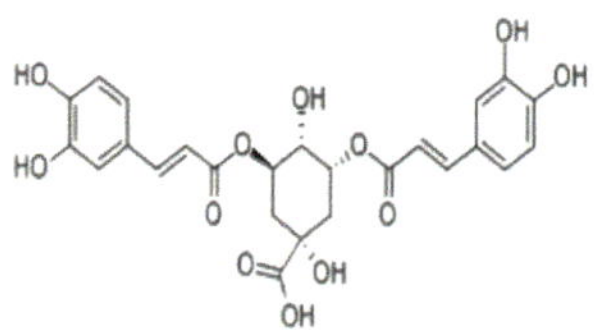

Iso-Chlorogenic Acid

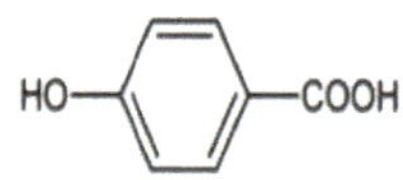

p-Hydroxy Benzoic Acid

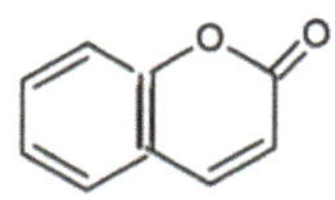

Coumarin

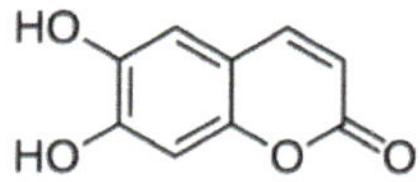

Aesculetin

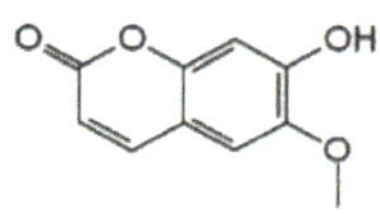

Scopoletin

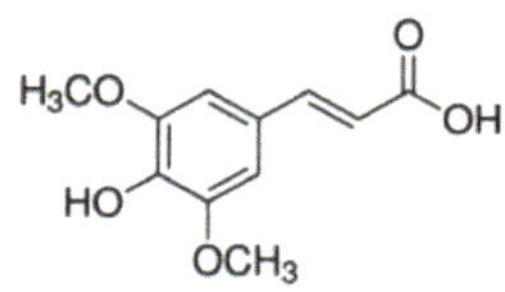

Trans-Sinapic Acid

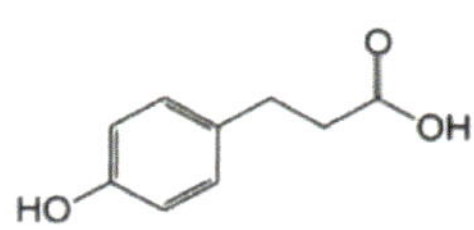

Phloretic Acid

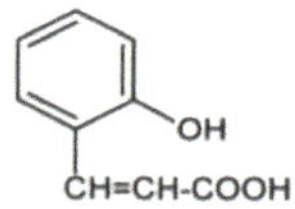

o-Coumaric Acid

Several flavonoid compounds have been analyzed and identified from the aerial parts of *A. aspera* and *A. indica.* A total of six were identified in leaves, flowers, and fruits. Myricetin, quercetin, kaempferol, apigenin, orientin, rutin, and only one unidentified compound in *A. indica* leaves were identified (Aruna, 2010, 2012b).

Anthocyanidins identified in *A. aspera* and *A. indica* are delphidin, malvidin, and peonidin from all parts (Aruna, 2010; Aruna et al., 2012b).

37.3 PHARMACOLOGY

37.3.1 Wound-Healing Activity

The role of *A. indica* leaf extracts in wound healing revealed positive results. In a study conducted by Lei et al. (2019), it was observed that the *A. indica*

Flavonoids in *Aeschynome aspera* and *A. indica*.

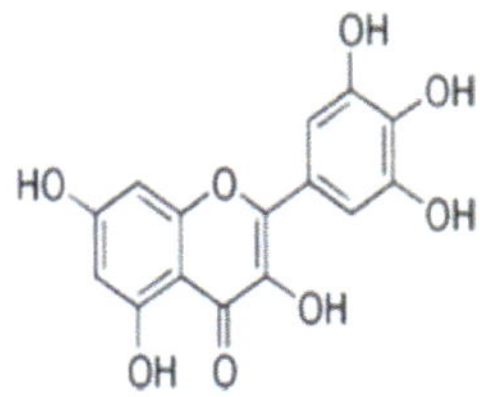

Myricetin

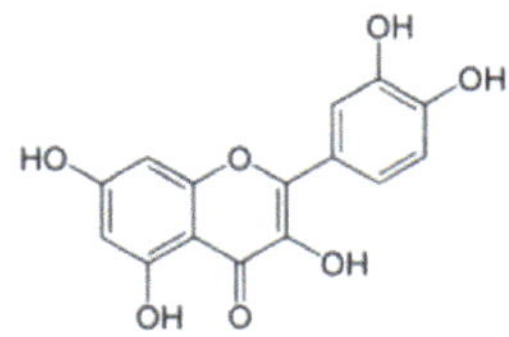

Quercetin

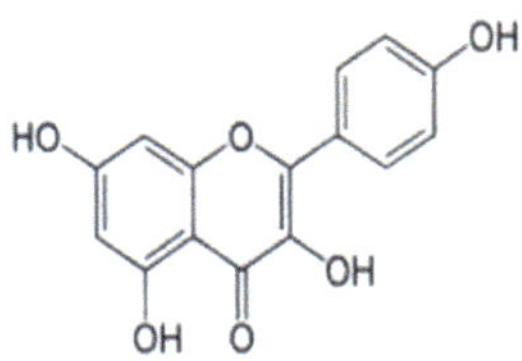

Kaempferol

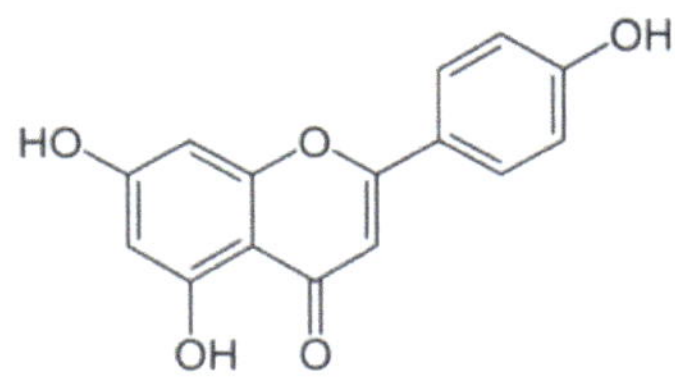

Apigenin

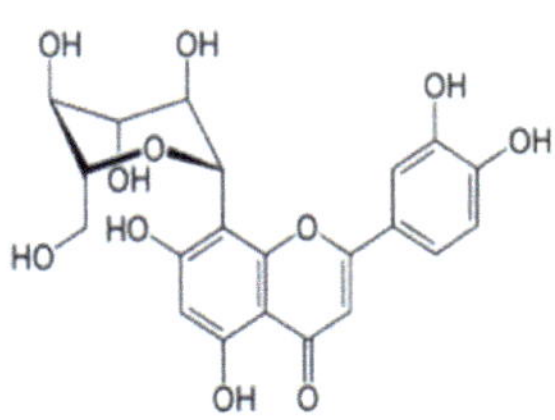

Orientin

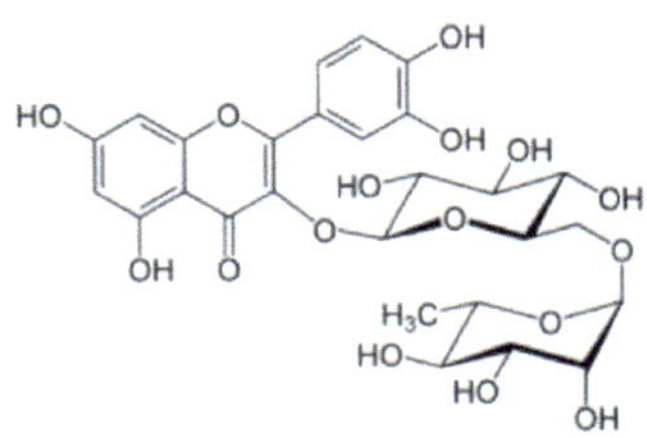

Rutin

Anthocynadins in *Aeschynome aspera* and *A. indica*.

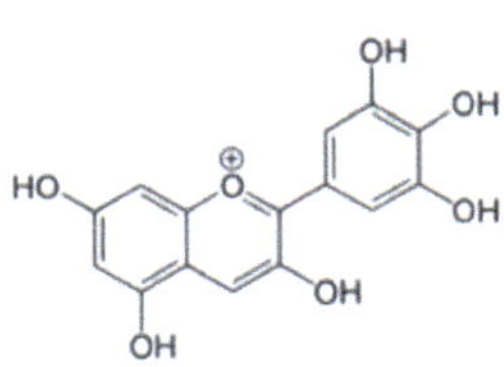

Delphidin

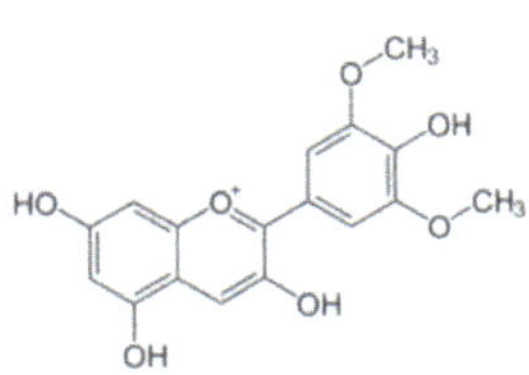

Malvidin

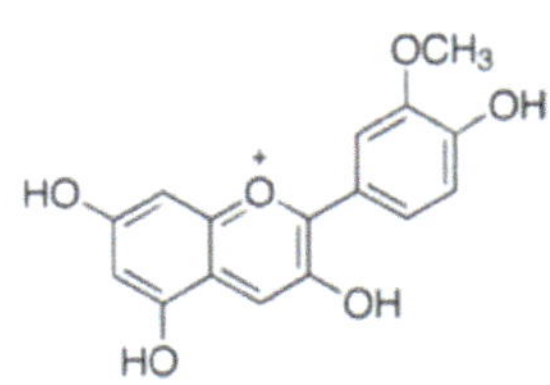

Peonidin

ethanol extract has showed a positive result in wound-healing activity (Lei et al., 2019). The extracts have shown epithelialization within 16 days whereas the vehicle groups have shown in 23 days. Other fractions such as the dichloromethane extract have showed no activity. No studies have been reported on the wound-healing activity of *A. aspera* species.

37.3.2 Antioxidant Activity

A. aspera and *A. indica* have shown effective antioxidant activity. The extracts of leaves, flowers, fruits, and roots from both species revealed the antioxidant activity. The activity is mainly due to the presence of phytoconstituents like phenols, flavanoids, alkaloids, steroids, and terpenoids. Furthermore, the silver nanoparticles of *A. aspera* leaf showed enhanced antioxidant activity when compared to leaf extracts alone (Bharathi and Rao 2015; Iswaryalakshmi and Akilabalamurugan, 2019; Vishnuvardhan et al., 2020a, 2020b).

37.3.3 Antihelmintic Activity

The antihelmintic activity of *Aeschynomene* species namely *A. indica* and *A. aspera* leaves of both species of 5, 10, and 15 mg along with standard drug Albendazole as positive control was studied. The antihelmintic activity of *A. indica* extracts showed paralysis time as 32–19 min and death time as 45–25 min, whereas the standard drug Albendazole showed paralysis time as 91–34 min and death time as 110–41 min. Similarly, *A. aspera* alcohol and methanol extracts revealed more effective results than the *A. indica*. The extracts of *A. aspera* showed paralysis time as 10–5 min and death time as 20–11 min compared to Albendazole. Therefore, the extracts of *A. aspera* and *A. indica* contain phytoconstituents that are effective in the killing of worms (Alekhya et al., 2013; Imtiaz et al., 2020).

37.3.4 Antibacterial Activity

Both the species *A. aspera* and *A. indica* on *Bacillus subtilis*, *Escherichia coli*, *Pseudomonas aeruginosa*, and *Staphylococcus aureus* with alcohol, methanol, cold water, and hot water extracts show antibacterial activity in the order of alcohol < methanol < cold water < hot water. More or less similar results have been obtained from both the species in terms of zone of inhibition. The silver nanoparticles of both the species exhibited antibacterial activity more effectively than the leaf extracts alone (Aruna et al., 2012a; Iswaryalakshmi and Akilabalamurugan, 2019; Vishnuvardhan et al., 2020a, 2020b).

37.3.5 Hepatoprotective Activity

The hepatoprotective activity of benzene and alcoholic root extracts of *A. aspera* was investigated in rats induced with carbon tetrachloride. The extracts did not produce any mortality even at 5000 mg/kg, while LD_{50} of benzene and alcoholic extracts were found to be safe at 100 mg/kg and 200 mg/kg, respectively. In benzene and alcoholic extracts-treated animals, the toxicity effect of carbon tetrachloride was controlled significantly by restoration of the serum bilirubin and enzyme levels as compared to the normal and standard drug silymarin-treated groups. Histology of liver in the treated animals with the extracts showed the presence of normal hepatic cords and absence of necrosis and fatty infiltration, which further evidence the hepatoprotective activity (Kumaresan and Pandae, 2011). There is no report on the hepatoprotective activity of *A. indica* species.

37.3.6 Antidiarrhoeal Activity

The antidiarrhoeal activity of *A. aspera* leaf extract showed significant results in the inhibition of defecation impulse by 46.01% and 71.68% at the doses of 250 mg/kg and 500 mg/kg, respectively. The positive control loperamide showed 88.59% inhibition of defecation at a dose of 25 mg/kg body weight (Imtiaz et al., 2020). No studies have been reported on the antidiarrhoeal activity of *A. indica* species.

37.3.7 Analgesic Activity

The analgesic activity of *A. aspera* with the leaf extract showed a significant inhibition of writhing impulse by 36.23 % and 61.32 % at the doses of 250 mg/kg and 500 mg/kg, respectively; whereas, it was observed that the positive control diclofenac Na exerted 70.03 % writhing inhibition at a dose of 25 mg/kg body weight on rats (Imtiaz et al., 2020). No studies have been reported on the analgesic activity of *A. indica* species.

KEYWORDS

- **Fabaceae**
- ***Aeschynome aspera***
- ***Aeschynomene indica***
- **bioactive phytoconstituents**
- **pharmacology**
- **in vitro studies**

REFERENCES

Aeschynomene indica. Germplasm Resources Information Network (GRIN). Agricultural Research Service (ARS), United States Department of Agriculture (USDA). Retrieved 21 Jan 2018.

Aeschynomene indica. Pacific Island Ecosystems at Risk (PIER). USFS.

Aeschynomene indica. USDA PLANTS.

AICAF. *Weeds in the tropics*. Association for International Cooperation of Agriculture and Forestry, Japan, 1997.

Alekhya, C.; Yasodamma, N.; Chaithra, D.; Job Roger Binny, A. Anthelmintic Activity of *Aeschynomene aspera* and *Aeschynomene indica*. *Int. J. Pharm. Pharma. Sci.* **2013,** *5* (2), 386–388.

Aruna, C. Studies on Phytochemical Analysis and Antimicrobial Activity of Leguminous Herbal Medicinal Plants [*Aeschynomene aspera* L., *A. indica* L., *Macroptilium atropurpureum* (DC.) Urban, and *Acacia caesia* (L.) Willd.] Ph.D. Thesis, Sri Venkateswara University, Tirupathi, 2010.

Aruna, C.; Suvarnalatha, A.; Alekhya, C.; Chaithra, D.; Yasodamma, N.; Meerasaheb, C. Phytochemical and Antimicrobial Studies of a Herbal Medicinal Plant *Aeschynomene aspera* L. leaf extracts. *J. Pharm. Res.* **2012a,** *5* (4), 1827–1837.

Aruna, C.; Chaithra, D.; Alekhya, C.; Yasodamma, N. Pharmacognostic Studies of *Aeschynomene indica* L. *Intern. J. Pharm. Pharma. Sci.* **2012b,** *4* (4), 76–77.

Asolkar, L. V.; Kakkar, K. K.; Charke, O. J. *Second Supplement to Glossary of Indian Medicinal Plants with Active Principles*; National Institute of Science Communication and Information Resources: New Delhi, 2010.

Ayyanar, M.; Ignacimuthu, S. Medicinal Plants Used by the Tribals of Tirunelveli Hills, Tamil Nadu to Treat Poisonous Bites and Skin Diseases. *Indian J. Trad. Knowl.* **2005,** *4* (3), 229–236.

Bharathi, M. P.; Rao, D. M. Evaluation of In Vitro Antioxidant Activity of *Aeschynomene indica*. *J. Pharm. Res.* **2015,** *9* (1), 21–26.

Burkill, H. M. *The Useful Plants of West Tropical Africa*, 2nd ed. Vol. 1 (Families A-D), Royal Botanic Gardens: Kew, 1985; pp 353–354.

Burkill, H. M. *The useful Plants of West Tropical Africa*; Royal Botanical Gardens: Kew, 1995.

Gillett, J. B.; Polhill, R. M.; Verdcourt, B. *Flora of Tropical East Africa. Leguminosae (Part 3) Subfamily Papilionoideae*; Royal Botanic Gardens: Kew, 1971.

http://www.plantsoftheworldonline.org/taxon/urn:lsid:ipni.org:names:472655-1.

http://www.theplantlist.org/tpl/record/ild-46362

https://indiabiodiversity.org/species/show/279283.

https://indiabiodiversity.org/species/show/279284.

https://www.cabi.org/ISC/datasheet/3451

Imtiaz, H.; Hossain, A.; Islam, F.; Sultana, R.; Rahman, M. M. Bioactivities of *Aeschynomene aspera* (Fabaceae) Leaf Extract. *Bangladesh Pharma. J.* **2020,** *23* (2), 109–116.

Iswaryalakshmi, K.; Akilabalamurugan, M. Phytochemical Screening Antioxidant and Antimicrobial Activity of *Aeschynomene aspera* Linn Root Extract. *J. Pharm. Biolog. Sci.* **2019,** *7* (1), 29–33.

Jha, V.; Goel, A. Utilization of Aquatic Biota—A Reference to the Folk Custom and Rituals in Mithila Region (North Bazar), India, 2004, p 370.

Kumaresan, P.; Pandae, V. Hepatoprotective Activity of *Aeschynomene aspera* Linn. *Pharmacologyonline* **2011,** *3*, 297–304.

Lei, Z. Y.; Chen, J. J.; Cao, Z. J.; Ao, M. Z.; Yu, L. J. Efficacy of *Aeschynomene indica* L. Leaves for Wound Healing and Isolation of Active Constituent. *J. Ethnopharmacol.* **2019,** *228* (10), 156–163.

Nadkarni, K. M. *Indian Materia Medica*; Popular Prakashan: Mumbai, 2003; pp 52–54.

Pal, D. C.; Jain, S. K. *Tribal Medicine*; Naya Prokash, 1998; p. 51.

Panda, A.; Misra, M. K. Ethnobotanical Survey of Some Wetland Plants of South Orissa and Their Conservation. *Indian J. Trad. Knowl.* **2011,** *10* (2), 296–303.

Rajan, B.G.; Chezhiyan, N. Strength and Wealth of Therapeutic Medicinal Plants in India. In Role of Biotechnology in *Medicinal and Aromatic Plants, Special Volume on Diseases. Vol. 6*; Khan, I. A.; Khanum, A., Eds.; Ukaaz Publications: Hyderabad, 2002; pp 286–293.

Rajendar, A. Phytotherapeutic Methods Used by Traditional Healers of Eturnagaram Mandal, Warangal, Andhra Pradesh, India. *Ethnobotanical Leaflets* **2010,** *14*, 361–365.

Sambamurthy, A. V. S. S. *Taxonomy of Angiosperms*; I. K. International Pvt. Ltd. S-25, 2005; p 337.

Vishnuvardhan, K.; Bommana, K.; Nimmanapalli, Y. Phytochemical Screening, Silver Nanoparticle Synthesis and Antibacterial Studies on the Leaves of *Aeschynomene aspera* L. *Int. J. Pharm. Sci. Res.* **2020a,** *11* (1), 451–463.

Vishnuvardhan, K.; Kavitha, B.; Nimmanapalli, Y. Synthesis of Silver Nanoparticles Using *Aeschynomene indica* L. Aqueous Leaf Extract and Evaluation of Its Antibacterial Activity. *J. Biochem. Technol.* **2020b,** *11* (1), 1–3.

Yasodamma, N.; Santhosh Kumar, M.; Paramageetham, C. H. Threat Assessment (IUCN Categorization) for Ethnomedicinal Plants Used by Chenchu Tribe of Gundlabramheswaram in Nallamalai Hills in Andhra Pradesh. *Ethnobotany* **2009,** *21* (1–2), 51–60.

Yoganarasimhan, S. N. *Medicinal Plants of India, Vol. 2 Tamilnadu.* Regional Science Institute (Ayur), Cyber Media: Bangalore, 2001.

Index

A

Abrus precatorius L.
  bioactives
    leaves, 80
    roots, 81
    seeds, 81–82
  pharmacology
    abortifacient activity, 90–91
    analgesic activity, 83
    antiallergic activity, 93
    antiarthritic activity and, 83
    anticancer activity, 88–89
    anticonvulsant effects, 83–84
    antidepressant efficacy, 84
    antidiabetic activity, 87
    antiepileptic activity, 85
    antifertility activity, 90
    antihelminthic activity, 92
    anti-implantation effect, 91
    anti-inflammatory activity, 82–83
    antimalarial activity, 93
    antimicrobial effects, 85–87
    antioxidant activity, 87–88
    antiserotonergic activity, 94
    antispermatogenic effect, 92
    antiviral activity, 87
    antiyeast activity, 85
    blood cells, 89–90
    cardiovascular protective activity, 92
    Estrous cycle disruption effect, 91
    hepatoprotective efficacy, 88
    immunomodulating activity, 89
    memory enhancer activity, 84
    nephroprotective activity, 85
    neuromuscular effects, 84
    neuroprotective effect, 83
    smooth muscle stimulant activity, 93
    wound healing activity, 94
Acacia ataxacantha DC
  bioactives, 38
  chemical structure, 39
  pharmacological activities
    antibacterial activities, 40
    antidiabetic activities, 41
    antifungal activities, 41
    antiinflammatory activities, 41–42
    antioxidant activities, 42
    antipyretic activities, 42
    gastro-intestinal role, 42–43
    toxicity study, 39–40
Acacia auriculiformis
  activity against worms, 58–59
  antibacterial activity, 56–57
  antidiabetic activity, 54
  antifungal activity, 57–58
  antiinflammatory activity, 56
  antilarvicidal activity/antimalarial
    activity, 59–60
  antimutagenic and anticarcinogenic
    activity, 55
  antioxidant activity, 53–54
  carbohydrates, 48
  CNS suppression activity, 55
  fatty acids, 48–49
  flavonoids and tannins, 49
    chemical structure, 50
  glycosides, 51
    chemical structure, 52
  hepatoprotective activity, 56
  memory enhancing activity, 55
  pharmacology, 51–53
  sperm immobilizing, 60
  spermicidal activity, 60
  wound healing activity, 55
Acacia dealbata, 373–374
  bioactives, 374–375
  pharmacology
    antimicrobial activity, 376–377
    antioxidant activity, 377
    cytotoxic activity, 377
Acacia ferruginea, 279–280
  bioactives, 280

pharmacology
antidiabetic activity, 283
anti-hemorrhoidal activity, 283
anti-inflammatory and analgesic effect, 282–283
antimicrobial and antifungal activity, 282
antioxidant activity, 280–282
antitumor activity, 282
Acacia leucophloea, 296
bioactives, 296–297
pharmacology
anti-inflammatory efficacy, 299
antioxidant and free radical scavenging, 298
gastrointestinal and respiratory activities, 298
Acacia macrostachya, 295–296
bioactives, 296–297
pharmacology
antidiabetic effectiveness, 299
antioxidant activity, 299
antiplasmodial activity, 299
antiradical activity, 299
Acacia mearnsii, 374
bioactives, 376
pharmacology, 378
Acacia modesta Wall., 287–288
bioactives
cyclitols, 288
nonprotein amino acids, 288
terpenoids, 288
pharmacology
analgesic activity, 291
antibacterial effects, 288–290
antifungal activity, 290
anti-hyperglycemic activity, 290–291
anti-inflammatory activity, 291
antioxidant activity, 292
antiplatelet activity, 291
brine shrimp cytotoxicity, 292
hemagglutination activity, 292
spasmolytic activity, 292–293
Adenanthera pavonina L.
bioactive phyto-constituents, 66–67
pharmacology
analgesic activity, 70–71
anthelmintic activity, 75
antibacterial and antifungal activity, 68
anticancer activity, 69–70
and anticonvulsant activity, 71
antidiarrheal activity, 72
and antihyperglycaemic activity, 72–74
anti-inflammatory and, 70–71
antinociceptive activity, 74–75
antioxidant and antimalarial activity, 68–69
antiviral activity, 71
blood cholesterol level, 75
CNS depressant, 71
hepatoprotective activity, 72
hypoglycaemic, 72–74
raw seeds, 65
seeds, 65
Aeschynome (Fabaceae), 485, 486
Aeschynome aspera, 485
Aeschynomene indica, 485, 486
bioactives, 488–490
Charak samhita, 487
pharmacology
analgesic activity, 493
antibacterial activity, 492–493
antidiarrhoeal activity, 493
antihelmintic activity, 492
antioxidant activity, 492
hepatoprotective activity, 493
wound-healing activity, 490–492
Albizia lebbeck (L.) Benth., 339–340
ethnobotany, 340
pharmacology, 341
antiallergic activity, 341–343
antianaphylactic activity, 348
antiarthiritic activity, 348–349
antiasthmatic activity, 347–348
antibacterial activity, 343
anti-inflammatory activities, 344
antioxidant activity, 344–345
anti-PCA activity, 345
antitumor activity, 345
antivenom properties, 343
immunomodulator activity, 345–346
nootropic activity, 347
ovicidal and adulticidal activity, 344
Parkinson's disease, 343
spermicidal activity, 346–347
phytochemical investigation, 340–341
Alhagi maurorum, 101
bioactives, 102–103

pharmacological activities
analgesic activity, 108
antibacterial and antiheamolytic activity, 107–108
antidepressant activity, 108
antidiarrhoeal activity, 109
antifungal activity, 108
antihyperglycemic effect, 109
anti-inflammatory activity, 108
antiplatelet activity, 109
antiproliferative activity, 109
anti-rheumatic activity, 107
anti-ulcer activity, 108
antiviral activity, 109
hepatoprotective activity, 103
phytoacaricidal activity, 107
wound healing activity, 107
Astragalus membranaceus (Fisch.), 113
bioactives
Astragalans I, II, and III, 119
chemical analysis, 119
dried sprouts, 118
ethanolic extract, 114
UPLC-QTOF-MS, 120
flavonoid derivatives, 118
pharmacological activities
allergic dermatitis, 124
antiarthritic effect, 123
anti-asthmatic effects, 124–125
anticancer, 121
antifatigue effect, 122–123
antimicrobial activity, 125
antioxidant effect, 124
anti-ulcerative colitis, 123
anxiolytic and anti-inflammatory effects, 123–124
cardioprotective effect, 121, 122
diabetic complications, 121–122
immunomodulatory activity, 120–121
neuroprotective activity, 121
reproductive toxicity, 124
toxicity of AS-IV, 122
saponins from, 115–117

B

Bauhinia purpurea L.
ayurvedic properties, 137
famous formulations, 138
bioactives, 130
bauhiniastatins (1–4), 131–132
purpuroside, 132
gastric cancer
predictor of, 138
pharmacology
and analgesic activity, 133–134
antiarthritic, 133–134
anticancer activity, 136–137
antifungal activity, 132–133
anti-inflammatory, 133–134
antimicrobial, 132–133
and antioxidant activity, 133, 134, 136
antiulcer activity, 134–135
cerebroprotective, 133
hepatoprotective, 136
nephroprotective activity, 135–136
wound healing, 134
Black wattle. *See Acacia dealbata, Acacia mearnsii*
Bowdichia virgilioides, 141
bioactives, 142–143
pharmacology
anticonvulsant activity, 146
antihyperglycemic effect, 145
anti-inflammatory activity, 143
antileishmanial, 146
antimalarial activity, 145
antimicrobial activity, 144
antineoplastic activity, 145
antinociceptive activity, 143
antioxidant efficacy, 144
anxiolytic activity, 145
immunomodulatory activities, 146
larvicidal activity, 145
wound healing properties, 144

C

Caesalpinia mimosides Lam., 151–152
bioactives, 153–154
pharmacology
analgesic activity, 158
antiarthritic activity, 158
anticancer activity, 156–157
anticholinesterase (AChE) inhibition activity, 158
anti-inflammatory property, 157
antimicrobial activity, 155

antioxidant activity, 155–156
neuritogenic and, 158–159
neuroprotective activity, 158–159
wound healing potential, 157
Cassia fistula L.
bioactives (phyotochemicals)
aliphatic compounds, 163–164
fistucacidin
(3,4,7,8,4)-pentahydroxyflavan, 164
fruit pulp of, 165
neutral lipids, 163
nonpolar compounds, 163
phytochemical analysis, 164–165
phytochemicals, 165
seeds, 162
twenty-seven compounds, 164
eukaryotic parasites
anthelmintic activity, 172
antiaging and, 180
anticancer effects, 178–179
antidiabetic activity, 174–175
antifertility effect, 180
anti-inflammatory activity, 175–176
antileishmaniasis activity, 181–182
antioxidant activity, 172–173
antiprotozoan effect, 171
antipyretic activity, 172
antitussive activity, 181
and diabetic conditions, 176
hepatoprotective activity, 176–177
hypocholesterolaemic effect, 177
immunomodulatory effect, 179
insecticidal activity, 181
laxative activity, 181
nephroprotective role, 178
neuroprotective effects, 178
oxidative stress, 176
toxicity studies, 179–180
tyrosinase inhibition studies, 180
wound healing property, 180
pharmacological activities
antibacterial and antifungal activity, 168–171
antiviral activity, 168
Clitoria ternatea L., 417–418
pharmacology
anesthetic activity studies, 427
antiasthmatic activity, 424
antibacterial activity, 420
anticancer effect, 426–427
antidiabetic activity, 423–424
antifungal activity, 419–420
antihelmintic and insecticidal effects, 421–422
anti-inflammatory, 421
antioxidant activity, 422–423
antipyretic and analgesic effects, 421
antiulcer capacity, 426
central nervous effect, 424–425
diuretic and anti-urolithiasis effect, 426
general behavior, effect on, 427
hepatoprotectivity activity, 423
hypolipidemic effect, 426
immunomodulatory impact, 427–428
wound-healing effects, 424
phytochemicals, 418–419

D

Derris scandens Benth.
bioactives
benzil derivatives, 195
benzoic acid derivatives, 195
flavanoids, 193–194
pterocarpene derivatives, 195
rotenoid, 195
terpenes and steroids, 195
dried stem of, 192
pharmacology
anticancer activity, 197–198
antifeedant activity and, 197
anti-inflammatory activity, 196
antimicrobial activity, 196–197
antioxidant activity, 198
biological activities, 199
and cancer treatment, 197–198
hypotensive activity, 198
immunostimulating activity, 198
toxicity against pest, 197

F

Flame thorn. *See Acacia ataxacantha* DC
Flemingia strobilifera, 303–304
bioactives, 304–305
pharmacology
analgesic activity, 305–307
anthelmintic activity, 307
antianxiety activity, 307

anticholesterol activity, 308
anticonvulsant activity, 308
antidiabetic activity, 308–309
anti-inflammatory activity, 309
antimicrobial activity, 310
antioxidant activity, 310–311
antiulcer activity, 311–312
estrogenic activity, 312
hepatoprotective activity, 312–313
Flemingia vestita, 317–318
bioactives, 318
pharmacology
anthelmintic activity, 319–322
anticancer activity, 322–323
antioxidant properties, 323
estrogenic action, 324
nematicidal properties, 323–324

I

Indigofera tinctoria, 461–463
pharmacological studies
anthelmintic activity, 468
anti-arthritic potential, 466
antibacterial activity, 468
anticancer activity, 467
antidyslipidemic activity, 468
antioxidant activity, 465
hepatoprotective activity, 465
immunomodulatory activity, 466
neuroprotective activity, 466
toxicity study, 469
phytochemical constituents, 463

K

Kidaichi, 487

L

Leguminosae. *See Psoralea corylifolia* L.
Luck Plant. *See Flemingia strobilifera*

M

Macrotyloma uniflorum (Lam.)
bioactive phytoconstituents
phenolic acids, 203
therapeutic potential, 202
pharmacological activities
analgesic and, 206–207
anti lithiatic activity, 207
antidiabetic activity, 206
antihelminthic activity, 207
antihypercholesterolemic effect, 205
anti-inflammatory activity, 206–207
antimicrobial activity, 206
antioxidant activity, 204–205
bioactive components, 204
seed extracts, 203
Mucuna pruriens L.
bioactive compounds
chemical structures, 25
flour and seeds, 26
GC-MS analysis data, 26
levodopa, 26
pharmacological activities, 23–24
phytochemicals present, 22
pharmacological activities
antidiabetic activity, 27
anti-helminthic activity, 31
anti-inflammatory activity, 30
antimicrobial activity, 28–29
antinutritional factors, 32
antioxidant activity, 30–31
antitumor activity, 29
antivenom activity, 29
antiviral activity, 28
aphrodisiac activity, 30
and DNA damage, 32
fertility enhancement, 27–28
food and feed supplement, 31
hepatoprotective activity, 28
mitochondrial dysfunction, 32
nutritional properties, 31–32
Parkinson's disease, 26–27
Mung bean. *See Vigna radiata* (L.) Wilczek

P

Prishnaparni. *See Uraria picta;* Uraria picta
Protein Tyrosine Phosphatase 1B (PTP 1B), 398
Psoralea corylifolia L., 327–328
bioactive constituents, 328–330
classification study, 329
list of metabolites, 330
loading plot of distinguishing metabolites, 329
pharmacology

anticancer property, 333
antidepressant effect, 332–333
antimicrobial activity, 330–331
antioxidant activity, 331–332
DNA polymerase inhibitory activity, 334
hepatoprotective activity, 332
photobiological activity, 330
topoisomerase inhibitory activity, 333–334
Pterocarpus marsupium Roxb., 213
bioactives
epicatechin, 215
isoflavonoids, 214
chemical structures of, 215–217
pharmacology
analgesic activity, 221
antibacterial activity, 220
anticancer activity, 220
antidiabetic activity, 217–218
antihyperlipidemic activity, 218
anti-inflammatory activity, 218
antimicrobial activity, 220
antioxidant activity, 218–219
cardiotonic activity, 220–221
hepatoprotective activity, 219
nootropic activity, 220
Pterolobium hexapetalum, 473, 474
bioactives, 475–479
pharmacological uses
acute toxicity study, 480
anticancer activity, 481
antidiarrhoel activity, 480
antimicrobial, 479
antioxidant activity, 479–480
antipyretic activity, 481
anti-ulcer activity, 480–481
ovicidal, larvicidal, and pupicidal activity, 481
wound-healing activity, 482
young pods and seed paste, 474

S

Saraca asoca, 1
pharmacological studies, 4
analgesic activity, 7
anthelmintic activity, 12
anticancer activity, 7–8
antidepressant activity, 11
antidiabetic activity, 10–11
anti-inflammatory activity, 6–7
antimicrobial activity, 12–13
anti-mutagenic, 8–9
anti-osteoporotic activity, 12
antioxidant activity, 5–6
antiproliferative, 8–9
antipyretic activity, 7
antiulcer activity, 9
aphrodisiac activity, 14
cardio-protective effect, 10
extracts, biological properties, 5
geno-protective activities, 8–9
hepatoprotective property, 14
immunomodulatory activity, 13
larvicidal activity, 14
neuroprotective activity, 14
radio-protective activity, 9–10
toxicity study, 15
uterotonic activity, 13
wound-healing effect, 14
phytochemical constituents
flavanol glycosides, 4
flavonoids, 3
Senna auriculata L., 381–382
medicinal plant, 382
pharmacological activities, 386
anthelmintic activity, 404
antibacterial/fungal activity, 387
anticancer effect, 402–403
anticlastogenic activity, 402
antidiabetic activity, 387
antidiabetic effect, 390–397
antihyperlipidemic efficiency, 403–404
anti-inflammatory activity, 401–402
antimicrobial activity, 388–389
antimutagenic capacity, 402
antinociceptive effect, 408
antioxidant studies, 404–406
antipyretic effect, 401
antivenom studies, 407
antiviral activity, 387
aphrodisiac property, 406
carbohydrate hydrolyzing enzyme inhibitor property, 406
cardioprotective effect, 407
hepatoprotective activity, 399–400

immunomodulatory activity, 406–407
laxative activity, 407
melanogenesis, effect on, 407–408
organ protective roles, 399
pharmacokinetic studies, 398–399
PTP 1B inhibitory activity, 398
in silico studies, 399
phytochemical constituents, 383–386
Senna hirsuta, 353–354
bioactives, 354–355
pharmacological activities
antibacterial activity, 356
antioxidant activity, 356
cytotoxicity assay, 356
hepatoprotective activity, 357
Senna occidentalis, 447
bioactives, 448–452
common names, 448
pharmacological activities
acute toxicity study, 456
anti-diabetic activity, 454–455
anti-inflammatory and antipyretic effects, 453–454
antimicrobial activity, 452–453
antioxidant activity, 454
antitrypanosomal activity, 455
hepatoprotective activity, 455
wound healing and sun protective effects, 455
substitute for coffee, 448
Senna siamea (family Fabaceae)
alkaloids, structure of, 227
bioactives, 226–227
cassibiphenol A and cassibiphenol B, 227
pharmacology
analgesic and anti-inflammatory activity, 229
antibacterial activity, 229
anticancer activity, 230
anti-hyperglycemic and, 230
anti-hyperlipidemia effect, 230
antioxidant efficacy, 229
antiplasmodial activity, 230
antiviral activity, 228
cardioprotective activity, 231
hepatoprotective activity, 229–230
mosquitocidal activity, 231
purgative activity, 231
tricyclic alkaloids, 228
Senna sophera (L.)
bioactives, 236
pharmacology
analgesic activity, 242
anti-asthmatic effects, 239
antibacterial activity, 241–242
anticancer activity, 240
antidiabetic/hypoglycemic activity, 236–237
antifungal activity, 241
anti-inflammatory activity, 239–240
antioxidant activity, 240
anti-ulcer activity, 241
diuretic activity, 242
hepatoprotective activity, 239
laxative activity, 242
sugar level, 238
Sesbania sesban (L.), 433–434
bioactive studies
anthelmintic activity, 440–441
antidiabetic activity, 441–442
antifertility activity, 442
anti-inflammatory activity, 438–439
antimicrobial activities, 435–437
antinociceptive activity, 440
antioxidant activity, 437–438
cytotoxic activity, 438
larvicidal activity, 438
molluscicidal activity, 442
spermicidal activity, 443
stimulatory effect on CNS, 442
ethnomedicine, 434
phytochemical constituents, 434–435, 436
Stryphnodendron adstringens (Mart.)
Coville, 247–248
bioactives, 248
pharmacology
acaricidal activity, 248
angiogenic properties, 248
anti-apoptotic activity, 249
antihelmintic effects, 249
anti-inflammatory properties, 249
antimicrobial activity, 249–250
anti-neoplastic activity, 250–251
antinociceptive activity, 251
antioxidant activity, 251
antivenom properties, 251

antiviral activities, 251
clinical trials, 255
cytoprotective effects, 252
genotoxicity and genoprotection, 252
immunomodulatory activities, 252
larvicidal activity, 253
molluscicidal activity, 253
protozoacidal activities, 253
toxicity, 253–254
wound-healing properties, 254

T

Tamarindus indica, 261–263
bioactive compounds, 263–264
pharmacological activity, 264–265
anticancer, 267
antidiabetogenic, 266
anti-inflammatory and analgesic, 267
antimicrobial, 265
antioxidant, 265–266
antivenom, 266–267
hepatotoxicity, 266
polyphenolic compounds from, 264
Tyrosinase, 368–369

U

Uraria picta, 271–272
bioactive phytoconstituents, 275
acaricidal activity, 275
antithrombotic property, 274
pharmacological significance, 273–275
phytochemicals
classes of, 273
source of important classes of, 272–273

V

Velvet bean. *See Mucuna pruriens* L.
Vigna radiata (L.) Wilczek, 359–360
bioactive components, 360–361, 362–363
biological activities
anticancer activity, 367
antidiabetic activity, 366
anti-hypertensive effects, 367–368
anti-inflammatory activity, 365–366
antimicrobial activity, 364–365
antioxidant activity, 360, 364
antisepsis effects, 368
myocardial preservation effects, 368
tyrosinase inhibition activity, 368–369